Smart Energy Management for Smart Grids

Smart Energy Management for Smart Grids

Special Issue Editors

Khmaies Ouahada
Omowunmi Mary Longe

MDPI • Basel • Beijing • Wuhan • Barcelona • Belgrade

Special Issue Editors
Khmaies Ouahada
University of Johannesburg
South Africa

Omowunmi Mary Longe
University of Johannesburg
South Africa

Editorial Office
MDPI
St. Alban-Anlage 66
4052 Basel, Switzerland

This is a reprint of articles from the Special Issue published online in the open access journal *Sustainability* (ISSN 2071-1050) from 2018 to 2019 (available at: https://www.mdpi.com/journal/sustainability/special_issues/Smart_Grids).

For citation purposes, cite each article independently as indicated on the article page online and as indicated below:

LastName, A.A.; LastName, B.B.; LastName, C.C. Article Title. *Journal Name* **Year**, *Article Number*, Page Range.

ISBN 978-3-03928-142-8 (Pbk)
ISBN 978-3-03928-143-5 (PDF)

© 2020 by the authors. Articles in this book are Open Access and distributed under the Creative Commons Attribution (CC BY) license, which allows users to download, copy and build upon published articles, as long as the author and publisher are properly credited, which ensures maximum dissemination and a wider impact of our publications.

The book as a whole is distributed by MDPI under the terms and conditions of the Creative Commons license CC BY-NC-ND.

Contents

About the Special Issue Editors . . . . **vii**

Preface to "Smart Energy Management for Smart Grids" . . . . **ix**

Oyeniyi Akeem Alimi, Khmaies Ouahada and Adnan M. Abu-Mahfouz
Real Time Security Assessment of the Power System Using a Hybrid Support Vector Machine and Multilayer Perceptron Neural Network Algorithms
Reprinted from: *Sustainability* **2019**, *11*, 3586, doi:10.3390/su11133586 . . . . **1**

Wilson Pavón, Esteban Inga and Silvio Simani
Optimal Routing an Ungrounded Electrical Distribution System Based on Heuristic Method with Micro Grids Integration
Reprinted from: *Sustainability* **2019**, *11*, 1607, doi:10.3390/su11061607 . . . . **19**

Jaehong Whang, Woohyun Hwang, Yeuntae Yoo and Gilsoo Jang
Introduction of Smart Grid Station Configuration and Application in Guri Branch Office of KEPCO
Reprinted from: *Sustainability* **2018**, *10*, 3512, doi:10.3390/su10103512 . . . . **37**

Christine Milchram, Geerten van de Kaa, Neelke Doorn and Rolf Künneke
Moral Values as Factors for Social Acceptance of Smart Grid Technologies
Reprinted from: *Sustainability* **2018**, *10*, 2703, doi:10.3390/su10082703 . . . . **55**

Jinchao Li, Tianzhi Li and Liu Han
Research on the Evaluation Model of a Smart Grid Development Level Based on Differentiation of Development Demand
Reprinted from: *Sustainability* **2018**, *10*, 4047, doi:10.3390/su10114047 . . . . **78**

Bharath Varsh Rao, Friederich Kupzog and Martin Kozek
Three-Phase Unbalanced Optimal Power Flow Using Holomorphic Embedding Load Flow Method
Reprinted from: *Sustainability* **2019**, *11*, 1774, doi:10.3390/su11061774 . . . . **103**

Jason Jihoon Ree and Kwangsoo Kim
Smart Grid R&D Planning Based on Patent Analysis
Reprinted from: *Sustainability* **2019**, *11*, 2907, doi:10.3390/su11102907 . . . . **119**

Sol Kim, Sungwon Jung and Seung-Man Baek
A Model for Predicting Energy Usage Pattern Types with Energy Consumption Information According to the Behaviors of Single-Person Households in South Korea
Reprinted from: *Sustainability* **2019**, *11*, 245, doi:10.3390/su11010245 . . . . **144**

Hanna Mela, Juha Peltomaa, Marja Salo, Kirsi Mäkinen and Mikael Hildén
Framing Smart Meter Feedback in Relation to Practice Theory
Reprinted from: *Sustainability* **2018**, *10*, 3553, doi:10.3390/su10103553 . . . . **168**

Aqdas Naz, Nadeem Javaid, Muhammad Babar Rasheed, Abdul Haseeb, Musaed Alhussein and Khursheed Aurangzeb
Game Theoretical Energy Management with Storage Capacity Optimization and Photo-Voltaic Cell Generated Power Forecasting in Micro Grid [†]
Reprinted from: *Sustainability* **2019**, *11*, 2763, doi:10.3390/su11102763 . . . . **190**

Ning Liang, Changhong Deng, Yahong Chen, Weiwei Yao, Dinglin Li, Man Chen and Peng Peng
Two-Stage Coordinate Optimal Scheduling of Seawater Pumped Storage in Active Distribution Networks
Reprinted from: *Sustainability* **2018**, *10*, 2014, doi:10.3390/su10062014 212

Simona-Vasilica Oprea, Adela Bâra, Adina Ileana Uță, Alexandru Pîrjan and George Căruțașu
Analyses of Distributed Generation and Storage Effect on the Electricity Consumption Curve in the Smart Grid Context
Reprinted from: *Sustainability* **2018**, *10*, 2264, doi:10.3390/su10072264 227

Ali Saleh Aziz, Mohammad Faridun Naim Tajuddin, Mohd Rafi Adzman, Makbul A. M. Ramli and Saad Mekhilef
Energy Management and Optimization of a PV/Diesel/Battery Hybrid Energy System Using a Combined Dispatch Strategy
Reprinted from: *Sustainability* **2019**, *11*, 683, doi:10.3390/su11030683 252

Seul-Gi Kim, Jae-Yoon Jung and Min Kyu Sim
A Two-Step Approach to Solar Power Generation Prediction Based on Weather Data Using Machine Learning
Reprinted from: *Sustainability* **2019**, *11*, 1501, doi:10.3390/su11051501 278

Claudia Zanabria, Filip Pröstl Andrén, Thomas I. Strasser
An Adaptable Engineering Support Framework for Multi-Functional Energy Storage System Applications
Reprinted from: *Sustainability* **2018**, *10*, 4164, doi:10.3390/su10114164 294

Byuk-Keun Jo, Seungmin Jung and Gilsoo Jang
Feasibility Analysis of Behind-the-Meter Energy Storage System According to Public Policy on an Electricity Charge Discount Program
Reprinted from: *Sustainability* **2019**, *11*, 186, doi:10.3390/su11010186 322

About the Special Issue Editors

Khmaies Ouahada (Prof) is a full Professor and head of the Department of Electrical and Electronics Engineering Science, the Faculty of Engineering and the Built Environment at the University of Johannesburg, Johannesburg, South Africa. He holds a doctorate in Information Theory and Telecommunications. His research interests are information theory; coding techniques; power-line communications; visible light communications; smart grid; energy demand management; renewable energy; wireless sensor networks; wireless communications; reverse engineering; and engineering education. He is the Founder and Chairman of the Centre for the Smart Systems (CSS) research group in the department. He serves as an Assistant Editor for the IEEE Access journal, USA and a member of the Editorial Board for Multidisciplinary Digital Publishing Institute (MDPI)—*Sustainability*, and (MDPI)—*Information and Digital Communications and Networks*, Elsevier journals. He has served as Guest Editor for special issues on Engineering Education in Multi-Disciplinary Publishing Institute (MDPI)—*Education Sciences*, Multi-Disciplinary Publishing Institute (MDPI)—*Energies and Coding and Modulation Techniques*, in the Multi-Disciplinary Publishing Institute (MDPI)—*Information* journals.

Omowunmi Mary Longe (Dr) received her D.Ing in 2017 from the University of Johannesburg, South Africa in Electrical and Electronic Engineering Science; and M.Eng and B.Eng degrees in Electrical and Electronics Engineering, in 2011 and 2001, respectively, from the Federal University of Technology, Akure, Ondo State, Nigeria. She is a senior member of the Institute of Electrical and Electronics Engineers (IEEE), and a member of the Society of Women Engineers (SWE), Organisation of Women in Science in Developing World (OSWD), Nigeria Society of Engineers (NSE), Association of Professional Women Engineers of Nigeria (APWEN), etc. She is also the IEEE PES Women in Power (WiP) Representative for South Africa. She has published more than thirty research and technical papers in peer-reviewed journals and professional conferences. She is a reviewer and guest editor for some ISI-listed journals, and also a member of Technical Programme Committees for local and international professional conferences. She is presently a Senior Lecturer in the department of Electrical and Electronic Engineering Science, University of Johannesburg, South Africa.

Preface to "Smart Energy Management for Smart Grids"

The main goal of developing a smart energy management system is to help corporations, organizations, groups, and individuals use energy consumption data wisely in order to maintain and improve energy consumption. This is, in fact, regarded as essential in the improvement of energy supplies, as it enhances access to more renewable energy usage at the household and community levels, which will also help with energy saving and CO_2 emission reduction.

Energy management can be defined as the conservation, control, and monitoring of energy in a smart grid. The main focus of this book is the investigation of the best energy management systems within micro-grids, households, and communities, in order to share energy in a very efficient way. The security and sustainability of smart grids are also discussed in this book. Three different energy sources will be discussed in this book: first, the energy generated and distributed by the utility, representing the main source of electricity to consumers; second, energy storage devices, which include battery storage, electric vehicles, and large-scale energy storage, where customers or third-party energy producers can store energy from the utility grid during lower price periods and use it during higher price periods; thirdly, renewable energy that is generated from natural sources, which are continuously replenished, thereby creating relative or total energy independence from the grid for customers.

Smart energy management can be carried out at two levels: household and community or household groupings. At the household or smart homes level, an energy management system should be installed according to consumer type and demand. Each consumer should be able to optimise its energy consumption and trading for comfort and profit. The energy management system can also offer smart, active consumers incentives above passive, smart consumers. Consumers should manage their internal energy from utility, storage and renewables. A smart energy management system should take control of these energy sources in order to maintain better consumption. At the community level, sharing available energy between households will help create a kind of independence that will help to sustain the grid.

This Book on "Smart Energy Management for Smart Grids" opens the door to research on the importance of developing smart energy management systems by designing smart techniques, mathematical approaches, and algorithms, in order to take control of energy consumption in smart homes and community households. The sharing of energy in a community can be modelled in a mathematical way, as in the case of many applications using game theory to create a demand–offer equilibrium that helps to manage and balance energy consumption.

Khmaies Ouahada, Omowunmi Mary Longe
Special Issue Editors

Article

Real Time Security Assessment of the Power System Using a Hybrid Support Vector Machine and Multilayer Perceptron Neural Network Algorithms

Oyeniyi Akeem Alimi [1,*], Khmaies Ouahada [1] and Adnan M. Abu-Mahfouz [2]

1. Department of Electrical & Electronic Engineering Science, University of Johannesburg, Johannesburg 2006, South Africa
2. Council for Scientific and Industrial Research (CSIR), Pretoria 0184, South Africa
* Correspondence: alimioyeniyi@gmail.com; Tel.: +27-73-802-9570

Received: 31 May 2019; Accepted: 24 June 2019; Published: 29 June 2019

Abstract: In today's grid, the technological based cyber-physical systems have continued to be plagued with cyberattacks and intrusions. Any intrusive action on the power system's Optimal Power Flow (OPF) modules can cause a series of operational instabilities, failures, and financial losses. Real time intrusion detection has become a major challenge for the power community and energy stakeholders. Current conventional methods have continued to exhibit shortfalls in tackling these security issues. In order to address this security issue, this paper proposes a hybrid Support Vector Machine and Multilayer Perceptron Neural Network (SVMNN) algorithm that involves the combination of Support Vector Machine (SVM) and multilayer perceptron neural network (MPLNN) algorithms for predicting and detecting cyber intrusion attacks into power system networks. In this paper, a modified version of the IEEE Garver 6-bus test system and a 24-bus system were used as case studies. The IEEE Garver 6-bus test system was used to describe the attack scenarios, whereas load flow analysis was conducted on real time data of a modified Nigerian 24-bus system to generate the bus voltage dataset that considered several cyberattack events for the hybrid algorithm. Sising various performance metricion and load/generator injections, en included in the manuscriptmulation results showed the relevant influences of cyberattacks on power systems in terms of voltage, power, and current flows. To demonstrate the performance of the proposed hybrid SVMNN algorithm, the results are compared with other models in related studies. The results demonstrated that the hybrid algorithm achieved a detection accuracy of 99.6%, which is better than recently proposed schemes.

Keywords: multilayer perceptron neural network; support vector machine; cyberattacks; optimal power flow; smart grid security; intruder detection system

1. Introduction

In recent times, rapid developments in technology have increased the rate of cyberattacks and cybercrimes on cyber-physical systems and institutions. Infrastructural security against these cyberattacks and cybercrimes have become increasingly important to individuals, organizations, and research centers. In a 2016 Global Economic Crime survey, cybercrime was ranked as the fourth most reported economic crime in South Africa, and the rate increased from 26% to 32% when compared to the reported cases in 2014 [1]. With regards to power systems and the electricity grid, the integration of the Internet of Things (IoT) and other technological tools have assisted in promoting grid efficiency and effectiveness. However, just like other important infrastructures, a plethora of new security concerns, such as cyberattacks, are becoming rampant on the power grid [2]. Moreover, the fact that the power grid is a vital asset among the country's various infrastructures makes it a highly attractive target for cyber-threats [2,3]. In the 2014 fiscal year, the Industrial Control Systems Cyber Emergency Response Team (ICS-CERT) announced that 79 of the 245 recorded cyber incidents on critical infrastructures targeted the energy sector [4]. Severe cyberattack examples, such as the Ukrainian power grid blackout in 2015 and the Israeli power grid in 2016, have shown that grid cyber-security is among the top priorities of national security [3,5]. Studies have shown that supervisory control and data acquisition (SCADA) systems, and other operational modules, including the State Estimation, Optimal Power Flow (OPF) can be successfully attacked [6–8]. Intruders take advantages of the various vulnerabilities in the grid network and modules to disrupt grid operation and stability, thereby causing blackouts and economic loss. These security issues have continuously necessitated attention from power system engineers and researchers into developing solutions.

Intruder detection schemes have been identified as a security solutions for power systems [3,9]. Intrusion detection systems (IDSs) in network processes aim to monitor, analyze, and react to any unauthorized and anomalous deviation from the normal profile of the network. Monitoring power system networks and module results in order to predict and detect intrusion and anomalies into the grid topology, database, and network data by adversaries is highly important for a reliable power system network. In recent times, various studies have proposed several formulations that focus on intrusion and anomaly detection for power systems [10–15]. The authors in [9,10] proposed an anomaly detection and correlation algorithm for substation cybersecurity using test systems as case studies. Further, machine learning techniques have been proposed as a viable option, as they are known to show tremendous performance in intrusion detections because of their accurate pattern recognition and learning abilities [16–18]. The authors in [16] ascertained that the machine learning approach is applicable to power system security. The authors successfully applied machine learning algorithms, including OneR, random forest, and Adaboost+JRipper, in classifying power system disturbances over a three-class (Attack, Natural Disturbance, and No Event) scheme. The authors in [18] developed different multi-model algorithms in order to find the best performer for voltage security monitoring and assessment. The authors used the IEEE 96 reliability test system as a case study and presented Random Forest as the best performer, with an accuracy of 99.89%. The authors in [19] proposed an artificial neural network algorithm (ANN) to detect power system cyberattacks on transmission network data. The authors evaluated their experiments on a 24-bus system and achieved a detection rate of 92–99.5% on the introduced anomalies. However, the consideration of scalability, demand, and generation uncertainty, which are highly common for power systems, were not considered. Further, the authors in [17] used some machine learning algorithms, involving a convolutional neural network, K-nearest neighbor, and XGBoost, to analyze raw data logs collected by phasor measurement units (PMUs) to detect intrusion into power systems. The authors achieved an average accuracy, precision, recall and F1 score of 0.9391, 0.938, 0.936, and 0.935 on 15 datasets, respectively. The authors in [2] also presented an IDS based on principal component analysis (PCA), whereby flow results are monitored and intrusion due to cyberattacks on transmission line parameters are detected. The authors used PCA to separate power flow variability into regular and irregular subspaces. They verified the performance of their algorithm using IEEE 24-bus and 118-bus reliability test systems and achieved good results. However, intrusions on several other input data such as the load, generator inputs, and network topology were not considered in their work. Furthermore, the authors in [20] presented a graph matching approach for power systems. The authors used IEEE 24-bus, 30-bus, and 118-bus benchmark test systems to implement their proposed scheme and achieved perfect scores. However, the proposed algorithm only considered the topological and configurational aspect of the power system database; intrusions into the power flow analysis were not considered.

We sought to improve the shortcomings in the above-mentioned literature, such as scalability, demand, and generation uncertainty, and topological and configurational intrusion of the power system. In this paper, a hybrid Support Vector Machine and Multilayer Perceptron Neural Network (SVMNN) algorithm, which involves a combination of Support Vector Machine (SVM) and feedforward Multilayer Perceptron Neural Network (MPLNN) algorithms, is developed for predicting and detecting power system cyber intrusion attacks. The key idea is to take advantage of two distinguished classifiers' abilities for predicting and detecting attacks on power systems. The logistic regression method is developed for the stacking process. The hybrid algorithm is modelled to evaluate a case study involving a 24-bus system AC power flow result dataset. This study made use of a real time generator and load data injections that showed the nonlinearity and uncertainty properties peculiar to power systems. Daily generator output profiles for a duration of twenty one (21) days and a load profile taken at an interval of thirty (30) minutes were used. The hypothesis is that at the end of each day, there will be ten (10) intrusive events involving simultaneous attacks, as described in [9]. The hypothesis of ten daily intrusive events was considered in order to have a balanced dataset for the prediction and detection algorithm. Feedforward MLPNN are known for their excellent learning abilities, especially in non-linear complex relationships and their good classification performance. With regards to its well-known flaw of non-optimal separation surfaces between classes, here, MLPNN is stacked with SVM, which is excellent in that regard. Further, unlike previous studies, the proposed scheme in this paper considered intrusions that affect the topological configuration, as well as intrusions on the load and generator output injections. High efficiency in precision and accuracy were achieved using the proposed scheme.

The specific novelties of this paper are stated briefly: (1) a description of power system cyber intrusion scenarios, involving topological modification and polluted data using a bus test system as a case study; (2) evaluating the effects of cyber intrusions on the AC power flow result of OPF and its relevant influences on voltage, power, and current flows; (3) load flow analysis using modified power system data and integrating various attack scenarios involving topological manipulation and load/generator injections; and (4) developing an effective hybrid scheme that involves taking advantage of two distinguished classifiers' abilities to evaluate the bus voltage dataset generated from the load flow results.

In this paper, two test bus systems were used as case studies. A modified IEEE Garver 6 bus test system was used in describing cyber intrusion scenarios, whereas a 24-bus system was used as the case study for the hybrid SVMNN prediction and detection scheme. The developed SVMNN algorithm presented 99.6% precision and accuracy rates in predicting and detecting the introduced attacks, which demonstrated the efficacy of the model in predicting and detecting both topological configurational intrusion as well as intrusions into the generator and load injections. All the simulations to generate the bus voltage dataset were conducted using the Electrical Transient Analyzer Program (ETAP) software. The ETAP was used to run the AC OPF processes, and the machine learning algorithms were designed, tested and evaluated using the Orange machine learning tool.

The rest of this paper is organized as follows. Section 2 presents the Materials and Methods while Section 3 presents the results and discussions. Section 4 presents the conclusions and recommendation for future work.

2. Materials and Methods

In this section, we describe the OPF processes and the mathematical formulations we used to generate our voltage dataset from the raw network data. We also discuss the methods used to develop the hybrid model and the case studies. All the simulations used to generate the voltage log dataset were generated using ETAP, while the classifiers' algorithms were implemented using the Orange machine learning tool. Both software packages were implemented on a 64-bit PC using an Intel Core i5-3340, 3.10 GHz CPU, with a total amount of 8.00 GB of RAM installed. In Section 2.1, we briefly discuss the optimal power flow, and in Section 2.2, we explain the mathematical formulations used in the paper. In Section 2.3, we discuss the prediction and detection model developed, and in Section 2.4, we present the case studies.

2.1. Optimal Power Flow

OPF modules are very vital in the operational decisions of the grid. They defines the steady state operation point, whereby the minimum generating cost is assured, and system operating constraints on quantities, such as real and reactive power, generator outputs, line flows, and voltage magnitudes, are maintained [2,19,21]. Grid control centers run multiple instances of the OPF module over regular time intervals so as to maintain the operational cost of the power system while ensuring its reliability despite variations in load requirement and available resources. It should be noted that some parameters and quantities, including line parameters and network topology, typically remain unchanged over time, unlike quantities like the load and the power dissipated by the generating units, which change often. Power flow equations can be determined through either AC or DC power flow calculations. Any error, wrong decision, or actions caused by an intrusion of the OPF modules can cause a series of operational failures, technical system instability, and huge financial losses.

Kirchhoff's law explains the theory of how power flows in an electrical network [22]. Using the node-voltage analysis explained in [22], provided the voltage outputs from the generating units, the load impedances, transmission line impedances, and susceptances for a network are given, the current and power flowing through the network can be computed. The current-voltage flow equation (IV equation) is derived in terms of the network admittance matrix, the current, and the voltage magnitudes. The network admittance matrix for an n-bus system has a relationship with the current matrix and voltage magnitude vector as presented in (1) [22]:

$$\begin{pmatrix} I_1 \\ I_2 \\ \vdots \\ I_n \end{pmatrix} = \begin{pmatrix} Y_{11} & Y_{12} & \cdots & Y_{1n} \\ Y_{21} & Y_{22} & \cdots & Y_{2n} \\ \vdots & \vdots & \cdots & \vdots \\ Y_{n1} & Y_{n2} & \cdots & Y_{nn} \end{pmatrix} \begin{pmatrix} V_1 \\ V_2 \\ \vdots \\ V_n \end{pmatrix} \quad (1)$$

where Y is the bus admittance matrix. The bus admittance matrix is given as $Y = G + jB$ [23,24]. The current vector is defined in (2), whereas (3) defines the voltage magnitude vector [22]:

$$I_n = [I_1, I_2, \ldots, I_n]^T \quad (2)$$

$$V_n = [V_1, V_2, \ldots, V_n]^T. \quad (3)$$

The state vector X for the n node system is given in terms of the voltage magnitude and voltage phase angle in (4) [25]:

$$X = [V_1 V_2 V_3 \ldots V_n \, \theta_2 \theta_3 \ldots \theta_{n-1}]^T \quad (4)$$

where θ is the $n-1$ dimensional vector representing voltage phase angle and V is the n-dimensional vector representing voltage magnitudes.

From the PV flow equations, the complex apparent power injection (+ve) or withdrawal (−ve) from bus n is defined in terms of p, q, and V as given in (5) [25]:

$$S_n = p_n + jq_n = V_n I_n^*. \quad (5)$$

Equations (6) and (7) express the current and voltage magnitude at bus n, respectively, in complex form as:

$$I_n = [I_n^r + jI_n^j] \quad (6)$$

$$V_n = [V_n^r + jV_n^j] \quad (7)$$

and substituting (6) and (7) into (5),

$$S_n = V_n I_n^* = (V_n^r + jV_n^j) \cdot (I_n^r - jI_n^j). \quad (8)$$

The real and reactive power from (5) is expressed in (9) as:

$$p_n = V_n^r \cdot I_n^r + V_n^j \cdot I_n^j \text{ and } q_n = V_n^j \cdot I_n^r - V_n^r \cdot I_n^j. \quad (9)$$

Note that j as a superscript refers to a complex number imaginary part, while r as a superscript refers to the real part. In terms of phasor angles, the real and reactive power at bus n can be expressed as (10) and (11), respectively [12,23,26,27]:

$$p_n = v_n \sum_{n' \in N} v_{n'} (G_{nn'} \cos(\theta_{nn'}) + B_{nn'} \sin(\theta_{nn'})) \quad (10)$$

$$q_n = v_n \sum_{n' \in N} v_{n'} (G_{nn'} \sin(\theta_{nn'}) - B_{nn'} \cos(\theta_{nn'})) \quad (11)$$

where $Y_{nn'} = G_{nn'} + jB_{nn'}$ is the line admittance between two buses n and n' and $\theta_{nn'}$ is the difference in phase angle between buses n and n'. The real and reactive power flowing from bus n and n' is given in (12) and (13), respectively as [6,12,27]:

$$p_{nn'} = v_n^2 (g_{sn} + g_{nn'}) - v_n v_{n'} (g_{nn'} \cos \theta_{nn'} + b_{nn'} \sin \theta_{nn'}) \quad (12)$$

$$q_{nn'} = -v_n^2 (b_{sn} + s_{nn'}) - v_n v_{n'} (g_{nn'} \sin \theta_{nn'} - b_{nn'} \cos \theta_{nn'}) \quad (13)$$

where $g_{sn} + jb_{sn}$ is the shunt branch admittance at bus n. The net apparent equation in (7) for bus n can be rewritten as (14):

$$s_n = p_n + jq_n = \begin{cases} (p_n^G - p_n^D) + j(q_n^G - q_n^D), & n \in \text{set of gens}, \\ -p_n^D - jq_n^D, & \text{otherwise}, \end{cases} \quad (14)$$

where p_n^G and q_n^G are defined as the controllable power injections/control input u. The power consumed at bus n, p_n^C, is related to the power flows of the lines connected to the bus n, as shown in (15):

$$p_n^C = \sum_{k \in K} p_{k,in}^L - \sum_{k \in K} p_{k,out}^L \quad \forall k \in K \quad (15)$$

where $L_{k,in}$ and $L_{k,out}$ represent set of incoming and outgoing lines of bus n, respectively, while the power flow via line k is denoted as p_k^L. The power consumed at bus n is related to the load power demand and power injection into the bus as expressed in (16):

$$p_n^C = p_n^D - p_n^G \quad \forall n \in N \quad (16)$$

where p_n^D and p_n^G are the load power demand and generated power at bus n, respectively.

2.2. Mathematical Formulation

OPF allows operators to specify a range of optimization criteria and some objective functions on quantities, including bus voltages and line flow. A mixed integer nonlinear programming problem AC OPF is formulated in the paper. There is an objective, and some constraints, that govern system performance. The objective is to find steady state operating points in terms of both state vectors and control inputs, whereby the power generated by the existing generators are optimally controlled to serve the load requirements and line flows in the network and minimize real and reactive power loss in the network. The objective function is subjected to the equality and inequality constraints in (17)–(21) [27,28]:

$$\sum_{n \in N} p_n^D - \sum_{n \in N} p_n^G - \sum_{k \in L_n,in} p_k^L + \sum_{k \in L_n,out} p_k^L = 0 \quad (17)$$

$$-p_k^{L,\,max} \leq p_k^L \leq p_k^{L,\,max} \quad \forall k \in K \tag{18}$$

$$0 \leq p_n^{G'} \leq p_n^{G,max} \quad \forall n \in N \tag{19}$$

$$\pi \leq \theta_n \leq \pi \quad \forall n \tag{20}$$

$$\theta_{Rn} = 0; \quad Rn: reference\ node. \tag{21}$$

The power balance equation to be solved is given in constraint (17), which gives the assurance that, at any node n, the summation of the total power dissipated by the generating unit n equals the summation of the power flowing in the lines and the total sum of the power demand. Constraint (18) defines the power flows via the lines, and the constraint limits the power flows via the lines within the network, with regards to their capacities. Constraint (19) is for the generator outputs' limits and ensures that the generator outputs' limits are not surpassed. The voltage phase angle limit constraint is shown in (20) and ensures that voltage angle limits are within the specified range. Constraint (21) is the constraint for the reference bus/node. Equation (21) ensures that the reference node has a voltage angle of 0 degrees [28].

2.3. Prediction and Detection Model

In this subsection, we describe the MLPNN, the SVM, and the hybrid SVMNN models that were employed in predicting and detecting the possibility of the power system network being compromised.

2.3.1. Multilayer Perceptron Neural Networks (MLPNN)

MLPNN is a feedforward neural network that uses backpropagation for its training process. Neural Network (NN) models are inspired and designed in a similar fashion to the human brain. However, unlike the brain, NNs utilize some mathematical functions that map input data to produce the output. A neural network operates in such a way that when data are presented at the input layer, the neural nodes (which are interconnected via respective weights and bias for each connections) execute some calculations using activation functions in all the successive layers until the input data reach the output nodes that produce the outputs. Typical activation functions used in neural networks include the sigmoid function and the Rectified linear units (ReLu), defined in (22) and (23), respectively [29]:

$$f(x') = \frac{1}{1+e^{-x'}} \tag{22}$$

$$R(x') = max(0, x') \tag{23}$$

Building a neural network algorithm begins with the simplest form, a 'single perceptron'. A perceptron is made up of a single McCulloch-Pitts neuron, which has modifiable weights and bias [30]. Figure 1a presents a perceptron process [30]. To create a multilayer perceptron, the perceptron is modified in such a way that it includes several layers of neurons with nonlinear activation functions, making it highly potent, as it can be implemented for nonlinear separable data. Considering the architectural model of a typical MLPNN presented in Figure 1b [31], the MLPNN has n inputs, one hidden layer with z' hidden neural nodes, and y output nodes.

Let us assume we have input data that is defined with the matrix [32]:

$$r = (r_1, r_2, \ldots\ldots, r_n). \tag{24}$$

Let us make the assumption that a vector r_1 that belongs to a class of the y output classes denotes the n feature values of case i'. Assuming α_g denotes the lower boundary limit and β_g denotes the upper limits of feature g, which equally relates to the minimum and maximum threshold values achievable for feature g, the mapping of the z' hidden layer neural node feedforward MLPNN process can be defined as [32]:

$$N : \{[\alpha_1, \beta_1], [\alpha_2, \beta_2], \ldots, [\alpha_n, \beta_n]\} \to [\gamma, \eta]^y : = N(r) = f(w_j f(w_i r - b_i) - b_j) \quad (25)$$

where w_i is the weight matrix that connects n input nodes to the z' hidden layer neural nodes, and w_j is the weight matrix that connects the z' neural nodes to output nodes y. Bias vectors b_i and b_j connect to the hidden and output layers, respectively. The function $f : \mathbb{R}^{\dim(a)} \to [\gamma, \eta]^{\dim(a)}$ defines the activation function that is fitted into individual nodes of the hidden layer's activation vector a, with γ and η being the lower and upper bounds. Each element in vector denotes the activation of each output layer node. Hence, classification is done based on the function class, which depends on the returned index of the maximum element in vector o.

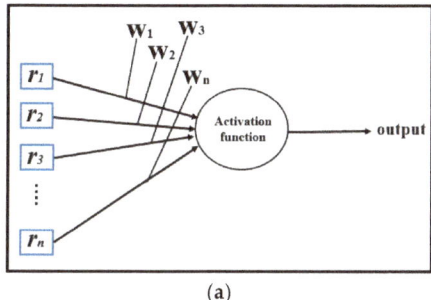
(a)

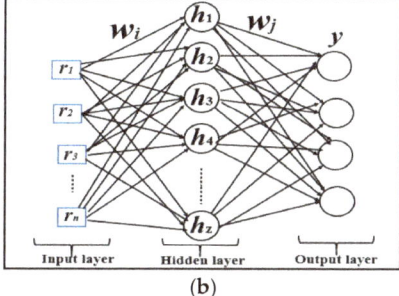
(b)

Figure 1. Neural Network models: (a) perceptron process; (b) multilayer perceptron neural network.

2.3.2. SVM Classifier

SVM is a dominant tool that is used in classification and regression problems. SVM was originally proposed for binary classifications, whereby the width of the margin between the two classes defines the optimization criterion. SVMs create a single hyper-plane, or sets of hyper-planes, in a high-dimensional feature space, which optimally separates the training patterns according to their classes. The efficient implementation of SVMs depends on the trade-off constant C and the kernel function K type, especially when it is required for nonlinear classification. Typical kernel functions include the linear, polynomial, sigmoid, and radial basis kernel function (RBF). The trade-off constant C is the soft margin parameter, which influences each individual support vector. Figure 2 [33] presents a linear SVM model showing the hyper-plane separation between the two classes.

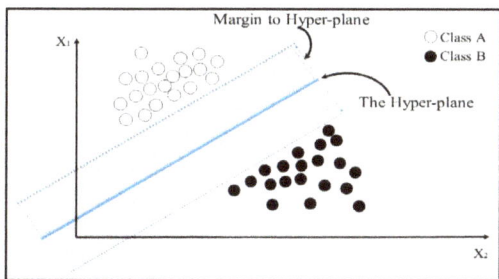

Figure 2. Support vector machine (SVM) hyper-plane separation of two class datasets.

As shown in Figure 2, the model presents the examples as space points, plotted such that the categories are kept apart by a distinct gap. Afterwards, new examples are plotted into the same space and predicted as either class depending on the side of the gap in which it is categorized.

Let us assume we have training data of n points $(\vec{x_1}, y_1), \ldots, (\vec{x_n}, y_n)$, where point x_i is a p-dimensional vector and $y_i = \pm 1$ labels the class to which point x_i belongs. SVMs tend to locate the maximum margin hyper-planes that split the group of points where x_i is for $y_i = +1$ from the groups where it is $y_i = -1$ [16]. The hyper-plane for the set of points $\vec{x}$ satisfies the equation $\vec{w}.\vec{x} - b = 0$, where $\vec{w}$ is the normal vector to the hyper-plane and b is the displacement term that determines the distance between the hyperplane and the origin [17].

2.3.3. Proposed Hybrid SVMNN

Hybrid learning methods are a process of combining two or more learning algorithms. This process is essential in achieving better accuracy and detection rates. A simplified flowchart of the hybrid SVMNN model is presented in Figure 3.

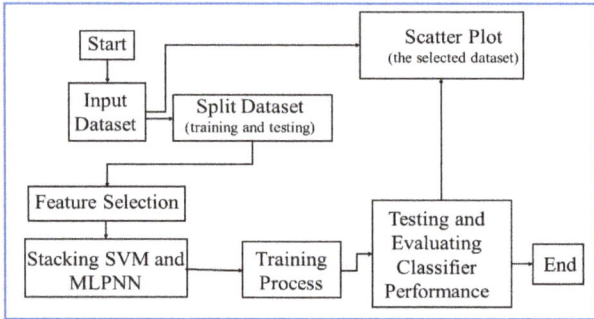

Figure 3. Flowchart of the Hybrid Support Vector Machine and Multilayer Perceptron Neural Network (SVMNN).

To evaluate the hybrid model's performance, it is required to have a sufficient experimental dataset with sensitive information for the algorithm's training and testing analysis. This is important for the effective performance of the algorithm. In this paper, bus voltage logs are captured as a dataset for the intrusion detection algorithm performance evaluation. The power system dataset used contains a total of 1218 training and testing dataset instances, with 24 features having binary targets. The scale and distribution of the dataset can produce a significant influence on the algorithm prediction and detection success. It should be noted that the bus voltage dataset that contains the poisoned datasets and the good datasets are combined and randomly split into two sets: the training and testing datasets. From the randomly organized data, 975 of the data samples were devoted to training, which is equivalent to approximately 80% of the dataset, whereas the remaining 20%, equaling 243 data samples, were dedicated to testing the trained model. The preprocessing stage of datasets, which includes transformation, normalization, discretization, and feature selection processes, are highly important for the efficiency of the machine learning algorithm. The feature selection process can vary based on the type of dataset being used. Since the dataset used in this paper uses numeric data, the data do not need any transformation. However, the dataset was normalized using the min-max scaling for effectiveness. For the developed hybrid model, the stacking utilized a back propagation MLPNN with three hidden layers of 30 neural nodes each. The L2 regularization parameter assists in reducing the generalization error as well as the overfitting problem. We varied the values of the L2 regularization parameters in order to achieve the best possible result from our developed MLPNN. Further, this study employed ReLu as the activation functions for the hidden layers. An Adam gradient-based optimizer was used as the solver for weight optimization. For the SVM, this study implemented the Library for

Support Vector Machines (LibSVM) package. The Cost C was chosen as 1.2. Three kernel functions (linear, polynomial, and RBF) were tested in order to find the best performer for our developed model. The gamma constant in kernel function was set at 0.25. This study used logistic regression for the stacking. The performance of the hybrid algorithm is evaluated and compared with the performance of individual classifiers (SVM and MLPNN) using a machine learning key performance indicator (KPI) confusion matrix. Popular classification performance measures, including precision, recall, and F1 score, will also considered for the evaluation. The metrics are discussed briefly [34,35]:

- Confusion Matrix

Confusion matrix refers to a table that is often used to explain and understand the performance of a classification model. The model evaluation metrics from a binary classifier confusion matrix typically have two dimensions: The actual class usually indexes one of the dimensions, whereas the other dimension is indexed by the classifier prediction.

- Precision

Precision presents how often the classifier model is correct. High precision correlates to a low false positive rate. Mathematically, precision is defined in (26) [34]:

$$\text{Precision} = TP/(TP + FP) \qquad (26)$$

where TP is the rate of true positives, defined as the correctly identified positives from the classifier model, and FP is the rate of false positives, which is defined as negative cases that have been wrongly identified/classified as positive ones.

- Recall (Sensitivity)

Recall is the measure that describes the ability of a prediction model to pinpoint cases of a particular class from a dataset. Mathematically, recall is defined in (27) [34]:

$$\text{Recall} = TP/(TP + FN) \; (27) \qquad (27)$$

where FN is the rate of false negative observations.

- F1 Score

F1 score is the harmonic average of Precision and Recall. The F1 score is considered to be a better metric compared to accuracy, especially in a classification involving uneven distribution:

$$\text{F1 Score} = 2 \times (\text{Recall} \times \text{Precision})/(\text{Recall} + \text{Precision}). \qquad (28)$$

2.4. Case Studies

In this paper, a modified version of the Garver IEEE 6 bus test system modelled in [28] was used in describing the attack scenarios, whereas a 24-bus system was used for the evaluation of the developed MLPNN algorithm. The Electrical Transient Analyzer Program (ETAP) Toolkit developed by ETAP, Operation Technology Inc, is a commercial software package that is widely used for power system design, simulation, monitoring operation, analysis, optimization, and stability studies. In this study, the ETAP version 16.0 was used to run the AC version of the OPF calculations. The implementation of the MLPNN algorithm used was based on the open source machine learning framework Orange (Orange 3.20.1). The SVM embedded in the Orange framework is from the LibSVM package, while the MLPNN uses the Sklearn Python Module.

2.4.1. Cyberattack Scenario Explanation Using a Modified Garver IEEE 6-Bus System

Figure 4 presents the one-line diagram of the modified Garver test system used for the cyberattack description and the consequences on the power system. The test system has six nodes consisting of three generating units, six loads, and seven lines connecting the nodes. The generating units are at node 1, node 3, and node 6, whereas the loads are on node 1, node 2, node 3, node 4, node 5, and node 6. The assumption is that the cyberattack only made changes to the topology and no changes were made to the physical parameters of the lines and load values. Table 1 provides the parameters used for the transmission lines of the modified test system. The parameters reflect the values of the impedances and susceptances of each of the transmission lines in the modified test system.

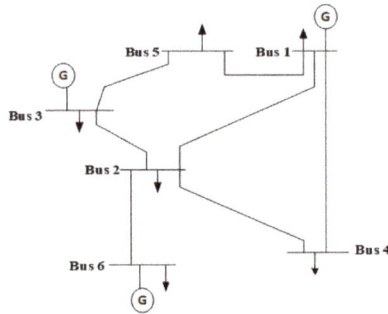

Figure 4. Modified IEEE Garver 6 bus test system.

Table 1. Transmission lines parameters.

S/N	$L_{k,out}$	$L_{k,in}$	R (ohm)	X (ohm)	$B_L(S)$
1	1	2	8.0×10^{-2}	1.05×10^{-1}	6.03
2	4	1	88.5×10^{-2}	4.0×10^{-2}	5.1×10^{-2}
3	1	5	88.5×10^{-2}	4.0×10^{-2}	5.1×10^{-2}
4	2	3	8.0×10^{-2}	1.21×10^{-1}	5.75
5	4	2	88.5×10^{-2}	4.0×10^{-2}	5.1×10^{-2}
6	2	6	8.0×10^{-2}	1.05×10^{-1}	6.03
7	3	5	88.5×10^{-2}	4.0×10^{-2}	5.1×10^{-2}

Attackers having prior knowledge about the network topology and/or having access to the grid network either through the help of an insider or via remote access may decide to slightly alter, isolate, or modify part of the network configuration or database. In the description of the attack, two scenarios are considered. In Scenario A, we assumed that the network is free of intrusion whereas in Scenario B, we assumed that the network is under attack and the grid operators are unaware of the cyber intrusion. For both scenarios, the grid is operational.

1. Scenario A

In Scenario A, an assumption was made that there was no manipulation or any attack intrusion on the network topology or data. Figure 4 presents the one-line diagram of the Scenario A test system. The generator power output data and load data for Scenario A are depicted in Table 2. The total load demand for the network without any intrusion is 255.74 MW, whereas the total generator power output from the three generators is 430.2 MW. Load flow was conducted on Scenario A using the load flow function in the ETAP program.

Table 2. Optimal generator power output and load data for Scenario A.

Bus/Node	Scenario A PG (MW)	Scenario A Load (MW)
1	240.2	42.5
2	-	24.74
3	150	51
4	-	42.5
5	-	70
6	40	25

2. Scenario B

In Scenario B, the assumption was made that there was intrusion, and the attacker(s) made some changes based on the simultaneous attack described in [6], whereby simultaneous attacks, which will not lead to a non-convergence simulation, are carried out on bus nodes. Attackers are aware that, for any type of attack, the isolation/de-energization of critical node(s), major sections, or the entire database will lead to non-convergence power flow computation. Therefore, the assumption in this paper is that attackers only make changes slight that affect the grid, but the network remains operational. Grid operators are unaware of the changes, and the network operates with the corrupted data. The one-line diagram for Scenario B is presented in Figure 5.

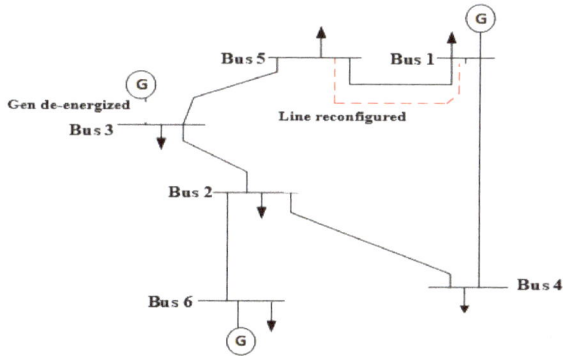

Figure 5. Scenario B test system.

As presented in Figure 5, the modified Garver 6-bus test system configuration topology has been altered due to the simultaneous attack. The simultaneous attack consists of the de-energization of the generator at node 3 and the manipulation of the sending and receiving buses of a line, such that intruders reconfigured the network by changing the origin and destination of line 1 from bus 1 to bus 5. The dashed line reflects the line that was attacked by the intruder.

The data for the generator units for Scenario B are depicted in Table 3. The total generator power output from the two supplying generators available in Scenario B is 280.2 MW, as the attack has already de-energized the generator at node 3.

Table 3. Optimal generator power output for Scenario B.

Bus/Node	PG (MW)
1	240.2
2	-
3	-
4	-
5	-
6	40

In both scenarios, the power flow computation converges. The Scenario B simulation results in the flows in the transmission lines using the poisoned data from the simultaneous attack, as depicted in Table 5.

2.4.2. Evaluation of the Prediction and Detection Algorithm Using the 24-Bus System

This paper made use of real time data. For simplicity, the network data used only covered the SouthWest and NorthWest geopolitical zone of the Nigerian grid's network topology. The 24-bus system used covered only some 330 kV stations across the geopolitical zones. Figure 6 depicts the one-line diagram of the modelled 24-bus system. The test system comprises 37 transmission lines, 17 loads, and 8 generators. The lines were modelled using their pi-equivalent circuits. The generators were modelled using steady state real and reactive powers limits. The loads were modelled using steady state real and reactive power consumption value limits.

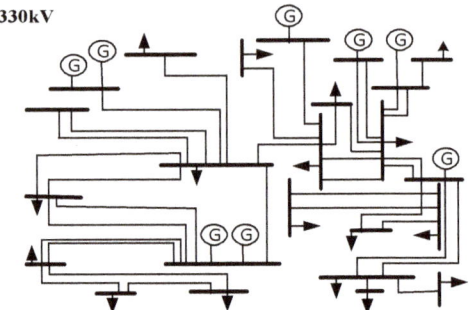

Figure 6. One-line diagram of the 24-bus system.

The modified generator data profile used in the study is presented in Figure 7, whereas the modified load data profile collected from the daily operational report used is presented in Figure 8. As shown in Figures 7 and 8, respectively, the generator data represent a daily generator data profile for a three weeks duration, whereas the load data profile has a time interval of thirty (30) minutes. Both the generator and load data used were for a one week duration, using modified data from the Nigerian Electricity Regulatory Commission daily operational report [36] from 1 to 21 February 2018.

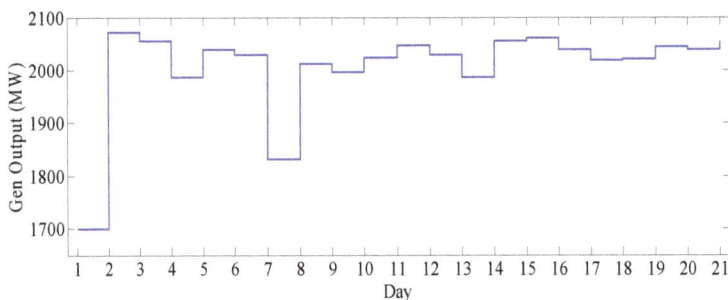

Figure 7. Generator output profile.

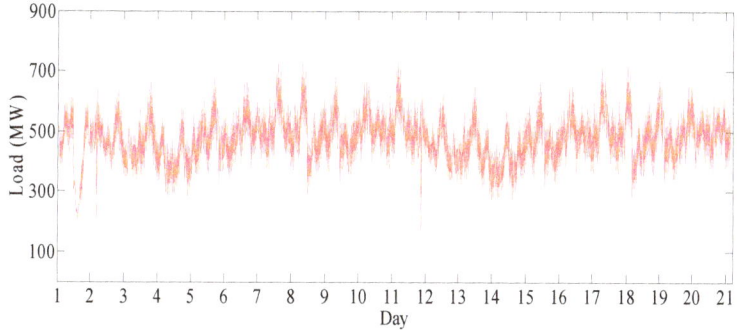

Figure 8. Load profile.

3. Results and Discussion

In this section, the results of the case studies are analyzed and presented.

3.1. Comparison of the Two Scenarios

The description of the cyber intrusion is explained with Scenarios A and B. Tables 4 and 5 present the flow simulation results of Scenario A and Scenario B respectively. Figure 9 presents the comparison of the bus voltage simulation result for Scenario A and B. Note that, for both scenarios, bus 1 has a bus voltage of 1 pu, as the bus is chosen as the slack/reference bus.

Table 4. Scenario A flow result.

S/N	$L_{k,out}$	$L_{k,in}$	Power Flow (MW)	Current Flow (Amp)
1	1	2	24.85	149.9
2	4	1	3.19	72.77
3	1	5	50.32	96.82
4	2	3	78.11	160.1
5	4	2	39.96	89.11
6	2	6	14.83	91.27
7	3	5	20.82	37.28

Table 5. Scenario B flow result.

S/N	$L_{k,out}$	$L_{k,in}$	Power Flow (MW)	Current Flow (Amp)
1	1	2	65.86	120.5
2	4	1	122.39	228.9
3	1	5	30.73	56.22
4	2	3	25.39	47.9
5	4	2	76.16	152.7
6	2	6	25.47	54.36
7	3	5	25.68	49.71

With intrusions, the normal operating limits of the grid can be above or below the limit. The upper and lower limits for voltage stability endorsed by American National Standards Institute (ANSI) are 1.05 pu and 0.95 pu, respectively [37]. As shown in Figure 9, bus 6 is clearly below the limit while bus 2 and bus 3 are very close to the lower standard limit for Scenario B. IEEE guidelines and operational safety require a response and action to be taken by the operators to rectify such situations. Hence, early detection of cyber intrusions into the power system network is highly important to grid operators.

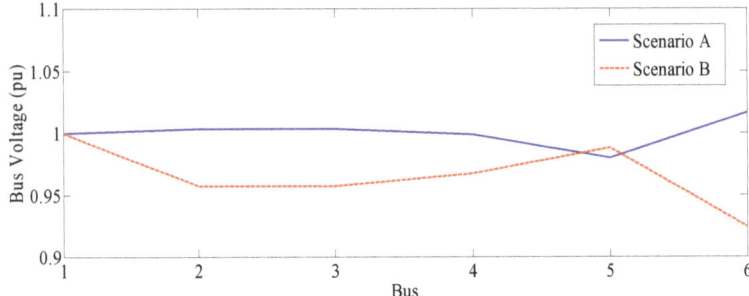

Figure 9. Scenario A and Scenario B bus voltage comparison results.

Figure 10 presents the comparison result of the current flows on the lines with and without intrusion (Scenario A and B respectively). Line 4 and line 2 have the highest magnitude for current flow for Scenario A and Scenario B, respectively. Despite the intrusion presence in Scenario B, Figure 10 shows a close relationship at line 7 for both scenarios in terms of currents flows on the lines, which typifies the fact that the presence of intrusion on a power system can be tedious to predict or pinpoint.

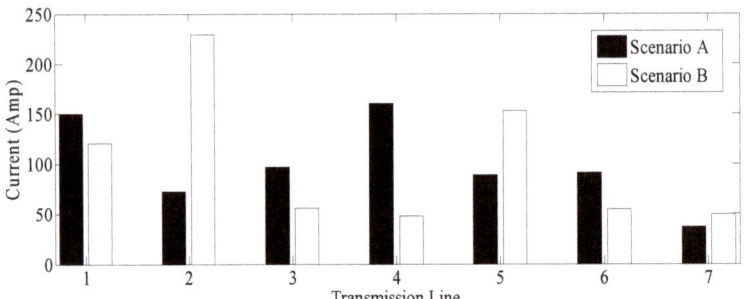

Figure 10. Scenario A and Scenario B current flow comparison results.

Moreover, Figure 11 depicts the comparison result of the power flows on the lines for both Scenario A and Scenario B. As shown in Figure 11, there is a significant difference in terms of the power flow on each individual line in the network. Line 2 has the highest magnitude of power flow for the intrusion-presence Scenario B, while the lowest magnitude of power flow occurred in the same line when there was no intrusion.

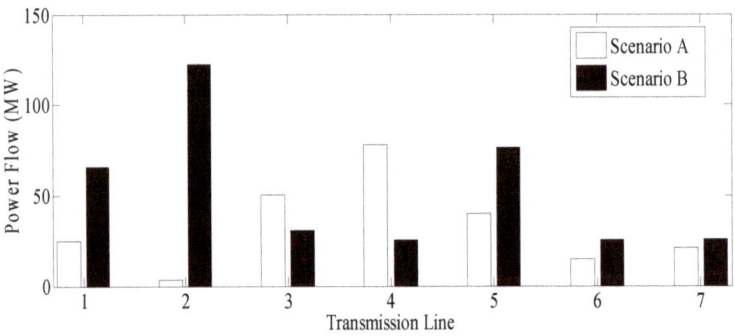

Figure 11. Scenario A and Scenario B power flow comparison results.

3.2. Hybrid SVMNN Classification Report

In order to obtain the best possible results from the developed SVM model, which will be stacked with the MLPNN, we experimented using three prominent kernel functions: RBF, Polynomial, and Sigmoid functions. The Cost C and gamma value were kept at constant values of 1.2 and 0.25, respectively. Some notable results from preliminary experiments are summarized in Table 6.

Table 6. Summarized classification results comparing the SVM kernel function's performance.

Kernel Function	Precision	Accuracy	Recall	F1 Score	Training Time (Second)
RBF	95.7%	95.5%	95.5%	95.2%	8.92
Polynomial	87.2%	85.6%	86.3%	86.5%	8.36
Sigmoid	81.6%	78.6%	78.6%	79.8%	5.83

As shown in Table 6, RBF presented the best result while the sigmoid kernel function gave the lowest accuracy. Hence, the RBF kernel function was stacked with the developed MLPNN. Also, after developing the MLPNN algorithm, in order to achieve a result with reduced generalization errors and overfitting problems, the L2 regularization parameter was varied. Table 7 presents the notable preliminary results achieved by varying the L2 regularization parameter.

Table 7. Summarized classification results varying the MLPNN L2 regularization parameter.

L2 Regularization Parameter	Precision	Accuracy	Recall	F1 Score	Training Time (Second)
100	87.5%	81.9%	81.9%	73.7%	18.38
85	94.3%	85.2%	85.2%	80.7%	17.83
65	94.6%	93.8%	93.8%	93.3%	16.35
50	97.6%	94.2%	94.2%	93.8%	14.25
35	96.2%	96.1%	96.1%	96.5%	9.8

Table 8 presents the hybrid SVMNN model's evaluation metrics from the confusion matrix. This model is a binary classifier, and the two classes targets are labelled non-intrusive data sample (NID) and intrusive data samples (ID). As shown in Table 8, the modelled SVMNN classifier predicted 203 non-intrusive data (NID) samples and 40 intrusive data (ID) samples from the testing data samples. Table 9 presents the classification result of the hybrid SVMNN compared with the best results from the standalone MLPNN and SVM classifiers.

Table 8. Model evaluation metrics.

No of Testing Data = 243	Predicted ID	Predicted NID
Actual ID	40	1
Actual NID	0	202

Non-intrusive data sample (NID); intrusive data samples (ID).

Table 9. Classification results.

Classifier	Precision	Accuracy	Recall	F1 Score
SVMNN	99.6%	99.6%	99.4%	99.6%
SVM alone	95.7%	95.5%	95.5%	95.2%
MLP alone	96.2%	96.1%	96.1%	96.5%

As presented in Table 9, the SVMNN algorithm showed precision and accuracy rates of 99.6%. The recall score is 99.4%, whereas the F1 score is 99.6%, which is much better than the best results achieved from the standalone SVM and standalone MLPNN methods. MLPNN has the ability to learn complex relationships and can easily generalize models and give efficient predictions. Thus, as expected, the standalone MLPNN gave a good result with an accuracy of 96.1%. However, the best result was achieved from the hybrid algorithm. Table 10 presents the comparison result from the paper with some proposed schemes in the literature.

Table 10. Classification result comparison with other schemes.

Classifier	Accuracy
SVMNN	99.6%
Mousavian et al. [19] (ANN model)	Average of 95.75%
Hink et al. [16] (Adaboost + JRipper model)	95%
Wang et al. [17] (AWV model)	93.91%
Valenzuela et al. [2] (PCA alone)	97%

As shown in Table 10, in a related model proposed by Mousavian et al. [19], the proposed ANN model was able to detect 92–99.5% (averaging 95.75% accuracy) involving a 24-bus system. Further, Hink et al. [16] compared several machine learning approaches and achieved an approximately 95% precision accuracy using Adaboost + JRipper for a binary classification. In a similar approach, the authors in [17] reported a detection accuracy result of 93.91% using a model that involved using a random forest as the basic classifier of AdaBoost and a weighted voting (AWV) model on PMU cyberattacks. Furthermore, the authors in [2] reported a detection accuracy result of 97% in the case of a severity class C attack involving an attack on only two lines at a time. Note that the authors in [2] did not consider intrusions into generator and load injections. In a similar article in [20], where the authors equally considered a 24-bus system in a graph matching approach and achieved a 100% result, it needs to be pointed out that the authors only considered cyberattacks on a topological power system configuration. However, in this paper, both topological configurational intrusions, as well as intrusions into the generator and load injections, were considered. The simulation results of the prediction and detection algorithm developed showed the effectiveness of the scheme, which can be employed for the effective protection of power systems.

4. Conclusions

Security threats, such as cyber intrusions into the power grid, necessitate responses from all stakeholders involved in the electricity grid. Detecting and preventing such cyber intrusions is important in current and future research. In this paper, power system cyber intrusion scenarios involving topological modifications and polluted data are described, and the effects of intrusions on the AC power flow result of OPF are discussed using a modified IEEE Garver 6 bus test system as a case study. A prediction and detection scheme based on a hybrid SVMNN was developed to predict and detect cyber intrusion attacks into the power system. The algorithm was developed to evaluate a bus voltage dataset. Several simultaneous attacks scenarios, including the removal of transmission lines and generators, were considered as cyber intrusions in the 24-bus case study. The proposed SVMNN method showed 99.6% precision and accuracy rates in predicting and detecting simultaneous attacks. However, despite showing tremendous accuracy in predicting and detecting cyber-intrusion, the developed algorithm cannot identify, locate, or eliminate present or future intrusion. Future work can focus on extending this work to developing avenues, to identify attacked stations and/or transmission lines.

Author Contributions: All the authors have contributed equally to this article.

Funding: This research was funded by the Council for Scientific and Industrial Research (CSIR), South Africa.

Conflicts of Interest: The authors declare no conflict of interest.

References

1. Economic Crime: A South African Pandemic. Global Economic Crime Survey 2016, 5th South African edition. Available online: www.pwc.co.za/en/assets/pdf/south-african-crime-survey-2016.pdf (accessed on 6 February 2019).
2. Valenzuela, J.; Wang, J.; Bissinger, N. Real-time intrusion detection in power system operations. *IEEE Trans. Power Syst.* **2013**, *3*, 1052–1062. [CrossRef]
3. Alimi, O.A.; Ouahada, K. Security Assessment of the Smart Grid: A Review focusing on the NAN Architecture. In Proceedings of the IEEE 7th International Conference on Adaptive Science & Technology (ICAST), Accra, Ghana, 22–24 August 2018.
4. The Industrial Control Systems Cyber Emergency Response Team (ICS-CERT). Incident Response Activity (September 2014–Febbruary 2015). Available online: https://goo.gl/9jGIjK (accessed on 13 June 2019).
5. Anwar, A.; Mahmood, A.N.; Pickering, M. Modeling and performance evaluation of stealthy false data injection attacks on smart grid in the presence of corrupted measurements. *J. Comput. Syst. Sci.* **2017**, *81*, 58–72. [CrossRef]
6. Anwar, A.; Mahmood, A.N. Vulnerabilities of smart grid state estimation against false data injection attack. In *Renewable Energy Integration*; Springer: Singapore, 2014; pp. 411–428.
7. Liu, Y.; Ning, P.; Reiter, M.K. False data injection attacks against state estimation in electric power grids. *ACM Trans. Inf. Syst. Secur. (TISSEC)* **2011**, *4*, 13. [CrossRef]
8. Yu, Z.H.; Chin, W.L. Blind false data injection attack using PCA approximation method in smart grid. *IEEE Trans. Smart Grid* **2015**, *6*, 1219–1226. [CrossRef]
9. Ten, C.W.; Hong, J.; Liu, C.C. Anomaly detection for cybersecurity of the substations. *IEEE Trans. Smart Grid* **2011**, *2*, 865–873. [CrossRef]
10. Talebi, M.; Wang, J.; Qu, Z. Secure power systems against malicious cyber-physical data attacks: Protection and identification. World Academy of Science, Engineering and Technology. *J. Comput. Syst. Sci. Eng.* **2012**, *6*, 757–764.
11. Tao, W.; Zhang, W.; Hu, C.; Hu, C. A Network Intrusion Detection Model Based on Convolutional Neural Network. In Proceedings of the International Conference on Security with Intelligent Computing and Big-data Services, Guilin, China, 14–16 December 2018; pp. 771–783.
12. Chaojun, G.; Jirutitijaroen, P.; Motani, M. Detecting false data injection attacks in ac state estimation. *IEEE Trans. Smart Grid* **2015**, *6*, 2476–2483. [CrossRef]
13. Kim, S.H.; Lim, S.C. Intelligent intrusion detection system featuring a virtual fence, active intruder detection, classification, tracking, and action recognition. *Ann. Nucl. Energy* **2018**, *112*, 845–855. [CrossRef]
14. Fang, J.; Qian, W.; Zhao, Z.; Yao, Y.; Wen, Z. Adaptively feature learning for effective power defense. *J. Vis. Commun. Image Represent.* **2019**, *60*, 33–37. [CrossRef]
15. Nishino, H.; Ishii, H. Distributed detection of cyberattacks and faults for power systems. *IFAC Proc.* **2014**, *47*, 11932–11937. [CrossRef]
16. Hink, R.C.B.; Beaver, J.M.; Buckner, M.A.; Morris, T.; Adhikari, U.; Pan, S. Machine learning for power system disturbance and cyber-attack discrimination. In Proceedings of the 7th IEEE International Symposium on Resilient Control Systems (ISRCS), Denver, CO, USA, 19–21 August 2014; pp. 1–8.
17. Wang, D.; Wang, X.; Zhang, Y.; Jin, L. Detection of power grid disturbances and cyber-attacks based on machine learning. *J. Inf. Secur. Appl.* **2019**, *46*, 42–52. [CrossRef]
18. Tomin, N.V.; Kurbatsky, V.G.; Sidorov, D.N.; Zhukov, A.V. Machine learning techniques for power system security assessment. *IFAC Pap.* **2016**, *49*, 445–450. [CrossRef]
19. Mousavian, S.; Valenzuela, J.; Wang, J. Real-time data reassurance in electrical power systems based on artificial neural networks. *Electr. Power Syst. Res.* **2013**, *96*, 285–295. [CrossRef]
20. Anwar, A.; Mahmood, A.N. Anomaly detection in electric network database of smart grid: graph matching approach. *Electr. Power Syst. Res.* **2016**, *133*, 51–62. [CrossRef]
21. Frank, S.; Steponavice, I.; Rebennack, S. Optimal power flow: a bibliographic survey I. *Energy Syst.* **2012**, *3*, 221–258. [CrossRef]
22. Dwivedi, A.; Yu, X. A maximum-flow-based complex network approach for power system vulnerability analysis. *IEEE Trans. Ind. Inform.* **2011**, *9*, 81–88. [CrossRef]

23. Bretas, A.S.; Bretas, N.G.; Braunstein, S.H.; Rossoni, A.; Trevizan, R.D. Multiple gross errors detection, identification and correction in three-phase distribution systems WLS state estimation: A per-phase measurement error approach. *Electr. Power Syst. Res.* **2017**, *151*, 174–185. [CrossRef]
24. Schavemaker, P.; Van der Sluis, L. *Electrical Power System Essentials*; John Wiley & Sons: Hoboken, NJ, USA, 2017.
25. Bolognani, S.; Dörfler, F. Fast power system analysis via implicit linearization of the power flow manifold. In Proceedings of the IEEE 53rd Annual Allerton Conference on Communication, Control, and Computing (Allerton), Montecello, IL, USA, 29 September–2 October 2015; pp. 402–409.
26. Mohseni-Bonab, S.M.; Rabiee, A.; Jalilzadeh, S.; Mohammadi-Ivatloo, B.; Nojavan, S. Probabilistic multi objective optimal reactive power dispatch considering load uncertainties using Monte Carlo simulations. *J. Oper. Autom. Power Eng.* **2015**, *3*, 83–93.
27. Zhu, J. *Optimization of Power System Operation*; John Wiley & Sons: Hoboken, NJ, USA, 2015.
28. Gbadamosi, S.L.; Nwulu, N.I.; Sun, Y. Multi-objective optimization for composite generation and transmission expansion planning considering offshore wind power and feed-in tariff. *IET Renew. Power Gener.* **2018**, *12*, 1687–1697. [CrossRef]
29. Baldi, P.; Vershynin, R. The capacity of feedforward neural networks. *arXiv* **2019**, arXiv:1901.00434. [CrossRef]
30. Khalil Alsmadi, M.; Omar, K.B.; Noah, S.A.; Almarashdah, I. Performance Comparison of Multi-layer Perceptron (Back Propagation, Delta Rule and Perceptron) algorithms in Neural Networks. In Proceedings of the IEEE International Advance Computing Conference (IACC), Patiala, India, 6–7 March 2009; pp. 296–299.
31. Abass, O.M. Neural networks in business forecasting. *Int. J. Comput.* **2015**, *19*, 114–128.
32. Egmont-Petersen, M.; Talmon, J.L.; Hasman, A.; Ambergen, A.W. Assessing the importance of features for multi-layer perceptrons. *Neural Netw.* **1998**, *11*, 623–635. [CrossRef]
33. Saleh, A.I.; Talaat, F.M.; Labib, L.M. A hybrid intrusion detection system (HIDS) based on prioritized k-nearest neighbors and optimized SVM classifiers. *Artif. Intell. Rev.* **2019**, *51*, 403–443. [CrossRef]
34. Ruuska, S.; Hämäläinen, W.; Kajava, S.; Mughal, M.; Matilainen, P.; Mononen, J. Evaluation of the confusion matrix method in the validation of an automated system for measuring feeding behaviour of cattle. *Behav. Process.* **2018**, *148*, 56–62. [CrossRef]
35. Deng, X.; Liu, Q.; Deng, Y.; Mahadevan, S. An improved method to construct basic probability assignment based on the confusion matrix for classification problem. *Inf. Sci.* **2016**, *340*, 250–261. [CrossRef]
36. Nigerian Electricity Regulatory Commission. Available online: https://nercng.org/index.php/library (accessed on 18 June 2018).
37. Ding, F.; Nagarajan, A.; Chakraborty, S.; Baggu, M.; Nguyen, A.; Walinga, S.; McCarty, M.; Bell, F. *Photovoltaic Impact Assessment of Smart Inverter Volt-Var Control on Distribution System Conservation Voltage Reduction and Power Quality*; National Renewable Energy Lab (NREL): Golden, CO, USA, 2016; NREL/TP-5D00-67296.

© 2019 by the authors. Licensee MDPI, Basel, Switzerland. This article is an open access article distributed under the terms and conditions of the Creative Commons Attribution (CC BY) license (http://creativecommons.org/licenses/by/4.0/).

Article

Optimal Routing an Ungrounded Electrical Distribution System Based on Heuristic Method with Micro Grids Integration

Wilson Pavón [1,*], Esteban Inga [1] and Silvio Simani [2]

1 Department of Electrical Engineering, Universidad Politécnica Salesiana, Quito EC170146, Ecuador; einga@ups.edu.ec
2 Department of Telecommunications, Università degli Studi di Ferrara, 050031 Ferrara, Italy; silvio.simani@unife.it
* Correspondence: wpavon@ups.edu.ec; Tel.: +593-098-768-1168

Received: 18 December 2018; Accepted: 14 March 2019; Published: 16 March 2019

Abstract: This paper proposes a three-layer model to find the optimal routing of an underground electrical distribution system, employing the PRIM algorithm as a graph search heuristic. In the algorithm, the first layer handles transformer allocation and medium voltage network routing, the second layer deploys the low voltage network routing and transformer sizing, while the third presents a method to allocate distributed energy resources in an electric distribution system. The proposed algorithm routes an electrical distribution network in a georeferenced area, taking into account the characteristics of the terrain, such as streets or intersections, and scenarios without squared streets. Moreover, the algorithm copes with scalability characteristics, allowing the addition of loads with time. The model analysis discovers that the algorithm reaches a node connectivity of 100%, satisfies the planned distance constraints, and accomplishes the optimal solution of underground routing in a distribution electrical network applied in a georeferenced area. Simulating the electrical distribution network tests that the voltage drop is less than 2% in the farthest node.

Keywords: electrical distribution system; graph theory; micro grids; heuristic; optimization; planning

1. Introduction

The unpredictable increasing in electricity demand has made challenging the design and planning of any electrical system in transmission or distribution level. The population growth, migration and city planning had reduced the performance of the Electric Distribution Systems (EDS) in large cities, especially in third world countries. The main reason for that is the conventional deployed EDS was designed without formal considerations of planning or projected demand. Consequently, the regular EDS are mainly unplanned and the electricity service throughout the networks are unsatisfactory with problems in the entire system, for instance reliability and stability.

Electricity transportation seriously concerns designers due to the large distance from generation to the final customer. Conversely, the generation in Micro Grid (MG) with Distribute Energy Resources (DER) is close to the end-user or is in the same Low Voltage (LV) network, therefore avoiding the power transmission [1]. Biomass, solar or wind power, and small hydro generators are some examples of DERs. Those alternatives are boosting the local generation, increasing the continuous electrical service, decreasing the fossil fuel dependency and can achieve a clean ecosystem by reducing emissions [2–4].

Nowadays, modern EDS must satisfy optimization, security, reliability, and energy efficiency requirements, which are considered as fundamental requirements in the design and implementation process. For instance, MG is the integration of optimal EDS with DER. In order to implement a Smart

Grid (SG), firstly the EDS should reach security and, reliability requirement via technical planning. Moreover, the EDS must be optimal and technically adequate because the end customer is close to that system, and due to its investment cost this is considerable compared to the entire network [5–7].

Furthermore, DERs are a promising solution for the implementation of Low Carbon (LC) Technologies in a conventional electrical system. Considering that the power generation industry is a considerable source of CO_2, a growing number of EDS have connected to DER in order to follow the LC policies [8]. The LC policies suggest countries adopt clear and measurable objectives to reduce emissions. There is some research which proposes an acceptable level of reduction, as is the case of [9], which proposed a model to reduce 80% of CO_2 emissions taking as based line 1990, and introduced the implementation of mitigation technologies, including DER in EDS.

The Figure 1 shows the percentages contributions of each technology in the reduction of emissions. Special attention is focused on the electricity decarbonisation, smart growth and rooftop PV. The first technology is mainly the integration of renewable energy, which is composed of 90% CO_2 free technologies. The second involves the optimal planning of EDS and the transportation systems. The third constitutes of rooftop PV implementation in residential and commercial buildings considering 10% of electricity demand should be reduced by the implementation of rooftop PV [9].

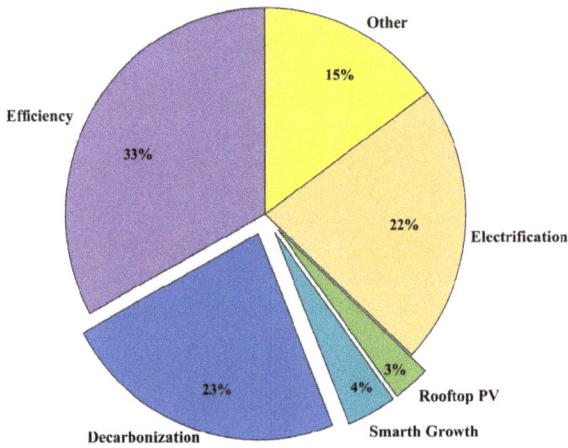

Figure 1. Percentage of CO_2 reduction contribution of Distribute Energy Resources (DER) implemented in a Electric Distribution Systems (EDS) [9].

The mathematical model proposes in this paper achieve the entire connectivity, in order to cope with this objective the minimum expansion tree algorithm was applied, and the radial topology of a georeferenced EDS was obtained. By this methodology, the power balance in the network is achieved automatically and guaranteed, as well as the scalability, including the case of whether further residential or industrial loads would be connected to the system. The model performance test was developed in a Geographic Information System (GIS), where all the elements of the network are represented as nodes. Aside from the map information like streets, roads, and natural features, this representation includes homes, LV transformers and substations of the selected region.

Several researchers have developed models to find the best topology for an optimal EDS planning. For instance, Ref. [10] is one of the first to have presented a detailed overview of expansion planning models, compared to the different mathematical techniques describing the objective functions, constraints, the programming technique, and the pros and cons associated with the model. On the other hand, the approach is commonly used in wireless communication, Inga et al. [6,7,11,12] proposed a hybrid wireless mesh network infrastructure considering a multi-hop system which is planned for electric consumption metering in a metropolitan area network, thereby performing an advanced metering infrastructure for use in MG.

Lavorato et al. [13] proposed a critical analysis to integrate the radially as a constraint in an optimization model of an EDS, and [14] proposed a mathematical procedure for modelling the radial networks. Both studies recognize that the radially constraint is a heavy burden to implement in any model. Other researches have proposed that the problem can be solved using a combination of algorithms, including heuristics to find a good initial solution and then apply the result to a deterministic mathematical optimization [14]. In [15–18] proposed implementation of the Minimal Spanning Tree (MST) to minimize the energy supplied by medium voltage (MV) in an EDS. In [15] algorithm allowed graphing compression, leading to savings in computing time. Ref. [19] also tackled the active power loss minimizing problem using MST.

The optimization algorithm for determining the route for MV feeders was developed using simulated annealing algorithm in [20], who proposed a three stages methodology. Additionally, researchers in [21] describe a heuristic with the objective of minimizing the loss of power applying EDS reconfiguration. Ref. [22] used the complex network analysis and graph theory to explain the properties and exposed the mathematical representation of the electrical topology that are implemented in the real EDSs. In [23] describes the network design problem using the cooperative Tabu research that is the first level of the capacitated multicommodity. Ref. [24] proposed a model, using the adapted genetic algorithm, to minimize the voltage drop in distribution transformers, considering size, quantity, and siting.

There are several heuristics methods that can be used to solve an optimisation problem, in the paper [25] a scheme is explained the pros and cons of the "best solvers", based on the analysis of a considerable amount of articles. The efficiency and closeness to the global optimum solution of some heuristic solvers are tested in [26], where implemented a Home Energy Management solved through five heuristic algorithms.

The heuristics methods applied GIS are investigated in several technological areas, for instance, the introduction of more flexible technologies in urban areas [27]. Whilst [28,29] study the DER penetration in an implemented photo-voltaic systems. The problem in [30] is solved through a modified Particle Swarm Optimization (PSO), which included a new mutation method to improve the global searching thereby avoiding the local optimum. Ref. [31] applied the local search heuristics representing the EDS as a spanning forest problem. The proposed algorithms are based on the research of the shortest spanning sub tree and connection network, originally proposed by [32,33].

Based on the extensive bibliographic research, a model of DER planning with MG integration deployed in a GIS is hardly resolved by linear programming, because it implies a large computational time due to the complexity and the massive amount of involved variables. The proposed problem represents a combinatorial problem, which includes the routing cost minimization as objective function and constraints of connectivity, radial, distance and voltage profiles. In conclusion, the problem is NP-Complete and as a result, lacks a globally optimal solution [30].

For the reasons exposed above, the raised problem is not trivial and it must be solved applying heuristic models. The solution of the mathematical model of the EDS planning is proposed as a routing problem which is approached through a complex network analysis and graph theory [34]. Hence, it is necessary to perform a heuristic model that can reach a near optimal solution or sub-optimal solution. The present paper presents a mathematical model that applies graph theory as a multi-layer algorithm; one of them addresses the problem of routing of a MV network, the second the LV network, and the third allocate the DER in the EDS.

The remainder of this article is organised as follow, in the Section 2 the problem formulation is presented, the simulation results are presented in Section 3. Finally, in Section 4 the conclusions, recommendations and future works.

2. Problem Formulation

The Optimal Routing of Electrical Distribution Networks is defined as a NP-complete problem, to deal with it a heuristic model is used. The model is divided into three algorithms; Algorithm 1 solves the problem in MV network, while Algorithm 2 works with the resolution in LV network, and Algorithm 3 determines the allocation of the rooftop PV in the scenario. In Table 1 are presented the variables used in the model.

Table 1. Parameters and variables.

Nomenclature	Description
X	Latitude element coordinate point or points
Y	Longitude element coordinate point or points
ij	Point to point search variables
X_s, Y_s	Residential customer location
X_{np}, Y_{np}	Street nearest point to any customer
X_{se}, Y_{se}	Substation location
X_{be}, Y_{be}	Streets intersection or candidate sites location
X_{tr}, Y_{tr}	MV to LV transformer final location
XL_{st}, YL_{st}	Member Points of L street
SH	End user location
Ind	Optimal transformer index
N	Number of residential customers
M	Number of LV transformers
S	Number of substations
P	Total Number of subscribers N+M+S
$demN_N$	Individual customer demand
$demM_M$	Individual LV transformer demand
G	PxP connectivity matrix
$dist$	PxP distance matrix
$dist_N$	Distance from N customer to corresponding transformer
Cap	Number capacity constraint for all LV transformer
R	Distance constraint (m) for all LV connections
$Path$	Network connectivity route
$Pred$	Association end-user transformer
PVs	PV amount in the network
PVC	PV rooftop location
PVP	PV power assignation
C	Total customer connectivity in percentage
$CostMV$	Total distance (m) cost of designed LV network
$CostLV_M$	Distance (m) cost of M tranformer
$CostLV$	Total distance (m) cost of desgined low voltage network
$Comp_E$	Computational cost (seg) for each experiment
i,j,k	Counter variables for control loops
$flag, used, z$	Temporal variables
$Loc1, Loc2$	Temporal variables

What the mathematical model accomplishes in the algorithm is represented by the next equation exposed below. The objective Function (1) finds the minimum length of path feeder, where C is the cost of distances and X represents the activation or deactivation of each node connection. The Equations (2) and (3) represent the radial nature for the network where the numbers of connections must be $n - 1$, where n is the number of nodes. Finally, the Equation (4) demonstrates that the connections have two states, like 0 or 1, whether there is disconnections or connection, respectively.

$$\text{Minimize} \sum_{ij \in E} C_{ij} X_{ij} \qquad (1)$$

$$\text{Subject to} \sum_{ij \in E} X_{ij} = n - 1 \qquad (2)$$

$$\sum_{ij \in E: i \in S, j \in S} X_{ij} \leq |S| - 1 \quad \forall S \subseteq V \qquad (3)$$

$$X_{ij} \in \{0, 1\} \quad \forall ij \in E \qquad (4)$$

Algorithm 1 Optimal location and routing a MV grid network.

1: **procedure**
2: *Step: 1 Variables*
3: $\quad P, distN, X, Y, Cap, R$
4: *Step: 2 Optimal transformer selection*
5: $\quad used \leftarrow prim(X, Y)$;
6: $\quad Ind \leftarrow find(sum(used) == 1)$;
7: $\quad X_{tr} \leftarrow X_{be}(Ind)$;
8: $\quad Y_{tr} \leftarrow Y_{be}(Ind)$;
9: *Step: 3 Find nearest street point to customer*
10: $\quad Loc1 \leftarrow [X_s Y_s]$;
11: $\quad Loc2 \leftarrow [XL_{st} YL_{st}]$;
12: $\quad \text{for } i \rightarrow 1 : N \text{ do}$
13: $\quad\quad \text{for } j \rightarrow 1 : length(XL_{st}) \text{ do}$
14: $\quad\quad\quad dist_{i,j} \leftarrow haversine(Loc1, Loc2)$;
15: $\quad\quad\quad z \leftarrow find(dist_{i,j} == min(min(dist_{i,j})))$;
16: $\quad\quad EndFor$
17: $\quad EndFor$
18: $\quad X_{np} \leftarrow Loc2(z, 1)$;
19: $\quad Y_{np} \leftarrow Loc2(z, 2)$;
20: *Step: 4 Optimal Routing MV grid*
21: $\quad X \leftarrow [X_{np} X_{tr} X_{se}]$;
22: $\quad Y \leftarrow [Y_{np} Y_{tr} Y_{se}]$;
23: $\quad dist_{i,j} = haversine(X, Y)$;
24: $\quad G(dist_{i,j} <= R) \leftarrow 1$;
25: $\quad path \leftarrow prim_{mst}(sparse(G))$;
26: *Step: 5 Determine the final cost of MV*
27: $\quad \text{for } i \rightarrow 1 : length(X) \text{ do}$
28: $\quad\quad \text{for } j \rightarrow 1 : length(X) \text{ do}$
29: $\quad\quad\quad costMV \leftarrow costMV + dist_{i,j}(path)$;
30: $\quad\quad EndFor$
31: $\quad EndFor$
32: **End procedure**

The Algorithm 1 has five steps. The first declares the variables, distance R and the capacity number Cap restriction to zero, or receives the georeference information from the map, including the latitude and longitude of end-user, candidate sites and substation location. The information was taken from an OpenStreetMap (OSM) file, including the georeferenced information about the houses' shape, main routes, streets, public spaces, and more. Step 2 determines the optimal transformer selection using Prim algorithm, which returns the number and transformer index of optimal configuration. Step 3 is responsible to find the nearest street point to customer, it is done through the distance calculation of each end-user to the each constituted point street, and determining the closest point to each home, this solution has the same number as end existing users. The fourth step searches the optimal routing of the MV grid, which used the haversine distance calculation to determine the distance between all elements in the network. After that, the connectivity matrix is calculated with the model restrictions, next the Prim minimal spanning tree is applied to find the minimum rout. The fifth step determines the cost that corresponds to the total distance of the elements of the MV network.

The Algorithm 2 determines the optimal routing of the LV grid network, which approaches the problem dividing the network in pieces of the transformer that serves to the end user customer. The solution is proposed in five steps as follows. Step 1 is similar as the Algorithm 1 and aim the initialization or complete the needed information. Step 2 determines the distance between each end user with all solution transformer of Algorithm 1. After that, the connectivity matrix is calculated, which considers the connectivity between the transformer and the substation as already done, and the connection from the substation to end-user as non available. Step 3 implements the Dijkstra algorithm calculation, which find the optimal LV connections. Step 4 calculates the optimal rout of the corresponded elements to the transformer, and the step individually considers the LV connections. Finally, step 5 calculates the final cost that correspond with the final distance of conductor in LV network.

Finally, the Algorithm 3 allows us to determine the allocation of the rooftop photo-voltaic panels in the houses. The houses percentage chosen is 10%, based on the contribution of PV in MG. The algorithm in step 1 gathers the end user coordinates in one array, after the PV amount is determined with the researched criteria and is stored in PVs, in step 2. In step 3 the center of mass is calculated though k-medoids algorithm, the scenario is divided into PVs variable clusters. In step 4, the electrical power is assigned for each end user, and the power for each rooftop is 10 KV, the same for all scenarios.

Algorithm 2 Optimal routing a LV grid network.

1: **procedure**
2: *Step: 1 Variables*
3:     $P, distN, X, Y, Cap, R$
4: *Step: 2 Determine the distance end user, transformer*
5:     $dist_{i,j} = haversine(X, Y)$;
6:     $G(dist_{i,j} <= R) \leftarrow 1$;
7:     $G(1 : N, N + M + 1 : N + M + S) \leftarrow inf$;
8:     $G(N + M + 1 : N + M + S, 1 : N) \leftarrow inf$;
9:     $G(N + 1 : N + M + S, N + 1 : N + M + S) \leftarrow inf$;
10: *Step: 3 Applying Dijkstra*
11:     $Pred \leftarrow dijkstra(G, P)$;
12: *Step: 4 Optimal Routing LV grid*
13:     **for** $Trans \rightarrow 1 : N$ **do**
14:         $X \leftarrow [X_{np}(Pred) X_{Trans}]$;
15:         $Y \leftarrow [Y_{np}(Pred) Y_{Trans}]$;
16:         $dist_{i,j} = haversine(X, Y)$;
17:         $G(dist_{i,j} <= R) \leftarrow 1$;
18:         $path \leftarrow prim_{mst}(sparse(G))$;
19:     *EndFor*
20: *Step: 5 Determine the final cost of LV*
21:     **for** $i \rightarrow 1 : length(X)$ **do**
22:         **for** $j \rightarrow 1 : length(X)$ **do**
23:             $costLV \leftarrow costLV + dist_{i,j}(path)$;
24:         *EndFor*
25:     *EndFor*
26: *End procedure*

Algorithm 3 Allocation of DER PV generator.

1: **procedure**
2: *Step: 1 Inizialization*
3:     $X \leftarrow [X_s]$;
4:     $Y \leftarrow [Y_s]$;
5:     $SH \leftarrow [XY]$;
6: *Step: 2 Determining PV amount*
7:     $PVs \leftarrow floor(length(SH) * 0.1)$;
8: *Step: 3 Determining the center of mass*
9:     $PVC \leftarrow kmedoids(SH, PVs)$;
10: *Step: 4 Power assignation*
11:     $PVP \leftarrow 10KV$;
12: **End procedure**

3. Analysis of Results

The case study is part of the EDS of the area of Tytherington in the north of Macclesfield in Cheshire, England. The limits in longitude in the present study are -2.1360 to -2.1270, meanwhile, the latitude starts from 53.2730 to 53.2810, the total area is 1.15 square kilometers. In the scenario, there are 813 loads with a total power of 5.4 MW. The presented model deploys the EDS, including the network planning expansion. Therefore, the model designs an efficient and reliable EDS, with the lowest investment cost. The network planning expansion allows us to use the initial configuration and expand the EDS with a short and medium time period. The model was developed with the algorithms one and two presented below, which was implemented in Matlab.

In the Table 2 are presented the simulation parameters used in the implementation. The selected area has a density of 700 end users per kilometer square, which is considered lower in comparison with the average density in the cities in Europe. The deployment requires a maximum distance of 100 m from an end user to transformer, with a coverage of 100% in the entire network. The installation type in both networks is under grounded and the configuration is radial in order to accomplish with the EDS requirements. The number of main feeders from the substation is one. Whilst, the voltage in the MV installation, between the substation and the transformers, is 11 KV, and the LV network voltage is 400 V. Finally, the concentrated load is balanced in all the experimental procedure.

The studied georeferenced scenario is shown in Figure 2. First, in order to analyze the designed network performance, the scenario was divided into six different clusters, the homes in the same cluster were outlining with the same colors. The division by clusters was made with the K-medoids algorithm, but any clustering algorithm could be used. The clusters are numbered from 1 to 6 in clockwise, starting with the left upper with the number 1 and the located in the middle left is the 6. The power consumption of each home depends on the cluster membership, in the cluster 1 the average consumption is 300 KVA, whilst the average power in the cluster 2 is 400 KVA and the houses of cluster 6 the consumption is 800 KVA, correspondingly. The power assignment is random normally distributed, depending on the cluster membership.

The substation location is aleatory, where must exist enough space for the implementation of this building. It can be changed, and the optimum substation allocation is proposed for future work. The transformer candidate sites are shown in the graph as well. These sites are called manhole or checkup points. To find these points are considered all the corners or bifurcation points in any street, in total there are 314 checkup points. These points are the input of the prim algorithm with the desired maximum distance, therefore the prim algorithm output is the final transformer allocation.

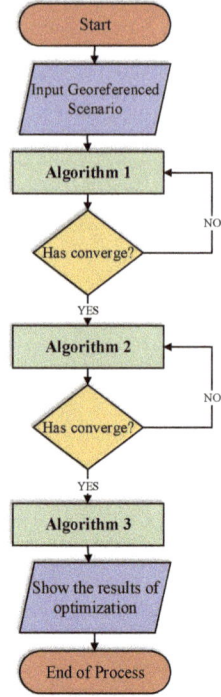

Figure 2. The flowchart of the ordinal interaction of the three algorithms proposed for the authors.

A constraint in the model is the maximum distance between the end user and their corresponding LV transformer. The distance restriction is an input parameter in the prim algorithm, that decided the final transformer allocation. Thus, based on this distance parameter two scenarios are proposed, the first scene takes the restriction of 80 m and the second 100 m, and are called A and B scenario, respectively.

The optimization applying graph theory is based on the connectivity matrix. The connectivity matrix of the presented scenario A is shown in Figure 3, where is seen a symmetrical square matrix of N+M+S elements. Where N is the number of end users, M is the number of activated transformers and S is the substations number. In order to find the connectivity matrix, the distance matrix is calculated, which as shown in the graph represents the distance between homes to homes, homes to transformers, homes to the substation and finally transformers to the substation. The color in the matrix represents the distance, for instance, a dark color means a closer distance compared with a light color. Moreover, the white dots illustrates the possible connections between nodes, the white dots are located whether the restriction distance is accomplished. The number nz in the bottom of the figure is 8426, which represents the number of total connections in the studied scenario. There are two extreme fringes in the figure, the right and the bottom one; those fringes represent the connection between transformers and end users. Notice that the form of the fringes changes respect to the rest of the figure, mainly that there are more white dots which means the higher connection possibility between transformers and end user. This is due the optimal transformer allocation. Besides, the principal diagonal consideration must be considered, because it represents the distance between the same node, and it must be changed for a greater distance in order to do not obtain erroneous model results.

Table 2. Parameter of Model Simulation Model.

Item	Parameter	Value
End user information	Density	700 per square kilometer
	Amount in study	813 in all study
	Location	Georeference
Deployment	Max transformer distance	100 m
	MV Network transformer coverage	100%
	LV Network end users coverage	100%
MV network parameters	Installation type	Undergrounded
	Network configuration	Radial
	Number of primary feeders number	1
	Voltage level	11 KV
	Total power demand	5.4 MVA
LV network parameters	Installation type	Undergrounded
	Network configuration	Radial
	Voltage level	400 V
	Concentrated load	balanced

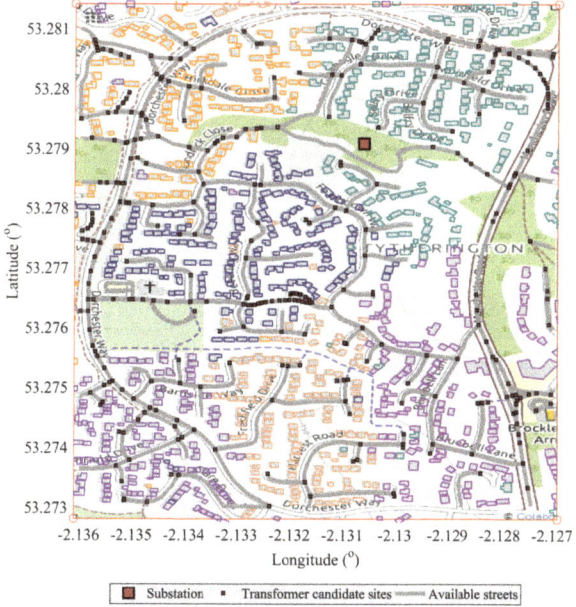

Figure 3. Studied scenario with the transformer candidate cites and substation localization. The end user power consumption is represent with different colors depending on the cluster.

The obtained result with the Algorithm 1 is the sub optimal MV network routing of scenario A is shown in Figure 4, which was generated with a distance constraint of 80 m and a connectivity of 100%. In this scenario, there are 76 transformers located in the candidate sites using the prim algorithm. Therefore, the distance and connectivity constraints are accomplished through the location of the transformers. Moreover, initially, by the MV network route origins in the substation and by means of one feeder deliveries power to all the MV transformers. The planned routing is radial, following the routes of the streets, consequently, the MV network can be implemented as an underground network. The MV network length is 14.05 km, connected by one conductor all the transformers through MST. The planned routing is an alternative method for resilience network in order to the designer can be planned optional routes in case of adverse operating conditions, this topic is proposed as future work. Scenario B is shown in Figure 5, which was generated with distance and connectivity constraints of 100 m and 100%, correspondingly. In the present scenario, there are 55 transformers located in the candidate sites, accomplishing the desired constraints. As well as the previous scenario, the planned routing is radial, following the routes of the streets, consequently, the MV network can be implemented as an underground network. The MV network length is 12.15 km, connected by one conductor all the transformers through MST. In scenario B, there are 21 transformers and 2 km of conductor less than the last analyzed scenario. However, those savings affects the LV distribution network design because the reduction in the transformers amount represents the overloading of them. Moreover, the transformers power capacity must be raised by reason that the corresponding demand will the higher. Under those circumstances, Scenario A henceforth will be called the suboptimal solution of the presented model. The presented results in Table 3 compare the data obtained from the A and B scenarios.

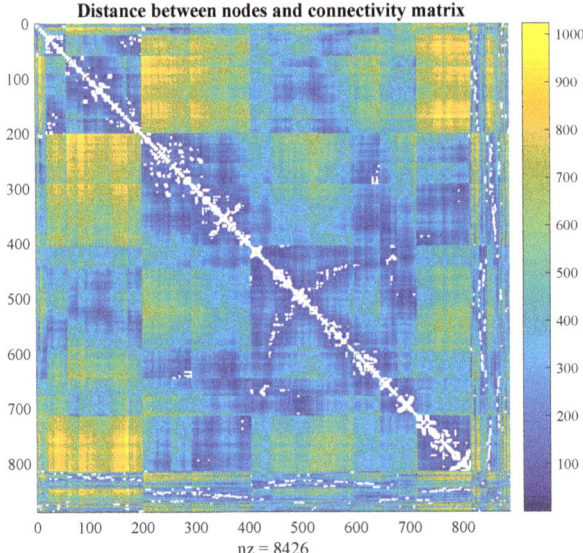

Figure 4. Distance and connectivity matrix of scenario A.

The LV network was designed through Algorithm 2, optimal routing of a LV grid network, explained in the sections below. The obtained result is presented in Figure 6, there is shown the georeferenced scenario with LV network implementation and irregular polygons sketched in the graph, delimiting the transformers area of service. The end users with 100% connectivity are connected to the LV network through the operator service cable. Those cables connect the home nearest point to the corresponding nearest street point; this calculation is included in the model. The distance in the LV network includes the length from house to the street and from that point down the street to the transformer. The model for the LV network design is subject to the application of distance restriction.

Figure 5. Sub optimal Routing of medium voltage (MV) Network, scenario A.

The connection representation between the transformer with their end users are shown in the Figure 7 through the irregular polygons, those indicate the service area of each activated transformer. The mentioned polygons gather the elements belonging to the individual LV EDS, the transformer normally is inside the polygon, but can be in the edge, the polygon joins all the connection house points including the transformer. If the transformer does not belong to any polygon, it means that it just delivers power to one end user, normally the closest one. There are some houses in the study that are not considered as nodes of the network, especially they are located within the map end limits. The end users connect to the corresponding transformer via under grounded electrical installation.

The result implementation of Algorithm 3 is shown in the Figure 8, the allocation of rooftop PV is. All the PV panels have a power of 10 KVA. The percentage of houses with PV panels are 10% of all the scenario, in total 79 houses, with a total power of 790 KV. As a result, the rooftop PV panels contribute to 14.5% to the total power deman. Notice that the distribution of the rooftop PV are in all the maps, showing the practical allocation of the PV panels. The PV will reduce the power consumption of the power delivered from the substation in approximately 10%. The MV transformer should be bidirectional for the implementation. The design should include the protection and control of the network.

The obtained results after the application of the three layers algorithm are shown in the figures below. The Figure 9 shows power (KVA) detailed in the 76 LV network, the purple bar represents the power consumption of the LV network, while the dark green bar is the representation of the power contribution of the installed rooftop PV, and the light green bar indicates the assigned MV transformer, taking on consideration standardized transformer values. Moreover, the pink area is the average power consumption that represents a power of 71.6 KVA. Analysing the power transformers, it is seen that the highest transformer has a power of 280 KVA, whereas the lowest assigned transformer is

10 KVA. The differences of each power LV network is due to the algorithm considerations that allow to taking the terrain characteristics respecting the imposed constraints.

The Figure 10 shows the number of end users connected to each LV transformer, where 32 is the maximum number, and 1 is the minimum end users, the average end users connected to each transformer is 11. It can be seen that there exists a direct relationship between the power in each LV network compared with the number of connected users, and the relation between those variables is considered as linear.

In addition, the proposed model allows planning of an MV and LV network in a georeferenced area, maintained under defined constraints and technical specifications with the minimal cost. Figure 11 presents the total number of transformer classified by the assigned electrical power. For instance, there are nine transformers of 10 KVA, four of 20 KVA, and 10 transformers with a power of 80 KVA, which is the highest amount of transformers. While the 120 KVA and 170 KVA transformers are quantitatively less than others, with one transformer, respectively. Furthermore, there are two 280 KVA transformers, which is the highest assigned power. Thus, it is demonstrated that there are areas in the scenario with high-end user density, this characteristic is a characteristic of the 30th and 39th transformers.

The distance constraint in the scenarios A and B, between end user to the corresponding transformer, are 80 and 100 respectively. The coverage is 100% and the % of drop voltage is less than 2% for both scenarios. The number of activated transformers for the scenario a is 76 with an MV grid length of 14.05 km, thus the average transformer distance to the user is 33 m. Compared with the scenario B, which has 55 transformers with a MV grid length of 12.15%, but the distance of the transformer to the end user is 40 m, higher than the case A. As a result, scenario A was selected for the sub-optimum scenario and the selected design to be implemented.

Figure 6. Sub optimal Routing of the MV Network, scenario B.

Sustainability **2019**, *11*, 1607

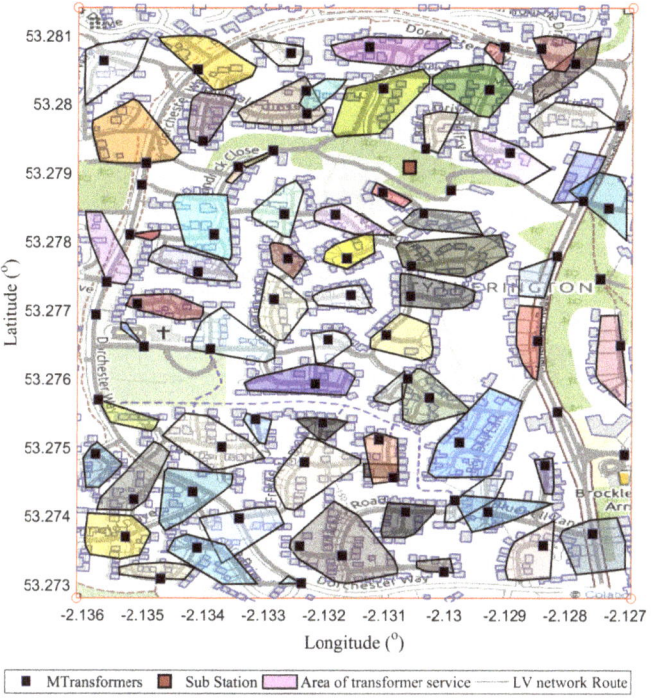

Figure 7. Suboptimal low voltage (LV) Network Routing. Transformer correspondence with end users.

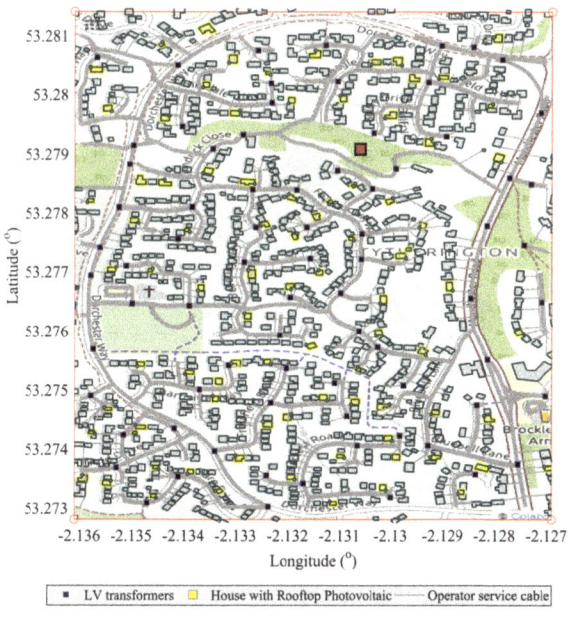

Figure 8. Photovoltaic (PV) rooftop on Distributed generation, considering the 10% of end users.

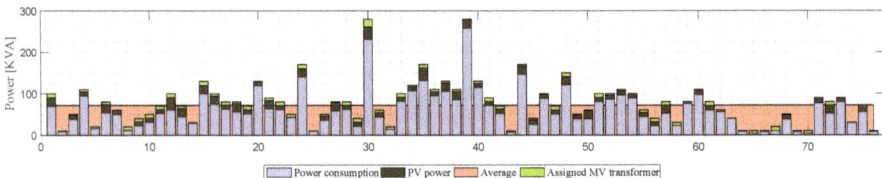

Figure 9. Obtained model results, power consumption and end users for each MV transformer. Cite: Author.

The results obtained in the scenario A were tested in electrical simulation software, the implementation is presented in the Figures 12 and 13, the numeric results are summarised in Table 2. The 76 transformers are in the exact location where the algorithm determined. The simulation was development taking on account the real distance of the feeders, thus the electrical analysis is close to the real implementation. Moreover, in the Figure 12 is shown the end user voltage compared in terms of distance from the sources, where it can see that the farthest have the higher drop voltage, but they are less than the 2% compared with the source.

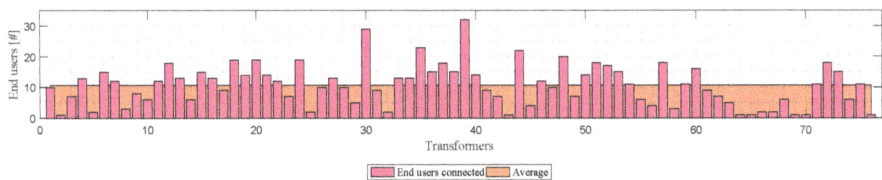

Figure 10. Obtained model results, power consumption and end users for each MV transformer.

Table 3. Implemented Results.

Specification	Scenario A	Scenario B
Max distance model constraint [m]	80	100
MV and LV Coverage [%]	100	100
Distribution transformers [number]	76	55
MV grid length [Km]	14.05	12.15
Voltage drop in [%]	Max. 2%	Max. 2%
LV Transformer to end user average distance [m]	33m	40m

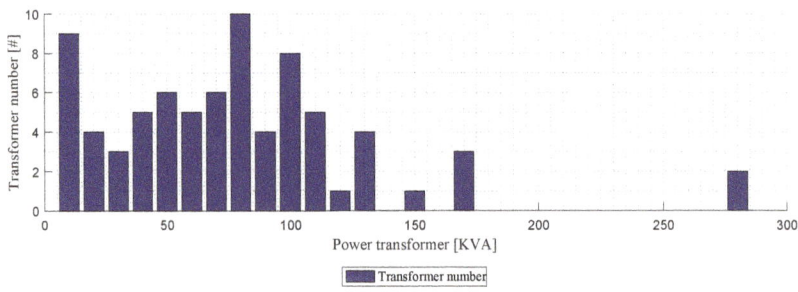

Figure 11. Amount of transformer depending on the electrical power assigned, considering standardised quantities.

Figure 12. The implemented MV network applied in electrical simulation software.

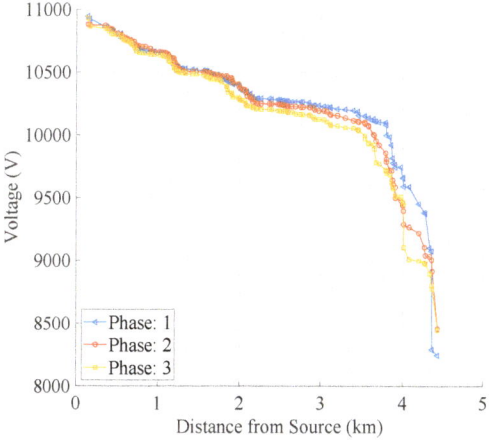

Figure 13. End user voltage (V) compared in terms of Distance from the source (m).

4. Conclusions and Future Works

This paper proposed a heuristic algorithm based model to solve the routing underground electrical networks problem in a georeferenced area. The model proposed a three layered algorithm; the first handles transformer allocation and routing of the MV network, the second algorithm works out the LV network and transformer sizing, and the third presents a method to allocate DER in an EDS. In this research, an array of rooftop photovoltaic panels with a specific criteria was allocated. The modelled networks were implemented in an electrical simulation software to demonstrate the feasibility of the proposed topology.

The proposed algorithm is capable of routing a network in a georeferenced area, taking into account the characteristics of the terrain, such as streets or intersections, including scenarios without squared streets. The modelled network achieves distance, and end user number constraints.

The suboptimal routing underground electrical networks were obtained, minimizing the implementation cost and maximizing the quality of electrical services and the reliability in the network, with a farthest node voltage drop of maximum 2%. The MV grid length was 14.05 km, with 76 activated transformers as a total number, an average of 71.6 KVA and 11 connected end users. Further, the allocated rooftop PV panels contributed 14.5% of the total demand of the network.

The optimum substation allocation, an alternative method for resilience network design in order to accommodate optional routes in case of adverse operating conditions, the application of control techniques and electrical protection in the EDS and the integration of the demand curve in the implementation of PV generation are proposed as future work.

Author Contributions: Conceptualization, methodology, software, resources, validation, formal analysis, W.P., E.I. and S.S.; investigation, writing—original draft preparation W.P.; writing—review and editing, supervision, E.I. and S.S.

Funding: This research was funded by Universidad Politécnica Salesiana-Ecuador. This work was supported by Universidad Politécnica Salesiana and GIREI—Smart Grid Research Group under the optimal model for control and operation of electrical distribution substations project—2018.

Conflicts of Interest: The authors declare no conflict of interest.

References

1. Wong, J.; Baroutis, P.; Chadha, R.; Iravani, R.; Graovac, M.; Wang, X. A methodology for evaluation of permissible depth of penetration of distributed generation in urban distribution systems. In Proceedings of the IEEE Power and Energy Society 2008 General Meeting: Conversion and Delivery of Electrical Energy in the 21st Century, PES, Pittsburgh, PA, USA, 20–24 July 2008; pp. 1–8. [CrossRef]
2. Vallejos, W.D.P. Standalone photovoltaic system, using a single stage boost DC/AC power inverter controlled by a double loop control. In Proceedings of the 2017 IEEE PES Innovative Smart Grid Technologies Conference—Latin America (ISGT Latin America), Quito, Ecuador, 20–22 September 2017; pp. 1–6. [CrossRef]
3. Editor, R.B. *Handbook of Distributed Generation Electric Power Technologies, Economics and Environmental Impacts*; Springer International Publishing: Cham, Switzerland, 2017; p. 819.
4. Zubo, R.H.; Mokryani, G.; Rajamani, H.S.; Aghaei, J.; Niknam, T.; Pillai, P. Operation and Planning of Distribution Networks with Integration of Renewable Distributed Generators Considering Uncertainties: A Review. 2017. Available online: https://doi.org/10.1016/j.rser.2016.10.036 (accessed on 14 March 2019).
5. Martinez-Pabon, M.; Eveleigh, T.; Tanju, B. Smart Meter Data Analytics for Optimal Customer Selection in Demand Response Programs. *Energy Procedia* **2017**, *107*, 49–59. [CrossRef]
6. Inga, E.; Céspedes, S.; Hincapié, R.; Cárdenas, A. Scalable Route Map for Advanced Metering Infrastructure Based on Optimal Routing of Wireless Heterogeneous Networks. *IEEE Wirel. Commun.* **2017**, *24*, 1–8. [CrossRef]
7. Peralta, A.; Inga, E.; Hincapié, R. FiWi Network Planning for Smart Metering Based on Multistage Stochastic Programming. *IEEE Latin Am. Trans.* **2015**, *13*, 3838–3843. [CrossRef]
8. Ge, S.; Wang, S.; Lu, Z.; Liu, H. Substation planning method in an active distribution network under low-carbon economy. *J. Modern Power Syst. Clean Energy* **2015**, *3*, 468. [CrossRef]

9. Williams, J.H.; DeBenedictis, A.; Ghanadan, R.; Mahone, A.; Moore, J.; Morrow, W.R.; Price, S.; Torn, M.S. The Technology Path to Deep Greenhouse Gas Emissions Cuts by 2050: The Pivotal Role of Electricity. *Science* **2011**, *335*, 53–59. [CrossRef] [PubMed]
10. Temraz, H.K.; Quintana, V.H. Distribution system expansion planning models: An overview. *Electr. Power Syst. Res.* **1993**, *26*, 61–70. [CrossRef]
11. Inga-Ortega, E.; Peralta-Sevilla, A.; Hincapie, R.C.; Amaya, F.; Monroy, I.T. Optimal dimensioning of FiWi networks over advanced metering infrastructure for the smart grid. In Proceedings of the 2015 IEEE PES Innovative Smart Grid Technologies Latin America (ISGT LATAM), Montevideo, Uruguay, 5–7 October 2015; pp. 30–35. [CrossRef]
12. Bader, D.A.; Cong, G. Fast shared-memory algorithms for computing the minimum spanning forest of sparse graphs. *J. Parallel Distrib. Comput.* **2006**, *66*, 1366–1378. [CrossRef]
13. Lavorato, M.; Franco, J.F.; Rider, M.J.; Romero, R. Imposing Radiality Constraints in Distribution System Optimization Problems. *IEEE Trans. Power Syst.* **2012**, *27*, 172–180. [CrossRef]
14. Ahmadi, H.; Martí, J.R. Distribution System Optimization Based on a Linear Power-Flow Formulation. *IEEE Trans. Power Deliv.* **2015**, *30*, 25–33. [CrossRef]
15. Carcamo-Gallardo, A.; Garcia-Santander, L.; Pezoa, J.E. Greedy Reconfiguration Algorithms for Medium-Voltage Distribution Networks. *IEEE Trans. Power Deliv.* **2009**, *24*, 328–337. [CrossRef]
16. Peralta, A.; Inga, E.; Hincapié, R. Optimal Scalability of FiWi Networks Based on Multistage Stochastic Programming and Policies. *J. Opt. Commun. Netw.* **2017**, *9*, 1172. [CrossRef]
17. Han, X.; Liu, J.; Liu, D.; Liao, Q.; Hu, J.; Yang, Y. Distribution network planning study with distributed generation based on Steiner tree model. In Proceedings of the 2014 IEEE PES Asia-Pacific Power and Energy Engineering Conference (APPEEC), Hong Kong, China, 7–10 December 2014; Volume 1, pp. 1–5. [CrossRef]
18. Oertel, D.; Ravi, R. Complexity of transmission network expansion planning NP-hardness of connected networks and MINLP evaluation. *Energy Syst.* **2014**, *5*, 179–207. [CrossRef]
19. Li, H.; Mao, W.; Zhang, A.; Li, C. An improved distribution network reconfiguration method based on minimum spanning tree algorithm and heuristic rules. *Int. J. Electr. Power Energy Syst.* **2016**, *82*, 466–473. [CrossRef]
20. Gouin, V.; Alvarez-Hérault, M.C.; Raison, B. Innovative planning method for the construction of electrical distribution network master plans. *Sustain. Energy Grids Netw.* **2017**, *10*, 84–91. [CrossRef]
21. De Oliveira, E.J.; Rosseti, G.J.; de Oliveira, L.W.; Gomes, F.V.; Peres, W. New algorithm for reconfiguration and operating procedures in electric distribution systems. *Int. J. Electr. Power Energy Syst.* **2014**, *57*, 129–134. [CrossRef]
22. Abeysinghe, S.; Wu, J.; Sooriyabandara, M.; Abeysekera, M.; Xu, T.; Wang, C. Topological properties of medium voltage electricity distribution networks. *Appl. Energy* **2018**, *210*, 1101–1112.. [CrossRef]
23. Crainic, T.G.; Li, Y.; Toulouse, M. A first multilevel cooperative algorithm for capacitated multicommodity network design. *Comput. Oper. Res.* **2006**, *33*, 2602–2622. [CrossRef]
24. Mendoza, J.E.; López, M.E.; Peña, H.E.; Labra, D.A. Low voltage distribution optimization: Site, quantity and size of distribution transformers. *Electr. Power Syst. Res.* **2012**, *91*, 52–60. [CrossRef]
25. Chicco, G.; Mazza, A. Heuristic optimization of electrical energy systems: Refined metrics to compare the solutions. *Sustain. Energy Grids Netw.* **2019**, *17*, 100197. [CrossRef]
26. Javaid, N.; Qureshi, T.N.; Khan, A.H.; Iqbal, A.; Akhtar, E.; Ishfaq, M. EDDEEC: Enhanced developed distributed energy-efficient clustering for heterogeneouswireless sensor networks. *Procedia Comput. Sci.* **2013**, *19*, 914–919. [CrossRef]
27. Alhamwi, A.; Medjroubi, W.; Vogt, T.; Agert, C. GIS-based urban energy systems models and tools: Introducing a model for the optimisation of flexibilisation technologies in urban areas. *Appl. Energy* **2017**, *191*, 1–9. [CrossRef]
28. Tang, Y.; Mao, X.; Ayyanar, R. Distribution system modeling using CYMDIST for study of high penetration of distributed solar photovoltaics. In Proceedings of the 2012 North American Power Symposium (NAPS), Champaign, IL, USA, 9–11 September 2012; pp. 1–6. [CrossRef]
29. Freitas, S.; Santos, T.; Brito, M.C. Impact of large scale PV deployment in the sizing of urban distribution transformers. *Renew. Energy* **2018**, *119*, 767–776. [CrossRef]

30. Aghaei, J.; Muttaqi, K.M.; Azizivahed, A.; Gitizadeh, M. Distribution expansion planning considering reliability and security of energy using modified PSO (Particle Swarm Optimization) algorithm. *Energy* **2014**, *65*, 398–411. [CrossRef]
31. Davidescu, G.; Stützle, T.; Vyatkin, V. Network planning in smart grids via a local search heuristic for spanning forest problems. In Proceedings of the 2017 IEEE 26th International Symposium on Industrial Electronics (ISIE), Edinburgh, UK, 19–21 June 2017; pp. 1212–1218. [CrossRef]
32. Kruskal, J.B. On the Shortest Spanning Subtree of a Graph and the Traveling Salesman Problem. *Proc. Am. Math. Soc.* **1956**, *7*, 48–50. [CrossRef]
33. Prim, R.C. Shortest Connection Networks And Some Generalizations. *Bell Syst. Tech. J.* **1957**, *36*, 1389–1401. [CrossRef]
34. Lezama, F.; Soares, J.; Vale, Z.; Rueda, J.; Rivera, S.; Elrich, I. 2017 IEEE competition on modern heuristic optimizers for smart grid operation: Testbeds and results. *Swarm Evolut. Comput.* **2019**, *44*, 420–427. [CrossRef]

© 2019 by the authors. Licensee MDPI, Basel, Switzerland. This article is an open access article distributed under the terms and conditions of the Creative Commons Attribution (CC BY) license (http://creativecommons.org/licenses/by/4.0/).

Article

Introduction of Smart Grid Station Configuration and Application in Guri Branch Office of KEPCO

Jaehong Whang [1], Woohyun Hwang [2], Yeuntae Yoo [3] and Gilsoo Jang [3,*]

1. KU-KIST GreenSchool, Graduate School of Energy and Environment, Korea University, Seoul 02841, Korea; jh_whang@korea.ac.kr
2. KEPCO Academy, Seoul 01793, Korea; hblue@kepco.co.kr
3. School of Electrical Engineering, Korea University, Seoul 02841, Korea; yooynt@korea.ac.kr
* Correspondence: gjang@korea.ac.kr; Tel.: +82-2-3290-3246

Received: 29 July 2018; Accepted: 28 September 2018; Published: 30 September 2018

Abstract: Climate change and global warming are becoming important problems around the globe. To prevent these environmental problems, many countries try to reduce their emissions of greenhouse gases (GHGs) and manage the consumption of energy. The Korea Electric Power Corporation (KEPCO) introduced smart grid (SG) technologies to its branch office in 2014. This was the first demonstration of a smart grid on a building, called the Smart Grid Station (SGS). However, the smart grid industry is stagnant despite of the efforts of KEPCO. The authors analyzed the achievements to date, and proved the effects of the SGS by comparing its early targets to its performance. To evaluate the performance, we analyzed the data of 2015 with the data of 2014 in three aspects: peak reduction, power consumption reduction, and electricity fee savings. Furthermore, we studied the economic analysis including photovoltaic (PV) and energy storage system (ESS) electricity fee savings, as well as running cost savings by electric vehicles. Through the evaluation, the authors proved that the performance surpassed the early targets and that the system is economical. With the advantages of the SGS, we suggested directions to expand the system.

Keywords: smart grid; Smart Grid Station; renewable energy sources; energy management system

1. Introduction

For many years, concerns about global warming and climate change have been growing. In response to these environmental problems, most developed and developing countries have held meetings and sought countermeasures. However, due to the expiration of the Kyoto Protocol in 2020 (Post2020), the Paris Agreement—a statement of intent to address climate change problems—was signed by 195 countries at the twenty-first Conference of the Parties of the United Nations Framework Convention on Climate Change (UNFCCC), held in Paris, France in 2015. The objective of the Paris Agreement was prevent the increase in the global average temperature from rising more than 2 °C above pre-industrial levels [1]. Each country set its own target with regard to greenhouse gas (GHG) emissions. Table 1 shows the goals for the GHG reduction of some selected countries.

Table 1. Goals of greenhouse gas (GHG) emissions reduction for selected countries [2–5].

Countries	Goals
China	To lower carbon dioxide emissions per unit of GDP by 60–65% from the 2005 level
EU	At least 40% domestic reduction in GHG emissions by 2030 compared to 1990
Japan	At the level of GHG emission reductions of 26% by 2030 compared to 2013
South Korea	GHG emission reduction by 37% from the business as usual (BAU) level by 2030

Through Intended Nationally Determined Contributions, the Korean government set a goal to reduce GHG emissions by 37% compared to business as usual (BAU) by 2030 [5]. In their effort, the Korean government has tried to expand renewable energy (RE) generation and developed new technologies. One of these technologies is the smart grid (SG), which is a new concept of an electrical grid integrated with information and communication technologies (ICT).

Since 2009, the Korea Electric Power Corporation (KEPCO, a public organization) has installed and demonstrated SG technologies. The Jeju Smart Grid Demonstration Project was the first test-bed built on Jeju Island in 2009. This project had five themes: smart place, smart transportation, smart renewable, smart power grid, and smart service. It included renewable energy sources, electric meters, electric vehicles, a battery system, demand responses, transmissions, communications, etc. Using the experience gained in that project, the Smart Grid Station (SGS) was built in the Guri branch office building of KEPCO in 2014 as the first demonstration. The "station" in SGS refers to a place or building that can provide various services. Therefore, the SGS is a place that provides intelligent electricity services to customers. This new business model is different from building energy management systems (BEMSs). The BEMS is used for minimizing energy costs by primarily managing HVAC (heating, ventilation, and air conditioning), lighting, and other systems [6], and Ock et al. have proposed a control system using building energy control patterns to adjust the energy use. The HVAC is regarded as an important portion in load demand in [7]. Ferro et al. [7] have suggested an architecture based on model predictive control to improve the operation efficiency of building energy consumption. A SG is generally composed of distributed energy resources (DERs), such as photovoltaics (PVs) and wind turbines (WTs), an operation system (OS) as an energy management system (EMS), an energy storage system (ESS), advanced metering infrastructure (AMI), and other smart devices [8]. References [9–11] are about smart zero-energy buildings that utilize internet of things technologies. Especially, Kolokotsa [9] have emphasized the importance of zero-energy buildings for the smart community, but the support basis is weak in that there is no case study. Wurtz et al. [10] have described a global research strategy to improve a smart software. The authors have also considered the strategy in a living lab. In [11], the authors have described smart buildings with a new technology. However, the proposed system is limited in that the system is focused on the internet of things. In [12], Kim et al. have suggested an EMS algorithm based on reinforcement learning to reduce energy cost. However, this research ignored charging and discharging loss of ESS, and the system is composed of simple devices. Barbato et al. [13] have focused on an energy management framework integrating renewable energy, storage bank, and demand response in a smart campus. The authors have discussed scenarios to minimize the energy cost. In [14], the authors have dealt with the lighting energy consumption in educational institutes, and have applied a data mining tool to reduce the energy waste of lighting. The recent research has described smart buildings, and has mostly focused on algorithms with improvements. On the other hand, we describe the first SGS demonstration, which is a more comprehensive solution than the smart buildings mentioned above, in that the SGS integrates software with hardware including electric vehicles (EVs), a building automation system (BAS), and a distribution automation system. We also conducted a performance evaluation and analyzed its economic feasibility. The goal of the SGS is to optimize energy consumption by utilizing various technologies, even though the SGS is connected to the power grid. Especially, the OS can balance supply and demand in real-time by monitoring and controlling the whole system. KEPCO has determined that the office could shave power peak and reduce power consumption by use of the SGS. As a result, the SGS has expanded to 121 of the branch offices. However, the expansion is limited to KEPCO's internal branch offices in Korea. Although it has been a few years since the first SGS was built, the smart grid industry is stagnant. The authors recognized that an analysis was needed to prove the performance of SGS to promote the SG industry. For this purpose, this paper presents the concepts and features in Section 2, the description of components in Section 3, and the analysis of the performance of SGS in Section 4. We calculated the performances about peak shaving, reducing power consumption, and saving electricity fees to prove the advantages of the technology. The authors also evaluate the economic feasibility by

PV, ESS, and EV in Section 5. Based on the results, a discussion is given in Section 6, and Section 7 concludes the paper.

2. SGS Concept and Features

The objectives of SGS are to optimize the usage of electricity and to reduce the electricity fee and consumption in a building, with the integration of various technologies.

When renewable energy sources are connected to both the grid and ESS-generated power, the power from renewable energy can be supplied to the load directly or can charge a battery of the ESS. Also, the battery is charged from the grid when the price of electricity on the grid is low and is discharged when the price is high. This allows a building to save money and reduce power consumption. In other words, less energy production is required from fossil fuel generators during peak load time. Consequently, the SGS benefits the environment by reducing CO_2 emissions. As a public organization, KEPCO has developed the SGS with small- and mid-sized businesses to grow together. Through the accompanied growth, the KEPCO has contributed to popularize components of the system. These effects are shown in Table 2.

Table 2. Details of the Smart Grid Station (SGS).

Objective	Reducing Consumption and Saving on Electricity Fees by Optimizing Usage in a Building
Expected Effects	<ul><li>Reduce 5.0% of electricity peak in a year</li><li>Reduce 9.6% of power consumption for a year</li><li>Save on electricity fees</li><li>Reduce 5.0% of CO_2 for a year</li><li>Accompanied growth with small- and mid-sized businesses</li></ul>
Features	<ul><li>Remote control of demand in a building</li><li>Energy management system construction based on SG</li><li>Information and communication technologies (ICT) convergence on intelligent energy management</li></ul>

The Guri SGS project consisted of two steps. The period of the first step was from October 2013 to February 2014. This step comprised the installation of PV, ESS, a slow-charging type of EV charger, AMI, a smart distribution board, and BAS components. A main goal of the first step was to optimize the building's energy consumption based on the SG. The second step comprised the addition of WT, an HVAC control system, and improving the operation system. The period of the second step was from December 2014 to June 2015.

Figure 1 is a diagram of SGS components. It shows power connections and communication connections. PV was connected to a power conversion system (PCS), which makes the power from the renewable energy stable. Because WT was installed at the second step, it was connected to the transformer (TR) room directly. This solution has the characteristics of a test-bed to optimize the operation of energy consumption.

The power from RE can charge batteries or supply loads including lights, outlets, HVAC, variable frequency drives (VFDs), and EVs. The equipment of the SGS is interconnected in the transformer (TR) room. Also, the KEPCO grid is directly connected to the TR room, and the power quality is checked by AMI. The operation system is a software program that plays a key role in integrating other technologies. This system gathers the various data, including voltage, current, frequency, communication status, and amount of generated power. This means that the OS can not only control each element remotely, but also maintain the power balance between various components. Each component will be described in Section 3.

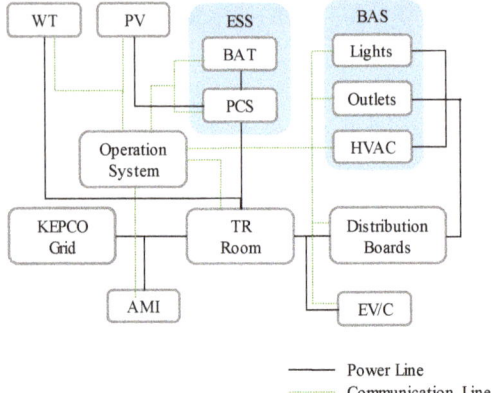

Figure 1. The main components of the SGS. AMI: advanced metering infrastructure; BAS: building automation system; BAT: battery; ESS: energy storage system; EV/C: electric vehicle chargers; HVAC: heating, ventilation, and air conditioning; KEPCO: Korea Electric Power Corporation; PCS: power conversion system; PV: photovoltaic; TR: transformer; WT: wind turbine.

3. Components Description

3.1. DERs and Operation System

3.1.1. Photovoltaic

The PV system was mounted at 30° on the rooftop. The maximum power of each module is 250 W, and the total capacity of the system is 20 kWp. The system is composed of 84 modules consisting of monocrystalline silicon cells, but four of them are dummies which are not generated and connected to the system. The dummies are decorations to make the shape rectangular in 4 by 21. The connection is 16 series by 5 parallel because of space restriction. The capacity was adopted at 5% of the contracted power (400 kW) of the Guri office to reduce 5% of the power peak. The PV system supplies the power to the building at peak load time and mid load time, and charges a battery at off-peak load time. By utilizing the PV system, KEPCO expected that 9.6% of power consumption would be possible. Table 3 provides the specifications of the PV system.

Table 3. Specifications of the PV system.

Device	Monocrystalline silicon
Max Power	250 Wp (60 cells)
Max Voltage	497.6 V (31.1 V × 16)
Efficiency	15.7%
Capacity	20 kWp (6 by 14, 84 modules, 4 dummies)
Connection	16 series by 5 parallel

The power from PV in summer season, from June to August, does not charge the battery but supplies to building loads directly to reduce the peak and the power consumption.

3.1.2. Wind Turbine

A WT system was mounted on the rooftop in March 2015. The PV and the WT installed were as in Figure 2. A vertical axis-type WT was selected for the SGS because this WT is suitable in urban areas since it is not influenced by the direction of the wind [15] and does not make noise when the turbine rotates. Also, its cut-in wind speed is a light wind of 3 m/s. These features make the WT system easy to install on buildings, but the rated power is 1.2 kW at 15 m/s. This WT was not custom-made, and there was a space restriction. To install the WT on the rooftop while making the best use of the space, the system designer could only choose the small size of WT rated at 1.2 kW. Although this WT can generate 53 kWh per year according to Weibull performance calculations, the system is not optimal for contributing to energy use reduction, because the wind does not blow fast enough in the urban area. Nevertheless, its presence is meaningful in that it is an attempt at WT installation on a building.

Because this vertical type of WT has unstable output at certain wind speeds and low efficiency [15,16], it has its own interconnected inverter to stabilize the output. Table 4 shows the features of the WT, and Table 5 shows the details of this inverter.

Figure 2. PV and WT system at the Guri office.

Table 4. Performance characteristics of the WT.

Type	Vertical axis
Size	1400 mm × 1800 mm
Rated power	1.2 kW at 15 m/s
Cut-in wind speed	3 m/s
Extreme wind speed	52.5 m/s

Table 5. Specifications of the WT inverter.

Input	Rated voltage	430 VDC
	Voltage range	60–600 VDC
Output	Max capacity	3kW (single phase)
	Rated voltage	220 VAC (±10%)
	Rated frequency	60 Hz (±0.5 Hz)
	Efficiency	98%
	Power factor	99%

3.1.3. Energy Storage System

The ESS is composed of a battery and a PCS. Figure 3 shows the battery and the PCS installed on the rooftop. The ESS can be used for either on-grid status or off-grid status. The ESS has various effects: peak shaving, load leveling, providing constant voltage and constant frequency (CVCF), cost reduction, load compensation, and so on [17]. In this paper, the authors focused on peak shaving and load shifting. Regarding the first function, the ESS charges power from renewable energy sources at the off-peak load time and discharges the power at the peak time. Concerning the second function, the ESS charges a battery with the power from the grid in the evening and discharges the battery in the afternoon. By peak shaving and load shifting, the electricity fee can be saved.

The PCS converts DC-to-AC and AC-to-DC. This means that it can function both as a converter and an inverter. PV systems generally have their own inverters, whereas the PCS used in the SGS is connected with the battery as well as the PV. Because the PCS is a hybrid type, it is possible to charge and discharge the battery simultaneously, making it possible to optimize the power from the PV and the battery. Figure 4 is the inner connection diagram of the PCS. The capacity of the PCS was determined as 30 kW by adding 20 kWp of the PV system and 10 kW of the expected peak reduction, which is 5% of the power peak (180 kW) at the Guri office.

Figure 3. Battery (**left**) and PCS (**right**) on rooftop.

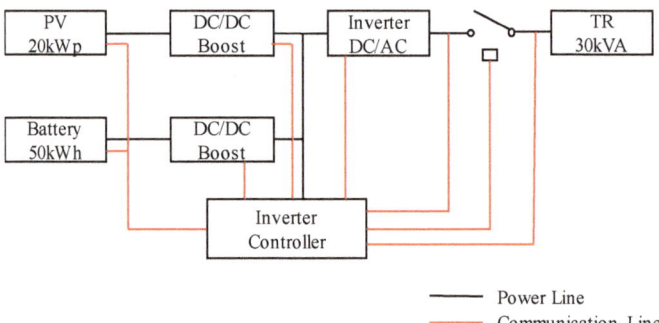

Figure 4. Inner connection diagram of PCS.

A lithium iron phosphate (LiFePO$_4$) battery was selected because this battery has better thermal and chemical safety than other types of batteries [18]. The life cycle is 4000 cycles at 80% of depth of discharge (DOD). Its size is 50 kWh, and the capacity was designed to discharge for five hours at 8 kW while considering 80% of DOD. The 4000 life cycles means that the battery can be used 4000 times if it charges-and-discharges power in the range of 20% to a full charge state. This range is established to prevent the battery from reaching a full discharge state. The specification of the ESS is in Table 6.

There are three discharge schedules that the operator adjusts, as follows:

1. Uniform discharge: discharges a uniform amount of power from the battery during peak and mid-load times continuously;
2. Continuous differential discharge: continuously discharges during peak and mid-load times, but the amount is different during peak load time;
3. Non-continuous differential discharge: discharges non-continuously during peak and mid-load times, and the amount is different during peak-load time.

In summary, the ESS charges the batteries at the off-peak load time and discharges them during the peak and mid-load times on weekdays. It is expected that a customer can reduce power peak by 5% with these schedules.

Table 6. Specifications of the ESS.

PCS	Capacity	30 kW/30 kVA, 60 Hz
	Control system	PWM converter (Pulse Width Modulation)
	Max efficiency	90%
	Power factor	Over 95%
	Max input voltage	800 VDC from Battery
		450–850 VDC from PV
	Output	45 A
BAT	Capacity	50 kWh (25 kWh × 2)
	Charge–discharge efficiency	95%
	Cell voltage	3.2 V, 20 Ah
	Rack voltage	422.4 V (11 modules, 396 cells)
	Nominal voltage	422.4 V (369.6–468.6 V)

3.1.4. SGS Operation System

In the SGS, the operation system plays the role of the energy management system (EMS) developed by KEPCO. It is a software program that can integrate the other components. The integration allows the OS to monitor the power consumption of all components in real-time. Moreover, it has a human–machine interface (HMI) that shows the details of the components. By monitoring, optimized management is possible. Specifically, the operator can set schedules for these devices. Regarding the PCS, the OS controls the charge–discharge operation mode as shown in Table 7. However, the WT is not considered in the modes, because the output of the WT is too small to contribute to the modes. The OS has three categories: system configuration, management, and statistics.

Table 7. Operation modes of PCS.

Status	Mode	Description
Charge	Full charge	Full Charge by PV + grid
	Only PV Charge	Charge only by PV
Discharge	Full discharge	Supply power to loads by PV + BAT
	Fixed PCS output	• Fixed PCS output • BAT output varies with PV output
	Fixed BAT output	• Fixed BAT output • PCS output varies with PV output
	Only PV discharge	Supply power to loads only by PV
Concurrent	Fixed BAT charge	• Some of PV output charges BAT • Other supplies to loads
	Fixed PCS output	• Some of PV output supplies to loads • Other charges BAT

The first category, system configuration, shows the real-time flow of power. In this section, users can check the status of components and monitor the general data, including electricity fee information, supplied power from each source, and the power consumption of the building. This helps users to understand the power flow. In this section, the operator monitors the overall status of the system, electricity fee information, real-time demand power, and supply power including the generation of RE and battery discharge. This section also includes information on the communication status of each device. The serial communications protocols used in the SGS are Modbus and Zigbee [19]. The components are connected in communication lines, and the data of the devices are gathered into the operation system. Especially, the PV, PCS, and battery data are sent by International Electrotechnical Commission (IEC) 61850. The IEC 61850 protocol standard is for substation to exchange data and enables the integration of control, measurement, and monitoring [20]. The used IEC 61850 models are the following: IEC 61850-7-420 is used to exchange of data with DERs, and IEC 61850-90-9 has functions for power converters focused on DC-to-AC and AC-to-DC conversions [19,21,22]. The PV data are voltage and current of generated power, and solar radiation. The PCS measures active power, reactive power, phase current, and phase voltage. Also, the battery sends the data of voltage, current, status of charge, temperature, and each cell's voltage, current, and temperature.

The management is for treating smart lights, smart outlets, VFDs, HVAC, and ESS. As an example of this section, the operator is not only able to monitor each smart outlet but also turn them on and off.

The statistics section is comprised of an overall analysis, DER analysis, and load forecasting. Overall analysis is for supplied power and peak per day, month, and year. This section shows the monthly analysis of supply/demand. The DER analysis shows the PV generation and the amount of battery discharge. One of the main functions of the OS is to forecast the demand of electricity by its own algorithm. Through this analysis, the OS controls the devices and power flow, and decides to charge or discharge the battery.

3.2. Other Ancillary Equipment

3.2.1. Advanced Metering Infrastructure

AMI installed in the transformer room of the SGS measures the amount of power supplied from the grid and checks the qualities of voltage, current, and frequency. Through the measures, the SGS can optimize the power supply. A general electricity meter monitors power quality every 15 min, whereas the AMI exports the data in real-time. The exported data are used to calculate the real-time electricity fee and analyze the operation status through the OS. By utilizing the data, the OS can operate the whole system flexibly. This AMI is connected in a series connection to current and in a parallel connection to voltage. Figure 5 shows a picture and the connection of the AMI, and Table 8 provides the detailed specifications of the AMI.

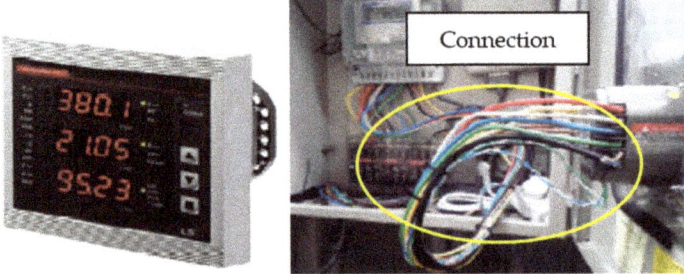

Figure 5. Pictures of AMI and connection line.

Table 8. Specifications of the AMI.

Potential transformer	AC 10–452 V/110 V
Current transformer	0.05–6 A (rated 5 A)
Measurement	Voltage, current, freq., etc.

3.2.2. Electric Vehicle Chargers

Outside of the office building, there were a few gasoline vehicles for outside work. Some of them were changed to electric vehicles, and six EV chargers were installed. Four of the chargers are a slow-charging type, and the others are a fast-charging type. The slow-charging type has an AC-type connector and charges the EV at 7–8 kW through a single phase of 220 VAC, and it takes about 5 to 6 h to reach a full charge. The fast-charging type has three kinds of socket: CHAdeMO, Combo, and 3-phase AC type. CHAdeMO and Combo supply power in DC. The fast-charging type chargers supply power at 50 kW by 380–450 VDC or 380 VAC. In fact, the EV chargers are considered as loads, while the EVs contribute to reduce the running cost of vehicles compared to gasoline vehicles, as described in Section 5. Through this section, it is proved that the EVs are more economical than gasoline vehicles. The EVs also have potential, in that they can be bridges to implement vehicle-to-grid (V2G) technology [23].

3.2.3. Building Automation System

For building automation systems, current transformers (CTs) were installed in each distribution board to measure the power quality and the consumed energy by time and by device. These CTs are solid-ring and split-core types, and they communicate with a multi-channel power meter by Modbus. Also, smart outlets and light switches were newly installed to reduce power peak and consumption by turning the devices on and off remotely or automatically. This reduction is directly reflected in a reduction in the amount of power that needs to be generated by the fossil fuel generators.

The outlets can cut off standby power. Their rated allowable current is 16 A, and their overload current is 20 A. For the smart lighting, gateways were installed to transmit control signals to the lighting from the OS.

The other controllable system of the BAS is the HVAC. The OS adjusts the air quality by controlling the frequency of the VFDs to reduce power consumption.

4. Performance Evaluation

To evaluate the performance of the Smart Grid Station, the authors analyzed the reduction of peak and consumption, as well as economic feasibility by comparison with the early targets. We acquired the real data of building demand, peak, and DERs measured in 2014 and 2015 from KEPCO.

The output data and the performance analysis were based on the real operation of the Smart Grid Station, following the algorithm shown in Figure 6. The algorithm was developed by KEPCO. Following the algorithm, grid power always supplies power to loads, and also optionally charges the battery at off-peak load times such as night time or on weekends. PV can generate when the sun shines, and WT can generate when the wind speed is over 3 m/s. If the generation of the PV and the WT exceeds the power demand, the extra power goes to charge the battery at off-peak load time. When the sources charge the battery, the power goes through PCS. At the peak load time or mid-load time, the grid power, renewable energy sources, and the battery ($P_{discharge}$) supply to loads to reduce the peak of the building. The blue line in Figure 6 shows the communication connection—all data gather into the OS. After gathering the generation, supply, and demand data from each device, the OS gives orders to the PCS.

In the SGS, peak power and consumption are reduced due to the DER. This makes it difficult to directly compare the decreased value with the unreduced value that could have been measured if not for the reduction. For this reason, the authors tried to compare the reduced peak and consumption with the values from 2014. However, new equipment and appliances were installed for the supervisory control and data acquisition (SCADA) room and the temporary office of Namyangju city in the Guri branch office, and we considered these changes in increment. Table 9 shows the details of increment in the building, and the authors assumed the usage time of the devices as in Table 10.

As the equipment for the SCADA room is ICT equipment, it is always used, even on weekends. Because the air conditioning system is an ice storage system, it does not contribute to a rise in the peak. Printers and cooling fans were considered to be unused during peak time to save energy.

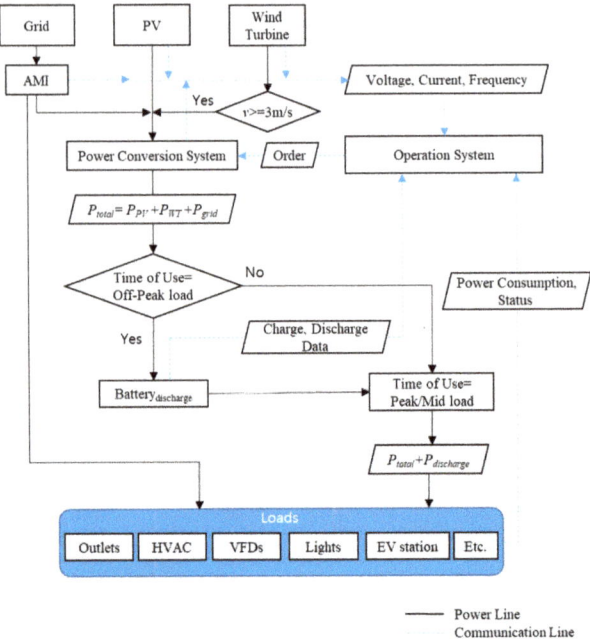

Figure 6. Operation algorithm of the Smart Grid Station.

Table 9. Newly installed equipment and appliances. SCADA: supervisory control and data acquisition.

	Installation Date	Items	Capacity (kW)	Hours Used		Consumption		
				Per a Day of Weekday	Per a Day of Weekend	Aug (kWh)	Sept (kWh)	Total (kWh)
SCADA	2014 10.30	Media rack	0.72	24	24	10,460.6	10,123.2	20,583.8
		Audio rack	0.5					
		6 DLP [1] Cube	1.32					
		4 LED TVs	0.52					
		Humidifier	11					
Temp. office	2015 03.24	Lights	1.8	10	-	378	360	738
		13 Computers (400 W)	5.2	10	-	1092	1040	2132
		13 Computers (300 W)	3.9	10	-	819	780	1599
		21 Monitors (170 W)	3.7	10	-	777	740	1517
		Hot & cold dispenser	0.85	24	24	632.4	612	1244.4
		Air handling unit	12.5	10	-	2625	2500	5125
		10 Cooling fans	0.3	6	-	37.8	36	73.8
		4 Printers (700 W)	2.8	1	-	58.8	56	114.8
		Ice storage AC sys.	15	10	-	3150	3000	6150
		Total consumption (kWh)				20,030.6	19,247.2	39,277.8
		The number of weekdays in each month				21 days	20 days	
		The number of weekends in each month				10 days	10 days	

[1] Display Lighting Projector Cube.

Table 10. Details of usage time.

	Usage Hours			Note
	Off-Peak Load	Mid-Load	Peak Load	
SCADA	6	8	10	All days
Office facilities [1]	6	3.5	0.5	08:30–18:30 on Weekdays
Hot & cold dispenser	6	8	10	All days
Ice storage AC system	-	-	10	Weekdays
Printers	1	-	-	Weekdays
Cooling fans	5	1	-	Weekdays

[1] Lights, computers, monitors, and air handling unit.

4.1. Peak Shaving

For a commercial building, once a peak power is measured, the peak is adopted for the electricity fee for the year. The maximum peak occurred in the summer. Thus, the authors compared the peak that occurred in August and September of 2014 with the peak in same months of 2015 as Table 11.

The peak shaving ratio (PSR) is the ratio between the maximum peak in 2015 and the maximum peak in 2014.

Table 11. Comparison of peaks.

	Day Peak (kW)	Evening Peak (kW)	Max Peak (kW)
2015.08	285.12	294.24	294.24
2015.09	255.84	213.12	
2014.08	271.08	259.56	271.08
2014.09	255.96	237.24	

$$Peak_{add} = 42.01 \text{ kW} \times 0.9 \cong 37.8 \text{ kW} \tag{1}$$

$$PSR(\%) = \frac{Peak_{max}^{2015} - Peak_{add}}{Peak_{max}^{2014}} \times 100 - 100 \tag{2}$$

$$PSR(\%) = \frac{294.24 - 37.8}{271.08} \times 100 - 100 = -5.40\% \tag{3}$$

In (1), 42.01 kW is the sum of rated power of all equipment except cooling fans, printers, and the ice storage AC system in Table 9. The value of added equipment capacity defined in (1) should be subtracted from max peak in 2015. Because the real contribution of the equipment to peak was unknown, the authors assumed the contribution by multiplying the rated capacity of devices by the power factor 0.9. Thus, the result of (1) is an estimate of the rated capacity multiplied by the power factor (0.9). A positive PSR value represents an increase of peak, whereas a negative value is a reduction. In (1) and (2), the result of the PSR was −5.40%, meaning that the peak was reduced by 5.40%.

4.2. Consumption Reduction

The second effect of the SGS is the reduction of power consumption. The data used to calculate peak reduction in Section 4.1 was also used in this section.

The consumption was separated into three time periods: off-peak load, mid-load, and peak load, as shown in Table 12. The added power consumption should be subtracted from the total consumption in 2015. However, because the contribution of the added devices to power consumption was unidentified, the power factor 0.9 was multiplied to assume the contribution in the same way as in (1). The consumption reduction ratio (CRR) is calculated in (5). A positive value of CRR represents an increase of power consumption, whereas a negative value means a reduction. In (4) and (5), the result of CRR was approximately −11.26%, which means that power consumption was reduced.

$$\text{Con}_{add} = 39,277.8 \text{ kWh} \times 0.9 \cong 35,350 \text{ kWh} \tag{4}$$

$$\text{CRR}(\%) = \frac{\text{Con}_{max}^{2015} - \text{Con}_{add}}{\text{Con}_{max}^{2014}} \times 100 - 100 \tag{5}$$

$$\text{CRR}(\%) = \frac{123,807 - 35,350}{99,686} \times 100 - 100 = -11.26\% \tag{6}$$

Table 12. Monthly consumption.

	Consumption (kWh)			
	Off-Peak Load	Mid-Load	Peak Load	Total
2015.08	24,299	20,703	19,128	123,807
2015.09	27,478	17,619	14,580	
2014.08	7844	21,322	18,891	99,686
2014.09	17,695	19,343	14,591	

4.3. Saved Electricity Fee

There are two kinds of electric rates: demand charge and energy charge. Demand charge is for the measured peak, and energy charge is different in each season. Time periods are divided into summer, spring/fall, and winter. Exact time periods are shown in Table 13.

Table 13. Segmentation by season and time [21].

Load Time	Summer	Spring/Fall	Winter
Off-peak load time	23:00–09:00		23:00–09:00
Mid-load Time		09:00–10:00 12:00–13:00 17:00–23:00	09:00–10:00 12:00–17:00 20:00–22:00
Peak load Time		10:00–12:00 13:00–17:00	10:00–12:00 17:00–20:00 22:00–23:00

Electric rate refers to each type of customer. The General Service rate is classified in General Service (A) I, General Service (A) II, and General Service (B) is for commercial building customers. These rates are subdivided into High-Voltage A for 3.3–66 kV and High-Voltage B ranged over 154 kV. Besides, customers can choose option I, option II, or option III depending on the customers' electricity use time for a month. The High-Voltage A option II of General Service (B) is for the customers who use electricity for 200–500 h per month, and whose contract demand of 300 kW or more. The details are shown in Table 14. In the table, the authors considered ₩1000 KRW as $1 USD for a convenience.

Table 14. Electric rates table for High-Voltage A Option II of General Service (B) [24].

	Demand Charge		$8.32/kW		
Energy charge ($/kWh)	Time Period		Off-Peak Load	Mid-Load	Peak Load
	Summer (1 Jun–31 Aug)		$0.0561	$0.109	$0.1911
	Spring/Fall (1 Mar–31 May/1 Sep–31 Oct)		$0.0561	$0.0786	$0.1093
	Winter (1 Nov–28 Feb)		$0.0631	$0.1092	$0.1667

The total power consumption and electricity fees in 2014 and 2015 are shown in Table 15. To calculate the saved fee ratio (SFR), the added fees were also considered. These fees are in Tables 16 and 17. By (7), the sum of added fees was $4127.892, and the SFR was calculated as −10.15%. This negative value means that the electricity fee was reduced by 10.15%, while a positive value indicates an increase.

$$Fee_{add} = Charge_{demand} + Charge_{energy} \tag{7}$$

$$\text{SFR}(\%) = \frac{Fee^{2015} - Fee_{add}}{Fee^{2014}} \times 100 - 100 \tag{8}$$

$$\text{SFR}(\%) = \frac{15,924 - 4,127.892}{13,128} \times 100 - 100 = -10.15\% \tag{9}$$

Although there was no early target for fee reduction, the analysis of electricity fees is enough to prove the effects of the Smart Grid Station.

Table 15. Total electricity fees in 2014 and 2015.

Date	Consumption (kWh)	Fee (USD)	Total (USD)
2015.08	64,130	$8659	$15,924
2015.09	59,677	$7265	
2014.08	48,057	$6424	$13,128
2014.09	51,629	$6704	

Table 16. Demand charge for added loads in 2015.

Month	Max Peak	Demand Charge	Charged Fee
Aug.	37.8 kW	$8.32/kW	$314.496
Sept.	37.8 kW	$8.32/kW	$314.496

Table 17. Energy charge for added loads in 2015.

2015		Consumption (kWh)	Fee [1] (USD)
Aug.	Off-peak Load	7744	$304.041
	Mid-load	5481.4	$387.755
	Peak load	6021.0	$761.778
	Total	19,247.2	$1453.574
Sept.	Off-peak Load	8056.7	$406.783
	Mid-load	5695.8	$558.758
	Peak load	6278.2	$1079.785
	Total	20,030.7	$2045.326

[1] Power factor of 0.9 was applied in the fee, and decimal point was rounded up.

5. Economic Analysis of Smart Grid Station

In Section 5, we studied the contribution with regard to the economic aspects of the contributions of by PV generation and EV, and the energy time shifting by the ESS. The monthly measurement period was from the first day of each month to the last day. The used data was measured by the OS.

5.1. Saved Electricity Fees by PV Generation

During August and September of 2015, PV generation was 2961.5 kWh in August and 2326.1 kWh in September. The total saved fee is the sum of saved demand charge (SDC) and saved energy charge (SEC) contributed by the PV system.

$$\text{SDC} = P_{PCS}^{Max}(\text{kW}) \times Charge_{demand} \tag{10}$$

$$SEC = W_{DER}(\text{kWh}) \times Charge_{energy} \quad (11)$$

It is difficult to know when the PV system generated power and how much the system generated. For this reason, the authors assumed the PV supplied power to loads at peak-load time and mid-load time in ratio of 8 to 2. Using this, the SDC and the SEC in August were found as follows:

$$SDC = 30 \text{ kW} \times \$8.32/\text{kW} = \$249.6 \quad (12)$$

$$SEC_{Aug} = 2961.5 \text{ kWh} \times (\$0.1911/\text{kWh} \times 0.8 + \$0.109/\text{kWh} \times 0.2) = \$517.374 \quad (13)$$

$$SEC_{Sept} = 2326.1 \text{ kWh} \times (\$0.1093/\text{kWh} \times 0.8 + 0.0786/\text{kWh} \times 0.2) = \$239.96 \quad (14)$$

The SDC values from September were the same as those from August. Likewise, the SEC values from September are in (14). In conclusion, the total saved fees were \$1256.534.

5.2. Running Cost Reduction by EV

There was one electric vehicle in 2015, and the running data is in Table 18. In August and September, the EV ran for 470 km and 345 km, respectively. We assumed the fuel efficiency of a gasoline-powered car is 10 km/L. Referring to the data, we compared the running cost (RC) of the EV with that of a gasoline-powered vehicle.

$$RC_{Gas} = \frac{\text{mileage (kW)}}{\text{km/L}} \times Price_{Gasoline} \quad (15)$$

$$RC_{EV} = W_{EVcharge} \times Price_W \quad (16)$$

By substituting figures, the results were as follows:

$$RC_{Gas}^{Aug} = \frac{470 \text{ kW}}{10 \text{ km/L}} \times \$1.56/\text{L} = \$73.32 \quad (17)$$

$$RC_{EV}^{Aug} = 105.8 \text{ kWh} \times \$0.109/\text{kWh} \cong \$11.533 \quad (18)$$

$$RC_{Gas}^{Sept} = \frac{345 \text{ kW}}{10 \text{ km/L}} \times \$1.59/\text{L} = \$54.855 \quad (19)$$

$$RC_{EV}^{Sept} = 68.4 \text{ kWh} \times \$0.786/\text{kWh} \cong \$5.377 \quad (20)$$

Table 18. Running data of EV in 2015.

Month	Mileage	EV Charge Amount (kWh)	Price of 1 kWh ($/kWh)	Price of Gasoline [1] ($/L)
Aug	470 km	105.8	$0.109	$1.56
Sept	345 km	68.4	$0.0786	$1.59

[1] The price of gasoline is the average value of the month.

According to (17) and (18), \$61.787 were saved, which means about 84.3% of the running cost was saved by the EV in August. Equations (19) and (20) also show that \$49.478—about 90.2% of the cost—was saved in September. Although the actual amount saved was small, this shows that the EV was much more effective than a gasoline-powered vehicle.

5.3. Saved Fee by ESS Scheduling

A customer charges the battery of the ESS at night, when the price of electricity is low, and discharges the power at the peak load time or mid-load time when the price is high. However, the power from the PV system does not charge the battery but supplies power to the building load directly in the summer to maximize the efficiency. In August, 758.9 kWh was charged to the battery, and the same amount was discharged. In September, 541.1 kWh was charged and discharged. We also adapted the same assumption that the ratio of 8 to 2 stated in Section 5.1. Equation (21) is a formula to calculate the fee reduction (FR). Time of use (TOU) applied in this equation is an electricity fee policy that varies with seasons and times, as shown in Table 14.

$$FR = \left(W_{discharge} \times TOU\right) - \left(W_{charge} \times fee_{off-peak}\right) \quad (21)$$

By substituting figures, the results were as follows:

$$FR_{Aug} = (758.9 \times (0.1911 \times 0.8 + 0.109 \times 0.2)) - (758.9 \times 0.0561) \cong \$89.991 \quad (22)$$

$$FR_{Sept} = (541.1 \times (0.1093 \times 0.8 + 0.786 \times 0.2)) - (541.1 \times 0.0561) \cong \$26.465 \quad (23)$$

As the calculations show, $116.456 was saved for two months. By load shifting, the cost of electricity was greatly reduced.

5.4. Economic Feasibility on Investment Cost

The economic benefits in 2015 are estimated in Table 19. The total construction cost was $174,542 which consisted of the purchasing, installation, and operation costs of all systems: PV, ESS, OS, BAS, AMI, smart outlets, smart lights, and smart distribution boards.

Table 19. Savings report in 2015.

	Investment	Expectation	Measured Amount	Saved Fee (USD)
PV generation	$51,000 USD	25,200 kWh	27,296 kWh	$6376
ESS discharge	$10,500 USD (PCS) $30,000 USD (BAT)	12,540 kWh	8430 kWh	

$$\text{ROI}(\%) = \frac{\text{Total Net Return}}{\text{Investment}} = \frac{\$6376 \times 20 \text{ years}}{\$174,542} \times 100\% \cong 73\% \quad (24)$$

The economic feasibility can be evaluated by calculating the return on investment (ROI). The net benefit in 2015 was $6376, and we assumed total net return based on the net benefit in 2015 as life expectancy of the installed devices was expected as 20 years. By (24), the ROI was calculated as approximately 73%. This value may seem that its benefit is low, but additional profits were not considered such as a CO_2 reduction, effect by BAS, a tax incentive, the renewable energy certificate, the avoided costs of the generation facilities and the transmission and distribution facilities, because there were constraints on the request for the indices to KEPCO. Thus, we simply appraised the benefits of the system. When the SGS in Guri branch office which was the first demonstration was constructed, the unit cost of the PV system was $2550/kW, that of the PCS was $350/kW, and that of the battery was $600/kWh. However, according to the National Renewable Energy Laboratory (NREL) [25], the unit cost of a commercial PV system was $1620/kW ($1.62 per watt) in 2017. Also, Bloomberg New Energy Finance in [26] described the unit cost of a Li-ion battery as $273/kWh in 2016. As a result, we expect that the SGS solution is much more economical than when the first demonstration was implemented. Based on the effects, the KEPCO had established 121 Smart Grid Stations as of 2016.

6. Discussion

In this paper, the authors studied the first demonstration of a Smart Grid Station in the Guri branch office of KEPCO to prove its effectiveness. The authors verified the performance and economic feasibility of the SGS to propose a future strategy. The performance was evaluated with regard to three aspects: peak shaving, reduction of power consumption, and electricity fee savings. The economic efficiency was feasible in terms of electricity fee and running cost of an EV. Measured values in 2015 were revised for objective comparisons with values in 2014 before the SGS was built.

The early main targets were 5% reduction of peak, 9.6% reduction of consumption, and savings in electricity fees. To evaluate the performance objectively, we compared the factors in 2015 with the values in 2014, while considering the increased loads in 2015. As described in Section 4, the performance for peak shaving was calculated as 5.40%. This means that the 5% of peak shaving as one of early targets was accomplished. Next, the power consumption was reduced by 11.26%. The savings in electricity fees did not have a specific target, but they were reduced by 10.15%. These benefits to early targets are arranged in Table 20.

Table 20. Comparisons of early targets with performance.

	Peak	Consumption	Electricity Fee
Early target	5.0%	9.6%	-
Performance	5.40%	11.26%	10.15%

An economic analysis was conducted in Section 5. We considered the saved electricity fees by the PV generation, the reduced running cost by the EV, and the saved fees by ESS scheduling. The saved electricity fees by PV were $1256.534, which is the sum of saved energy charge and demand charge for two months. The reduced running cost by the use of an EV was calculated by comparing an EV to a gasoline vehicle. The running cost for the EV was cheaper than the cost of the gasoline vehicle by about 90.2%. Additionally, the ESS contributed to savings in the electricity fees of $116.456 for two months by load shifting. This means that the greater the capacity of ESS, the greater the savings. Based on the savings by PV and ESS, we calculated the ROI as approximately 73% which may seem low. However, we considered only few benefits because of the constraints. Moreover, due to the gradually decreasing unit price of each system annually, the ROI may be higher at present.

7. Conclusions

According to our analysis, the early targets were accomplished, and the effectiveness of the SGS was proven. Considering the proved effectiveness, the KEPCO has already installed SGSs in 121 of its branch offices, and the possibility of applying the technology to other buildings was also proved. We also suggest the commercialization of the SGS. By expanding, the SGS will contribute to encouraging the industry and to build a smart city, which is a city-sized energy solution. This could come from supporting price policies for devices, and private businesses will participate in the industries actively. To support the expansion of the SG, the convenience, safety, and efficiency of the SGS should be also improved for customers. Also, improvement of the OS would allow the system to integrate and control more and various devices and technologies. As a next step, the authors will study the Smart Town, which is a town-sized energy solution composed of various kinds of buildings founded in the KEPCO Academy in 2016.

Author Contributions: J.W. designed and wrote the paper. W.H. supported and helped to collect the data. Y.Y. and G.J. conducted the review. G.J. contributed to the modification. All the authors have read and approved the final manuscript.

Funding: This research received no external funding.

Acknowledgments: This work was supported by Korea Electric Power Corporation (R17XA05-4) and under the framework of international cooperation program managed by National Research Foundation of Korea (No. 2017K1A4A3013579).

Conflicts of Interest: The authors declare no conflict of interest.

Nomenclature

GHG	Greenhouse Gas
SG	Smart Grid
PV	Photovoltaic
WT	Wind Turbine
AMI	Advanced Metering Infrastructure
BAS	Building Automation System
SGS	Smart Grid Station
RE	Renewable Energy
HVAC	Heating, Ventilation and Air Conditioning
VFD	Variable Frequency Drive
TOU	Time of Use
PSR	Peak Shaving Ratio
CRR	Consumption Reduction Ratio
SFR	Saved Fee Ratio
SDC	Saved Demand Charge
SEC	Saved Energy Charge
RC	Running Cost
FR	Fee Reduction

References

1. UNFCCC. *Adoption of the Paris Agreement*; UNFCCC: Paris, France, 2015.
2. Enhanced Actions on Climate Change. Available online: http://www4.unfccc.int/Submissions/INDC/Submission%20Pages/submissions.aspx (accessed on 20 March 2018).
3. Submission by Latvia and the European Commission on Behalf of the European Union and Its Member States. Available online: http://www4.unfccc.int/Submissions/INDC/Submission%20Pages/submissions.aspx (accessed on 20 March 2018).
4. Submission of Japan's Intended Nationally Determined Contribution. Available online: http://www4.unfccc.int/Submissions/INDC/Submission%20Pages/submissions.aspx (accessed on 20 March 2018).
5. Submission by the Republic of Korea Intended Nationally Determined Contribution. Available online: http://www4.unfccc.int/Submissions/INDC/Submission%20Pages/submissions.aspx (accessed on 20 March 2018).
6. Ock, J.; Issa, R.R.A.; Flood, I. Smart building energy management systems (BEMS) simulation conceptual framework. In Proceedings of the 2016 Winter Simulation Conference, Arlington, VA, USA, 11–14 December 2016; pp. 3237–3245.
7. Ferro, G.; Laureri, F.; Minciardi, R.; Robba, M. Optimal Integration of Interconnected Buildings in a Smart Grid: A Bi-Level Approach. In Proceedings of the 2017 UKSim-AMSS 19th International Conference on Computer Modelling & Simulation (UKSim), Cambridge, UK, 5–7 April 2017; pp. 155–160.
8. Borlase, S. Smart Grid Technologies. In *Smart Grids Infrastructure, Technology, and Solutions*; CRC Press: Boca Raton, FL, USA, 2015; pp. 79–125.
9. Kolokosta, D. The role of smart grids in the building sector. *Energy Build.* **2016**, *116*, 703–708.
10. Wurtz, F.; Delinchant, B. Smart buildings integrated in "smart grids": A key challenge for the energy transition by using physical models and optimization with a "human-in-the-loop" approach. *C. R. Phys.* **2017**, *18*, 428–444. [CrossRef]
11. Sidid, S.; Gaur, S. Smart Grid Building Automation Based on Internet of Things. In Proceedings of the International Conference on Innovations in Power and Advanced Computing Technologies, Vellore, India, 21–22 April 2017; pp. 1–4.
12. Kim, S.; Lim, H. Reinforcement Learning Based Energy Management Algorithm for Smart Energy Buildings. *Energies* **2018**, *11*, 2010. [CrossRef]
13. Barbato, A.; Bolchini, C.; Geronazzo, A.; Quintarelli, E.; Palamarciuc, A.; Piti, A.; Rottondi, C.; Verticale, G. Energy Optimization and Management of Demand Response Interactions in a Smart Campus. *Energies* **2016**, *9*, 398. [CrossRef]

14. Cabrera, D.F.M.; Zareipour, H. Data association mining for identifying lighting energy waste patterns in educational institutes. *Energy Build.* **2013**, *62*, 210–216. [CrossRef]
15. Ragheb, M. *Vertical Axis Wind Turbines*; University of Illinois at Urbana-Champaign: Champaign, IL, USA, 2011.
16. Soedibyo, J.F.; Ashari, M. Performance Comparison of Vertical Axis and Horizontal Axis Wind Turbines to Get Optimum Power Output. In Proceedings of the 2017 15th International Conference on Quality in Research: International Symposium on Electrical and Computer Engineering, Bali, Indonesia, 24–27 July 2017; pp. 429–433.
17. Mohd, A.; Ortjohann, E.; Schmelter, A.; Hamsic, N.; Morton, D. Challenges in Integrating Distributed Energy Storage Systems into Future Smart Grid. In Proceedings of the IEEE ISIE, Cambridge, UK, 30 June–2 July 2008; pp. 1627–1632.
18. Alegria, E.; Brown, T.; Minear, E.; Laseter, R.H. CERTS Migcrogrid Demonstration with Large-Scale Energy Storage and Renewable Generation. *IEEE Trans. Smart Grid* **2014**, *5*, 937–943. [CrossRef]
19. Lee, K.; Lee, Y.; Seo, J.; Lee, S.; Seo, D. Case study on Smart Grid Station using IEC 61850. In Proceedings of the KIEE Summer Conference, Pyeongchang, Korea, 13–15 July 2016; pp. 24–25.
20. ABB. *650 Series IEC 61850 Communication Protocol Manual*; ABB: Vasteras, Sweden, 2011.
21. Communication Networks and Systems for Power Utility Automation—Part 7-420. Available online: https://infostore.saiglobal.com/preview/is/en/2009/i.s.en61850-7-420-2009.pdf?sku=1138048 (accessed on 9 May 2018).
22. Communication Networks and Systems for Power Utility Automation—90-7. Available online: https://webstore.iec.ch/preview/info_iec61850-90-7%7Bed1.0%7Den.pdf (accessed on 9 May 2018).
23. Tan, K.M.; Ramachandaramurthy, V.K.; Yong, J.Y. Integration of electric vehicles in smart grid: A review on vehicle to grid technologies and optimization techniques. *Renew. Sustain. Energy Rev.* **2016**, *53*, 720–732. [CrossRef]
24. KEPCO General Service Electricity Rate Table. Available online: http://cyber.kepco.co.kr/ckepco/front/jsp/CY/E/E/CYEEHP00202.jsp (accessed on 9 May 2018).
25. Fu, R.; Feldman, D.; Margolis, R.; Woodhouse, M.; Ardani, K. *U.S. Solar Photovoltaic System Cost Benchmark: Q1 2017*; NREL: Golden, CO, USA, 2017.
26. Curry, C. *Lithium-ion Battery Costs and Market*; Bloomberg New Energy Finance Report; BNEF: New York, NY, USA, 2017.

© 2018 by the authors. Licensee MDPI, Basel, Switzerland. This article is an open access article distributed under the terms and conditions of the Creative Commons Attribution (CC BY) license (http://creativecommons.org/licenses/by/4.0/).

Review

Moral Values as Factors for Social Acceptance of Smart Grid Technologies

Christine Milchram *[ID], Geerten van de Kaa, Neelke Doorn[ID] and Rolf Künneke

Faculty of Technology, Policy and Management, Delft University of Technology, Jaffalaan 5, 2628 BX Delft, The Netherlands; g.vandekaa@tudelft.nl (G.v.d.K.); n.doorn@tudelft.nl (N.D.); r.w.Kunneke@tudelft.nl (R.K.)
* Correspondence: c.milchram@tudelft.nl; Tel.: +31-15-27-85297

Received: 5 June 2018; Accepted: 23 July 2018; Published: 1 August 2018

Abstract: Smart grid technologies are considered an important enabler in the transition to more sustainable energy systems because they support the integration of rising shares of volatile renewable energy sources into electricity networks. To implement them in a large scale, broad acceptance in societies is crucial. However, a growing body of research has revealed societal concerns with these technologies. To achieve sustainable energy systems, such concerns should be taken into account in the development of smart grid technologies. In this paper, we show that many concerns are related to moral values such as privacy, justice, or trust. We explore the effect of moral values on the acceptance of smart grid technologies. The results of our systematic literature review indicate that moral values can be both driving forces and barriers for smart grid acceptance. We propose that future research striving to understand the role of moral values as factors for social acceptance can benefit from an interdisciplinary approach bridging literature in ethics of technology with technology acceptance models.

Keywords: smart grid; smart energy; sustainability; values; technology acceptance; technology adoption

1. Introduction

Driven by climate change mitigation and transition to low carbon energy systems, governments worldwide have set targets to increase the use of renewable energy sources. The 2030 European energy targets include a minimum 27% share of renewable energy consumption [1]. Growing shares of renewables, particularly from wind and solar energy, lead to rising intermittencies of energy supply and to a larger number of small and decentralized generation sites. Growing intermittencies and decentralization, however, lead to challenges for balancing supply and demand in networks that were designed for relatively few large and controllable power plants [2,3].

Smart grid technologies are praised as one solution to support the integration of rising shares of renewable energy sources into power networks and are thus seen as essential in the transition to sustainable energy systems [2,4]. They allow accounting for higher supply intermittencies and decentralization by using innovative information and communication technologies (ICT). For consumers, they contribute to increased information and awareness of energy use, potentially enabling energy savings [5]. As such, smart grids can be a promising solution to reducing greenhouse gas emissions in the electricity system while at the same time dealing with rising energy costs [6]. Although the concept of smart grid technologies comprises many technological applications and lacks a single definition, widely accepted definitions include efficient management of intermittent supply, two-way communication between producers and consumers, and the use of innovative ICT solutions [7,8].

In spite of their promising benefits for low-carbon energy systems, several challenges are associated with smart grid technologies. Next to concerns about high costs as well as uncertain investment and regulatory environments, moral values underlie many societal concerns [9]. Concerns about data privacy and security have already delayed smart meter introductions in Europe and the US [10,11]. The possibility to share end-users' energy consumption data automatically and in (near) real time with grid operators and store these data in central databases raises concerns that energy companies could use this data to get insight into activities in a household that are considered as private [12]. Related to storing sensitive data in central databases are fears that these data could be threatened by cyberattacks and used in a harmful way. Additionally, consumer fears of reduced autonomy are reflected in concerns that smart meters or smart household appliances might give energy companies more control over a household's electricity use [13]. Further concerns that energy suppliers will not be transparent about benefits and pass financial savings on to their customers relate to the values of trust and a fair distribution of costs and benefits [14].

Challenges in the smart grid development which are related to moral values need to be addressed to achieve sustainable energy systems and might hinder the wider acceptance and adoption of smart grid technologies. There is an extensive literature on factors influencing technology acceptance and adoption in the field of technology and innovation management [15,16], and social psychology [17,18]. Innovation management scholars emphasize market acceptance, which is determined largely by environmental and market-specific factors, the characteristics of the technology itself, and firm-level characteristics [19–21]. Theories in social psychology, on the other hand, concentrate on individual user acceptance, with models stressing the importance of technology specific beliefs, social influences, and personality beliefs as factors for acceptance [17,18,22]. Although these bodies of literature focus on a wide range of potential factors for acceptance, moral values—characteristics of a technology with ethical importance [23]—are typically not included in these factors. Given that moral values underlie societal concerns uttered in public debates, there is a need for research that addresses how moral values impact the acceptance of smart grid technologies. This paper therefore aims at exploring this relationship. It addresses the questions which moral values are relevant for the acceptance of smart grid technologies and how these values influence smart grid acceptance. The paper contributes to the development of sustainable smart grid technologies. To achieve sustainability, it is important not only to consider environmental impacts such as carbon emissions but also social and ethical impacts such as privacy and justice. We stress the importance of social and ethical aspects for sustainability by emphasizing the role of moral values for smart grid technologies.

The paper is structured as follows: The next section provides a theoretical background on moral values drawing from the field of ethics of technology, as well as factors for technology acceptance and adoption drawing from technology and innovation management, and social psychology. Sections 3 and 4 contain the methodology and results of a systematic literature review on values associated with the acceptance of smart grid technologies. The two final sections are devoted to discussions and conclusions.

2. Theoretical Perspectives

2.1. Ethics of Technology

Moral values are evident in societal concerns about smart grid technologies. Ethics of technology is the major field concerned with moral values and technologies. Moral values are used to make statements about ethical and social consequences of technologies. Although an unanimously agreed upon definition of the term 'moral values' is lacking, they often refer to abstract principles and "general convictions and beliefs that people should hold paramount if society is to be good" [24] (p. 1343). They are considered to be intersubjectively shared, which means they are principles that different individuals can relate to and generally hold important [24,25]. As such, moral values relate to convictions of what is perceived as good and bad that are shared by members of a society [26]. Typical examples of importance for technologies are health, well-being, safety, or justice [23,27].

Evaluations of technologies with respect to ethical and social consequences are grounded in the understanding that technologies are not neutral objects, but value-laden [28,29]. That means that they are capable of endorsing or harming values [30]. Winner [30] gives the much-cited example of very low overpasses over the only highway connecting New York with Long Island Beach, thereby hindering public busses (the main method of transportation for less well-off societal groups including racial minorities) to access the beach. The example is often used to illustrate the moral importance of technological design [29,31]. Moral considerations of technological design are especially relevant as technologies do usually not only fulfill the specific function they are designed for, but also have positive and negative side effects [32].

For the design of technologies, moral values are (perceived) technology characteristics that go beyond functional requirements and address requirements of ethical importance such as justice, trust, privacy and more [28,33]. They are seen as identifiable entities that should be considered in design or be embedded in technologies. To embed value in technologies through design choices, Value Sensitive Design (VSD) scholars follow a tripartite approach [28,33]. The approach consists of iterative conceptual, empirical, and technical investigations (for a detailed description of the approach, see for example [27,34,35]). Conceptual investigations are applied to find out what values are relevant, and to identify indirect and direct stakeholders as well as reflections on how to deal with value conflicts. Empirical investigations focus on the stakeholders as unit of analysis in order to get insights into their interpretation and prioritization of different values. Technical investigations focus on the technology itself to identify which technological features support or harm which values. They refer to the "translation" of abstract values into concrete design requirements of the technology.

VSD scholars strive for an in-depth understanding of moral values and the design of technologies that are "better" from an ethical standpoint. Their research aim is focused on integrating convictions "that people should hold paramount if society is to be good" [24] (p. 1343) into the design of technologies. Hence, their research aim does typically not include testing effects of their design on social acceptance of technologies.

2.2. Technology Acceptance and Adoption

Acceptance of novel energy technologies is typically defined in terms of perceptions of stakeholders involved in energy projects [36]. Acceptance can range from passive consent with novel technologies to more active approval such as taking action to promote a technology [37]. Adoption of technologies is defined as the behavior to purchase and use a technology [38]. Adoption can therefore be measured through e.g., market share. Some scholars include behavior towards energy technologies in their definition of "acceptance." When acceptance is defined as purchase/use, "acceptability" is sometimes used to refer to positive attitudes towards technologies (e.g., [39–41]). For the purpose of this research, the definition of acceptance includes the purchase or use of a technology.

Various scholars have focused on factors that affect acceptance of technologies, particularly in the fields of technology and innovation management, and social psychology (Table 1).

2.2.1. Technology and Innovation Management

Scholars in the area of technology and innovation management take a market and firm perspective towards factors for technology acceptance and adoption [15,16,19–21,42,43]: Factors pertain to environmental and market-specific factors, the characteristics of the technology itself, and firm-level characteristics [16,44].

Within environmental and market-specific factors, a strong emphasis is put on network effects. Network effects are positive consumption externalities that occur when the utility of a technology for one consumer increases with the number of other consumers that have adopted the technology [15,19,20,44]. In addition, a high diversity in the inter-organizational network, which is the extent to which stakeholders from different industries are involved in developing and marketing a technology, is beneficial for technology adoption [16,45,46].

Table 1. Overview of factors for technology acceptance/adoption.

Type of Factors	Factors (Examples)	Technology & Innovation Management	Social Psychology
Environmental and market characteristics	Network effects, switching costs, installed base, regulators, suppliers	✓	
Technology-specific characteristics	Technological superiority, complementary goods, compatibility	✓	
Firm-level characteristics	Financial strength, brand reputation, pricing strategy, time of market entry	✓	
Perceived technology-specific characteristics	Performance and effort expectancy, cost-benefit perceptions, hedonic motivations		✓
Perceived social influences	Subjective norm, image		✓
Perceived personality characteristics	Personal norms, ecological worldviews, innovativeness		✓
Others	Experience, habit		✓

Related to characteristics of the technology, the extent to which a given technology performs superior to competing technologies (i.e., its technological superiority) is generally regarded as beneficial for its adoption [20]. In addition, a greater availability and variety of complementary goods has a positive effect on adoption [15,19,47].

In addition, firm-level characteristics are found to impact technology adoption. The financial strength of the firm in terms of the availability of appropriate financial resources to develop and market the technology [48], the brand reputation and credibility [16], and a strong learning orientation from past experiences [15] are beneficial for the firm's specific technology to become adopted. Several factors are related to the firms' strategic choices connected to the introduction of the technology, such as the pricing strategy and timing of market entry [15,16].

2.2.2. Social Psychology

Whereas technology management scholars focus on a firm or market perspective, social psychologists concentrate on individual user acceptance. Among the most prominent theories are the Theory of Planned Behavior (TPB) [22], the Technology Acceptance Model (TAM) [49], and its advancements to the Unified Theory of Acceptance and Use of Technology (UTAUT) [17,50], the Norm Activation Model (NAM) [51,52], or the Value-Belief-Norm theory (VBN) [18]. (Note that the term "values" in this context needs differentiation from moral values in an ethics of technology context. Value orientations or values are referred to in social psychology as individuals' personality characteristics [53]. Moral values in an ethics of technology context are perceived characteristics of the technology [33].)

Factors for technology acceptance can be categorized as technology-specific beliefs, social influences, and personality beliefs. Technology-specific beliefs include beliefs that a technology will be useful and enhance the achievement of a consumer's goal (performance expectancy) and perceptions of the ease of use associated with a technology (effort expectancy) [17,49]. Consumers are also more likely to adopt a technology if they perceive facilitating conditions, including the support available to use a technology [17,22]. Monetary aspects are considered in terms of the perceived trade-off between costs and gains. Finally, hedonic motivations (expected fun, enjoyment) are also found to positively impact acceptance [17,54].

Social influences—interchangeably used with subjective norm [22], and image [50]—cover perceptions that important others such as family and friends believe they should use a technology and the belief that the use will enhance their social status [17].

Personality-specific beliefs mostly refer to the role of personal norms as factors for pro-environmental behavior. They play a prominent role in the Norm Activation Model (NAM) [51,52] and the Value-Belief-Norm theory (VBN) [18]. Personal norms are perceptions about one's moral obligation to take pro-environmental actions [18,40]. They are shaped by ecological worldviews, which are general beliefs about the relationship between humans and the environment [40].

Scholars also combine models focusing on technology-specific beliefs such as TPB and TAM with models focusing on personality-beliefs such as NAM. Broman Toft et al. [38] for example combine TAM with NAM and show that if smart grid technologies are perceived as useful and easy to use, consumers are likely to show stronger personal norms to use the technology. Huijts et al. [41] posit that perceived costs and benefits—elements from TBP—impact personal norms, which is a concept from NAM.

3. Method

To understand the role of moral values for the acceptance of smart grid technologies in greater details, we conducted a systematic literature review. We analyzed journal articles reporting the results of empirical studies to ensure capturing original research results. Articles were retrieved from the databases Scopus and Web of Science (see Table 2 for the full search queries). To capture a diverse range of smart grid technologies, search terms included smart grid, smart energy, smart metering, smart home, home energy management, energy and digitalization, and smart technology. Acceptance, acceptability, and adoption were used as search terms, because, as outlined in Section 2.2, these are common concepts which are often used interchangeably to study social acceptance of emerging technologies (e.g., [36–41]). An initial screening of relevant publications revealed that the term "values" is often not mentioned explicitly, even when moral values were included as factors for smart grid acceptance [2,55–58]. To ensure capturing all relevant publications, the term "values" was therefore not included in our search terms.

Table 2. Search queries used in the systematic literature review.

Database	Search Query	# of Results	Date
Scopus	((TITLE-ABS-KEY (smart AND grid) OR TITLE-ABS-KEY (smart AND meter*) OR TITLE-ABS-KEY (smart AND energy) OR TITLE-ABS-KEY (smart AND home*) OR TITLE-ABS-KEY (home AND energy AND management) OR TITLE-ABS-KEY (smart AND technology) OR TITLE-ABS-KEY (energy AND digital*)) AND (TITLE-ABS-KEY (acceptance) OR TITLE-ABS-KEY (acceptability) OR TITLE-ABS-KEY (adoption))) AND (LIMIT-TO (DOCTYPE, "ar ") OR LIMIT-TO (DOCTYPE, "ip")) AND (LIMIT-TO (SUBJAREA, "ENER ") OR LIMIT-TO (SUBJAREA, "ENVI") OR LIMIT-TO (SUBJAREA, "OCI") OR LIMIT-TO (SUBJAREA, "BUSI")) AND (LIMIT-TO (LANGUAGE, "English"))	444	5 January 2018
Web of Science	(TS = (smart grid OR smart energy OR smart meter* OR smart home* OR home energy management OR smart technology OR energy digital*) AND TS = (acceptance OR acceptability OR adoption)) AND LANGUAGE: (English) AND DOCUMENT TYPES: (Article) Refined by: WEB OF SCIENCE CATEGORIES: (ENVIRONMENTAL SCIENCES OR ECONOMICS OR ENVIRONMENTAL STUDIES OR PSYCHOLOGY APPLIED OR BUSINESS OR SOCIOLOGY OR GREEN SUSTAINABLE SCIENCE TECHNOLOGY OR URBAN STUDIES OR PSYCHOLOGY MULTIDISCIPLINARY OR PSYCHOLOGY EXPERIMENTAL OR SOCIAL SCIENCES INTERDISCIPLINARY)	262	5 January 2018

The database search resulted in 706 articles, which were screened for inclusion in the detailed review (see Figure 1 for flow diagram of systematic literature review). After removing duplicates, the 532 unique search results were screened based on their abstracts. Articles that solely focused on technical issues or did not report results of empirical studies were excluded. As a result, for example, a study by Park et al. [59] was eligible for further analysis because it investigated consumer acceptance of a home energy management system. In contrast, a study by Vagropoulos et al. [60] was excluded because it presented an optimization model and did not empirically assess the acceptance of smart grid technologies. This abstract screening resulted in a total of 103 relevant articles, which were subsequently analyzed with respect to moral values as factors for smart grid acceptance. In the analysis, we searched for values of ethical importance often mentioned in the VSD literature. In addition, we aimed to find additional values that were reported in empirical smart grid studies but not included in prior literature. Apart from identifying values, we analyzed their conceptualizations, the relevant stakeholder group, the technical context, and applied methodologies. The analysis resulted in a group of 49 papers that reported moral values as factors for smart grid acceptance (see Appendix A) and a group of 54 studies that did not include moral values as factors for smart grid acceptance (for example a study by Kobus et al. [61] focusing on the role of smart appliances to bring about electricity demand shift by residential households).

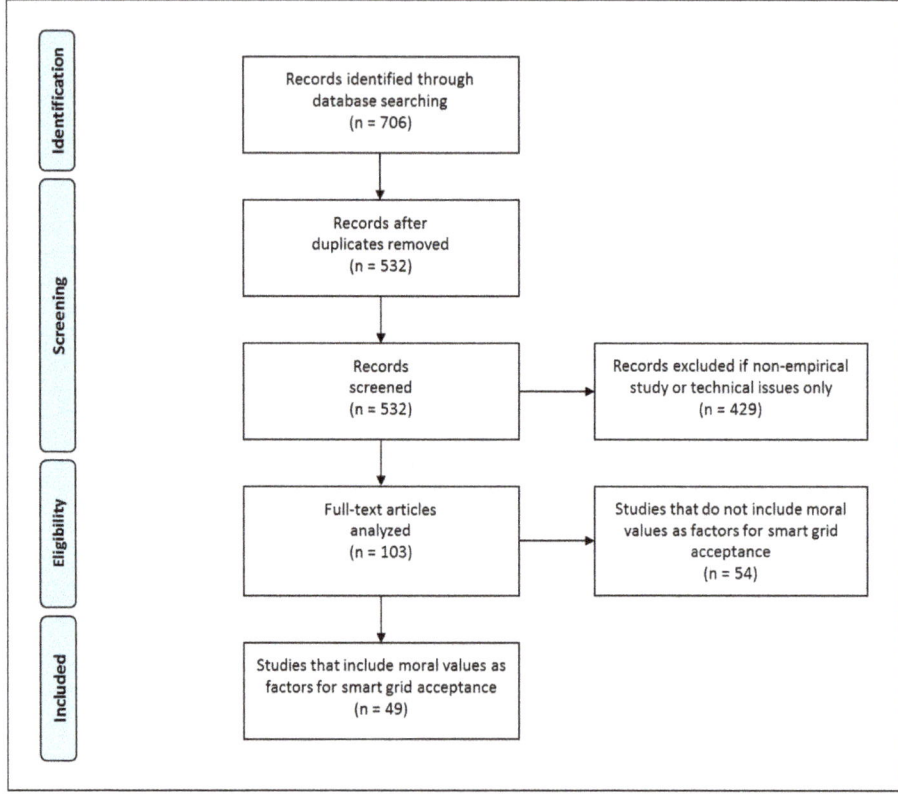

Figure 1. Flow diagram for systematic literature review (Based on [62]).

4. Results

Our literature review reveals that moral values can act as factors for smart grid acceptance; moral values were found in 49 articles on smart grid acceptance (see Appendix A). These articles were published in 23 different journals. However, more than 50% are concentrated in four journals: *Energy Research & Social Science* and *Energy Policy* were the most frequent journals, with 10 and nine publications respectively, followed by four publications in *Energy Efficiency* and three publications in *Renewable and Sustainable Energy Reviews*. The journals cover a large diversity of subject areas, including energy research, environmental science, engineering, business and management research, computer science, psychology, and philosophy. (Journals have been mapped to subject areas based on their categorizations in Scopus and Web of Science.)

The most prevalent subject area was energy research: 32 out of 49 articles were published in this field. A smaller number of articles were published in the three subject areas that can provide the theoretical background to understand the role of values for social acceptance and were reviewed earlier. First, this concerns ethics of technology: two articles were published in journals within the subject area of philosophy (*Journal of Information, Communication and Ethics in Society*, and *Public Understanding of Science*). Second, three articles were published in journals that contribute to the field of technology and innovation management, such as *Technological Forecasting and Social Change*. Third, a total of 18 publications are from journals where theories on technology acceptance from social psychology are widely used, for example the *Journal of Consumer Policy* and *Psychology & Marketing*, but also *Energy Research & Social Science*.

Twenty-five studies with qualitative approaches exploring smart grid acceptance used predominantly expert interviews, focus groups, public workshops, and in-depth interviews, while 27 studies used quantitative methodologies to test the impact of various values on acceptance or adoption (three publications rely both on qualitative and quantitative methodologies). Twelve publications tested consumer acceptance of smart grid technologies based on technology acceptance models used in social psychology. The other 14 articles using quantitative methods derived their own antecedents of smart grid acceptance.

In the 49 publications, a range of moral values have emerged as factors for acceptance or adoption of various smart grid technologies (Table 3). These values were reported either as drivers or barriers of smart grid acceptance/adoption. A value is classified as a "driver" if it provides impulse, motivation, or reason for smart grid introduction or if smart grid technologies are perceived to have a positive influence on these values. A value is identified as a "barrier" if it is expressed as concerns or if there is a perceived fear that the technology might have adverse consequences for this specific value.

The drivers of smart grid acceptance were environmental sustainability, security of supply, and transparency. Data privacy, data security, (mis)trust, health, justice, and reliability were found as barriers to smart grid acceptance. Control, inclusiveness, quality of life, and affordability were partly identified as driver and partly as barrier. All of these values emerged in studies using inductive qualitative approaches. Most of them were also included in quantitative studies, with the exception of distributive justice, inclusiveness, quality of life, and transparency.

The majority of these values are relevant for citizen or consumer acceptance. Only seven articles report values relevant for office workers, manufacturing companies, energy companies, or the society at large. While values for office workers are similar to consumers' concerns (trust and quality of life or comfort), the values reported for companies and the societies in general are the main drivers for smart grid development: environmental sustainability and security of supply.

Table 3. Values relevant for the acceptance of smart grid technologies.

Values	# of Articles (N = 49)	Smart Grid	Smart Metering	Smart Home	DSM	Household Storage	Smart EV Charging	Sources
Environmental Sustainability	22	+	+	+	+		+	[11,56–59,63–79]
Security of Supply	7		+	+	+	+	+	[3,67–69,72,78,79]
Transparency and Accuracy	6		+	+	+			[56,58,67,76,77,80]
Privacy	24	−	−	−	−			[2,11,14,55,56,58,63,74–77,79,81–92]
Security	15	−	−	−	−			[2,11,55,68,74–77,79,81,85,86,89,92,93]
(Mis)Trust	14			−	−			[14,63,75,76,81,83,90–92,94–98]
Health	5		−					[11,56,68,86,91]
Distributive and Procedural Justice	5							[14,56,57,75,96]
Control and Autonomy	14	−	−	+/−	+/−			[14,55,56,66,67,73,79,81,88,93,94,99–101]
Inclusiveness	7		−	+/−				[14,75–77,79,81,93]
Quality of Life	7			+	−			[58,66,73,75–77]
Reliability	5			+/−				[75,77,79–81]
Affordability of energy	4		+/−	+/−	−			[11,71,76,99]

+: Driver; −: Barrier; +/−: mentioned both as driver and barrier depending on study; refer to text for details; DSM: Demand-side management; EV: Electric vehicle.

4.1. Moral Values That Act as Drivers of Smart Grid Acceptance

The most often cited positive driving force (22 publications, [11,56–59,63–79]) for the acceptance of various smart grid technologies was their contribution to the environmental sustainability of energy systems. Environmental sustainability refers to the reduction of emissions from the electricity sector, thereby contributing to climate change goals [75,78]. Smart grid technologies contribute to environmental sustainability by facilitating the integration of renewable energy sources and electric vehicles [63,69,72]. In addition, smart metering and smart home networks are perceived by consumers to enable them to save energy through better visualization of the energy consumption of various household appliances, thereby lowering not only energy costs but also emissions [56,57,75].

Another key factor positively related to the acceptance of smart grid technologies was the security of electricity supply (seven publications, [3,67–69,72,78,79]). "Security of supply" in the context of electricity systems is defined as a low risk of interruptions in the supply [3]. Given that the electricity system is vital for the functioning of modern societies, a high security of supply is one of the central values in any debate on changing energy systems. Smart meters were perceived to enhance the security of supply, because they allow detection and reduction of power outages faster than conventional meters [68]. Household electricity storage systems allow to reduce the risk of supply interruptions because they can serve as a buffer for excess energy and allow to decouple electricity generation from consumption [3]. Smart charging systems allow to shift the charging time of electric vehicles and thereby can help to avoid grid overload problems [69,72].

In the context of smart metering, smart home, and demand-side management, transparency and accuracy were found to be further values motivating the acceptance of such technologies (six publications, [56,58,67,76,77,80]). Greater accuracy and a better overview of energy consumption data as well as transparency in the impact of consumption patterns on cost and the environment, which are enabled through smart meters and in-home displays, contributed positively to the acceptance these technologies [56,58].

4.2. Moral Values That Form Barriers for Smart Grid Acceptance

Privacy was by far the most prevalent moral value reported as a perceived barrier, mentioned in 24 publications [2,11,14,55,56,58,63,74–77,79,81–92]. Concerns about privacy are related to the increased collection and transmission of information on energy consumption compared to traditional meters [2]. Triggered by the possibility to share end-users' energy consumption data automatically and in real time with grid operators and store these data in central databases, consumers are concerned that energy companies could use these data to get insight into activities in a household that are considered as private [55,56]. Explicitly mentioned was the fear that smart grid technologies could allow identification of the type and time of use of household appliances [86]. In addition, consumers were concerned that their personal data could be sold commercially [91]. One study also reported the perceived danger in the effect of combining different pieces of data to reveal more information or patterns about consumer behavior that could be extracted from single pieces [75].

Concerns about data and cyber security were the second most often reported barrier to smart grid, smart metering, and smart home acceptance (15 publications, [2,11,55,68,74–77,79,81,85,86,89,92,93]). Security refers to the existence of mechanisms that ensure that personal data is protected from outside, malicious attacks [2,68]. The increased collection and transmission of more energy consumption data than with "dumb" systems are at the core of security concerns. Consumers are concerned that their consumption data, which is transmitted to e.g., grid operators, might fall into the wrong hands due to cyberattacks. They stress the importance of ensuring that personal data is adequately protected and encrypted [81,85,89]. In addition, and specifically connected to smart home platforms, consumers uttered the fear that outsiders could get more easy access to their private spaces/homes [55].

Trust, or rather the lack of trust by consumers in organizations charged with the implementation and management of smart grid technologies (e.g., electric utilities, governmental authorities), was reported as one of the key barrier values for smart grid acceptance (14 publications, [14,63,75,76,81,83,90–92,94–98]).

While trust was mainly relevant in consumer acceptance studies, one study from the perspective of US utilities revealed that utilities were aware of the problematic (mis)trust by consumers towards their companies [63]. Consumers' lack of trust is reflected in concerns that the utilities industry and the government (a) are not open about their benefits and (b) will not pass any financial savings on to customers. Consumers also found it difficult to understand why utilities would promote energy-saving messages while they are perceived to increase profits with an increased energy consumption [76,91,94]. Additionally, concerns were related to the degree of trust that the personal data shared through smart meters with energy companies is protected [90,98].

In the context of smart metering, consumers perceived health risks were found to be negatively connected to the acceptance and use of smart meters (five publications, [11,56,68,86,91]). Perceived health risks refer to the subjective evaluation of potential health threats resulting from an event or an activity [56]. Health risks were connected with exposure to electromagnetic radiation from smart meters [68,86,91]. Whether or not radiation poses objective threats to consumers' health, the fact that smart meters are perceived as health risks in studies on consumer acceptance indicates that such concerns should be taken seriously by utilities and governmental authorities when introducing smart metering.

Concerns about the fairness of smart metering and demand-side management reflected the values of distributive and procedural justice as a barrier for smart grid acceptance from the perspective of energy consumers (five publications, [14,56,57,75,96]). Distributive justice refers to a fair distribution of costs and benefits among the key stakeholders involved in these technologies [14,56,57]. Consumers feared that they will have to bear the costs for the introduction of smart metering without receiving apparent benefits while energy providers would profit from financial savings [57]. In addition, there was a perception that the responsibility for saving energy would be pushed on consumers while supplier obligations to ensure low consumer prices would be neglected [14]. Procedural justice refers to fairness in decision making processes, often based on the fact that all relevant stakeholders are able to participate in the process. Although this concern was less prevalent than distributive justice, it yielded interesting results in a study by Guerreiro et al. [56]. The authors were interested in the use of smart meters combined with an in-home display and found that increased perceptions of procedural justice let to decreased use in the devices. It might be that respondents who perceived the process of introduction as being fair felt a lower need to control the equipment.

4.3. Moral Values with Ambiguous Effects on Smart Grid Acceptance

Control or autonomy—defined in this context as the perception that one can direct events in life free of outside influence [100]—was related to consumers concerns about loss of control and autonomy with the introduction of smart metering and the installation of smart home platforms. They feared losing control to ICT systems and perceived the monitoring of daily behavior as too intrusive and restrictive [55,81]. Concerns were also directed to a fear of loss of control towards energy suppliers, who might manage their energy consumption for them [14,56]. While control was mostly perceived as a barrier (12 publications, [14,55,56,67,73,79,81,88,93,94,99,101]), a later study reported a positive effect of control on the acceptance of an automated demand-side response tariff [100]. This suggests that concerns about the loss of control play a more ambivalent role than previously assumed. The authors explain the effect with two reasons. Firstly, the tariff's impact was clearly defined (e.g., the room temperature was only allowed to shift by 1 °C). Secondly, the option of overriding the automation was presented, which might have restored perceptions of self-control [14,100].

Inclusiveness was both seen as a barrier and a driver for smart grid acceptance. Inclusiveness refers to giving all different societal groups the possibility to be included in the technological development. On the one hand, six studies revealed that consumers were concerned that elderly people, disabled people, and people with less affinity to computers and IT systems would be systematically excluded from the smart grid development [14,75,77,79,81,93]. In another study, however, consumers expressed positive views about the benefits, the support, and the additional services that smart homes could offer in assisted living for the elderly and people with disabilities [76].

Increased quality of life was seen as a driver for smart home technologies in six publications [58,66,73,75–77]. Smart home services such as health monitoring or a remote control of security are perceived as practical and automation is seen as enhancing convenience and comfort [73,76]. However, it was reported as a barrier in one study, in which building occupants were concerned with reductions in their living quality as a consequence of demand-side management [102]. When building equipment such as ventilation fans or cooling systems have communication and control capabilities to steer the energy demand of the building automatically, the effects on the perceived thermal comfort of building occupants was reported as a major concern and barrier for the implementation of such a DSM measure [102].

The reliability of novel smart home technologies was questioned and reported as barrier by consumers in four publications [75,77,79,81]. The adoption of non-mainstream technology was seen as risky with respect to the malfunctioning of the system, such as a break-down of communication systems or room sensors being triggered unintentionally [75,77]. Consumers felt unease at becoming reliant on computer systems they might not fully understand. In addition, concerns were reported that innovations, once adopted, would not widely spread or become rapidly obsolete due to fast technological progress. This was especially seen problematic when smart home technologies were seen as a costly and long-term investment [79]. However, one publication found that in-home displays have the ability to enhance the reliability of an entire home energy management system because such displays support in discovering system failures or underperformance [80].

Future affordability of energy was found to be both a driver and a barrier for the acceptance of smart metering, smart home platforms, and demand-side management. Affordability is the availability of financial means to be able to pay for energy. In two studies [11,99], the potential of smart meters and smart home platforms to save energy and prevent energy poverty were seen as reasons to accept these technologies. In two different studies [71,76], however, consumers were concerned about hidden costs and were generally skeptical whether smart grid technologies will indeed reduce their energy bills.

5. Discussion

Our literature review on the role of moral values for the acceptance of smart grid technologies showed that values are indeed discussed in the literature on smart grid acceptance and adoption. However, their relationship with acceptance is not always clear. Whereas certain values are always seen as either drivers or barriers, others could be seen as having an ambiguous effect on acceptance. We turn to a more detailed discussion of our findings.

5.1. Values as Factors for Consumer and Citizen Acceptance

In general, our results show that moral values can act as important factors for consumer and citizen acceptance of smart grid technologies. The fact that all the values we found have emerged from inductive, qualitative studies indicates that consumers expressed values in an unprompted way as both drivers for smart grid development and concerns around these technologies. Thus, values were not a priori introduced into these studies by researchers but were expressed by consumers independently. In addition, quantitative studies confirmed for almost all reported values that they influence consumer or citizen acceptance. Distributive justice, inclusiveness, quality of life, and transparency were the exceptions which were only reported in qualitative studies.

However, our results also show that there are two aspects of values which pose additional complexities to their investigation as factors for acceptance. First, some values were found to have an ambiguous effect on acceptance. More specifically, whereas some values were clearly positive forces driving smart grid development (e.g., environmental sustainability) and some were clearly consumer concerns around the technology (e.g., privacy, justice), some were mentioned both as drivers and barriers. For example, studies mentioned the potential of smart grid technologies to save energy and thus save costs as perceived benefits with regards to energy affordability [11,99]. However, consumers were also concerned that they will have to bear the costs for the introduction of smart grid technologies through higher electricity bills. Another example is inclusiveness; whereas there are concerns that several societal groups (e.g., the elderly, disabled) would be systematically excluded because of the focus on novel ICT [14], benefits that smart homes in particular could offer in assisted living for the elderly and disabled are expressed [76]. Additionally, the value of control was mostly perceived as a barrier due to a perceived loss of consumers' control to electronic devices or energy suppliers. Automated demand-response tariffs were particularly in focus of this concern. However, once the impact of such a tariff was clearly defined, the degree of external control through the tariff was very small, and consumers had the option to override the automation, the perceived loss of control was no longer a problem [100].

These examples illustrate the importance of the detailed technological and regulatory context for the effect direction on acceptance. In the example of control, the way an automated demand-response tariff was structured with respect to definition of boundaries of the automation or overriding possibilities was decisive whether control was seen as a barrier or not. The debate to what extent smart metering impacts energy affordability depends on the regulation of electricity prices: if smart meters enable consumers to save costs by using less energy, these savings might be offset because costs for the smart metering infrastructure are socialized, i.e., paid by consumers through the network tariffs on electricity bills.

The examples also illustrate that whether certain values have a positive or negative impact on smart grid acceptance depends on their interpretation by consumers. Values can therefore be characterized as "contestable concepts," having two levels of meaning [103]. The first level is expressed in a short definition; for example, energy affordability is generally defined as having the financial means to be able to pay for energy. The second level of meaning refers to the value's conception. Here, contestation occurs over how the concept should be interpreted and whether a technology contributes to the value or endangers it [103]. It is thus important to understand values at the level of conception, since this is the level where controversies arise and the way values impact technology acceptance might depend on their conception [104]. In the example of affordability, the debate is not about the definition or importance of affordability, the debate is whether certain features of smart grid technologies are perceived to contribute to energy affordability while others do not. As a consequence, future research should carefully consider different potential conceptions of values when testing their effect on acceptance.

Second, certain values are closely interrelated, increasing the complexity in deriving their separate effects on smart grid acceptance. Probably the most prevalent relationship could be observed between data privacy and security. Both concerns are related to the increased transmission and storing of personal data. They are frequently mentioned in context with each other [2,55,56,76] or even measured as one construct (e.g., [84,85]). However, they are different concepts. Privacy refers to the concern that individuals' personal data can be used externally to infer information about activities that are considered as private [12]. Security concerns on the other hand are defined in terms of the risk that personal data is subject to malicious external attacks, e.g., through hacking [2]. Their conceptual differentiation means on the one hand that different measures need to be taken by policy makers and industry actors responsible for smart grid introduction to protect consumers' privacy and data security. On the other hand, their conceptual differentiation could imply different effects on consumer acceptance. They should therefore be treated as separate concepts in academic studies on smart grid acceptance.

Distributive justice is connected to affordability concerns. Consumers were concerned that they will have to bear the costs for e.g., the smart meter introduction, whereas energy providers would profit from financial savings [57]. Consumers perceived an unfairness that smart grid technologies might lead to higher energy costs and a lower affordability of energy [56]. As a consequence, concerns about fairness and affordability might reinforce each other in their negative effect on smart grid acceptance.

In addition, several values were connected with the perceived trust of consumers in energy companies and government authorities. Concerns about distributive justice were connected with the lack of trust that energy companies are not open about their benefits and would not pass on financial savings to consumers [57,76]. Also, trust was related to privacy and security concerns: Perceived consumer trust about the protection of personal data [98]. This points to the central importance of trust between consumers and authorities or organizations charged with the implementation and management of smart grid technologies as potential antecedent for several other values; a relationship that is worth considering in smart grid acceptance studies. Trust is also suggested as antecedent for consumer beliefs by Huijts et al. [41] in their conceptual development of a framework for acceptance of energy technologies. Trust is suggested as influencing positive and negative affect, perceived costs, risks and benefits, and also procedural justice.

5.2. Combining Insights from Ethics with Technology Acceptance Literature

In contrast to our results, current theoretical frameworks for technology acceptance and adoption do not seem to pay attention to moral values as factors for acceptance (see Section 2.2). Frameworks for technology acceptance and adoption in technology and innovation management fields focus on market-, firm-, and technology-specific characteristics [16,21,44]. In social psychology, technology acceptance models focus on factors pertaining to technology beliefs, social influences, and personality beliefs [17,18].

Therefore, we propose that moral values should be included more systematically in studies on the acceptance or adoption of smart grid technologies, and potentially technology acceptance in general. Scientific understanding of the role of values for technology acceptance can be gained by combining insights from ethics of technology with literature on technology acceptance.

Ethics of technology and particularly VSD approaches can be beneficial for the identification and conceptualizations of relevant values for a particular technological context. In their tripartite approach, VSD scholars place great emphasis on identifying relevant values. They do this both from an ethical normative perspective and a descriptive perspective relying on the opinions of key stakeholders involved with a technology [27,33]. In addition, they acknowledge that values can be interpreted and prioritized differently by different stakeholder groups and therefore integrate considerations around conceptions of values explicitly in their empirical approaches [35,104]. Their in-depth understanding of different conceptualization of values can contribute to the two complexities about the relationships between values and social acceptance we encountered in our results, namely that these relationships hinge on detailed interpretations of values and that there are mutual interdependencies between different values. Methods of elicitation of technology specific values from VSD can be used by researchers studying smart grid acceptance. This includes what VSD researchers call conceptual investigations, philosophically informed considerations of how stakeholders might be affected by the technology. It also includes empirical investigations, in which VSD scholars use the entire range of qualitative and quantitative empirical methods to answer questions such as how stakeholders interpret different values for the given technological context or which values are prioritized by different stakeholder groups affected by the technology [34].

Ethicists and VSD scholars focus on the understanding of values and possibilities to integrate them into technological design. However, their research aims do not include testing whether a design for values increases the acceptance and adoption of technologies. Their approach seems to underlie the implicit proposition that a proper integration of values that are judged as important for the context of a specific technology will contribute to enhancing acceptance in society [28].

The literature on technology acceptance is complementary to that because it does study the impact of a diverse range of factors on technology acceptance and adoption. Thus, it provides not only rigorous quantitative methods to test relationships but also measurement scales for values and acceptance in surveys or experiments [17,40,41].

More specifically, our results indicate that adaptations of technology acceptance models from social psychology might be suitable to include moral values (see Section 2.2.2 for a review). Half of the publications in our systematic literature review including values as factors and using deductive theory testing approaches investigate smart grid acceptance based on models used in social psychology (e.g., [3,56,85,98,100]). Although they only include a sub-set of relevant values in their models, these studies provide first indications how to integrate values in acceptance models and which other model variables values might be related to.

Most of these scholars study values as direct antecedents of intentions to use or use of smart grid technologies. For example, Fell et al. [100] find that control over comfort and timing of activities are related to intentions to adopt a demand-side management scheme and Römer et al. [3] relate security of supply concerns to purchase intentions of household storage systems.

A number of studies show effects of values on several different variables in technology acceptance models, particularly perceived risk and perceived usefulness or ease of use, concepts that are used in both UTAUT and TAM. Chou et al. [85] find that concerns on data privacy and security impact perceived risk. In a similar vein, Guerreiro et al. [56] stress the connection between health concerns and perceived risk. Park et al. [68] find that perceived security of supply and environmental sustainability impact perceived usefulness, and perceived security and health concerns affect perceived risk. Perceived usefulness and risk impact in turn impacts intentions to use smart grid technologies.

The indication from our results that technology acceptance models from social psychology might be suitable to include moral values is in line with a proposed framework for public acceptance of sustainable energy technologies such as wind mills or hydrogen vehicles by Huijts et al. [41]. The authors stress the importance of procedural and distributive justice measured as perceived fairness of the decision process leading up to the technology's introduction as well as the perceived fair distribution of costs and benefits, affecting attitudes toward the technologies. Additionally, they hypothesize that the degree of trust in actors that are responsible for the technology is seen as influencing positive and negative affect, perceived costs, risks and benefits, which in turn affect attitudes toward the technologies. Positive attitudes toward technologies are then related to intentions to accept and technology acceptance.

6. Conclusions

Smart grid technologies are seen as an important enabler in the transition to more sustainable energy systems, but the development has been challenged among others by societal concerns [2,11]. In this paper, we showed that societal concerns about smart grid technologies reflect moral values, which are (perceived) technology characteristics about ethical and social consequences of technologies such as justice, trust, or privacy. We proposed that concerns related to moral values might hinder the wider acceptance and adoption of smart grid technologies. The paper set out to address the questions which moral values are relevant for smart grid technologies and how they influence smart grid acceptance.

Our results show that moral values can act as drivers and barriers for consumer and citizen acceptance of smart grid technologies. On the one hand, values such as environmental sustainability and security of supply positively influence smart grid acceptance. On the other hand, concerns about privacy, security, or health negatively impact their acceptance. In addition, several values were mentioned both as driving factors for smart grid acceptance and as concerns (e.g., affordability, inclusiveness). Studying the impact of values on acceptance is not only made complex by these ambiguous interpretations, but also by instrumental relationships between certain values such as affordability and distributive justice. It is thus important to consider the detailed technological and regulatory context, the nature of values as contestable concepts, and interdependencies between them.

Based on our results, we propose that future research should strive for a better understanding of the role of moral values as factors for smart grid acceptance in order to contribute to embedding values in smart grid design. This can be done by bridging literature from ethics of technology with technology acceptance. Ethicists study in depth which values are implied in certain technologies. In their focus on a normative perspective, however, they do not relate values to the empirical acceptance of technologies [28]. Technology acceptance studies provide a complementary perspective because they test the impact of a wide range of factors on acceptance, yet typically without considering values as factors [15–18]. The results of our systematic literature review show that especially acceptance models widely used in social psychology such as TAM, TPB, or UTAUT offer a good foundation to study the effect of values as perceived technology characteristics on smart grid acceptance.

Author Contributions: The article's concept and main message were developed in joint discussions. C.M. conducted the review and wrote the paper. G.v.d.K. and N.D. contributed to Sections 2.1 and 2.2.1. R.K. gave input on several drafts. All authors read, revised, and approved the final manuscript.

Funding: Funding for work on this article has been provided by the Netherlands Organization for Scientific Research (NWO) under the Responsible Innovation Program [Grant No. 313-99-305], the Amsterdam Institute for Advanced Metropolitan Solutions (AMS), and TFECo B.V.

Conflicts of Interest: The authors declare no conflict of interest. The funders had no role in the design of the study; in the collection, analyses, or interpretation of data; in the writing of the manuscript, and in the decision to publish the results.

Appendix A

Table A1. Overview of articles considering values for smart grid acceptance.

Authors	Year	Journal	Citations *	Main Contribution	Methodology	Technology
Aduda et al.	2016	Sustainable Cities and Society	18	Investigate effect of demand-side management on building performance indicators	Field study with follow-up survey	DSM
Balta-Ozkan et al.	2013	Energy Policy	88	Explore key barriers to smart home adoption in the UK	Expert interviews, deliberative public workshops	Smart Home
Balta-Ozkan et al.	2013	Energy	28	Explore key barriers to smart home adoption in the UK	Expert interviews, deliberative public workshops	Smart Home
Balta-Ozkan et al.	2014	Technology Analysis and Strategic Management	11	Explore technical and economic drivers and barriers to smart home market development in three European countries (UK, DE, IT)	Deliberative public workshops	Smart Home
Balta-Ozkan et al.	2014	Energy Research & Social Science	22	Explore drivers and barriers to smart home market development in three European countries (UK, DE, IT)	Deliberative public workshops	Smart Home
Barnicoat & Danson	2015	Energy Research & Social Science	17	Explore how older tenants in rural Scotland interact with technology	In-depth interviews	Smart Home
Begier	2014	Journal of Information, Communication and Ethics in Society	0	Explore strategies to build relationships with energy consumers during exchange of energy meters	Focus groups, survey	Smart Metering
Berry et al.	2017	Energy Efficiency	0	Explore residential consumers' attitudes towards and experiences with an in-home display and energy management system	In-depth interviews	Smart Home
Buchanan et al.	2016	Energy Policy	6	Explore opportunities and threats of smart metering initiatives	Focus groups	Smart Metering/Smart Services
Buryk et al.	2015	Energy Policy	11	Investigate impact of disclosing environmental benefits on DSM adoption	Choice experiment	DSM
Chen et al.	2017	Energy Research & Social Science	8	Investigate social-psychological factors affecting smart meter support and adoption intention	Survey	Smart Metering
Cherry et al.	2017	Energy Research & Social Science	6	Explore experts' and public's visions of smart homes	Semi-structured interviews	Smart Home
Chou & Yutami	2014	Applied Energy	16	Investigate antecedents of willingness to adopt smart meter	Survey	Smart Metering
Chou et al.	2015	Renewable and Sustainable Energy Reviews	6	Investigate antecedents of willingness to adopt smart meter	Survey	Smart Metering

Table A1. *Cont.*

Authors	Year	Journal	Citations *	Main Contribution	Methodology	Technology
Dedrick et al.	2015	Electronic Markets	3	Examine factors influencing smart grid adoption among US utilities	Semi-structured interviews	Smart Grid
Ehrenhard et al.	2014	Technological Forecasting and Social Change	19	Explore acceptance of smart home among the elderly	In-depth interviews	Smart Home
Fell et al.	2015	Energy Research & Social Science	18	Investigate factors for acceptance of different demand-side response tariffs	Experiment	DSM
Gerpott & Paukert	2013	Energy Policy	27	Investigate factors for willingness-to-pay for smart meters	Survey	Smart Metering
Ghazal et al.	2015	Renewable and Sustainable Energy Reviews	3	Investigate factors for consumer acceptance of a smart plug system	Survey	Smart Home
Goulden et al.	2014	Energy Research & Social Science	90	Explore perceptions of centralized and decentralized smart grid platforms	Focus groups	Smart Grid
Guerreiro et al.	2015	Energy Efficiency	3	Understand socio-psychological and technological aspects that influence use of smart meters	Survey, discourse analysis	Smart Metering
Hall et al.	2016	Energy Policy	6	Explore consumer interest and responses to the concept of cost-reflective pricing	Focus groups	DSM
Hammer et al.	2015	User Modelling and User-Adapted Interaction	5	Build user-trust model for decision making on energy management systems in office buildings	Survey experiment, (Living Lab) model	Energy management systems
Hess & Coley	2014	Public Understanding of Science	16	Explore complaints in the public debate on wireless smart meters in California	Discourse analysis	Smart Metering
Kahma & Matschoss	2017	Energy Research & Social Science	4	Investigate the non-adoption of smart energy services through focus on non-users	Survey	Smart Home
King & Jessen	2014	International Journal of Law and Information Technology	5	Explores the key privacy and data protection concerns for both the EU and USA consumers related to data sharing in smart metering systems	Secondary data analysis (of legal regimes)	Smart Metering
Krishnamurti et al.	2012	Energy Policy	93	Explore consumer beliefs about smart meters in the US	In-depth interviews, survey	Smart Metering
Li et al.	2017	Applied Energy	1	Investigate user perception of smart grids and energy flexible buildings to identify suitable user groups	Survey	Smart Grid
Luthra et al.	2014	Renewable and Sustainable Energy Reviews	61	Explore barriers to smart grid adoption	Expert interviews	Smart Grid
Matschoss et al.	2015	Energy Efficiency	4	Identify pioneering customers for novel energy efficiency services enabled by smart grid technologies	Survey	DSM
Mesarić et al.	2017	Sustainability	2	Explore the influence of users' energy-related behavior on smart grid processes	Focus groups	DSM

Table A1. Cont.

Authors	Year	Journal	Citations *	Main Contribution	Methodology	Technology
Michaels & Parag	2016	Energy Research & Social Science	7	Investigated perceptions of demand reduction, load shifting, and energy storage as prosumer activities in Israel	Survey	DSM
Moser	2017	Energy Efficiency	1	Investigate factors for social acceptance of load-shifting programs for smart appliances	Experiment	DSM
Muench et al.	2014	Energy Policy	21	Explore barriers to smart grid implementation	Expert interviews	Smart Grid
Ornetzeder et al.	2009	WIT Transactions on Ecology and the Environment	1	Explore public's opinion on future sustainable energy technology research	Participatory technology assessment workshop	Smart Metering Smart Home
Paetz et al.	2012	Journal of Consumer Policy	85	Explore behavioral aspects, motives, and barriers for smart home acceptance	Focus groups	Smart Home
Park et al.	2014	Energy Policy	19	Tested factors for consumer acceptance of smart meters	Survey	Smart Metering
Park et al.	2017	Sustainability	0	Investigate consumer acceptance of a home energy management system	Survey	Smart Home
Raimi & Carrico	2016	Energy Research & Social Science	4	Examine the American lay public's level of knowledge about smart meters	Survey	Smart Metering
Römer et al.	2015	Electronic Markets	4	Investigate factors for household acceptance of electricity storage systems	Survey	Household Storage
Sandström & Keijer	2010	OPEN HOUSE INTERNATIONAL	0	Explore attitudes and acceptance of residents towards smart homes	Survey	Smart Home
Schmalfuß et al.	2015	Energy Research & Social Science	3	Investigate user experience with smart charging system	Field study with follow-up interviews	Smart Charging
Schweitzer et al.	2016	Psychology & Marketing	2	Investigate impact of perceived disempowerment on adoption intention of smart home applications	Experiment	Smart Home
Shrouf & Miragliotta	2015	Journal of Cleaner Production	54	Explore experts view on energy-efficient production management practices supported by the Internet of Things	Expert interviews	Smart Metering and appliances in factory production processes
Spence et al.	2015	Nature Climate Change	14	Investigate public perceptions of different demand-side management possibilities in the UK	Survey	DSM
Will & Schuller	2016	Transportation Research Part C: Emerging Technologies	8	Investigate factors for the acceptance of smart charging	Survey	Smart Charging
Wilson et al.	2017	Energy Policy	12	Identify perceived benefits and risks of smart home technologies	Survey	Smart Home
Yang et al.	2017	Industrial Management and Data Systems	4	Investigate customers' adoption intentions of smart home services	Survey	Smart Home
Zhou & Brown	2017	Journal of Cleaner Production	10	Compare factors for smart metering penetration rates across five European countries	Case study research (secondary data)	Smart Metering

* Number of citations according to Scopus/Web of Science.

References

1. European Commission 2030 Energy Strategy. Available online: https://ec.europa.eu/energy/en/topics/energy-strategy/2030-energy-strategy (accessed on 16 December 2016).
2. Muench, S.; Thuss, S.; Guenther, E. What hampers energy system transformations? The case of smart grids. *Energy Policy* **2014**, *73*, 80–92. [CrossRef]
3. Römer, B.; Reichhart, P.; Picot, A. Smart energy for Robinson Crusoe: An empirical analysis of the adoption of IS-enhanced electricity storage systems. *Electron. Mark.* **2015**, *25*, 47–60. [CrossRef]
4. Lund, H.; Hvelplund, F.; Østergaard, P.; Möller, B.; Mathiesen, B.V.; Connolly, D.; Andersen, A.N. Analysis: Smart Energy Systems and Infrastructures. In *Renewable Energy Systems*; Lund, H., Ed.; Elsevier Inc.: Oxford, UK; Waltham, MA, USA, 2014; pp. 131–184, ISBN 9780124104235.
5. Fallah, S.; Deo, R.; Shojafar, M.; Conti, M.; Shamshirband, S. Computational Intelligence Approaches for Energy Load Forecasting in Smart Energy Management Grids: State of the Art, Future Challenges, and Research Directions. *Energies* **2018**, *11*, 596. [CrossRef]
6. Pooranian, Z.; Abawajy, J.; P, V.; Conti, M. Scheduling Distributed Energy Resource Operation and Daily Power Consumption for a Smart Building to Optimize Economic and Environmental Parameters. *Energies* **2018**, *11*, 1348. [CrossRef]
7. Xenias, D.; Axon, C.J.; Whitmarsh, L.; Connor, P.M.; Balta-Ozkan, N.; Spence, A. UK smart grid development: An expert assessment of the benefits, pitfalls and functions. *Renew. Energy* **2015**, *81*, 89–102. [CrossRef]
8. Pooranian, Z.; Nikmehr, N.; Najafi-Ravadanegh, S.; Mahdin, H.; Abawajy, J. Economical and environmental operation of smart networked microgrids under uncertainties using NSGA-II. In Proceedings of the 2016 24th International Conference on Software, Telecommunications and Computer Networks (SoftCOM), Split, Croatia, 22–24 September 2016.
9. Sintov, N.D.; Schultz, P.W. Adjustable green defaults can help make smart homes more sustainable. *Sustainability* **2017**, *9*, 1–12. [CrossRef]
10. Cuijpers, C.; Koops, B.-J. Smart Metering and Privacy in Europe: Lessons from the Dutch Case. In *European Data Protection: Coming of Age*; Gutwirth, S., Leenes, R., de Hert, P., Poullet, Y., Eds.; Springer: Dordrecht, The Netherlands, 2013; pp. 269–293, ISBN 978-94-007-5170-5.
11. Raimi, K.T.; Carrico, A.R. Understanding and beliefs about smart energy technology. *Energy Res. Soc. Sci.* **2016**, *12*, 68–74. [CrossRef]
12. McKenna, E.; Richardson, I.; Thomson, M. Smart meter data: Balancing consumer privacy concerns with legitimate applications. *Energy Policy* **2012**, *41*, 807–814. [CrossRef]
13. Ligtvoet, A.; Van de Kaa, G.; Fens, T.; Van Beers, C.; Herder, P.; Van den Hoven, J. Value Sensitive Design of Complex Product Systems. In *Policy Practice and Digital Science: Integrating Complex Systems, Social Simulation and Public Administration in Policy Research*; Janssen, M., Wimmer, A.M., Deljoo, A., Eds.; Springer International Publishing: Cham, Switzerland, 2015; pp. 157–176, ISBN 978-3-319-12784-2.
14. Buchanan, K.; Banks, N.; Preston, I.; Russo, R. The British public's perception of the UK smart metering initiative: Threats and opportunities. *Energy Policy* **2016**, *91*, 87–97. [CrossRef]
15. Schilling, M.A. Technology Success and Failure in Winner-Take-All Markets: The Impact of Learning Orientation, Timing, and Network Externalities. *Acad. Manag. J.* **2002**, *45*, 387–398.
16. Suarez, F.F. Battles for technological dominance: An integrative framework. *Res. Policy* **2004**, *33*, 271–286. [CrossRef]
17. Venkatesh, V.; Thong, J.; Xu, X. Consumer Acceptance and Use of Information Technology: Extending the Unified Theory of Acceptance and Use of Technology. *MIS Q.* **2012**, *36*, 157–178. [CrossRef]
18. Stern, P.C. Toward a Coherent Theory of Environmentally Significant Behavior. *J. Soc. Issues* **2000**, *56*, 407–424. [CrossRef]
19. Cusumano, M.A.; Mylonadis, Y.; Rosenbloom, R.S. Strategic Maneuvering and Mass-Market Dynamics: The Triumph of VHS over Beta. *Bus. Hist. Rev.* **1992**, *66*, 51–94. [CrossRef]
20. Katz, M.L.; Shapiro, C. Network externalities, competition, and compatibility. *Am. Econ. Rev.* **1985**, *75*, 424–440.
21. Schilling, M.A. Technological Lockout: An Integrative Model of the Economic and Strategic Factors Driving Technology Success and Failure. *Acad. Manag. Rev.* **1998**, *23*, 267–284. [CrossRef]
22. Ajzen, I. The Theory of Planned Behavior. *Organ. Behav. Hum. Decis. Process.* **1991**, *50*, 179–211. [CrossRef]

23. Shrader-Frechette, K.S.; Westra, L. Overview: Ethical Studies about Technology. In *Technology and Values*; Shrader-Frechette, K.S., Westra, L., Eds.; Rowman & Littlefield Publishers: Lanham, MD, USA, 1997; pp. 3–10.
24. Taebi, B.; Kadak, A.C. Intergenerational considerations affecting the future of nuclear power: Equity as a framework for assessing fuel cycles. *Risk Anal.* **2010**, *30*, 1341–1362. [CrossRef] [PubMed]
25. Van de Poel, I. Values in engineering design. In *Handbook of the Philosophy of Science*; Meijers, A.W.M., Ed.; Oxford University Press: New York, NY, USA, 2009; Volume 9, pp. 973–1006.
26. Künneke, R.W.; Mehos, D.C.; Hillerbrand, R.; Hemmes, K. Understanding values embedded in offshore wind energy systems: Toward a purposeful institutional and technological design. *Environ. Sci. Policy* **2015**, *53*, 118–129. [CrossRef]
27. Friedman, B.; Kahn, P.H., Jr.; Borning, A.; Huldtgren, A. Value sensitive design and information systems. In *Early Engagement and New Technologies: Opening Up the Laboratory*; Doorn, N., Schuurbiers, D., Van de Poel, I., Gorman, M., Eds.; Springer: Dordrecht, The Netherlands, 2013; pp. 55–95, ISBN 9400778430.
28. Manders-Huits, N. What values in design? The challenge of incorporating moral values into design. *Sci. Eng. Ethics* **2011**, *17*, 271–287. [CrossRef] [PubMed]
29. Albrechtslund, A. Ethics and technology design. *Ethics Inf. Technol.* **2007**, *9*, 63–72. [CrossRef]
30. Winner, L. Do Artifacts Have Politics? *Daedalus* **1980**, *109*, 121–136.
31. Shilton, K.; Koepfler, J.A.; Fleischmann, K.R. Charting Sociotechnical Dimensions of Values for Design Research. *Inf. Soc.* **2013**, *29*, 259–271. [CrossRef]
32. Barry, C. The Ethical Assessment of Technological Change: An overview of the issues. *J. Hum. Dev.* **2001**, *2*, 167–189. [CrossRef]
33. Flanagan, M.; Howe, D.C.; Nissenbaum, H. Embodying values in technology: Theory and practice. In *Information Technology and Moral Philosophy*; Van Den Hoven, J., Weckert, J., Eds.; Cambridge University Press: New York, NY, USA, 2008; pp. 322–353.
34. Friedman, B.; Kahn, P.; Borning, A. Value sensitive design: Theory and methods. In *University of Washington Technical Report*; University of Washington: Washington, DC, USA, 2002; pp. 2–12.
35. Davis, J.; Nathan, L.P. Value Sensitive Design: Applications, Adaptations, and Critiques. In *Handbook of Ethics, Values, and Technological Design: Sources, Theory, Values and Application Domains*; van den Hoven, J., Vermaas, P.E., van de Poel, I., Eds.; Springer: Dordrecht, The Netherlands, 2015; pp. 11–40.
36. Wüstenhagen, R.; Wolsink, M.; Bürer, M.J. Social acceptance of renewable energy innovation: An introduction to the concept. *Energy Policy* **2007**, *35*, 2683–2691. [CrossRef]
37. Sauter, R.; Watson, J. Strategies for the deployment of micro-generation: Implications for social acceptance. *Energy Policy* **2007**, *35*, 2770–2779. [CrossRef]
38. Broman Toft, M.; Schuitema, G.; Thøgersen, J. Responsible technology acceptance: Model development and application to consumer acceptance of Smart Grid technology. *Appl. Energy* **2014**, *134*, 392–400. [CrossRef]
39. Schuitema, G.; Steg, L.; Forward, S. Explaining differences in acceptability before and acceptance after the implementation of a congestion charge in Stockholm. *Transp. Res. Part A Policy Pract.* **2010**, *44*, 99–109. [CrossRef]
40. Steg, L.; Dreijerink, L.; Abrahamse, W. Factors influencing the acceptability of energy policies: A test of VBN theory. *J. Environ. Psychol.* **2005**, *25*, 415–425. [CrossRef]
41. Huijts, N.M.A.; Molin, E.J.E.; Steg, L. Psychological factors influencing sustainable energy technology acceptance: A review-based comprehensive framework. *Renew. Sustain. Energy Rev.* **2012**, *16*, 525–531. [CrossRef]
42. Suarez, F.F.; Utterback, J.M. Dominant designs and the survival of firms. *Strateg. Manag. J.* **1995**, *16*, 415–430. [CrossRef]
43. Gallagher, S.; Park, S.H. Innovation and competition in standard-based industries: A historical analysis of the US home video game market. *Eng. Manag. IEEE Trans.* **2002**, *49*, 67–82. [CrossRef]
44. Van de Kaa, G.; Van den Ende, J.; de Vries, H.J.; Van Heck, E. Factors for winning interface format battles: A review and synthesis of the literature. *Technol. Forecast. Soc. Chang.* **2011**, *78*, 1397–1411. [CrossRef]
45. Van de Kaa, G.; De Vries, H.J.; Rezaei, J. Platform selection for complex systems: Building automation systems. *J. Syst. Sci. Syst. Eng.* **2014**, *23*, 415–438. [CrossRef]
46. Van de Kaa, G.; De Vries, H.J. Factors for winning format battles: A comparative case study. *Technol. Forecast. Soc. Chang.* **2015**, *91*, 222–235. [CrossRef]

47. Van de Kaa, G.; De Vries, H.J.; Van den Ende, J. Strategies in network industries: The importance of inter-organisational networks, complementary goods, and commitment. *Technol. Anal. Strateg. Manag.* **2015**, *27*, 73–86. [CrossRef]
48. Teece, D.J. Profiting from technological innovation: Implications for integration, collaboration, licensing and public policy. *Res. Policy* **1986**, *15*, 285–305. [CrossRef]
49. Davis, F.D.; Bagozzi, R.P.; Warshaw, P.R. User Acceptance of Computer Technology: A Comparison of Two Theoretical Models. *Manag. Sci.* **1989**, *35*, 982–1003. [CrossRef]
50. Venkatesh, V.; Davis, F.D. A theoretical extension of the technology acceptance model: Four longitudinal Studies. *Manag. Sci.* **2000**, *46*, 186–205. [CrossRef]
51. Schwartz, S.H. Normative Influences on Altruism. In *Advances in Experimental Social Psychology*; Berkowitz, L., Ed.; Advances in Experimental Social Psychology; Academic Press: New York, NY, USA, 1977; Volume 10, pp. 221–279.
52. Schwartz, S.H.; Howard, J.A. A normative decision-making model of altruism. In *Altruism and Helping Behavior: Social, Personality and Developmental Perspective*; Rushton, J.P., Sorrentino, R.M., Eds.; Lawrence Erlbaum: Hillsdale, NJ, USA, 1981; pp. 198–211.
53. Schwartz, S.H. Are There Universal Aspects in the Structure and Contents of Human Values? *J. Soc. Issues* **1994**, *50*, 19–45. [CrossRef]
54. Ahn, M.; Kang, J.; Hustvedt, G. A model of sustainable household technology acceptance. *Int. J. Consum. Stud.* **2016**, *40*, 83–91. [CrossRef]
55. Ehrenhard, M.; Kijl, B.; Nieuwenhuis, L. Market adoption barriers of multi-stakeholder technology: Smart homes for the aging population. *Technol. Forecast. Soc. Chang.* **2014**, *89*, 306–315. [CrossRef]
56. Guerreiro, S.; Batel, S.; Lima, M.L.; Moreira, S. Making energy visible: Sociopsychological aspects associated with the use of smart meters. *Energy Effic.* **2015**, *8*, 1149–1167. [CrossRef]
57. Hall, N.L.; Jeanneret, T.D.; Rai, A. Cost-reflective electricity pricing: Consumer preferences and perceptions. *Energy Policy* **2016**, *95*, 62–72. [CrossRef]
58. Paetz, A.-G.; Dütschke, E.; Fichtner, W. Smart Homes as a Means to Sustainable Energy Consumption: A Study of Consumer Perceptions. *J. Consum. Policy* **2012**, *35*, 23–41. [CrossRef]
59. Park, E.-S.; Hwang, B.; Ko, K.; Kim, D. Consumer Acceptance Analysis of the Home Energy Management System. *Sustainability* **2017**, *9*, 2351. [CrossRef]
60. Vagropoulos, S.I.; Balaskas, G.A.; Bakirtzis, A.G. An Investigation of Plug-In Electric Vehicle Charging Impact on Power Systems Scheduling and Energy Costs. *IEEE Trans. Power Syst.* **2017**, *32*, 1902–1912. [CrossRef]
61. Kobus, C.B.A.; Klaassen, E.A.M.; Mugge, R.; Schoormans, J.P.L. A real-life assessment on the effect of smart appliances for shifting households' electricity demand. *Appl. Energy* **2015**, *147*, 335–343. [CrossRef]
62. Moher, D.; Liberati, A.; Tetzlaff, J.; Altman, D.G.; Altman, D.; Antes, G.; Atkins, D.; Barbour, V.; Barrowman, N.; Berlin, J.A.; et al. Preferred reporting items for systematic reviews and meta-analyses: The PRISMA statement. *PLoS Med.* **2009**, *6*. [CrossRef] [PubMed]
63. Dedrick, J.; Venkatesh, M.; Stanton, J.M.; Zheng, Y.; Ramnarine-Rieks, A. Adoption of smart grid technologies by electric utilities: Factors influencing organizational innovation in a regulated environment. *Electron. Mark.* **2015**, *25*, 17–29. [CrossRef]
64. Ghazal, M.; Akmal, M.; Iyanna, S.; Ghoudi, K. Smart plugs: Perceived usefulness and satisfaction: Evidence from United Arab Emirates. *Renew. Sustain. Energy Rev.* **2015**, *55*, 1248–1259. [CrossRef]
65. Li, R.; Dane, G.; Finck, C.; Zeiler, W. Are building users prepared for energy flexible buildings?—A large-scale survey in the Netherlands. *Appl. Energy* **2017**, *203*, 623–634. [CrossRef]
66. Mesarić, P.; Đukec, D.; Krajcar, S. Exploring the Potential of Energy Consumers in Smart Grid Using Focus Group Methodology. *Sustainability* **2017**, *9*, 1463. [CrossRef]
67. Moser, C. The role of perceived control over appliances in the acceptance of electricity load-shifting programmes. *Energy Effic.* **2017**, 1–13. [CrossRef]
68. Park, C.K.; Kim, H.-J.; Kim, Y.-S. A study of factors enhancing smart grid consumer engagement. *Energy Policy* **2014**, *72*, 211–218. [CrossRef]
69. Schmalfuß, F.; Mair, C.; Döbelt, S.; Kämpfe, B.; Wüstemann, R.; Krems, J.F.; Keinath, A. User responses to a smart charging system in Germany: Battery electric vehicle driver motivation, attitudes and acceptance. *Energy Res. Soc. Sci.* **2015**, *9*, 60–71. [CrossRef]

70. Shrouf, F.; Miragliotta, G. Energy management based on Internet of Things: Practices and framework for adoption in production management. *J. Clean. Prod.* **2015**, *100*, 235–246. [CrossRef]
71. Spence, A.; Demski, C.; Butler, C.; Parkhill, K.; Pidgeon, N. Public perceptions of demand-side management and a smarter energy future. *Nat. Clim. Chang.* **2015**, *5*, 550–554. [CrossRef]
72. Will, C.; Schuller, A. Understanding user acceptance factors of electric vehicle smart charging. *Transp. Res. Part C Emerg. Technol.* **2016**, *71*, 198–214. [CrossRef]
73. Wilson, C.; Hargreaves, T.; Hauxwell-Baldwin, R. Benefits and risks of smart home technologies. *Energy Policy* **2017**, *103*, 72–83. [CrossRef]
74. Zhou, S.; Brown, M.A. Smart meter deployment in Europe: A comparative case study on the impacts of national policy schemes. *J. Clean. Prod.* **2017**, *144*, 22–32. [CrossRef]
75. Balta-Ozkan, N.; Davidson, R.; Bicket, M.; Whitmarsh, L. The development of smart homes market in the UK. *Energy* **2013**, *60*, 361–372. [CrossRef]
76. Balta-Ozkan, N.; Amerighi, O.; Boteler, B. A comparison of consumer perceptions towards smart homes in the UK, Germany and Italy: Reflections for policy and future research. *Technol. Anal. Strateg. Manag.* **2014**, *26*, 1176–1195. [CrossRef]
77. Balta-Ozkan, N.; Boteler, B.; Amerighi, O. European smart home market development: Public views on technical and economic aspects across the United Kingdom, Germany and Italy. *Energy Res. Soc. Sci.* **2014**, *3*, 65–77. [CrossRef]
78. Buryk, S.; Mead, D.; Mourato, S.; Torriti, J. Investigating preferences for dynamic electricity tariffs: The effect of environmental and system benefit disclosure. *Energy Policy* **2015**, *80*, 190–195. [CrossRef]
79. Cherry, C.; Hopfe, C.; MacGillivray, B.; Pidgeon, N. Homes as machines: Exploring expert and public imaginaries of low carbon housing futures in the United Kingdom. *Energy Res. Soc. Sci.* **2017**, *23*, 36–45. [CrossRef]
80. Berry, S.; Whaley, D.; Saman, W.; Davidson, K. Finding faults and influencing consumption: The role of in-home energy feedback displays in managing high-tech homes. *Energy Effic.* **2017**, *10*, 787–807. [CrossRef]
81. Balta-Ozkan, N.; Davidson, R.; Bicket, M.; Whitmarsh, L. Social barriers to the adoption of smart homes. *Energy Policy* **2013**, *63*, 363–374. [CrossRef]
82. Begier, B. Effective cooperation with energy consumers: An example of an ethical approach to introduce an innovative solution. *J. Inf. Commun. Ethics Soc.* **2014**, *12*, 107–121. [CrossRef]
83. Chen, C.; Xu, X.; Arpan, L. Between the technology acceptance model and sustainable energy technology acceptance model: Investigating smart meter acceptance in the United States. *Energy Res. Soc. Sci.* **2017**, *25*, 93–104. [CrossRef]
84. Chou, J.-S.; Yutami, G.A.N. Smart meter adoption and deployment strategy for residential buildings in Indonesia. *Appl. Energy* **2014**, *128*, 336–349. [CrossRef]
85. Chou, J.-S.; Kim, C.; Ung, T.-K.; Yutami, G.A.N.; Lin, G.-T.; Son, H. Cross-country review of smart grid adoption in residential buildings. *Renew. Sustain. Energy Rev.* **2015**, *48*, 192–213. [CrossRef]
86. Hess, D.J.; Coley, J.S. Wireless smart meters and public acceptance: The environment, limited choices, and precautionary politics. *Public Underst. Sci.* **2014**, *23*, 688–702. [CrossRef] [PubMed]
87. King, N.J.; Jessen, P.W. Smart metering systems and data sharing: Why getting a smart meter should also mean getting strong information privacy controls to manage data sharing. *Int. J. Law Inf. Technol.* **2014**, *22*, 215–253. [CrossRef]
88. Krishnamurti, T.; Schwartz, D.; Davis, A.; Fischhoff, B.; de Bruin, W.B.; Lave, L.; Wang, J. Preparing for smart grid technologies: A behavioral decision research approach to understanding consumer expectations about smart meters. *Energy Policy* **2012**, *41*, 790–797. [CrossRef]
89. Luthra, S.; Kumar, S.; Kharb, R.; Ansari, M.F.; Shimmi, S.L. Adoption of smart grid technologies: An analysis of interactions among barriers. *Renew. Sustain. Energy Rev.* **2014**, *33*, 554–565. [CrossRef]
90. Matschoss, K.; Kahma, N.; Heiskanen, E. Pioneering customers as change agents for new energy efficiency services???an empirical study in the Finnish electricity markets. *Energy Effic.* **2015**, *8*, 827–843. [CrossRef]
91. Michaels, L.; Parag, Y. Motivations and barriers to integrating 'prosuming' services into the future decentralized electricity grid: Findings from Israel. *Energy Res. Soc. Sci.* **2016**, *21*, 70–83. [CrossRef]
92. Yang, H.; Lee, H.; Zo, H. User acceptance of smart home services: An extension of the theory of planned behavior. *Ind. Manag. Data Syst.* **2017**, *117*, 68–89. [CrossRef]

93. Ornetzeder, M.; Bechtold, U.; Nentwich, M. Participatory assessment of sustainable end-user technology in Austria. *WIT Trans. Ecol. Environ.* **2009**, *121*, 269–278. [CrossRef]
94. Goulden, M.; Bedwell, B.; Rennick-Egglestone, S.; Rodden, T.; Spence, A. Smart grids, smart users? the role of the user in demand side management. *Energy Res. Soc. Sci.* **2014**, *2*, 21–29. [CrossRef]
95. Hammer, S.; Wißner, M.; André, E. Trust-based decision-making for smart and adaptive environments. *User Model. User-Adapt. Interact.* **2015**, *25*, 267–293. [CrossRef]
96. Kahma, N.; Matschoss, K. The rejection of innovations? Rethinking technology diffusion and the non-use of smart energy services in Finland. *Energy Res. Soc. Sci.* **2017**, *34*, 27–36. [CrossRef]
97. Sandström, G.; Keijer, U. Smart home systems—Accessibility and trust. *Open House Int.* **2010**, *35*, 6–14.
98. Gerpott, T.J.; Paukert, M. Determinants of willingness to pay for smart meters: An empirical analysis of household customers in Germany. *Energy Policy* **2013**, *61*, 483–495. [CrossRef]
99. Barnicoat, G.; Danson, M. The ageing population and smart metering: A field study of householders' attitudes and behaviours towards energy use in Scotland. *Energy Res. Soc. Sci.* **2015**, *9*, 107–115. [CrossRef]
100. Fell, M.J.; Shipworth, D.; Huebner, G.M.; Elwell, C.A. Public acceptability of domestic demand-side response in Great Britain: The role of automation and direct load control. *Energy Res. Soc. Sci.* **2015**, *9*, 72–84. [CrossRef]
101. Schweitzer, F.; Van den Ende, E. To Be or Not to Be in Thrall to the March of Smart Products. *Psychol. Mark.* **2016**, *33*, 830–842. [CrossRef] [PubMed]
102. Aduda, K.O.; Labeodan, T.; Zeiler, W.; Boxem, G.; Zhao, Y. Demand side flexibility: Potentials and building performance implications. *Sustain. Cities Soc.* **2016**, *22*, 146–163. [CrossRef]
103. Jacobs, M. Sustainable Development as a Contested Concept. In *Fairness and Futurity: Essays on Environmental Sustainability and Social Justice*; Dobson, A., Ed.; Oxford University Press: Oxford, UK, 1999; pp. 21–45.
104. Dignum, M.; Correljé, A.; Cuppen, E.; Pesch, U.; Taebi, B. Contested Technologies and Design for Values: The Case of Shale Gas. *Sci. Eng. Ethics* **2016**, *22*, 1171–1191. [CrossRef] [PubMed]

© 2018 by the authors. Licensee MDPI, Basel, Switzerland. This article is an open access article distributed under the terms and conditions of the Creative Commons Attribution (CC BY) license (http://creativecommons.org/licenses/by/4.0/).

Article

Research on the Evaluation Model of a Smart Grid Development Level Based on Differentiation of Development Demand

Jinchao Li [1,2,*], **Tianzhi Li** [1] **and Liu Han** [3]

1 School of Economics and Management, North China Electric Power University, Beijing 102206, China; 1172206115@ncepu.edu.cn
2 Beijing Key Laboratory of New Energy and Low-Carbon Development, North China Electric Power University, Beijing 102206, China
3 State Grid Economic and Technological Research Institute CO., LTD., Beijing 102209, China; hanliu@chinasperi.sgcc.com.cn
* Correspondence: lijc@ncepu.edu.cn; Tel.: +86-159-0116-1636

Received: 22 September 2018; Accepted: 1 November 2018; Published: 5 November 2018

Abstract: In order to eliminate the impact of inter-regional differentiation of development demand on the objective evaluation of the development level of smart grid, this paper establishes the evaluation model of weight modification, transmission mechanism and combination of subjective and objective weights. Firstly, the Analytic Hierarchy Process method is used to calculate the weights of evaluation indices of effect layer and then the indices of development demand are used to modify the weights of them. The association analysis and the correlation coefficient are used to establish the weights conduction coefficient between the effect level and the base level. Then the subjective weights of the indices of the base layer are calculated. The objective weights of the indices of the base layer are obtained by using the entropy method. The subjective weights of the base layer and the objective weights obtained by the entropy method are averagely calculated, and the comprehensive weights of the evaluation indices of the base layer are obtained. Then each index is scored according to the weights and index values. Finally, the model is used to quantitatively inspect the level of development of smart grid in specific regions and make a horizontal comparison, which provides a useful reference for the development of smart grids. The relevant examples verify the correctness and validity of the model.

Keywords: smart grid; differentiation; development demand; comprehensive evaluation

1. Introduction

Based on an integrated and high-speed bi-directional communication network, smart grid is designed to be reliable, safe, economical, efficient, and environment-friendly through advanced sensing and measurement technologies, equipment technologies, control methods, and decision support system technologies. Key features of it include self-healing, motivating and engaging users, defending against attacks, providing power quality that meets 21st century user needs, allowing access to a variety of power generation forms, activating power markets, and optimizing asset applications for efficient operation. As for its application range, it is more and more extensive. For example, in recent years, some areas have combined smart grids with intelligent transportation to build new smart cities [1]. As an important part of the energy internet, it has drawn wide attention from all of the world and has now become a new trend in the development of the world's power grid [2–4].

Investment is the economic foundation for the development of smart grids, but due to the different driving forces of smart grid development in different countries, the focus of investment in the construction of smart grids is also different. In 2010, the top ten countries that the central government invested in the smart grid are shown in the Figure 1. Their total investment has reached $18.4 billion and will continue to grow in the future. For Europe, its development focus is on the optimal operation of the power grid, the optimization of power grid infrastructure, and the development of communications and information technology. For the United States, its development and construction focus is on low-carbon and energy efficiency. For Japan, its construction focus is on the green economy. In China, its construction focus at present is to improve the resource allocation capability, safety level, and operating efficiency of the power grid. The development of smart grids in China is divided into three stages: pilot stage for planning, stage of comprehensive construction, and stage of guiding and improving. The situation of smart grid investment in each stage is shown in Figure 2 below.

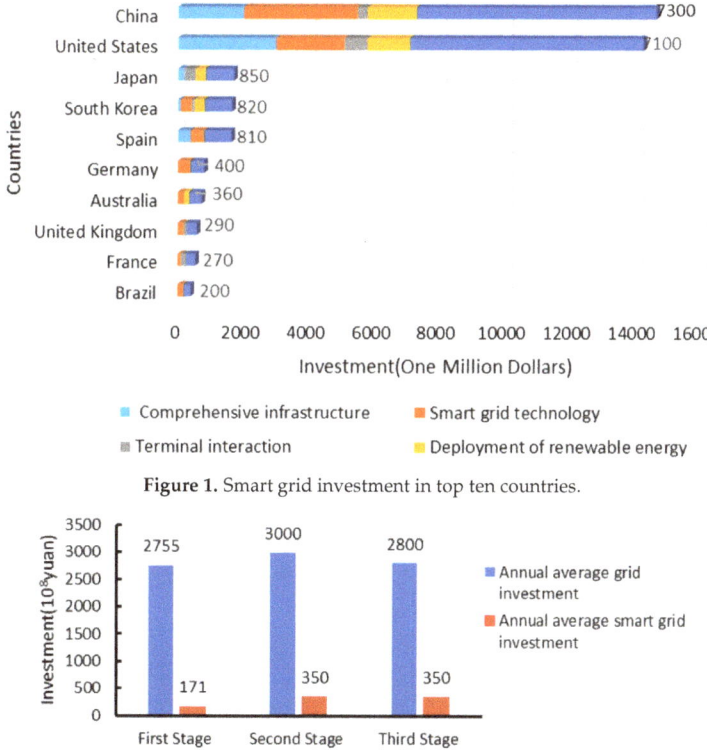

Figure 1. Smart grid investment in top ten countries.

Figure 2. Investment status of China at each stage.

After high-speed construction in recent years, the development of smart grid at abroad has entered a stage of normalization. As developed countries such as Europe and the United States have a high level in the development, construction, and operation management of power grids, a great deal of research work has been carried out on the assessment of smart grids. The experience has been accumulated and relatively rich achievements have been achieved [5–7]. For China, the development of the smart grid has also entered a critical stage. Under the layout of the State Grid Corporation on smart grids, provincial power grid companies have responded to the call to speed up the pace of development and construction. Therefore, it is urgent to establish a sound evaluation system and mechanism to evaluate the level of smart grid development to guide its direction of development. Based on this background, this paper establishes an index system that combines the effect layer and the base layer, evaluates the development level of the smart grid in a specific region, looks for its weak links, and gives corresponding optimization suggestions.

The paper is organized as follows. The second part serves as a literature review. The third part introduces the establishment of the index system. The fourth part and the fifth part respectively introduce the effect layer index system and the base layer index system. In the sixth part, the paper gives the evaluation process of the development level of the smart grid. The seventh part analyzes the examples. Finally, the eighth part offers conclusion.

2. Literature Review

At present, evaluations of smart grids have been conducted by scholars at home and abroad. The specific literatures are shown in the Table 1 below.

Table 1. Research on smart grid evaluation.

Author	Evaluation Content	Indices/Dimensions	Method/Model
Yu et al. (2018) [8]	Power quality (PQ) coupling of smart grid	Pattern construction, pattern representation, and time series pattern matching	Time series pattern
Park et al. (2018) [9]	Intelligent demand management of the micro grid	High-Power LED, System, Demand Resource Management, Micro-Distributed ESS	A micro-distributed ESS-based smart LED streetlight system
Jesus et al. (2018) [10]	Investments of smart grid	Definitions and Assumptions, parameter specification, economics of the smart grid, statement of the optimization problem and solution approach.	Multi-level optimization model
Peng et al. (2018) [11]	Reliability and cascading risk of a smart grid system	Theoretical analysis, Numerical simulations	Model based on complex network theory
Leszczyna (2018) [12]	Cyber security of smart grid	Reviews, Vulnerability identification, Vulnerability analysis	Systematic analysis
Cacciatore et al. (2017) [13]	Cost Analysis of Smart Lighting for Smart Cities	Delay-based (DEL), Encounter-based (ENC), Dimming (DIM)	Heuristics for smart lighting based on the peculiar characteristics of the employed technology
Hashemi-Dezaki et al. (2017) [14]	Reliability of smart grids	The uncertainties of power systems, the stochastic output generation of renewable resources, the behaviors of PHEV owners, availability of physical elements, cyber elements	A new reliability evaluation method simultaneously considering the DCPIs, DGs, and PHEVs
Munshi et al. (2017) [15]	Smart grids	Data acquisition, data storing and processing, data querying, data analytics components	A comprehensive big data framework
Woo et al. (2017) [16]	Cyber Security of smart grid	Information systems, Power Systems	Optimal power flow (OPF), power flow tracing, Analytic hierarchy process
Lloret-Gallego et al. (2017) [17]	Resilience of ICT platforms in Smart distribution grids.	Reliability, Adaptation Capacity, Elasticity, Plasticity, Evolvability	EMPOWER Resilience Evaluation Framework
Vazquez et al. (2017) [18]	Smart Grid Demonstration Project	Mean Absolute Error (MAE), Mean, Absolute Percentage Error, (MAPE)	Adaptive load forecasting methodology
Rossebø et al. (2017) [19]	Risk assessment of Smart	Impact assessment, Threat and vulnerability, Assessment, Risk estimation and prioritization, Risk treatment, Risk acceptance	SEGRID Risk Management Methodology (SRMM)

Table 1. Cont.

Author	Evaluation Content	Indices/Dimensions	Method/Model
Coppo et al. (2015) [20]	The Italian smart grid pilot projects	System average interruption frequency index, system average interruption duration index, customer average interruption duration index, customer average interruption frequency index, customers experiencing multiple interruptions	Numerical simulations
Xenias et al. (2015) [21]	UK smart grid	Standards, Technical issues, Data handling, Market structure, Regulation, Co-ordination, Customer engagement, Investment	Policy Delphi
Liu et al. (2015) [22]	Risk of transmission lines in smart grid	Primary Filtering Technique, Secondary Filtering Technique	Bi-level model
Personal et al. (2014) [23]	The degree of goal achievement of Smart Grid	Improve of Energy Efficiency, Increase of Renewable Energy Use, Reduction of Emissions, Secondary Objectives	Hierarchical metric/a set of KPIs
Dong et al. (2014) [24]	Technological Progress of Smart Grid	Investment, labor inputs, technology	Production function theory, DEA, RRA
Hu et al. (2014) [25]	Technology maturity of Smart Grid	Time, production processes and technical features	A model include Time Production Processes, Time Technical Features and Processes Technical Features
Song et al. (2014) [26]	Smart Distribution Grid	Strong degree of the network, facilities intelligence, supply reliability, power quality, operational efficiency, grid interactivity, development coordination	Hierarchical optimization model and DEMATEL-ANP-counter entropy method
Song et al. (2014) [27]	Reliability of Smart Grids	Information subsystem failure, Communication subsystem failure, Intelligent substation failure, Protection subsystem failure, Power supply failure, Failures of other devices depending on the architecture	Layered Fault Tree Model
Bracco et al. (2014) [28]	SG (Smart Grid)/Smart Microgrid	Technical, economic and environmental performance indicators	A mathematical model that the minimize the SPM daily operational costs
Wang et al. (2013) [29]	Operation performance of smart grid	Economic operation, supply quality and services (distribution line length, substation capacity, net assets, loss rate, electricity quantity, supply area).	Optimal fuzzy, algorithm and data envelopment analysis
Niu et al. (2013) [30]	Regional Grid	Safety, Economy, Quality, Efficiency	Hierarchical optimized combination evaluation
Li et al. (2012) [31]	Smart Distribution Grid	The model of two-level index synthesized cloud and remarks cloud	Cloud model
Bilgin et al. (2012) [32]	Performance of ZigBee in smart grid environments	Network throughput, End-to-end delay, Delivery ratio, Energy consumption	Wireless sensor network-based smart grid applications
Xie et al. (2012) [33]	Safety of Smart Grid	Structural safety of transmission network, structural safety of distribution network, high-efficient system and equipment support, operational safety and stability, adequacy and resilience	AHP-Entropy combined, Method
Sun et al. (2011) [34]	Smart grid	IBM smart grid maturity model, The DOE smart grid development evaluation system, the EPRI smart grid construction assessment indicators, The EU smart grid benefits assessment system	Comparative analysis

3. Construction of the Index System

The comprehensive evaluation model of the level of development of the smart grid aims to achieve systematic evaluation of the overall level of the smart grid. Therefore, this paper establishes an index system from the effect layer and the base layer and builds the relationship between the two layers. The effect layer reflects the inherent nature of the development of smart grid and is dedicated to meeting the development needs of smart grid, while the base layer is the focus of smart grid construction. The index system structure of this paper shown in Figure 3.

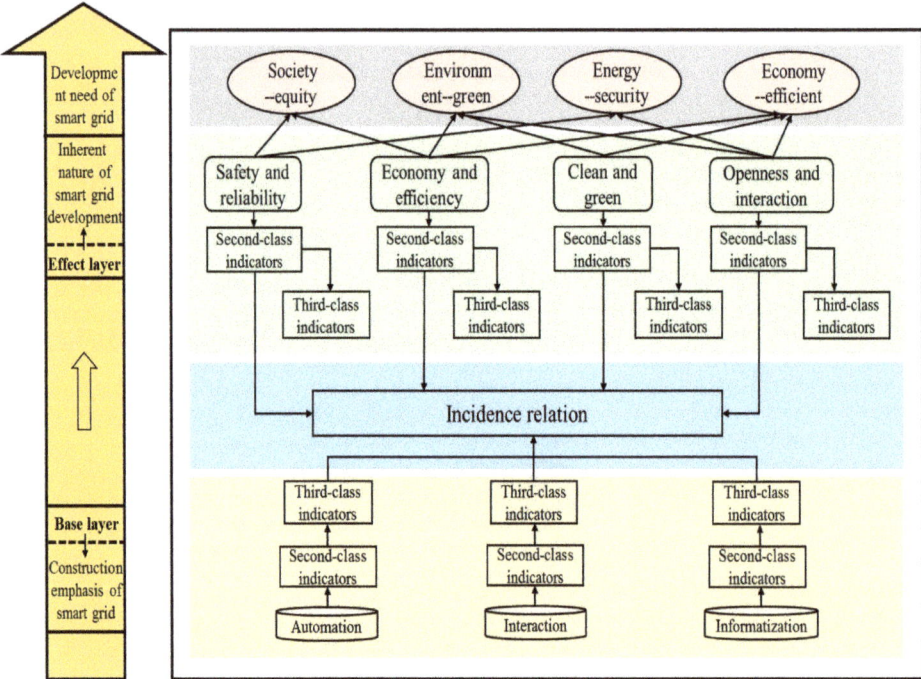

Figure 3. The structure of index system.

4. Index System of Effect Layer

4.1. Safety and Reliability

The safe and reliable operation are the key tasks for the future development of power grid. It involves the power supply security and reliability of power quality [35] and the ability to enhance the safety, stability and accident prevention capability of large power grids. The construction of communication information network is an important part of the intelligent construction of China's power grid, so the safety of communication information is equally worth noting. Therefore, the index system of the safety and reliability of the power grid is mainly established from two aspects: the safety and reliability of power grid and the safety of communication information, as shown in Table 2 [36].

Table 2. Index system of safety and reliability.

Second-Level Indicators	Third-Level Indicators	Code
The safety and reliability of power grid	The number of power transmission accident	E1
	The number of power transformation accident	E2
	The self-healing speed of the distribution network [37]	E3
	The self-healing rate of the distribution network	E4
	The reliability of power supply(urban user) [38]	E5
	The reliability of power supply (rural user)	E6
The safety of communication information	The index of the safe operation of information and communication system	E7
	The number of information events	E8

4.2. Economy and Efficiency

Economy and efficiency is to improve the grid operation and transmission efficiency, reduce operating costs and promote the efficient use of energy resources and power assets, so the index system of economics and efficiency of the power grid is mainly established from three aspects: economic benefits, grid efficiency and staff efficiency, as shown in Table 3.

Table 3. Index system of economics and efficiency.

Second-Level Indicators	Third-Level Indicators	Code
Economic benefits	The revenue of value-added services [39]	E9
	The recovery of electricity	E10
	The fair coefficient of electricity consumption	E11
Grid efficiency [40]	The annual maximum load utilization	E12
	The maximum load rate of power lines	E13
	The annual average equivalent load rate of line operation	E14
	The annual maximum load rate of main transformer	E15
	The annual average equivalent load rate of main transformer operation	E16
Staff efficiency	The efficiency of transmission staff	E17
	The efficiency of transformation staff	E18
	The efficiency of urban distribution network staff	E19
	Overall labor productivity	E20

4.3. Clean and Green

Clean and green means to improve the energy structure, improve the level of electrification, realize the large-scale development of clean energy and optimize the configuration of it in a wide range, replace the fossil energy with clean energy, and make the clean energy gradually become the dominant energy in the future. Therefore, the index system of the clean and green of the power grid is mainly established from three aspects: green power generation, green power grid, and green electricity, as shown in Table 4.

Table 4. Index system of cleanliness and green.

Second-Level Indicators	Third-Level Indicators	Code
Green power generation	The proportion of renewable energy power generation	E21
	The realization ratio of annual utilization hours of renewable energy	E22
	Abandoned wind ratio	E23
	Distributed power energy permeability	E24
Green power grid	The land disturbance area of unit quantity of electricity	E25
	The floor area saved by smart substation	E26
	Comprehensive line loss rate	E27
Green electricity	The electricity saved by demand-side management	E28
	The proportion of electricity in the terminal energy consumption	E29
	Power replacement ratio	E30

4.4. Openness and Interaction

Openness and interaction means that based on the platform of intellectualized service which built by smart grid to adapt to the connection and interaction of various types of power supply and load flexibly to meet the diverse needs of customers. Therefore, the index system of the openness and interaction of the power grids mainly established from four aspects: the transparency of power grid, the openness of power grid, quality service, and interactive effect, as shown in Table 5.

Table 5. Index system of openness and interaction.

Second-Level Indicators	Third-Level Indicators	Code
The transparency of power grid	The depth of information disclosure	E31
	The speed of information update	E32
	The convenience of getting information	E33
The openness of power grid	the growth rate of electric quantity in electric power market transaction	E34
	The investment in the open area of the grid business	E35
	The scale and proportion of the direct power-purchase for the large user	E36
	The completeness that all kinds of users access the standard system	E37
Quality service	The evaluation index of quality service	E38
Interactive effect	The year-on-year growth rate of the grid's annual maximum load utilization	E39
	The proportion of electricity of implementing peak and valley time price [41]	E40
	The power saved by demand-side management	E41
	The capability of load monitoring and control	E42
	The utilization rate of electric vehicles	E43

5. Index System of Base Layer

Based on the basic requirements of the construction of smart grid, this paper divides the basis of intelligent grid construction into three aspects: automation, interaction and information, and takes them as first-level index to establish the evaluation index system of the base layer.

5.1. Automation

Power network automation mainly refers to the automated operation of the power system. By running modern communication technology, network technology and automatic control technology, it reaches the automatic detection and control of grid operation, enhances the ability of online monitoring and self-protection operation, and effectively improves the efficiency of grid operation, to ensure reliable and efficient operation of the power grid. Therefore, the index system of power network automation is mainly established from four aspects: transmission automation, substation automation, distribution automation, and dispatching automation, as shown in Table 6.

Table 6. Index system of automation.

Second-Level Indicators	Third-Level Indicators	Code
Transmission automation	The total capacity of flexible AC transmission device	B1
	The proportion of energy-saving wire	B2
	The application of disaster prevention and reduction technology	B3
	The proportion of the lines applying condition monitoring technology	B4
	The proportion of the lines applying intelligent inspection technology	B5
Substation automation	The proportion of smart substation	B6
	The coverage of the patrol robot of substation	B7
	The coverage of condition monitoring of transformer equipment	B8
Distribution automation	The coverage of distribution automation	B9
	The coverage of feeder automation	B10
	Coverage of the command platform of power distribution repairs in a rush	B11
	Coverage of distribution power automation terminal	B12
Dispatching automation [42]	The coverage of provincial/prefecture (county) level smart grid dispatching control system	B13
	The coverage of provincial/prefecture (county) level standby scheduling	B14
	The coverage of dual access of dispatch data net	B15
	The access rate of station terminal dispatch data network	B16
	the coverage of secondary security system	B17

5.2. Interaction

Interactive technology of the smart grid is a key technology and development direction which can improve the capacity of the grid to carry new energy and ensure the power quality of the grid. It can achieve the multi-directional interaction among the power supply, power grid and users, and allows users to participate more in the process of power balance by changing users' electricity behavior and developing the access of distributed energy. Therefore, the index system of interaction is mainly established from four aspects: interaction of electricity use, electric vehicles, large-scale access to new energy sources, and distributed power supply, as shown in Table 7.

Table 7. Index system of interaction.

Second-Level Indicators	Third-Level Indicators	Code
Interaction of electricity use	The coverage of electricity information collection system	B18
	The coverage of intelligent ammeter	B19
	The coverage of power service management platform	B20
	The method of demand-side response to electricity prices	B21
	The area density of the interactive business hall	B22
Electric vehicles	The area density of city charge (change) power station	B23
	The linear density of highway filling (change) power station	B24
	The matching degree of electric vehicle and charger	B25
Large-scale access to new energy sources	The coverage of new energy power forecasting system	B26
	The completion rate of wind and PV power grid detection	B27
	The proportion of new energy installed capacity	B28
Distributed power supply	The proportion of distributed power installed capacity	B29
	The realization rate of distributed generation forecast	B30

5.3. Informatization

Grid informatization refers to the process of cultivating and developing new productivity represented by intelligent tools such as computers and network communication technologies in the power grid and improving the operation and management of the power grid. It is reflected in the construction of communication network and information construction index system as shown in Table 8.

Table 8. Index system of informatization.

Second-Level Indicators	Third-Level Indicators	Code
Construction of communication network	The optical fiber coverage of substations(35 kV and above)	B31
	The cable coverage of backbone communication network	B32
	The bandwidth capacity of communication transmission network platform	B33
	The fiber coverage of 10 kV communication access network	B34
	The rate of PFTTH	B35
Information construction	The coverage of SG-ERP system	B36
	The automatic monitoring rate of information communication equipment	B37
	The availability rate of information network	B38
	The availability rate of business systems	B39

6. Evaluation Process of Smart Grid Development Level

6.1. Implementation Path of Evaluation Model

The comprehensive evaluation model of the development level of smart grid is based on the theory of system evaluation and can accurately evaluate the overall development level of smart grid. By decomposing and refining the smart grid, it deepens its understanding of the smart grid, enhances the specificity and representativeness of the evaluation index, and improves the accuracy of the evaluation results. Through the research on the coordinative relationship among the indicators, a dynamic weight calculation method is designed to realize the two-way interaction between the effect layer and base layer.

When choosing the evaluation method of smart grid, this paper select the appropriate evaluation method based on the characteristics of each attribute and index, and combine with the application scope of the method, so as to obtain a more accurate and reasonable evaluation result. The evaluation model process of this paper as shown in Figure 4.

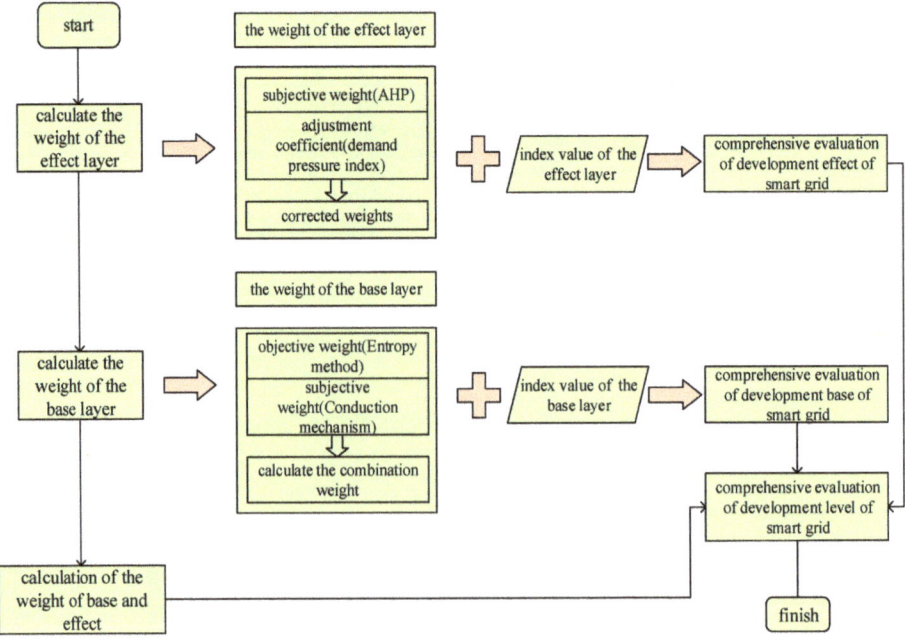

Figure 4. Technical path of comprehensive evaluation model.

6.2. Method of Evaluation Model

6.2.1. Subjective Weights of the Effect Level Indicators by the AHP Method

The Analytic Hierarchy Process (AHP) is a practical multi-objective decision-making method. When AHP is used to analyze the decision-making problem, first of all, we need to rationalize and stratify the issue so as to construct a hierarchical structural model. The basic steps are as follows:

(1) Establish a hierarchical structure.
(2) Construct a judgment matrix.

Hierarchies reflect the relationship between the factors, but the criteria of the criterion layer do not necessarily share the same weight in the target measure. This article uses the numbers 1–9 and their reciprocal as a scale. Table 9 lists the meaning of 1–9 scale:

Table 9. The judgment basis of scale value and related description.

Scale Value	Description
1	Indicates that elements i and j are of equal importance
3	Representing the elements i and j, the former is slightly more important than the latter
5	Representing the elements i and j, the former is significantly more important than the latter
7	Representing the elements i and j, the former is awfully more important than the latter
9	Representing the elements i and j, the former is perfectly more important than the latter
2, 4, 6, 8	The importance is between the above two
Reciprocal	Representing the importance of elements i and j in contrast to the above

(3) Hierarchical single arrangement and consistency checking

Hierarchical single arrangement is based on the judgment matrix, calculating the target element in the previous level, and determining the importance (weight) of level and its associated elements. The method of solving the largest eigenvector of the judgment matrix is used to obtain the weight of single arrangement. The formula is:

$$CW = \lambda_{max} W \qquad (1)$$

where λ_{max} and W denote the maximum eigenvalue of the judgment matrix C and the corresponding eigenvector.

In order to avoid the contradictory judgment result in the process of expert judgment, it is necessary to check consistency of hierarchical single arrangement. Check the consistency of the judgment matrix by calculating the CR value:

$$CR = CI/RI \qquad (2)$$

$CI = (\lambda_{max} - n)/(n - 1)$ is the dimension of the judgment matrix, RI is the corresponding random value, as shown in Table 10.

Table 10. RI value that correspond to n.

N	1	2	3	4	5	6	7	8	9
RI	0	0	0.58	0.9	1.12	1.24	1.32	1.41	1.45

If CR < 0.1, then the judgment matrix can be regarded as satisfactory consistency. The judgment matrix can be used as a hierarchical analysis. If CR ≥ 0.1, the judgment matrix is not satisfactory, and the judgment matrix needs to be adjusted and corrected.

(4) Hierarchical total ordering and consistency checking

Through the above steps, a set of weight vectors can be obtained. Ultimately, we should obtain the weight of sorting the goals in each element, especially in the lowest level, so as to make a choice of solutions. The total sequencing weight will synthesize the weights of the single criteria from top to bottom.

Suppose that the upper level (level A) contains m factors $A_1, \ldots, A_m$, and the total weight of their levels is $a_1, \ldots, a_m$. The next level (B level) contains n factors $B_1, \ldots, B_n$, whose rank ordering weights for A_j are respectively $b_{1j}, \ldots, b_{nj}$ ($b_{ij} = 0$ when B_i is unassociated with A_j). We now ask for the weight of each factor in the B-layer about the total goal, that is, find the total weight $b_1, \ldots, b_n$ of the hierarchy of each factor in the B-tier. The calculation is performed in the following way:

$$b_i = \sum_{j=1}^{m} b_{ij} a_j, i = 1, \ldots, n \qquad (3)$$

The hierarchical total ordering also needs to be checked for consistency. The test is still performed from the high level to the low level layer by layer like the total level of the hierarchy. The pairwise comparison judgment matrix of factors related to A_j in layer B is checked for consistency in a single ranking, and the single-order consistency index is obtained as $CI(j)$, $(j = 1, \ldots, m)$, and the corresponding average When the random consistency index is $RI(j)$ ($CI(j)$ and $RI(j)$ have been obtained when the levels are single-ordered), the proportion of random coherence of the total order of the B-level is:

$$CR = \frac{\sum_{j=1}^{m} CI(j) a_j}{\sum_{j=1}^{m} RI(j) a_j} \qquad (4)$$

when $CR < 0.10$, it is considered that the hierarchical total ordering results have a satisfactory consistency and accept the analysis result.

6.2.2. Correcting the Weight of the Effect Layer in the Direction of Development Demand

Based on the basic cluster analysis of the development of the provincial power grids in the country, the provincial power grids can be divided into three categories. Provincial power grids A, B, and C are selected from each of them, their development demand index values separately calculated, the development demand index values of the provinces where they are located as the target value averaged, and the demand pressure index calculated separately. The first-level indicators at the effect level are revised to meet the demand-oriented goal. The specific process is as follows in Figure 5.

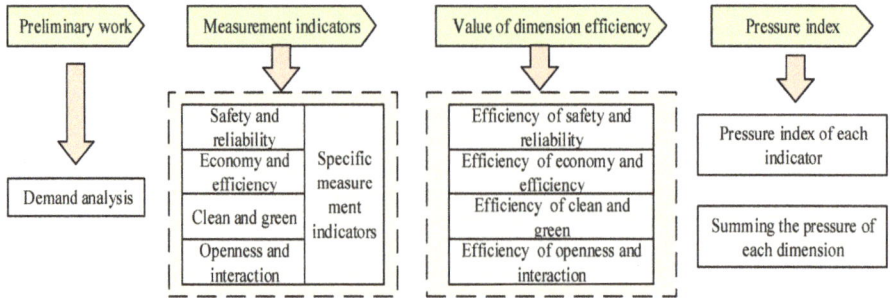

Figure 5. Process of weight modification.

(1) The measurement index of smart grid development demand

Based on the availability of the current indicator data, the measurement indicators of smart grid development demand are shown in Table 11.

Table 11. The measurement indicators of smart grid development demand.

Dimensions of Demand	Quantitative Measurement Indicators	Code of Demand Pressure Indicators
Safety and reliability (D_1)	The proportion of a type of load	DC_1
	The proportion of secondary industry production GDP	DC_2
	Load density	DC_3
	Capacity-load ratio	DC_4
	Urbanization rate	DC_5
Economy and efficiency (D_2)	Return on assets	DC_6
	Overall labor productivity	DC_7
	Electricity sale of unit assets	DC_8
	Energy intensity	DC_9
	Ratio of power generation and electricity	DC_{10}
Clean and green (D_3)	The proportion of clean energy production	DC_{11}
	Air-quality index	DC_{12}
	Carbon dioxide emissions per unit area	DC_{13}
	Carbon intensity	DC_{14}
	The proportion of electrical energy in terminal energy consumption	DC_{15}
Openness and interaction (D_4)	Reasonable degree of utilization hours of power generation equipment	DC_{16}
	Ratio of urban-rural power supply reliability	DC_{17}
	Quality service evaluation index	DC_{18}
	Per capita electricity consumption	DC_{19}
	The proportion of tertiary industry production GDP	DC_{20}

The demand pressure index formula is shown in Equation (5).

$$demand\ pressure\ index = \frac{Max(actual\ value, target\ value)}{Min(actual\ value, target\ value)} \quad (5)$$

Demand pressure index of safety and reliability (D_1)
= $DC_1 + DC_2 + DC_3 + DC_4 + DC_5$
Demand pressure index of economy and efficiency (D_2)
= $DC_6 + DC_7 + DC_8 + DC_9 + DC_{10}$
Demand pressure index of clean and green (D_3)
= $DC_{11} + DC_{12} + DC_{13} + DC_{14} + DC_{15}$
Demand pressure index of openness and interaction (D_4)
= $DC_{16} + DC_{17} + DC_{18} + DC_{19} + DC_{20}$

(2) After the normalization process as the requirement four-dimensional weight value

The above calculation results and the AHP method are used to correct the weight of the first-level indicators of the effect layer. The two mean values are taken as the final weight of the indicator, and the second-level and third-level indicators' weights of the effect layer are corrected in order.

6.2.3. Determination of Objective Weights of Base Layer Indicators

In this paper, the Entropy Method [43] is used to calculate the objective weights of the base layer indicators. It is a method to determine the weights based on the amount of information provided by the observations of each index. It is an objective method of empowerment that embodies the size of the evaluation of indicators in objective information. The basic implementation steps are as follows:

(1) Evaluation index membership degree matrix standardization

The n object to be evaluated corresponds to the index values of the m evaluation indices and constitutes a membership evaluation standard R.

$$R = \begin{pmatrix} r_{11} & r_{12} & \cdots & r_{1m} \\ r_{21} & r_{22} & \cdots & r_{2m} \\ \cdots & & & \cdots \\ r_{n1} & r_{n2} & \cdots & r_{nm} \end{pmatrix}$$

In this evaluation index system, there are differences in the dimension, content, merits and demerits of each indicator, etc. Therefore, it is necessary to standardize the value of each indicator. There are two kinds of standardized processing methods: The larger the indicator data is, the better, that is, the positive indicator. The standard formula is:

$$r_{ij} = \frac{x_{ij} - \min x_{ij}}{\max x_{ij} - \min x_{ij}} \qquad (6)$$

when the indicator data is smaller, the better, that is, the inverse indicator, the standard formula is:

$$r_{ij} = \frac{\max x_{ij} - x_{ij}}{\max x_{ij} - \min x_{ij}} \qquad (7)$$

(2) Normalize each indicator value and calculate the proportion of the indicator value of the ith evaluation object under the jth indicator:

$$P_{ij} = \frac{r_{ij}}{\sum_{j=1}^{n} r_{ij}} \qquad (8)$$

(3) Calculate the entropy of the jth indicator:

$$H_j = -K \sum_{i=1}^{n} P_{ij} \ln P_{ij} (j = 1, 2, \ldots, m) \qquad (9)$$

Among them:

$$K = 1/\ln n (K > 0, 0 \leq P_{ij} \leq 1) \qquad (10)$$

and assume that:

$$P_{ij} = 0, P_{ij} \ln P_{ij} = 0 \qquad (11)$$

(4) Calculate the difference coefficient of the jth indicator:

$$\alpha_j = 1 - H_j \qquad (12)$$

(5) Calculate the weight of the jth indicator:

$$w_j = \frac{\alpha_j}{\sum_{j=1}^{m} \alpha_j} \qquad (13)$$

6.2.4. Relationship among the Effect Layer and the Base Layer Indicators

Until now, the smart grid construction period is not long, and there are few index data, the correlation analysis based on the index data may have errors. Therefore, this paper first uses the expert scoring method to judge the correlation degree between two-level indicators of effect layer and the key indicators of the basic layer. The specific process is as follows in Figure 6.

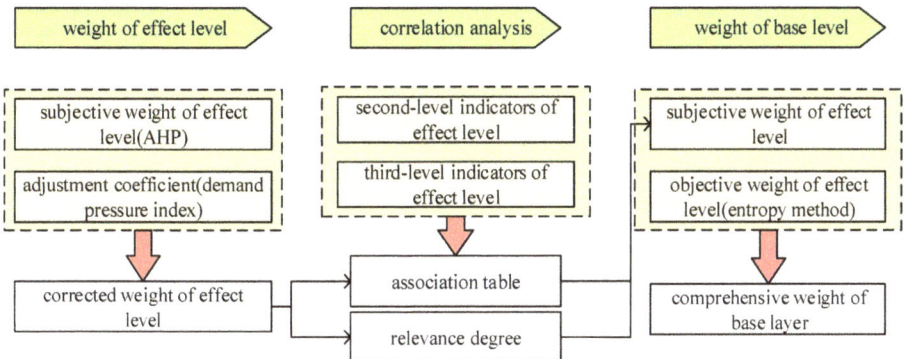

Figure 6. Process of correlation analysis.

The association table between effect layer and base layer indicators is showed in Table 12.

Table 12. Relationship between effect layer and base layer indicators.

Second-Level Indicators of Effect Level	Third-Level Indicators of Base Level Which Associated with It
The safety and reliability of power grid	B1, B3~B14, B26, B27, B30
The safety of communication information	B15~B17, B31~B34, B37~B39
Economic benefits	B18~B21, B23~B25, B35
Grid efficiency	B1, B13, B21
Staff efficiency	B5~B9, B11~B13, B36
Green power generation	B13, B26~B28
Green power grid	B2, B6
Green electricity	B20, B21, B23~B25
Transparent grid	B21, B22
Open grid	B23~B25, B29
Quality service	B11, B20, B22
Interactive effect	B18, B21~B25

There are two factors that affect the subjective weights of the basic layer indicators: one is the weight of the related effect level indicators, and the other is the size of the correlation between them. Therefore, this article uses the multiplication of these two as the subjective weights of the base layer indicators.

The subjective weights and objective weights of the basic layer are arithmetically averaged to obtain the comprehensive weight of the basic layer evaluation indicators.

Through the correlation analysis of qualitative and quantitative analysis between the second-level indicators of the effect layer and the third-level indicators of the base layer, the subjective weights of the effect layer are transmitted to the third-level indicators of the basic level, and the guiding effect of the effect on the foundation is achieved.

7. Case Study

Using the above-mentioned index system and evaluation method, three provincial power grids are selected, and scores are assigned to each aspect of smart grid development in combination with the weights and index values, thereby assessing the development level of smart grids. The results are as follows.

7.1. Province A

The relevant data (for example, the reliability of power supply, overall labor productivity, the proportion of renewable energy power generation, the evaluation index of quality service) that can reflect the first-level indicators of the effect layer in A province is used as a reference. Ten experts are hired to score the importance level of the first-level indicators in the effect layer, and the weights of the first-level indicators are calculated by the judgment matrix given by the experts. Finally, the weight result obtained by the AHP method is the average value of the calculation results of the ten expert judgment matrix, and then the weight is corrected by the indices of development demand to obtain the final weight of the first-level index, and so on, and the weights of the indicators at all levels are calculated.

The weights calculated using the judgment matrix given by one of the experts is showed in Table 13, and has passed the consistency test.

Table 13. The judgment matrix.

	Safety and Reliability	Economy and Efficiency	Clean and Green	Openness and Interaction	Weight
Safety and reliability	1	2	3	4	0.4285
Economy and efficiency	1/2	1	5	6	0.3810
Clean and green	1/3	1/5	1	2	0.1170
Openness and interaction	1/4	1/6	1/2	1	0.0735

Therefore, the average value calculated by the judgment experts is given by the ten experts and then corrected to the final weight of the effect layer index. The weight of the first-level index of the base layer is calculated by the entropy weight method. The final result is shown in Table 14.

Table 14. The weight of the first-level indices.

Index	Weight
Safety and reliability	0.3707
Economy and efficiency	0.2444
Clean and green	0.2517
Openness and interaction	0.1332
Automation	0.5039
Interaction	0.2811
Informatization	0.2150

The second-level indicator "Green Power Generation", and its associated indicators, are analyzed and the correlations are shown in Table 15.

Table 15. The relevance of the "green power generation" indicators.

Relevance of the "Green Power Generation" Indicators				
Associated indicators	B13	B26	B27	B28
Degree of association	0.4555	0.1289	0.8795	0.4537

According to the above introduction, the subjective weights of base layer and objective weights of base layer which obtained by using the entropy method are arithmetically averaged to obtain the comprehensive weight of the corresponding base layer evaluation indicators. The results are shown in Table 16.

Table 16. The weight of the corresponding base layer index.

Index of Base Layer	Subjective Weight	Objective Weight	Comprehensive Weight
B13	0.3175	0.0298	0.1737
B26	0.2030	0.0322	0.1176
B27	0.1322	0.0298	0.0810
B28	0.1184	0.0236	0.0710

(1) Evaluation results of effect level

Province A's evaluation results of effect level are showed in Figures 7 and 8.

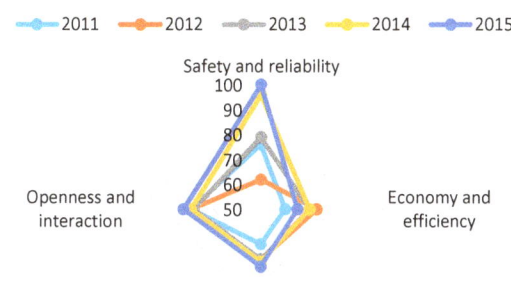

Figure 7. Radar map of construction effect evaluation of Province A.

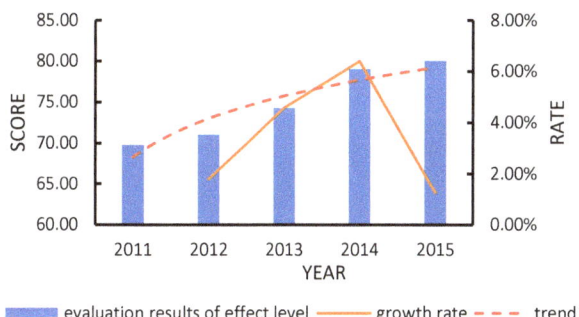

Figure 8. Evaluation results of effect level of Province A.

As can be seen from the above figure, the radar area is increasing year by year, and the score of the effect level development over the years is also gradually increasing, but the growth rate has a certain fluctuation.

In terms of safety and reliability, the power grid company of the province A actively responded to the call and during the "Twelfth Five-year Plan" period [44], it accelerated the construction of a strong smart grid including the ultra-high voltage (UHV), built a comprehensive demonstration project of the eco-city smart grid and promoted its application, ensuring the province's reliable supply of power energy and greatly increasing the safety of the power grid. In terms of clean and green, the province is committed to improving the efficiency of thermal power energy use and promoting energy conservation and emission reduction, therefore, the level of it has been improved to some extent. On the whole, the development level of smart grid effect level of province A should be fully promoted through the two main lines of technological progress and management improvement.

(2) Evaluation results of base level

Province A's evaluation results of base level are showed in Figures 9 and 10.

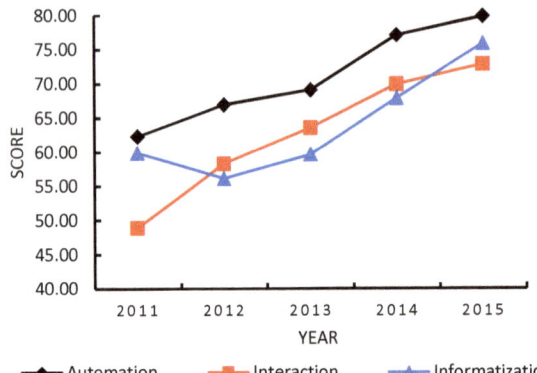

Figure 9. Line chart of construction base evaluation of Province A.

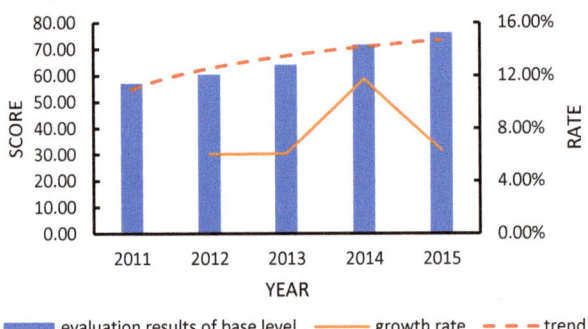

Figure 10. Evaluation results of base level of Province A.

The level of automation, interaction, and informatization of the smart grid in the province has been gradually improved, so the overall level of its base layer is on the rise. The provincial power company's smart grid construction plan was completed in 2010 and entered the full-scale construction phase of the smart grid in 2011. During the "Twelfth Five-Year Plan" period, the provincial electric power company increased its investment in the construction of a smart grid, and extensively adopted modern technology and automation equipment. As a result, the level of the base level of power grid has been comprehensively improved.

7.2. Province B

As mentioned above, the weights of corresponding indicators of province B are showed in Tables 17–19.

Table 17. The weight of the first-level indicators.

Layer	Indicators	Weight
Effect layer	Safety and reliability	0.3961
	Economy and efficiency	0.2352
	Clean and green	0.2367
	Openness and interaction	0.1319
Base layer	Automation	0.5039
	Interaction	0.2811
	Informatization	0.2150

Table 18. The relevance of the "green power generation" indicators.

Relevance of the "Green Power Generation" Indicators				
Associated indicators	B13	B26	B27	B28
Degree of association	0.7212	0.3892	0.4712	0.4807

Table 19. The weight of the corresponding base layer indicators.

Indicators of Base Layer	Subjective Weight	Objective Weight	Comprehensive Weight
B13	0.0464	0.0298	0.0381
B26	0.0179	0.0322	0.0250
B27	0.0107	0.0298	0.0202
B28	0.0135	0.0236	0.0186

7.2.1. Evaluation Results of Effect Level

Province B's evaluation results of effect level are showed in Figures 11 and 12.

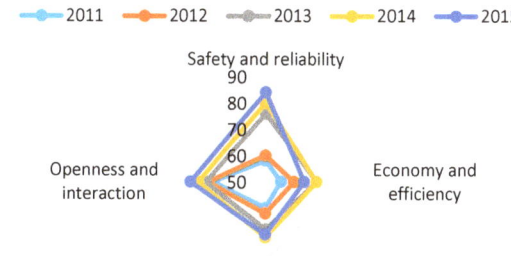

Figure 11. Radar map of construction effect evaluation of Province B.

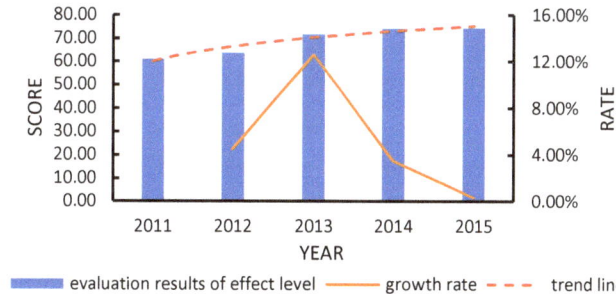

Figure 12. Evaluation results of effect level of Province B.

As can be seen from the radar map, the radar area is increasing year by year, and the improvement of the intelligent effect tends to be flat. Among them, the progress of safety and reliability is relatively fast, indicating that the Province B's smart grid construction has a good effect on the construction of power grids and power supplies. In terms of economy and interaction, it may not perform well because related projects are mostly piloted or promoted. From the above figure, it can also be seen that the level of the power grid effect of the province is slowly growing, and the growth rate is fluctuating.

In the aspect of safety and reliability, it is indicated that the construction of the power grid is under the background of UHV AC and DC landing in the Central Plains, and priority is given to ensuring a wide range of optimal allocation of energy resources. In terms of cleanness and green and openness and interaction, the company is a power grid based on thermal power, and marketing and interactive services are starting. In terms of economic and efficiency, due to the large number of historical problems in the grid, the overall weak distribution network and low operating efficiency have not yet been fundamentally reversed.

7.2.2. Evaluation Results of Base Level

Province B's evaluation results of base level are showed in Figures 13 and 14.

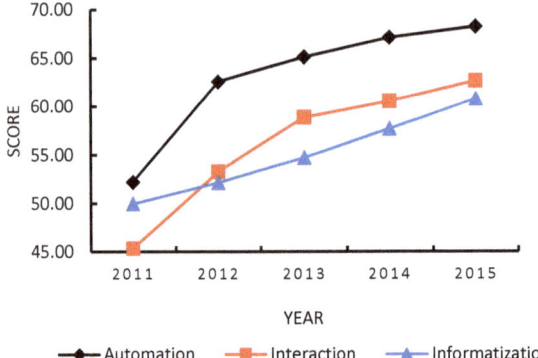

Figure 13. Line chart of construction base evaluation of Province B.

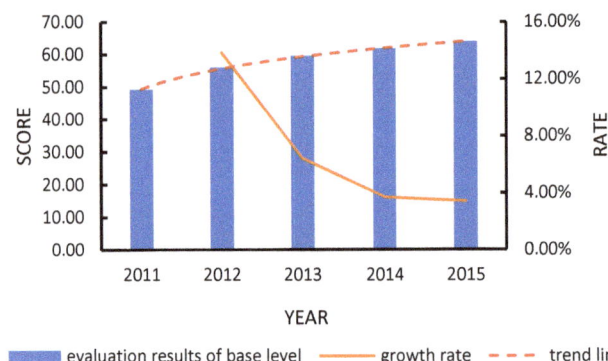

Figure 14. Evaluation results of base level of Province B.

The scores of various indicators have increased year by year, indicating that the basic level of Province B's power grid has become better year by year, among them, the progress of automation and interaction has been greater, indicating that the company's smart grid construction has achieved significant improvement in technology.

In terms of automation, based on the status of the company's balanced power grid, the company's grid security and resource allocation capabilities have been significantly improved through strong cooperation with UHV AC/DC interconnected power grid construction in such areas as power generation, transmission, and dispatch. In terms of informatization and interaction, it can be seen from the above figure that its level is increasing year by year. This is because of the development of related technologies such as measurement, communications, information, and control.

7.3. Province C

As mentioned above, the weights of corresponding indicators of province C are showed in Tables 20–22.

Table 20. The weights of the first-level indicators.

Layer	Indicators	Weight
Effect layer	Safety and reliability	0.2580
	Economy and efficiency	0.2699
	Clean and green	0.2845
	Openness and interaction	0.1876
Base layer	Automation	0.5039
	Interaction	0.2811
	Informatization	0.2150

Table 21. The relevance of the "green power generation" indicators.

Relevance of the "Green Power Generation" Indicators				
Associated indicators	B13	B26	B27	B28
Degree of association	0.3115	0.1919	0.3038	0.5434

Table 22. The weights of the corresponding base layer indicators.

Indicators of Base Layer	Subjective Weight	Objective Weight	Comprehensive Weight
B13	0.0152	0.0298	0.0225
B26	0.0030	0.0322	0.0176
B27	0.0742	0.0298	0.0520
B28	0.0762	0.0236	0.0499

7.3.1. Evaluation Results of Effect Level

Province C's evaluation results of effect level are showed in Figures 15 and 16.

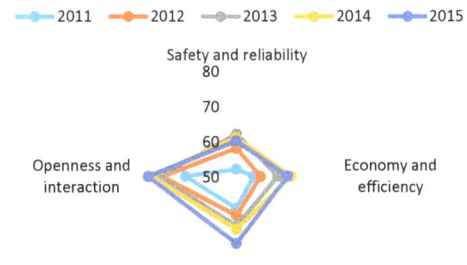

Figure 15. Radar map of construction effect evaluation of Province C.

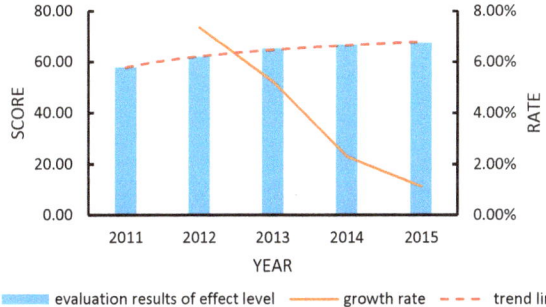

Figure 16. Evaluation results of effect level of Province C.

As can be seen from the above figure, the overall level of the effect layer is increased year by year, but the growth rate fluctuates.

Due to the abundance of wind resources in the province C, ten million kilowatts of wind power bases were built during the period of the Twelfth Five-Year Plan. Some wind power bases are centralized renewable energy generation (CRG) in terms of access methods. After the CRG is connected to the power grid, it has an important and positive effect on energy conservation, emission reduction and energy structure optimization, but it has affected the security and stable operation of the power grid to some extent. At the same time, clean energy alternative projects have been carried out in some areas of the province, which has made great progress in cleaning and environmental protection.

7.3.2. Evaluation Results of Base Level

Province C's evaluation results of base level are showed in Figures 17 and 18.

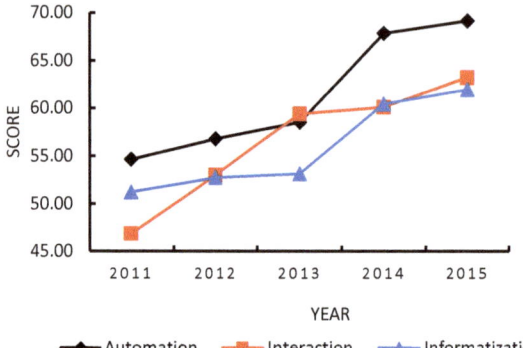

Figure 17. Line chart of construction base evaluation of Province C.

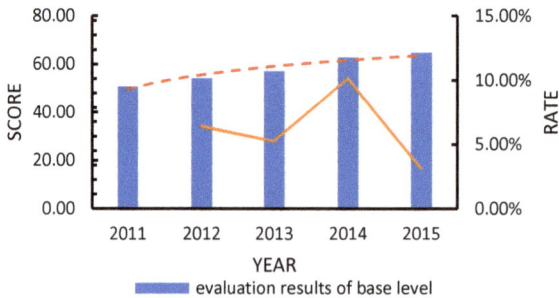

Figure 18. Evaluation results of base level of Province C.

From the above figure, we can see that the level of automation, interaction and informatization of the smart grid in the province has been increasing year by year, but the growth rate is different.

In terms of automation and informatization, the growth rate accelerated in 2013. This is because in 2013, the provincial power grid company carried out all-round power grid geographic information system collection work. As of June 2015, with the province's total 35 kV, 110 The GIS data collection work of the KV transmission line was fully completed. The power company of the province realized visualization, space, and automation management of the power grid through the power grid GIS "big data", thus greatly improving the automation and informatization level of the entire power grid. As for interaction, its growth rate has been relatively stable.

7.4. Comparison

The above method can be used to compare the development level of smart grids in the three provinces. The results are shown in the Figures 19–21.

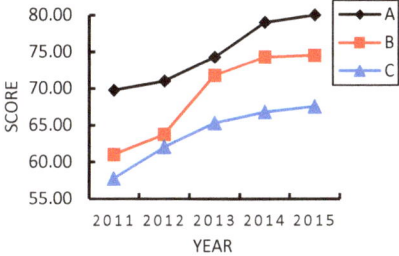

Figure 19. Evaluation results of effect level.

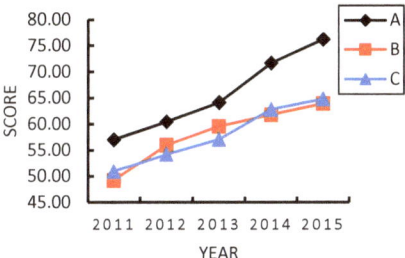

Figure 20. Evaluation results of base level.

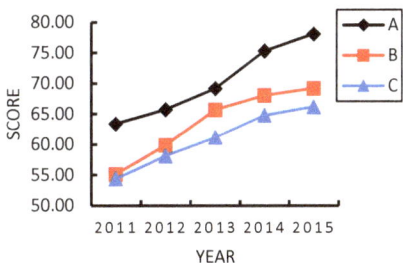

Figure 21. Comprehensive evaluation results.

As can be seen from above figures, in the early stage of smart grid construction in 2011, although the scores of the base layer were low, the construction achievements were remarkable, and the scores in the effect layer were relatively high. With the promotion and construction of the smart grid, although the levels of intelligence of the grid infrastructure keep growing at a certain rate, the speed of the improvement of the effect layer has decreased year by year and tends to be flat; by 2014–2015, although the score of the base layer continues to increase, the score of the effect layer has grown little, which fully reflects the development rule that the smart grid has been in an all-round construction phase.

The final comprehensive evaluation results show that Province A is better than Province B in the overall level of smart grid development, and Province B is better than Province C. Therefore, the power companies in Province B and Province C need to further strengthen the construction and operation management of the smart grid.

8. Conclusions

Based on the difference of demand for the development of smart grid, this paper first establishes its own index system. Subsequently, this paper proposes the implementation path of the evaluation model. Finally, three typical provinces were selected to evaluate and compare the level of smart grid development.

We know that as the smart grid is written into the "Twelfth Five-Year Plan", its status in the country's strategic emerging industries gradually emerged, and the nation's smart grid construction was fully launched. The year 2011 is the first year for the smart grid to enter the comprehensive construction phase. It is also the starting point for the smart grid to achieve leapfrogging from pilot construction to comprehensive construction. By 2015, a strong smart grid operation control and two-way interactive service system has been formed to basically achieve friendly access and coordinated control of renewable energy such as wind power and solar power generation.

Based on this background, this paper proposes to evaluate the development level of smart grid using weight modification, transmission mechanisms, and evaluation methods combining subjective and objective weights, and selects three typical provinces for case demonstration.

For Province A, its development goal in the Twelfth Five-Year Plan is to build a strong, self-reliant, economical, compatible, flexible, and integrated urban power grid that matches the orientation of urban development and is characterized by information, automation, and interaction. Therefore, the government has increased investment, promoted energy-saving construction, and adopted modern technology and automation equipment. As a result, the level of the effect layer and base layer of smart grid is relatively high, and its overall level is also relatively high compared to other provinces. For provinces B and C, due to historical issues and different stages of development, the overall level of smart grid development is relatively low compared to province A.

Through the analysis of relevant examples, it can be proved that the evaluation model can make an association analysis between the construction foundation and the construction effect, and make a comprehensive and deep evaluation of the development level of the smart grid in our country, which is of guiding significance to the future intelligent construction of the power grid.

When using this evaluation method to calculate the weight, the general indicators such as the number of transmission and transformation accidents, overall labor productivity, the proportion of renewable energy power generation, the rate of electricity market transaction power growth, the ratio of intelligent substation, the coverage of smart electric meter, and so on, can be directly applied to different regions and countries, but it should be noted that because different metrics are used around the world, such as assessing the reliability of power systems, there are different indicators, such as SAIDI/SAIFI, and the indicators proposed in this paper are not all versatile, therefore, when applying this method, some substitutions can be made appropriately without affecting the content embodied in the indicator.

Author Contributions: J.L. contributed to the conception and design. T.L. contributed to the computation and English editing. L.H. contributed to the analysis of calculation results and English checking. All of the authors drafted and revised the manuscript together and approved its final publication.

Funding: This work has been supported by "Ministry of Education, Humanities and Social Science Fund, Nos. 15YJC630058", "Beijing Social Science Fund, Nos. 18GLB023", "the Fundamental Research Funds for the Central Universities, Ns. 2017MS083", and "the Science and Technology Project of SGCC".

Acknowledgments: The authors thank the reviewers and editors whose comments improved the quality of this paper.

Conflicts of Interest: The authors declare no conflict of interest.

References

1. Ringenson, T.; Eriksson, E.; Wangel, J. *The Limits of the Smart Sustainable City*; The Workshop on Computing within Limits; ACM: New York, NY, USA, 2017; pp. 3–9.
2. Chen, S.Y.; Song, S.F.; Li, L.X.; Shen, J. Summary of Smart Grid Technology. *Power Syst. Technol.* **2009**, *33*, 1–7.

3. Chen, S.Y.; Song, S.F.; Li, L.X.; Shen, J. Research on Intelligent Grid Evaluation Index System. *Power Syst. Technol.* **2009**, *33*, 14–18.
4. Yu, Y.; Luan, W. Review of Smart Grid. *Proc. CSEE* **2009**, *29*, 1–8.
5. Ni, J.; He, G.; Shen, C.; Deng, Y.; Deng, Z.; Huang, W. A review of assessment of smart grid in America. *Autom. Electr. Power Syst.* **2010**, *34*, 9–13.
6. *Smart Grid Method and Model*; IBM Corporation: Beijing, China, 2010.
7. European Smart Grids Technology Platform. *Strategic Deployment Document for Europe's Electricity Networks of the Future*; European Commission: Brussels, Belgium, 2008.
8. Yu, H.; Jia, Q.; Wang, N.; Dong, H. A Data-Driven Modeling Strategy for Smart Grid Power Quality Coupling Assessment Based on Time Series Pattern Matching. *Math. Probl. Eng.* **2018**, *2018*, 1–12. [CrossRef]
9. Park, S.; Kang, B.; Choi, M.I.; Jeon, S.; Park, S. A micro-distributed ESS-based smart LED streetlight system for intelligent demand management of the micro grid. *Sustain. Cities Soc.* **2018**, *39*, 801–813. [CrossRef]
10. Jesus, O.D.; Antunes, C.H. Economic valuation of smart grid investments on electricity markets. *Sustain. Energy Grids Netw.* **2018**, *16*, 70–90. [CrossRef]
11. Peng, H.; Kan, Z.; Zhao, D.; Han, J.; Lu, J.; Hu, Z. Reliability analysis in interdependent smart grid systems. *Phys. A Stat. Mech. Its Appl.* **2018**, *500*, 50–59. [CrossRef]
12. Leszczyna, R. Standards on cyber security assessment of smart grid. *Int. J. Crit. Infrastruct. Prot.* **2018**, *22*, 70–89. [CrossRef]
13. Cacciatore, G.; Fiandrino, C.; Kliazovich, D.; Granelli, F.; Bouvry, P. Cost analysis of smart lighting solutions for smart cities. In Proceedings of the IEEE International Conference on Communications, Paris, France, 21–25 May 2017.
14. Hashemi-Dezaki, H.; Askarian-Abyaneh, H.; Shams-Ansari, A.; DehghaniSanij, M.; Hejazi, M.A. Direct cyber-power interdependencies-based reliability evaluation of smart grids including wind/solar/diesel/distributed generation and plug-in hybrid electrical vehicles. *Int. J. Electr. Power Energy Syst.* **2017**, *93*, 1–14. [CrossRef]
15. Munshi, A.A.; Mohamed AR, I. Big data framework for analytics in smart grids. *Electr. Power Syst. Res.* **2017**, *151*, 369–380. [CrossRef]
16. Woo, P.S.; Kim, B.H. Methodology of Cyber Security Assessment in the Smart Grid. *J. Electr. Eng. Technol.* **2017**, *12*, 495–501. [CrossRef]
17. Lloret-Gallego, P.; Aragüés-Peñalba, M.; Van Schepdael, L.; Bullich-Massagué, E.; Olivella-Rosell, P.; Sumper, A. Methodology for the Evaluation of Resilience of ICT Systems for Smart Distribution Grids. *Energies* **2017**, *10*, 1287. [CrossRef]
18. Vazquez, R.; Amaris, H.; Alonso, M.; Lopez, G.; Moreno, J.I.; Olmeda, D.; Coca, J. Assessment of an Adaptive Load Forecasting Methodology in a Smart Grid Demonstration Project. *Energies* **2017**, *10*, 190. [CrossRef]
19. Rossebø, J.E.; Wolthuis, R.; Fransen, F.; Björkman, G.; Medeiros, N. An Enhanced Risk-Assessment Methodology for Smart Grids. *Computer* **2017**, *50*, 62–71. [CrossRef]
20. Coppo, M.; Pelacchi, P.; Pilo, F.; Pisano, G.; Soma, G.G.; Turri, R. The Italian smart grid pilot projects: Selection and assessment of the test beds for the regulation of smart electricity distribution. *Electr. Power Syst. Res.* **2015**, *120*, 136–149. [CrossRef]
21. Xenias, D.; Axon, C.J.; Whitmarsh, L.; Connor, P.M.; Balta-Ozkan, N.; Spence, A. UK smart grid development: An expert assessment of the benefits pitfalls functions. *Renew. Energy* **2015**, *81*, 89–102. [CrossRef]
22. Liu, X.; Liu, X.; Li, Z. Cyber Risk Assessment of Transmission Lines in Smart Grids. *Energies* **2015**, *8*, 13796–13810. [CrossRef]
23. Personal, E.; Guerrero, J.I.; Garcia, A.; Peña, M.; Leon, C. Key performance indicators: A useful tool to assess Smart Grid goals. *Energy* **2014**, *76*, 976–988. [CrossRef]
24. Dong, H.; Zheng, Y. Measuring Technological Progress of Smart Grid Based on Production Function Approach. *Math. Probl. Eng.* **2014**, *2014*, 861820. [CrossRef]
25. Hu, D.G.; Zheng, J.; Zhang, Y.H.; Xu, H.L.; Zhou, H.M.; Yang, J.K.; Pu, T.J. Study on Technological Evaluation Modeling of Smart Grid. *Appl. Mech. Mater.* **2014**, *494–495*, 1747–1752. [CrossRef]
26. Xu, Q.; Li, Y.; Yang, X. Comprehensive assessment system and method of smart distribution grid. *Power Syst. Technol.* **2014**, *860–863*, 1901–1908.
27. Song, G.; Chen, H.; Guo, B.A. Layered fault tree model for reliability evaluation of smart grids. *Energies* **2014**, *7*, 4835–4857. [CrossRef]

28. Bracco, S.; Delfino, F.; Pampararo, F.; Robba, M.; Rossi, M.A. Mathematical model for the optimal operation of the University of Genoa Smart Polygeneration Microgrid: Evaluation of technical economic environmental performance indicators. *Energy* **2014**, *64*, 912–922. [CrossRef]
29. Wang, Q.R. Performance evaluation of smart grid based on optimal fuzzy algorithm method. *Appl. Mech. Mater.* **2013**, *482*, 346–349. [CrossRef]
30. Niu, D.X.; Liu, J.P.; Liu, T.; Guo, H.C. Comprehensive evaluation of regional grid development level of China under the smart grid construction. *Appl. Mech. Mater* **2013**, *300–301*, 640–644. [CrossRef]
31. Li, L.; Liu, L.; Yang, C.; Li, Z. The comprehensive evaluation of smart distribution grid based on cloud model. *Energy Procedia* **2012**, *17*, 96–102. [CrossRef]
32. Bilgin, B.E.; Gungor, V.C. Performance evaluations of ZigBee in different smart grid environments. *Comput Netw.* **2012**, *56*, 2196–2205. [CrossRef]
33. Xie, C.; Dong, D.; Hua, S.; Xu, X.; Chen, Y. Safety evaluation of smart grid based on AHP-entropy method. *Syst. Eng. Procedia* **2012**, *4*, 203–209.
34. Sun, Q.; Ge, X.; Liu, L.; Xu, X.; Zhang, Y.; Niu, R.; Zeng, Y. Review of smart grid comprehensive assessment systems. *Proc. Chin. Soc. Univ. Electr. Power Syst. Its Autom.* **2011**, *12*, 219–229. [CrossRef]
35. Caciotta, M.; Leccese, F.; Trifiro, A. From power quality to perceived power quality. In Proceedings of the IASTED International Conference on Energy and Power Systems, Rhodes, Greece, 26–28 June 2006; pp. 94–102.
36. Song, Y.; Yang, X.; Sun, J. Low carbon, efficient, safe and reliable smart grid. *Energy China* **2009**, *31*, 23–27.
37. Li, T.; Xu, B. Self-healing and its benchmarking of smart distribution grid. *Power Syst. Prot. Control* **2010**, *38*, 105–108.
38. Wang, Z.; Su, H. Cost-benefit analysis model for reliability of distribution network automaton system. *Power Syst. Prot. Control* **2014**, *42*, 98–103.
39. Wang, C.; Xu, Y.; Wei, Q.; Zhao, D.; Liu, D. Analysis of intelligent community business model and operation mode. *Power Syst. Prot. Control* **2015**, *43*, 147–154.
40. Han, L.; Zhuang, B.; Wang, Z.; Zhang, L. The research about power grid efficiency indexes. *East China Electr. Power* **2011**, *39*, 850–854.
41. Cheng, Y.; Zhai, N. Evaluation of TOU price oriented to smart grid. *Power Syst. Prot. Control* **2010**, *38*, 196–201.
42. Xin, Y.; Shi, J.; Zhou, J.; Gao, Z.H.; Tao, H.Z.; Shang, X.W.; Zhai, M.Y.; Guo, J.C.; Yang, S.C.; Nan, G.L.; et al. Technology development trends of smart grid dispatching and control system. *Autom. Electr. Power Syst.* **2015**, *39*, 2–8.
43. Feng, Y.; Li, X.; Li, X. Comprehensive evaluation of railway safety based on entropy weight method and grey relational analysis. *J. Saf. Environ.* **2014**, *14*, 73–79.
44. Xinhua Net. Outline of the Twelfth Five-Year Plan for National Economic and Social Development of the People's Republic of China. Available online: http://www.china.com.cn/policy/txt/2011-03/16/content_22156007.htm (accessed on 5 November 2018). (In Chinese)

© 2018 by the authors. Licensee MDPI, Basel, Switzerland. This article is an open access article distributed under the terms and conditions of the Creative Commons Attribution (CC BY) license (http://creativecommons.org/licenses/by/4.0/).

Article

Three-Phase Unbalanced Optimal Power Flow Using Holomorphic Embedding Load Flow Method

Bharath Varsh Rao [1,*], Friederich Kupzog [1] and Martin Kozek [2]

[1] Electric Energy Systems—Center for Energy, AIT Austrian Institute of Technology, 1210 Vienna, Austria; friederich.kupzog@ait.ac.at
[2] Institute of Mechanics and Mechatronics—Faculty of Mechanical and Industrial Engineering, Vienna University of Technology, 1060 Vienna, Austria; martin.kozek@tuwien.ac.at
* Correspondence: bharath-varsh.rao@ait.ac.at; Tel.: +43-664-8825-6043

Received: 16 February 2019; Accepted: 21 March 2019; Published: 24 March 2019

Abstract: Distribution networks are typically unbalanced due to loads being unevenly distributed over the three phases and untransposed lines. Additionally, unbalance is further increased with high penetration of single-phased distributed generators. Load and optimal power flows, when applied to distribution networks, use models developed for transmission grids with limited modification. The performance of optimal power flow depends on external factors such as ambient temperature and irradiation, since they have strong influence on loads and distributed energy resources such as photo voltaic systems. To help mitigate the issues mentioned above, the authors present a novel class of optimal power flow algorithm which is applied to low-voltage distribution networks. It involves the use of a novel three-phase unbalanced holomorphic embedding load flow method in conjunction with a non-convex optimization method to obtain the optimal set-points based on a suitable objective function. This novel three-phase load flow method is benchmarked against the well-known power factory Newton-Raphson algorithm for various test networks. Mann-Whitney U test is performed for the voltage magnitude data generated by both methods and null hypothesis is accepted. A use case involving a real network in Austria and a method to generate optimal schedules for various controllable buses is provided.

Keywords: unbalanced three-phase distribution networks; optimal power flows; genetic algorithm; holomorphic embedding load flow method; simulation

1. Introduction

In recent years, with the integration of distributed generators, electric storage, electrical vehicles, and demand response units, the role of distribution systems is changing. Distributed energy units (DERs) are posing problems mainly in the low-voltage networks with their intermittency and uncontrollability. New innovative solutions are required to maintain grid security. Management of low-voltage distribution networks are challenging since they contain large array of devices which need to be controlled, and monitoring systems are limited. The above DERs along with loads should be run in a sustainable fashion since it is one of the biggest challenges. Various methods to control the DERs are presented in [1].

A so-called advanced distribution management system (ADMS) has come into existence, evolving from the transmission network's supervisory control and data acquisition systems (SCADA). This is possible with the increase in smart meters and monitoring devices in the network which provides data acquisition abilities [2]. ADMS provides functionalities such as load flow analysis, optimal power flow, monitoring and control capabilities similar to SCADA systems [3]. This must in theory, host advance functionalities such as adaptive protection leading to self-healing, real-time monitoring, dynamic network reconfiguration and control [4–8]. This will provide intelligence to the grid with topology processor, state estimation, load and generation modeling [3]. The grid needs to be operated optimally and the power flows should be optimal to reduce losses, increase security, and maximize economic benefit. Energy balance should be maintained for secure operation of the network to maintain the frequency and voltage within its limits.

Optimal Power Flow (OPF) is one of the most fundamental functionalities of ADMS. In the literature, various OPF algorithms can be found. The authors in [9] describe an OPF algorithm to control active, reactive power, and transformer taps. The objective is to minimize system costs and losses. This method is based on Newton-Raphson load flow. Feasible power flow is solved, and the optimum is close to the load flow solution. Therefore, Jacobian information is used to calculate the optimum in a linear fashion. In [10], a non-linear programming technique is used to provide solution to OPF problem the objectives being economic dispatch and generation cost minimization. Same as before, load flow is performed to determine a feasible solution. Fletcher-Powell method is used to minimize the objective function. A general economic dispatch problem is implemented in [11]. This approach is similar to that in [9,10]. In [12], an OPF method for power system planning is provided. It used generalized reduced gradient technique to find the optimum. Hessian Matrix-based OPF method is illustrated in [13]. It combines non-linear programming, Newton-based methods and uses Hessian matrix load flow to minimize the quadratic objective. In [14], an OPF algorithm using Newton's method with Hessian in place of Jacobian matrix and Lagrangian multipliers is provided. It provides good convergence when compared to its predecessors. An OPF problem which includes steady-state security is presented in [15]. It is an extension of [9] which includes exact constraints on outage contingencies. In [16], a solution to the optimal dispatch problem using Jacobian matrix is implemented. It provides rapid convergence which can be used in online control. An OPF algorithm based on reduced gradient method is proposed in [17]. It is used to minimize generator loading and optimize voltage levels. Load flow equations are represented as equality constraints. The authors in [18] have described an OPF method using reduced Hessian matrix with systematic constraint handling. It provides accurate solution, good convergence, and description about acceleration factor is provided. In [19], modified recursive quadratic programming (MRQP)-based OPF is implemented. MRQP is based on [11]. An algorithm to solve large OPF problem is presented in [20]. It decomposes the original large problem into set of subproblems which are constrained linearly using augmented Lagrangian.

In 1991, a landmark paper [21] classified various OPF techniques into two categories. Class A describes a series of algorithms which uses ordinary load flow to get an intermediate solution and this solution is under normal circumstances is close to the optimal load flow solutions. Using Jacobian matrix and various other sensitivity relations, optimization is performed iteratively. At each iteration, new load flow is performed. The optimal solution of this class strongly depends on the accuracy of load flow solution. With a load flow solution, set of voltages, and phase angles, Jacobian matrix and set of incremental power flow equations are available or can be extended. If a load flow solution exists, it already satisfies all the constraints. The optimization problem is solved separately by incorporating the sensitivity relation from before to arrive at an optimal one. In [9] an implementation of Class A algorithm is provided. Another example of such implementation is provided in parts one and two in [22] and [23] respectively using linear programming.

Class B refers to the class of algorithms which depend on exact optimal conditions and therefore use load flow equations as equality constraints. The optimal solution is dependent on detailed formulation of the OPF problem with the entire search space. This does not need a load flow solution. However, these kinds of problems are non-convex in nature. Therefore, convex relaxation or non-convex solvers are needed to compute the optimum which poses its own difficulties. It deals with the optimality conditions from Lagrangian function and comprises of derivatives of constraints and objective functions. Since the Hessian matrix is sparse and remains constant, it further increases the simplicity of this method and ease at which the optimum is achieved. Constraint handling is one of the biggest challenges of this class of algorithms. Using a heuristic method, constraints are handled as penalty terms which requires refactoring at every step and therefore, leads to degradation of the performance.

The above two classes of algorithms has various advantages and disadvantages. Performance of Class A directly depends in the performance of load flow techniques such as Newton-Raphson, Gauss-Seidel and widely used Fast Decoupled Method. It is shown in [24] that the above-mentioned methods have convergence and robustness issues. This may result in inaccurate load flow solutions. If the load flow does not result in a so-called high voltage or operable solution, Class A algorithms fails. In Class B algorithms, getting a global operable solution is challenging since it needs convex relaxation or heuristic techniques and the operable solution is difficult to achieve by respecting all the constraints.

The authors in this paper present a third class of algorithms, a Class C. This class combines Class A and Class B. It uses a reliable load flow described in Section 2 method wrapped around a heuristic to determine the optimal solution. The load flow provides accurate operable voltage and phase angle solution at every step and the heuristic uses this as equality constraints as described in Class B. Class C algorithms present various advantages. Operable voltage and phase angle solution is obtained at each iteration with the help of THELM. THELM always finds a solution, if it exists, irrespective of initial conditions whereas, Newton-Raphson load flow method leads to a non-convergent solution at very low or high loading conditions [24]. Since THELM is used in Class C method, the results are high voltage and operable. Global OPF solution can be obtained with a non-convex solver.

The following contributions and structure of the paper is as follows,

1. Load Flow Solution to three-phase unbalanced distribution network using Holomorphic Embedding Load Flow Method (HELM) is described in Section 2.
2. Benchmarking of HELM against established Newton-Raphson load flow solver from DIgSILENT PowerFactory [25] which is a well-known power system simulation and analysis software. This is discussed in Section 2.
3. OPF using Distributed Genetic Algorithm, a Class C algorithm is described in detail in Section 3
4. Simulation of OPF is performed to generate active and reactive power schedules at controllable nodes (see Section 5). This algorithm is applied to a real network in Austria.

2. Three-Phase Unbalanced Load Flow Method

A solution to the load flow problem is mostly obtained using numerical iterative methods such as Gauss-Seidel with its slow convergence and improved Newton-Raphson method, which provides better convergence [26,27]. Newton-Raphson method is computationally expensive since it must calculate Jacobean matrix at each iteration step in-spite of using sparse matrix techniques [28]. Various decoupled methods have been implemented which exploits the weak link between active power and voltage, in which Jacobian matrix needs to be calculated only once. One such method is Fast Decoupled Load Flow method which is widely used in the community [29]. The above-mentioned iterative techniques face similar problems with no guaranteed convergence since it depends on the initial conditions. This is due to the fact that load flow equations are non-convex in nature with multiple solutions. It is difficult to control the way these iterative methods converge to an operable solution [24]. In the literature, multiple implementations to improve the convergence of such traditional algorithm have been illustrated with limited success [30–36].

To use load flow methods in near or real-time applications, the physical models should fully deterministic and solved with reliability. HELM is one such candidate which can full fill these requirements [24].

Three-Phase Holomorphic Embedding Load Flow Method

Power flow equations, for example, the load bus equation described in Equation (1) is inherently non-analytical. Holomorphic principles can be applied to such equations by means of embedding a complex variable α such that the resulting problem is analytic in nature.

$$\sum_{k \in \Omega} Y_{ik} V_k = \frac{S_i^*}{V_i^*}, \quad i \in \Omega_{PQ} \tag{1}$$

Voltage of the slack bus is assumed to be $V_0 = 1.0$ pu. and Bus 00 (see Appendix A) is always set to be slack bus.

Holomorphic embedding can be done in various methods. Equation (2) represents the simplest form. Bus voltages are the functions of the demand scalable complex variable α.

$$\sum_{k \in \Omega} Y_{ik} V_k(\alpha) = \frac{\alpha S_i^*}{V_i^*(\alpha^*)}, \quad i \in \Omega_{PQ} \tag{2}$$

The research work in [24] suggests that the operable voltage solution can be obtained by analytic continuum of Equation (2) at $\alpha = 1$ using the unique solution which exists when $\alpha = 0$

$$\sum_{k \in \Omega} Y_{ik} V_k(\alpha) = \frac{\alpha S_i^*}{\overline{V}_i(\alpha^*)}, \quad i \in \Omega_{PQ} \tag{3}$$

$$\sum_{k \in \Omega} Y_{ik}^* \overline{V}_k(\alpha) = \frac{\alpha S_i}{V_i(\alpha)}, \quad i \in \Omega_{PQ} \tag{4}$$

Equations (3) and (4), represent a set of polynomial equations and by using the Grobner bases, V_i and $\overline{V}_i$ are holomorphic except for finite singularities.

$$\overline{V}_i(\alpha) = (V_i(\alpha^*))^*, \quad i \in \Omega \tag{5}$$

According to [24], if Equation (5) holds good, then Equations (3) and (4) can be reduced to Equation (2). Equation (5) is referred to as reflecting condition.

Since voltages of from Equation (2) for $\alpha = 0$ as discussed above, it can be extended to power series described in Equation (6) and (7) at $\alpha = 0$.

$$V_i(\alpha) = \sum_{n=0}^{\infty} V_i[n] \alpha^n, \quad i \in \Omega \tag{6}$$

$$\frac{1}{V_i(\alpha)} = W_i(\alpha) = \sum_{n=0}^{\infty} W_i[n] \alpha^n, \quad i \in \Omega \tag{7}$$

Equation (9) is obtained by substituting 7 into 2 and power series coefficients can be calculated to a desired degree.

$$\sum_{k \in \Omega} Y_{ik}^* \sum_{n=0}^{\inf} V_k[n](\alpha^n) = \alpha S_i^* W_i^*[n] \alpha^n \tag{8}$$

The following steps are involved to calculate voltages.

1. For $\alpha = 0$, solve Equation (9) to obtain a linear equation where the left-hand side of the equation represents the slack bus at which $V_0[\alpha] = 1$.

$$\sum_{k \in \Omega} Y_{ik} V_k[0] = 0, \quad i \in \Omega_{PQ} \tag{9}$$

2. The reduced Y bus matrix is assumed to be non-singular. Equation (10) can be obtained based on the non-singularity assumption.

$$W_i[0] = \frac{1}{V_i[0]} \quad (10)$$

3. Remaining power series coefficients can be obtained to the desired n^{th} degree by equating the coefficients from Equation (11)

$$\sum_{k \in \Omega} Y_{ik} V_k[0] = S_i^* W_i^*[n-1], \quad i \in \Omega_{PQ} \ n \geq 1 \quad (11)$$

$W_i[n-1]$ are calculated using the lower order coefficients described in Equation (12).

$$W_i[n-1] = -\frac{\sum_{m=0}^{n-2} V_i[n-m-1] W_i[m]}{V_i[0]} \quad (12)$$

4. Pade approximations which are particular kind of rational approximations are used for analytical continuum to determine the voltages at $\alpha = 1$.

Based on the fundamentals of HELM discussed above, various research work dealing with enhancing or improving the method is available. One of the major deficiencies of the HELM described in [24] is that the PV/Generator bus is not defined. A PV bus model was presented in [37]. Ref. [38] presents an improved PV bus model and the major contribution of this paper is to provide alternative models capable of solving general networks. The authors have provided four methods with various parameters for PV bus. In the literature, three-phase formulation of HELM is lacking. In this paper, method four developed in [38] is extended to a novel three-phase unbalanced formulation which can be seen below. Equation (13) represents a general form of three-phase unbalanced HELM. Network models including various device models such as loads, generators, transformers are derived from the models developed in [39]. The seed solution, non-singularity of matrix A in Equation (14) and the reflective conditions of holomorphic functions are taken, as is, from [38]. Three-phase unbalanced form for a multi-bus system for PQ and PV bus types is presented below.

$$\begin{bmatrix} A_1^a & A_1^b & A_1^c & A_2^a & A_2^b & A_2^c \\ A_{PQ_3}^a & A_{PQ_3}^b & A_{PQ_3}^c & A_{PQ_4}^a & A_{PQ_4}^b & A_{PQ_4}^c \\ A_{PV_3}^a & A_{PV_3}^b & A_{PV_3}^c & A_{PV_4}^a & A_{PV_4}^b & A_{PV_4}^c \end{bmatrix} \begin{bmatrix} Re\{V^a[n]\} \\ Re\{V^b[n]\} \\ Re\{V^c[n]\} \\ Im\{V^a[n]\} \\ Im\{V^b[n]\} \\ Im\{V^c[n]\} \end{bmatrix} = \begin{bmatrix} r_{1,n-1} \\ r_{PQ2,n-1} \\ r_{PV2,n-1} \end{bmatrix} \quad (13)$$

Is of the form,

$$Ax = b \quad (14)$$

where the matrix A can be further clarified as,

$$\begin{aligned} A_{1_{ij}}^P &= G_{ij}^P + \delta i, j Re\{y_i^P\}, & i,j \in \Omega, P \in a,b,c \\ A_{2_{ij}}^P &= B_{ij}^P - \delta i, j Im\{y_i^P\}, & i,j \in \Omega, P \in a,b,c \\ A_{PQ3_{ij}}^P &= B_{ij}^P - \delta i, j Im\{y_i^P\}, & i,j \in \Omega_{PQ}, P \in a,b,c \\ A_{PV3_{ij}}^P &= 2\delta i, j, & i,j \in \Omega_{PV}, P \in a,b,c \\ A_{PQ4_{ij}}^P &= G_{ij}^P + \delta i, j Im\{y_i^P\}, & i,j \in \Omega_{PQ}, P \in a,b,c \\ A_{PV4_{ij}}^P &= 0, & i,j \in \Omega_{PV}, P \in a,b,c \end{aligned} \quad (15)$$

where G and B are the conductance and susceptance, respectively. $\delta i, j = 1$ if $i = j$, else 0.
The right-hand side matrix elements are defined as follows,

$$r_{1,n-1,i} = \delta_{n,1}(P_i - Re\{y_i\}) - Re\left\{\sum_{m=1}^{n-1} V_i^*[m] \sum_{k \in \Omega} \sum_{p \in P} Y_{ik} V_k[n-m]\right\}, \quad i \in \Omega \quad (16a)$$

$$r_{PQ2,n-1,i} = \delta_{n,1}(-Q_i - Im\{y_i\}) - Im\left\{\sum_{m=1}^{n-1} V_i^*[m] \sum_{k \in \Omega} \sum_{p \in P} Y_{ik} V_k[n-m]\right\}, \quad i \in \Omega_{PQ} \quad (16b)$$

$$r_{PV2,n-1,i} = -\sum_{m=1}^{n-1} \sum_{p \in P} V_i^*[m] V_i[n-m] + (1 + \alpha(M_i - 1)^2)[n], \quad i \in \Omega_{PV} \quad (16c)$$

where, M_i is the target voltage magnitude for PV bus.

The power series were calculated for using the above equations and Viskovatov Pade approximant algorithm is used to determine the voltages and phase angles similar to the ones in [24,37].

3. Optimal Power Flow Model

As described in Section 1, OPF algorithms can be classified under three classes. There have been a lot of research on OPF and this can be seen in the vast array of work available in the literature.

Type C algorithms requires non-convex solvers to perform optimization problems. Non-convex solvers have been previously used to solve OPF problems but, they are used in the context of Class B algorithms. The authors in [40] have provided a method to plan reactive power flows optimally using generic algorithm as it provides optimum which is a global one. The proposed method is applied to two 51 and 224 bus networks. An OPF problem is solved using generic algorithm in [41] as a unified power flow controller to regulate branch voltages with respect to both angle and magnitude. It minimizes real power losses and security limits of power flows are maintained. Reactive power planning using hybrid genetic algorithm is presented in [42]. It uses genetic algorithm at the highest level and linear programming to get the optimum sequentially. This can be considered as a modified version of Class A. It uses genetic algorithm instead of just load flow to determine the initial converged solution to the OPF problem.

In [43], a feeder reconfiguration technique is presented. It uses genetic algorithm in the context of OPF to reduce losses in a distribution system. Switches are opened to determine the initial population. The authors in [44], have presented a hybrid evolutionary algorithm with multi-objective OPF. It is used to minimize losses, voltage, and power flow deviations and generator costs. In [45], optimal placement and sizing of capacitor banks in distributed networks using genetic algorithm is presented. The objective is to simultaneously improve the power quality and sizing of fixed capacitor banks.

In this paper, the OPF problem is formulated as follows,

$$\begin{aligned}
\underset{x}{\text{minimize}} \quad & F(x,u) \\
\text{subject to} \quad & u \in U \\
& G(x,u) = 0, \\
& H(x,u) \leq 0
\end{aligned} \quad (17)$$

where,

x and u represent sets of state and input variables.

$F(x,u)$ is the objective function for the OPF problem. Typical objectives are total generator cost, loss minimization in network and in this paper, the objective function chosen is the three-phase unbalance minimization (see Section 4).

$G(x,u)$ and $H(x,u)$ represents the equality and inequality constraints of the OPF problem.

In the context of type C algorithms, accurate and reliable load flow is used as equality constraints. In this case, THELM is used.

Typical inequality constraints for a three-phase unbalanced distribution grids are enlisted below,

Limits on active power (kW) of a (generator) PV node:	$P_{Low_i} \leq P_{PV_i} \leq P_{High_i}$
Limits on voltage (V (pu.)) of a PV or PQ node:	$|V_{Low_i}| \leq |V_i| \leq |V_{High_i}|$
Limits on tap positions of a transformer:	$t_{Low_i} \leq t_i \leq t_{High_i}$
Limits on phase shift angles of a transformer:	$\theta_{Low_i} \leq \theta_i \leq \theta_{High_i}$
Limits on shunt capacitances or reactances:	$s_{Low_i} \leq s_i \leq s_{High_i}$
Limits on reactive power (kVAr) generation of a PV node:	$Q_{Low_i} \leq Q_{PV_i} \leq Q_{High_i}$
Upper limits on active power flow in transmission lines or transformers:	$P_{i,j} \leq P_{High_{i,j}}$
Upper limits on MVA flows in lines or transformers:	$P_{i,j}^2 + Q_{i,j}^2 \leq S_{High_{i,j}}^2$
Upper limits on current magnitudes in lines or transformers:	$|I_{i,j}| \leq |I_{High_{i,j}}|$
Limits on voltage angles between nodes:	$\Theta_{Low_i} \leq \Theta_i - \Theta_j \leq \Theta_{High_i}$

In this paper, the non-convex solver used is a genetic algorithm, to minimize the objective function. Genetic algorithm is chosen due to its wide use in OPF techniques, ease of parallelizability to handle large networks and its probabilistic transition rule. The authors have used the method developed in [46]. Genetic algorithms of the kind, mixed integer non-linear non-convex is used to include all the constraints mentioned above. To accommodate THELM in genetic algorithm, it is included in the fitness function and penalty functions are used to include constraints.

4. Three-Phase Unbalance Minimization

As mentioned in Section 1, it is essential to manage the distribution network optimally and in a balanced fashion. Various methods have been presented in the literature to minimize three-phase unbalance. In [47], a method to minimize three-phase unbalance is presented. Reactive power compensation is performed using flexible AC transmission system (FACTS) devices to minimize the three-phase unbalance. It is applied to a simple study case of four bus system. This method does not provide optimal scheduling of loads and does not include all the buses in the network. It is applicable only to local grid where the FACTS devices are located. The authors in [48] have provided a method to minimizing network unbalance using phase swapping. A genetic and greedy algorithm is used to optimally swap the phases to generate a convenient solution, leading to a minimum number of swaps to minimize network unbalance. In [49], plug-in hybrid electric vehicles are used to minimize local three-phase unbalance. It does not include a grid perspective and is done only on the point of common coupling.

In this paper, OPF from Section 3 is applied to a real network in Austria. Figure 1 represents a real low-voltage distribution network. In this use case, the objective function is to minimize three-phase voltage unbalance which can be seen in Equation (18). This objective can be realized in multiple ways and in this paper, reference balanced voltages are used.

$$\text{minimize } J = \sum_{k \in \Omega} \sum_{p \in P} (real(V_{k,balanced}^p) - real(V_k^p))^2 + (imag(V_{k,balanced}^p) - imag(V_k^p))^2 \quad (18)$$

where, $P \in phases(a,b,c)$ and B represented all the buses in the network. The voltages are represented in rectangular coordinate system with real part of voltage being the magnitude and imaginary part being the phase angle. Both real and imaginary value are considered because both phase and angle of voltages need to be balanced.

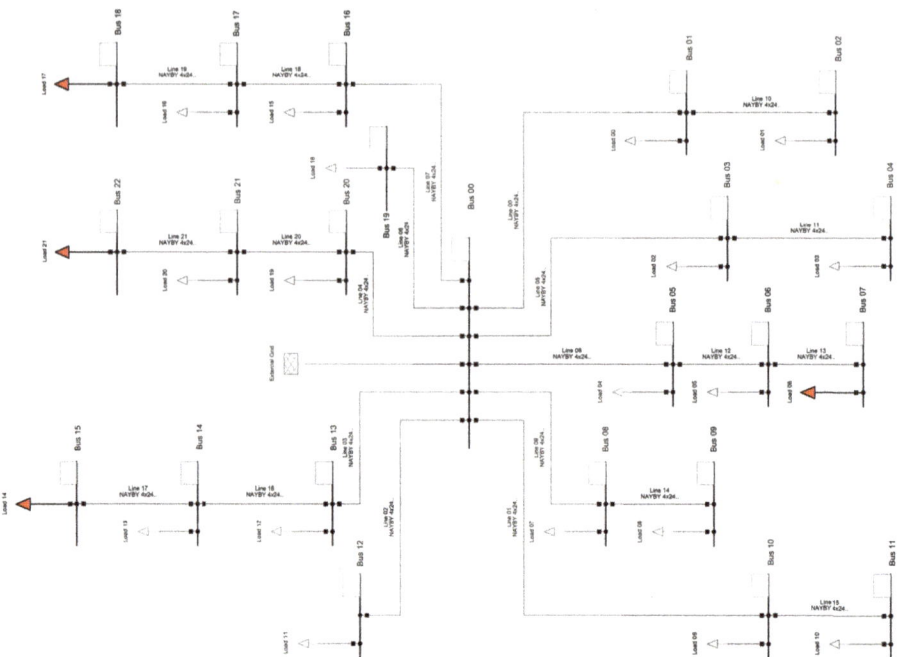

Figure 1. Topology of a real network in Austria with controllable loads at Bus 07, Bus 15, Bus 18, Bus 22. It represents a three-phase unbalance low-voltage distribution network with bus voltages rated at 0.4 kV.

The controllable variables are per phase active (P) and reactive powers (Q) at buses 07, 15, 18, and 22. The single-phase loads are replaced with three-phase loads (see Figure 2).

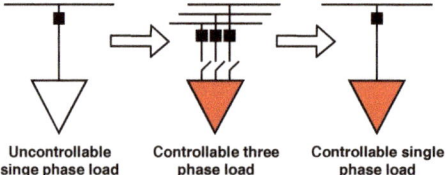

Figure 2. Single-phase loads are replaced by three-phase ones which can take both positive and negative values.

For simplicity of representation, these three-phase loads are represented as single-phase loads with red coloring. This can be observed in Figure 1. P and Q on individual loads can be modulated by the OPF algorithm and can take values which are both positive and negative essentially, acting as a prosumer node.

5. Simulation Results

This section provides simulation results to the concepts presented in the previous sections. In Section 5.1, THELM described in Section 2 is validated against DIgSILENT PowerFactory Newton-Raphson algorithm. In Section 5.2, simulation results for three-phase unbalanced optimal power is presented with the three-phase unbalance minimization objective presented in Section 4.

5.1. Validation of THELM

THELM is benchmarked against load flow solver in an established power system analysis tool, DIgSILENT PowerFactory. Various simple networks are drawn with increased level of complexity (see Appendix A).

Voltages from 1000 random load flows by varying active and reactive power at load buses from ±10 kW and ±0.8 kVAr (which accounts for power factor 0.9) are generated and tabulated below.

Mann-Whitney U test is used to compare the sample means of voltage magnitudes from the two methods, since the samples are non-parametric in nature. It checks whether to accept or reject the null hypothesis. Mann-Whitney U test is similar to student's T test but is suitable for non-parametric samples. A sample of 100 voltages from both THELM and Power Factory NR methods are used. Various statistical information and test results are tabulated in Table 1. Columns mean, standard deviation, min, 25%, 50%, 75%, max are calculated by taking the absolute difference between their respective voltages. All the data above is calculated by taking the average between various buses and phases. From the column statistic and p-value, it can be observed that their means are statistically insignificant, and the null hypothesis is accepted. The results from the test suggests that THELM produces results which are acceptable for load flow analysis with lower deviations from one another.

Table 1. Benchmarking THELM against Power Factory NR method.

	Mean	Std	Min	25%	50%	75%	Max	Statistic	p Value
Test 00	0.00489	0.00350	1.14×10^{-5}	0.00194	0.00430	0.00724	0.01591	2.38719	0.016977
Test 01	0.00572	0.00414	1.01×10^{-5}	0.00237	0.00496	0.00846	0.01938	3.23993	0.001195
Test 02	0.00209	0.00027	0.00167	0.00186	0.00205	0.00228	0.00287	5.47075	4.48×10^{-8}
Test 03	0.00015	0.00028	9.38×10^{-8}	4.57×10^{-5}	9.85×10^{-5}	0.00018	0.00529	−4.69619	2.65×10^{-6}
Test 04	0.00343	0.00246	1.08×10^{-5}	0.00144	0.00296	0.00508	0.01125	−5.6259	1.84×10^{-8}
Test 05	0.00178	0.00023	0.00142	0.00159	0.00175	0.00195	0.00241	−5.83359	5.42×10^{-9}
Test 06	0.05586	0.01114	0.02464	0.04779	0.05540	0.06339	0.09113	5.69187	1.25×10^{-8}

5.2. Simulation Results for Three-Phase Unbalanced Optimal Power Flow

Simulation is performed for the real grid detailed in Figure 1 using OPF algorithm described in Section 3. It is performed for one day from 2018-8-31 00:00:00 to 2018-9-01 00:00:00 with the sampling time of 15min (96 intervals). Load profiles are from smart-meter data, from real households and are acquired from all the buses in the network updating a database. Forecasted profiles are inputs to the OPF algorithm and optimal schedules are generated based on it. Load forecasting is performed for this time horizon using convolutional neural networks, using data until 2018-8-30 23:45:00. It is performed for one day (day ahead forecast) and more details able it is not provided since it is out of scope.

Load flow solution is non-causal in nature and to generate an optimal schedule for controllable buses, it must be run for all 96 intervals. OPF is performed using Class C algorithm presented in Section 3 for controllable buses described in Section 4. It can also be observed that the optimal schedules are generated for all the three phases and can take both positive and negative values. Real profile is recorded during day for uncontrollable loads at the buses.

Forecasted, optimal and real active and reactive power consumption profiles at one of the controllable buses (Bus 15) can be seen in Figures 3 and 4 respectively. At Bus 15, all the loads are single phased (connected to phase C).

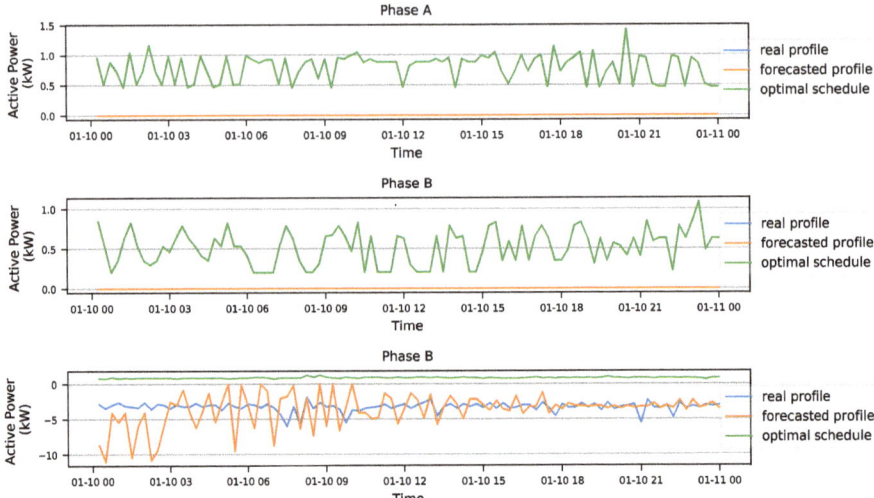

Figure 3. Active power of real, forecasted, and optimal profiles at Bus 15. It can be observed that the real and forecasted data is zero for phases A and B. This is for to the fact that the loads are single phased and connected only to phase C. During the OPF, they are replaced with three-phase controllable loads. On the x-axis, data time format is MM-dd HH. Data is from 2018-8-31 00:00:00 to 2018-9-01 00:00:00.

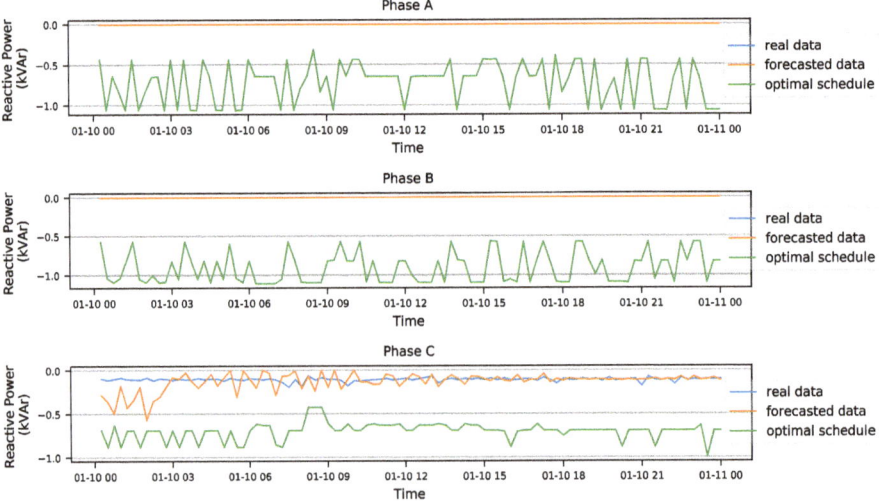

Figure 4. Reactive Power of real, forecasted, and optimal schedules at Bus 15. On the x-axis, data time format is MM-dd HH. Data is from 2018-8-31 00:00:00 to 2018-9-01 00:00:00.

Active and reactive power for all the phases can be observed in Figures 3 and 4.

Using the three schedules shown in Figures 3 and 4, load flows are performed using THELM described in Section 2. Loads flows are performed for all intervals and are represented using box-plots.

Figure 5 describes the averaged objective function values based on Equation (18). It can be observed that the three-phase unbalance has been reduced from 0.879 for real and forecasted profiles to 0.529 for optimal profiles which accounts for 39% unbalance minimization based on the defined objective function (see Section 4).

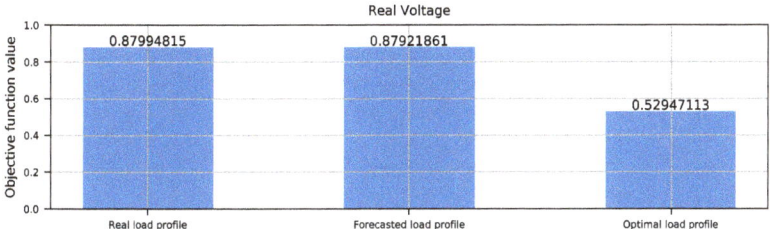

Figure 5. Average values of optimal power flow objective for real, forecasted, optimal voltage and phase angles based on Equation (18).

From Figure 6, it can be observed that the voltages are indeed balanced, and the average values are close to balanced voltages. Additionally, the nature of the objective function used has also caused the voltages to cluster around 1 pu. since the balanced real part of the balanced voltage is exactly 1 pu.

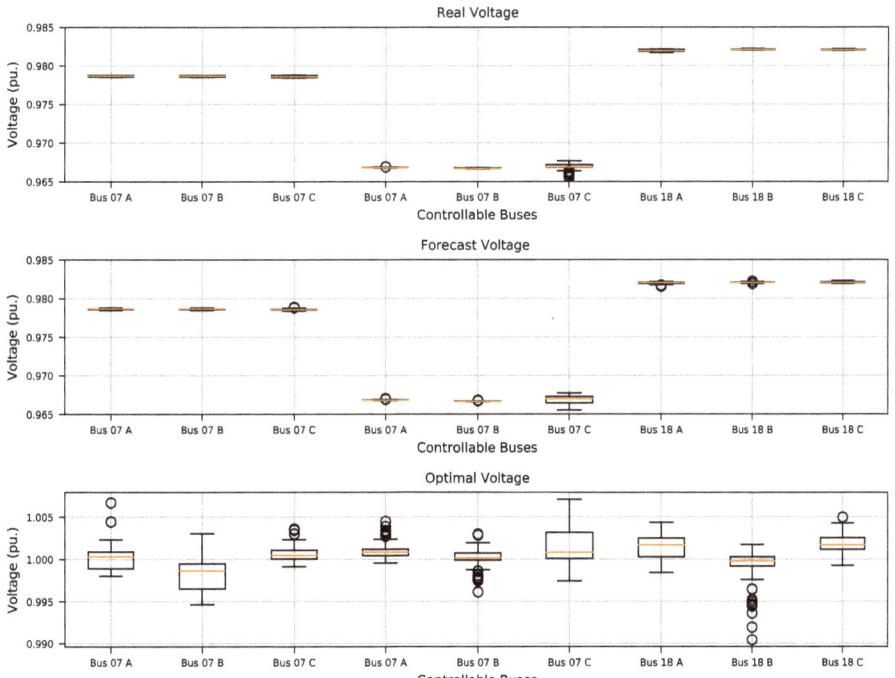

Figure 6. Voltage profiles generated from real, forecasted and optimal schedules from Figures 3 and 4.

6. Conclusions and Outlook

In this paper, a novel class of OPF algorithm is presented. It uses a novel three-phase unbalanced HELM presented in Section 2. Benchmarks are performed to test the performance of THELM and DIgSILENT Power Factory Newton-Raphson method described in Section 5.1. These benchmarks were performed on various test networks. Mann-Whitney U test was performed, and it was concluded that the results from both load flow methods were statistically indistinguishable and null hypothesis was accepted. Using THELM, optimal power flow method was developed using genetic algorithm in Section 3, describing the type C class of algorithms. The novel Class C algorithm provides various advantages over Class A and B OPF algorithms as discussed in Section 1. A use case with an objective function to minimize three-phase unbalance was applied to a real network in Austria in Section 5. The reason for choosing this objective is motivated by the requirements of the network operator and to handle the unbalance locally. It involves the generation of active and reactive power schedules for four controllable buses using smart-meter forecasts from other uncontrollable loads in the network (see Figure 1). Optimal schedules for these buses were generated and used to produce voltages using THELM and the results were described in Figure 6. It can be observed that the three-phase voltage unbalance has reduced up to 39% and the optimal average objective function values can be observed in Figure 5.

In future work, the scalability and replicability of the method needs to be analyzed. The method needs to be applied to various larger networks with large number of nodes. Simulation time and code optimization is not considered a priority for this study. To use this method in a real-time or near-real-time operation, the algorithm needs to be optimized. In this work, only three-phase unbalance minimization is used. OPF with various other objective functions need to be considered.

Author Contributions: Conceptualization, B.V.R., F.K. and M.K.; Formal analysis, B.V.R. and M.K.; Investigation, B.V.R.; Methodology, B.V.R., F.K. and M.K.; Supervision, F.K. and M.K.; Validation, B.V.R. and F.K.; Visualization, B.V.R. and F.K.

Funding: This research received no external funding.

Conflicts of Interest: The authors declare no conflict of interest.

Appendix A. Test Networks

Various test networks used for the analysis described in Section 5.1.

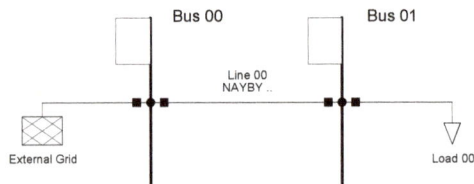

Figure A1. Test 00.

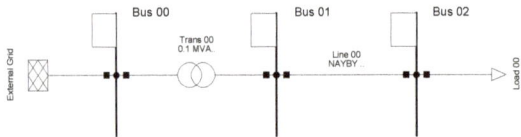

Figure A2. Test 01.

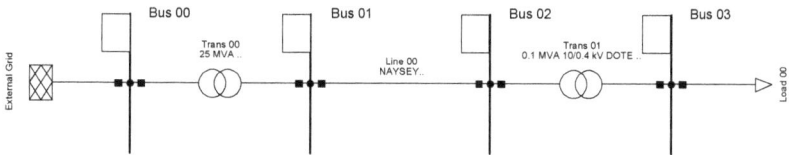

Figure A3. Test 02.

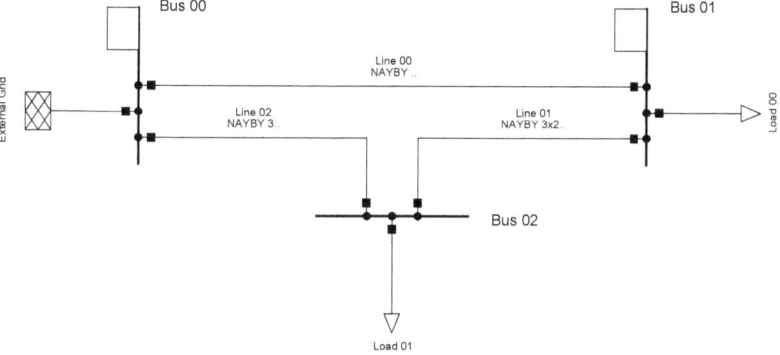

Figure A4. Test 03.

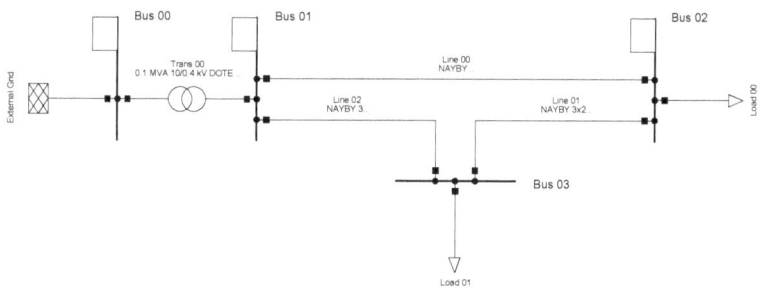

Figure A5. Test 04.

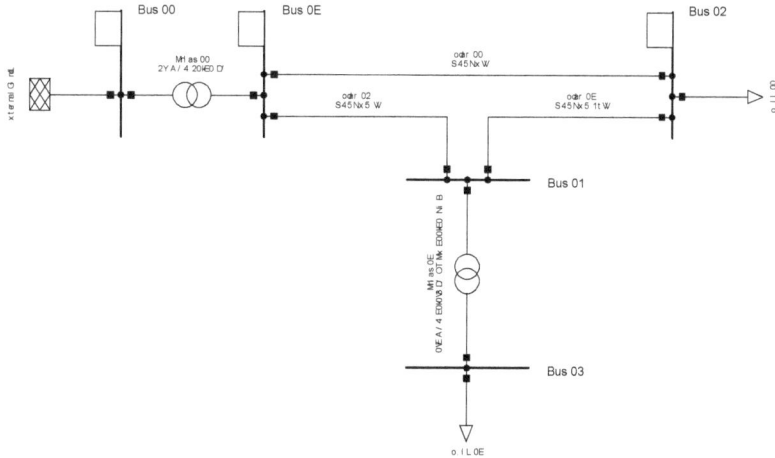

Figure A6. Test 05.

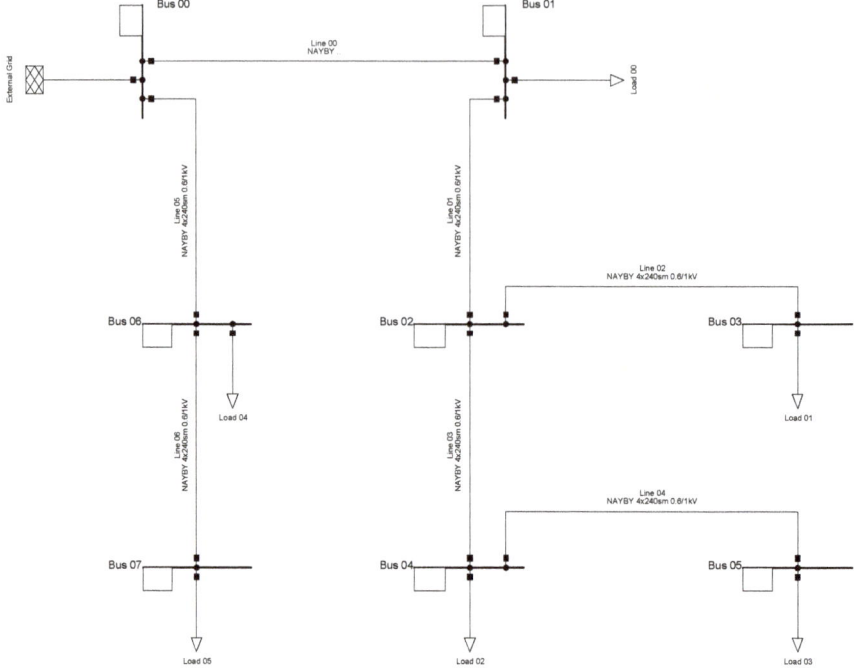

Figure A7. Test 06.

References

1. Serițan, G.; Porumb, R.; Cepișcă, C.; Grigorescu, S. Integration of Dispersed Power Generation. In *Electricity Distribution: Intelligent Solutions for Electricity Transmission and Distribution Networks*; Energy Systems; Karampelas, P., Ekonomou, L., Eds.; Springer: Berlin/Heidelberg, Germany, 2016; pp. 27–61. [CrossRef]
2. Campos, F.; Marques, L.; Silva, N.; Melo, F.; Seca, L.; Gouveia, C.; Madureira, A.; Pereira, J. ADMS4LV #8211; Advanced Distribution Management System for Active Management of LV Grids. *CIRED Open Access Proc. J.* **2017**, *2017*, 920–923. [CrossRef]
3. Fan, J.; Borlase, S. The Evolution of Distribution. *IEEE Power Energy Mag.* **2009**, *7*, 63–68. [CrossRef]
4. Horowitz, S.H.; Phadke, A.G. Boosting Immunity to Blackouts. *IEEE Power Energy Mag.* **2003**, *1*, 47–53. [CrossRef]
5. Novosel, D.; Begovic, M.M.; Madani, V. Shedding Light on Blackouts. *IEEE Power Energy Mag.* **2004**, *2*, 32–43. [CrossRef]
6. Taylor, C.W.; Erickson, D.C.; Martin, K.E.; Wilson, R.E.; Venkatasubramanian, V. WACS-Wide-Area Stability and Voltage Control System: R D and Online Demonstration. *Proc. IEEE* **2005**, *93*, 892–906. [CrossRef]
7. Ilic, M.D.; Allen, H.; Chapman, W.; King, C.A.; Lang, J.H.; Litvinov, E. Preventing Future Blackouts by Means of Enhanced Electric Power Systems Control: From Complexity to Order. *Proc. IEEE* **2005**, *93*, 1920–1941. [CrossRef]
8. Santo, M.D.; Vaccaro, A.; Villacci, D.; Zimeo, E. A Distributed Architecture for Online Power Systems Security Analysis. *IEEE Trans. Ind. Electron.* **2004**, *51*, 1238–1248. [CrossRef]
9. Dommel, H.W.; Tinney, W.F. Optimal Power Flow Solutions. *IEEE Trans. Power Appar. Syst.* **1968**, *PAS-87*, 1866–1876. [CrossRef]
10. Sasson, A.M. Combined Use of the Powell and Fletcher—Powell Nonlinear Programming Methods for Optimal Load Flows. *IEEE Trans. Power Appar. Syst.* **1969**, *PAS-88*, 1530–1537. [CrossRef]
11. El-abiad, A.H.; Jaimes, F.J. A Method for Optimum Scheduling of Power and Voltage Magnitude. *IEEE Trans. Power Appar. Syst.* **1969**, *PAS-88*, 413–422. [CrossRef]

12. Peschon, J.; Bree, D.W.; Hajdu, L.P. Optimal Power-Flow Solutions for Power System Planning. *Proc. IEEE* **1972**, *60*, 64–70. [CrossRef]
13. Sasson, A.M.; Viloria, F.; Aboytes, F. Optimal Load Flow Solution Using the Hessian Matrix. *IEEE Trans. Power Appar. Syst.* **1973**, *PAS-92*, 31–41. [CrossRef]
14. Rashed, A.M.H.; Kelly, D.H. Optimal Load Flow Solution Using Lagrangian Multipliers and the Hessian Matrix. *IEEE Trans. Power Appar. Syst.* **1974**, *PAS-93*, 1292–1297. [CrossRef]
15. Alsac, O.; Stott, B. Optimal Load Flow with Steady-State Security. *IEEE Trans. Power Appar. Syst.* **1974**, *PAS-93*, 745–751. [CrossRef]
16. Happ, H.H. Optimal Power Dispatch. *IEEE Trans. Power Appar. Syst.* **1974**, *PAS-93*, 820–830. [CrossRef]
17. Mukherjee, P.K.; Dhar, R.N. Optimal Load-Flow Solution by Reduced-Gradient Method. *Proc. Inst. Electr. Eng.* **1974**, *121*, 481–487. [CrossRef]
18. Bala, J.L.; Thanikachalam, A. An Improved Second Order Method for Optimal Load Flow. *IEEE Trans. Power Appar. Syst.* **1978**, *PAS-97*, 1239–1244. [CrossRef]
19. Lipowski, J.S.; Charalambous, C. Solution of Optimal Load Flow Problem by Modified Recursive Quadratic-Programming Method. *Transm. Distrib. IEE Proc. C Gener.* **1981**, *128*, 288–294. [CrossRef]
20. Burchett, R.C.; Happ, H.H.; Wirgau, K.A. Large Scale Optimal Power Flow. *IEEE Trans. Power Appar. Syst.* **1982**, *PAS-101*, 3722–3732. [CrossRef]
21. Glavitsch, H.; Bacher, R. Optimal Power Flow Algorithms. *Control Dyn. Syst.* **1991**, *41*, 135–205. [CrossRef]
22. Stott, B.; Hobson, E. Power System Security Control Calculations Using Linear Programming, Part I. *IEEE Trans. Power Appar. Syst.* **1978**, *PAS-97*, 1713–1720. [CrossRef]
23. Stott, B.; Hobson, E. Power System Security Control Calculations Using Linear Programming, Part II. *IEEE Trans. Power Appar. Syst.* **1978**, *PAS-97*, 1721–1731. [CrossRef]
24. Trias, A. The Holomorphic Embedding Load Flow Method. In Proceedings of the 2012 IEEE Power and Energy Society General Meeting, San Diego, CA, USA, 22–26 July 2012; pp. 1–8. [CrossRef]
25. Home—DIgSILENT. Available online: https://www.digsilent.de/en/ (accessed on 1 February 2019).
26. Ward, J.B.; Hale, H.W. Digital Computer Solution of Power-Flow Problems [Includes Discussion]. *Trans. Am. Inst. Electr. Eng. Part III Power Appar. Syst.* **1956**, *75*, 398–404. [CrossRef]
27. Tinney, W.F.; Hart, C.E. Power Flow Solution by Newton's Method. *IEEE Trans. Power Appar. Syst.* **1967**, *PAS-86*, 1449–1460. [CrossRef]
28. Tinney, W.F.; Walker, J.W. Direct Solutions of Sparse Network Equations by Optimally Ordered Triangular Factorization. *Proc. IEEE* **1967**, *55*, 1801–1809. [CrossRef]
29. van Amerongen, R.A.M. A General-Purpose Version of the Fast Decoupled Load Flow. *IEEE Trans. Power Syst.* **1989**, *4*, 760–770. [CrossRef]
30. Stott, B. Effective Starting Process for Newton-Raphson Load Flows. *Proc. Inst. Electr. Eng.* **1971**, *118*, 983–987. [CrossRef]
31. Tripathy, S.C.; Prasad, G.D.; Malik, O.P.; Hope, G.S. Load-Flow Solutions for Ill-Conditioned Power Systems by a Newton-Like Method. *IEEE Power Eng. Rev.* **1982**, *PER-2*, 25–26. [CrossRef]
32. Schaffer, M.D.; Tylavsky, D.J. A Nondiverging Polar-Form Newton-Based Power Flow. *IEEE Trans. Ind. Appl.* **1988**, *24*, 870–877. [CrossRef]
33. Tylavsky, D.J.; Crouch, P.E.; Jarriel, L.F.; Chen, H. Advances in Fast Power Flow Algorithms. *Control Dyn. Syst.* **1991**, *44*, 295–343. [CrossRef]
34. Crouch, P.E.; Tylavsky, D.J.; Chen, H.; Jarriel, L.; Adapa, R. Critically Coupled Algorithms for Solving the Power Flow Equation. *IEEE Trans. Power Syst.* **1992**, *7*, 451–457. [CrossRef]
35. Tylavsky, D.J.; Crouch, P.E.; Jarriel, L.F.; Singh, J.; Adapa, R. The Effects of Precision and Small Impedance Branches on Power Flow Robustness. *IEEE Trans. Power Syst.* **1994**, *9*, 6–14. [CrossRef]
36. Tylavsky, D.J.; Schaffer, M.D. A Nondiverging Power Flow Using a Least-Power-Type Theorem. *IEEE Trans. Ind. Appl.* **1987**, *IA-23*, 944–951. [CrossRef]
37. Subramanian, M.K.; Feng, Y.; Tylavsky, D. PV Bus Modeling in a Holomorphically Embedded Power-Flow Formulation. In Proceedings of the 2013 North American Power Symposium (NAPS), Manhattan, KS, USA, 22–24 September 2013. [CrossRef]
38. Wallace, I.; Roberts, D.; Grothey, A.; McKinnon, K.I.M. Alternative PV Bus Modelling with the Holomorphic Embedding Load Flow Method. *arXiv* **2016**, arXiv:1607.00163.

39. Bazrafshan, M.; Gatsis, N. Comprehensive Modeling of Three-Phase Distribution Systems via the Bus Admittance Matrix. *IEEE Trans. Power Syst.* **2018**, *33*, 2015–2029. [CrossRef]
40. Iba, K. Reactive Power Optimization by Genetic Algorithm. *IEEE Trans. Power Syst.* **1994**, *9*, 685–692. [CrossRef]
41. Lai, L.L.; Ma, J.T. Power Flow Control with UPFC Using Genetic Algorithms. In Proceedings of the International Conference on Intelligent System Application to Power Systems, Orlando, FL, USA, 28 January–2 February 1996; pp. 373–377. [CrossRef]
42. Urdaneta, A.J.; Gomez, J.F.; Sorrentino, E.; Flores, L.; Diaz, R. A Hybrid Genetic Algorithm for Optimal Reactive Power Planning Based upon Successive Linear Programming. *IEEE Trans. Power Syst.* **1999**, *14*, 1292–1298. [CrossRef]
43. Lin, W.M.; Cheng, F.S.; Tsay, M.T. Distribution Feeder Reconfiguration with Refined Genetic Algorithm. *Transm. Distrib. IEE Proc. Gener.* **2000**, *147*, 349–354. [CrossRef]
44. Das, D.B.; Patvardhan, C. Useful Multi-Objective Hybrid Evolutionary Approach to Optimal Power Flow. *Transm. Distrib. IEE Proc. Gener.* **2003**, *150*, 275–282. [CrossRef]
45. Masoum, M.A.S.; Ladjevardi, M.; Jafarian, A.; Fuchs, E.F. Optimal Placement, Replacement and Sizing of Capacitor Banks in Distorted Distribution Networks by Genetic Algorithms. *IEEE Trans. Power Deliv.* **2004**, *19*, 1794–1801. [CrossRef]
46. Back, T.; Fogel, D.B.; Michalewicz, Z. (Eds.) *Basic Algorithms and Operators*, 1st ed.; IOP Publishing Ltd.: Bristol, UK, 1999.
47. Korovkin, N.V.; Vu, Q.S.; Yazenin, R.A. A Method for Minimization of Unbalanced Mode in Three-Phase Power Systems. In Proceedings of the 2016 IEEE NW Russia Young Researchers in Electrical and Electronic Engineering Conference (EIConRusNW), St. Petersburg, Russia, 2–3 February 2016; pp. 611–614. [CrossRef]
48. Fernandes, C.M.M. Unbalance between Phases and Joule's Losses in Low Voltage Electric Power Distribution Networks. p. 9. Available online: https://fenix.tecnico.ulisboa.pt/downloadFile/395142112117/Resumo%20Alargado%20Carlos%20Fernandes.pdf (accessed on 15 February 2019).
49. Fernandez, J.; Bacha, S.; Riu, D.; Turker, H.; Paupert, M. Current Unbalance Reduction in Three-Phase Systems Using Single Phase PHEV Chargers. In Proceedings of the 2013 IEEE International Conference on Industrial Technology (ICIT), Cape Town, South Africa, 25–28 February 2013; pp. 1940–1945. [CrossRef]

© 2019 by the authors. Licensee MDPI, Basel, Switzerland. This article is an open access article distributed under the terms and conditions of the Creative Commons Attribution (CC BY) license (http://creativecommons.org/licenses/by/4.0/).

Article

Smart Grid R&D Planning Based on Patent Analysis

Jason Jihoon Ree and Kwangsoo Kim *

Department of Industrial and Management Engineering, Pohang University of Science and Technology, Pohang 37673, Korea; jjree@postech.ac.kr or jasonree32@gmail.com
* Correspondence: kskim@postech.ac.kr

Received: 29 March 2019; Accepted: 19 May 2019; Published: 22 May 2019

Abstract: A smart grid employs information and communications technology to improve the efficiency, reliability, economics, and sustainability of electricity production and distribution. The convergent and complex nature of a smart grid and the multifarious connection between its individual technology components, as well as competition between private companies, which will exert substantial influences on the future smart grid business, make a strategic approach necessary from the beginning of research and development (R&D) planning with collaborations among various research groups and from national, industry, company, and detailed technological levels. However, the strategic, technological, business environmental, and regulatory barriers between various stakeholders with collaborative or sometimes conflictive interests need to be clarified for a breakthrough in the smart grid field. A strategic R&D planning process was developed in this study to accomplish the complicated tasks, which comprises five steps: (i) background research of smart grid industry; (ii) selection of R&D target; (iii) societal, technological, economical, environmental, and political (STEEP) analysis to obtain a macro-level perspective and insight for achieving the selected R&D target; (iv) patent analysis to explore capabilities of the R&D target and to select the entry direction for smart grid industry; and, (v) nine windows and scenario planning analyses to develop a method and process in establishing a future strategic R&D plan. This R&D planning process was further applied to the case of a Korean company holding technological capabilities in the sustainable smart grid domain, as well as in the sustainable electric vehicle charging system, a global consumer market of smart grid. Four plausible scenarios were produced by varying key change agents for the results of this process, such as technology and growth rates, policies and government subsidies, and system standards of the smart grid charging system: Scenario 1, 'The Stabilized Settlement of the Smart Grid Industry'; Scenario 2, 'The Short-lived Blue Ocean of the Smart Grid Industry'; Scenario 3, 'The Questionable Market of the Smart Grid'; and, Scenario 4, 'The Stalemate of the Smart Grid Industry'. The R&D plan suggestions were arranged for each scenario and detailed ways to cope with dissonant situations were also implied for the company. In sum, in this case study, a future strategic R&D plan was suggested in regard to the electric vehicle charging technology business, which includes smart grid communication system, battery charging duration, service infrastructure, public charge station system, platform and module, wireless charging, data management system, and electric system solution. The strategic R&D planning process of this study can be applicable in various technologies and business fields, because of no inherent dependency on particular subject, like electric vehicle charging technology based on smart grid.

Keywords: R&D planning; patent analysis; sustainable smart grid technology; R&D strategy; STEEP analysis; scenario planning; electric vehicle charging technology

1. Introduction

Smart grid is an electrical grid that uses information and communications technology to gather information regarding the behaviors of suppliers and consumers in an automated fashion to improve the efficiency, reliability, economics, and sustainability of electricity production and distribution [1–12]. Through the construction of smart grid, the bi-directional transfer of electrical information allows for increases in rational energy consumption and provides high quality energy and various supplementary services. Furthermore, new business opportunities for green energy and electric automobiles are expected to be established through the convergence of sustainable green technologies. Accordingly, technologically-developed countries, such as the United States of America (USA), Japan, China, and European countries, have established smart grid agencies to fund research on smart grid [1–4,13–17].

For the purposes of a convergent and complex system, like smart grid, a strategic approach is necessary from the beginning of the research planning phase. Investment and research in various fields are essential due to the smart grid being a large-scale system on the national level. Additionally, collaborations in research are mandatory, because the connection between the different technologies is significantly sizable. While the current linear economy utilizes resources to manufacture products and disposes of them after consumption, the smart grid industry also meticulously aligns to the recent growth in interests of the circular economy framework, which aims to achieve a desired zero-waste society [18] through reusable, restorative and regenerative measures of wastes to retain as much value in producing new and renewed products or services [19–21]. Development in smart grid optimization is imperative due to its potential and innate aptitude to improve both the economic and environmental positions in support of the circular economy concept [22–24]. Furthermore, turbulent competition between numerous companies can exert substantial influences on the future smart grid business. Research and development (R&D) planning for smart grid must be established in detail and ensued from national, industry, and company levels due to such diverse reasons.

Currently, strategic planning and research investments are being conducted in regard to smart grid R&D on a national level [1–4,13–17]. However, such strategic R&D planning exclusively depends on groups of experts from various technology disciplines pertaining to the smart grid. Through heavy reliance on expert opinions, a problem arises, as the research plan can be susceptible to skewed biases [25,26]. For example, smart grid government funding can be misallocated toward certain technologies in favor of the interests of these experts for personal or political benefits, rather than for the purposes of advancements of a smart grid. Thus, a systematic and methodological approach is vital to support expert opinions and further increase the legitimacy of the R&D planning process. Through systematic R&D planning, rapid changes in technology, society, and environment can be coped with, both dexterously and continuously.

In this study, a R&D planning process was developed based on patent analysis combined with scenario planning analyses and nine windows from the Theory of Inventive Problem Solving (Russian acronym: TRIZ) [27] in order to construct decisive and strategic suggestions for the competitiveness of the smart grid R&D business. With the developed process, strategic R&D planning was performed on a smart grid company in conjunction to electric vehicle charging technology as a case study. Furthermore, this study has offered a proposition that can cover: (i) short-term strategies utilizing current technologies or technological shortcomings and (ii) intermediate-term and long-term strategies employing future prediction for new business opportunities. The resulting strategies of this study also stretch beyond the classical reliance on the perspectives of experts by using a methodology-based approach, which can provide new, useful, and generally overlooked insights and advices. In addition, the R&D planning process that is proposed in this study is not strictly tailored only for the purposes of smart grid, and it thus may be malleable and suitable for usages in other business R&D planning and strategizing activities for company R&D managers, technology policy and decision makers, and innovative individuals.

The rest of this paper consists of the sections, as follows: Section 2 describes research background, Section 3 depicts strategic R&D planning process, Section 4 portrays the application of the proposed method in a case study of Smart Grid R&D Planning for Hyosung Corporation, and Section 5 describes the conclusion.

2. Research Background

2.1. Strategic R&D Planning

In general, one needs to establish strategic R&D planning ahead of discussing its methodology. R&D refers to "creative work that is undertaken on a systematic basis in order to increase the stock of knowledge, including knowledge of man, culture, and society, and the use of knowledge to devise new applications" [28]. R&D can be frameworked in a systematic manner, as shown in Figure 1. The evolution of R&D can be classified into several generations, as follows. The first generation of R&D comprises of technologies that are developed through the plethora of investments of past research, which originated from the intellectual curiosity of researchers of various disciplines. The R&D driven by marketing for the purposes of marketability categorizes the second generation. As a result, a trend of cooperation between the R&D and marketing departments and the individual projects for new business starts to develop. The third generation is progressed from R&D projects that are instigated via the strategies of enterprise corporation. Currently, the majority of global companies are stressing the importance and practice of the third generation R&D efforts. Since the beginning of the 1990s, in particular, as companies' R&D activities have been combined with the enterprise corporation strategy, establishing the direction of R&D investment according to strategic goals and markets has become more essential. Finally, the fourth generation extends beyond the third generation R&D by forecasting the future and reconfiguring the business strategy, which accentuates and centers around the R&D strategy.

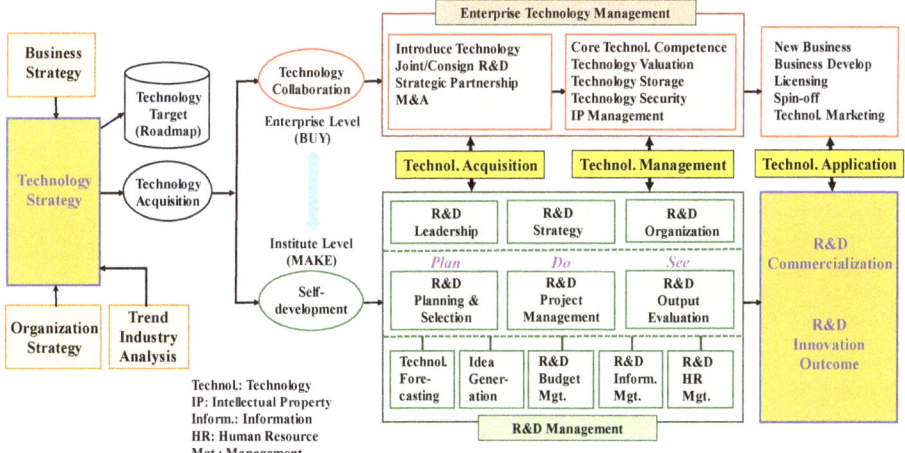

Figure 1. Overview of research and development (R&D) Framework.

2.2. Patent Analysis

Patents are reports of the latest information, which reliably reflects contemporary technological developments and changes. Thus, patent analysis is recognized as an essential process that is required in all assessments of technology trends and technology planning activities [29]. Furthermore, information that was obtained from patent analysis has been utilized in a variety of instances with the advancements of R&D, in particular, for strategic R&D planning.

Strategic R&D planning may be categorized into three stages [30]. As illustrated in Table 1, the first stage of R&D planning is limited to simply patenting the developed technologies without the consideration or application of available patent information. Typically starting from the second stage, patent information is used strategically and, however, the second stage of R&D strategies does not extend beyond an investigation of the prior similar technologies and the defensive strategies regarding patent infringement. Beginning from the third stage of R&D planning, patent information is deemed to be significant and is extensively utilized to its overall potential by comparison to the activities of the first and second stage R&D planning. The patent information is manipulated for various purposes, including merger and acquisition (M&A) technology procurement, new business and product planning, patent valuation, patent licensing, and patent portfolios. Furthermore, the use of patent information for such purposes has become the core trend for the R&D planning activities of technology-based companies.

Table 1. Utilization of patent analysis for each stage of R&D planning.

1st Stage	2nd Stage	3rd Stage
No Strategy	Defensive Strategy	Offensive Strategy
Data, Information	Information, Knowledge	Knowledge, Intelligence

Patent analysis can be divided into qualitative and quantitative patent analyses. Qualitative patent analysis mainly focuses on the technological information of patents in order to discover the core patents and assess technological trends. In addition, patent citation analysis, patent network analysis, and patent map analysis are conducted in qualitative patent analysis to extensively survey not only patents as individual documents, but also as connected entities, for a deeper perception of the technological domain. On the other hand, quantitative patent analysis generally involves the organization and illustration of bibliographical and technological data into graphical format. Through such quantitative analysis, the industry and domain trends of inventors, priority, nation, assignees, and citations can be apprehended.

2.3. TRIZ: The 9 Windows Analysis

Generalizing technologies after the extensive analysis of 40,000 patents developed the Theory of Inventive Problem Solving, or Teoriya Reshniya Izobretatelskikh Zadatch (TRIZ) [27]. Altshuller noted that technology has followed certain patterns and rules in creating new inventive patentable ideas [31]. TRIZ is a useful tool for analyzing technology and countless researchers from various disciplines have applied and validated it. Additionally, TRIZ techniques have been used in the case of technology forecasting and R&D planning activities [32–34].

The problem solving process of TRIZ, as illustrated in Figure 2a, converts a specific problem into one of TRIZ's generic problems, which then is used to apply a generic solution. With a given generic solution, this process is able to find a specific solution to the original specific problem. At a glance, this problem solving process seems to be exceptionally elementary; however, innumerable complex TRIZ tools are being developed to reveal solutions to problems, and substantial research is continuously required to acquire the ability to convert unique problems into TRIZ's general problems. Despite such lingering difficulties, TRIZ remains as a dominant source in coping with problems that involves particular aspects of technology, as well as diverse business problems at a company level.

Other than TRIZ's widely known techniques, such as the 40 Inventive Principles, 76 Standard Solutions and Algorithm for Inventive Problem Solving, or Algoritm Resheniya Izobretatelskikh Zadatch (ARIZ), process, there exists a vast collection of methodologies that have been developed. These methodologies include Technology Evolution Patterns [35–37], Ideal Final Result technique [38], and nine windows technique [39], and more. The 40 Inventive Principles of Problem Solving method solves the technical and physical contradictions [40] and it is "closer to application than the laws of technological systems evolution" [41]. The 76 Standard Solutions help to solve system problems without identifying contradictions and they are applied to alleviate unwanted exchanges between two parts of a system [42], and ARIZ is a logical structure aimed to simplify a complex problem to be easier to solve [43].

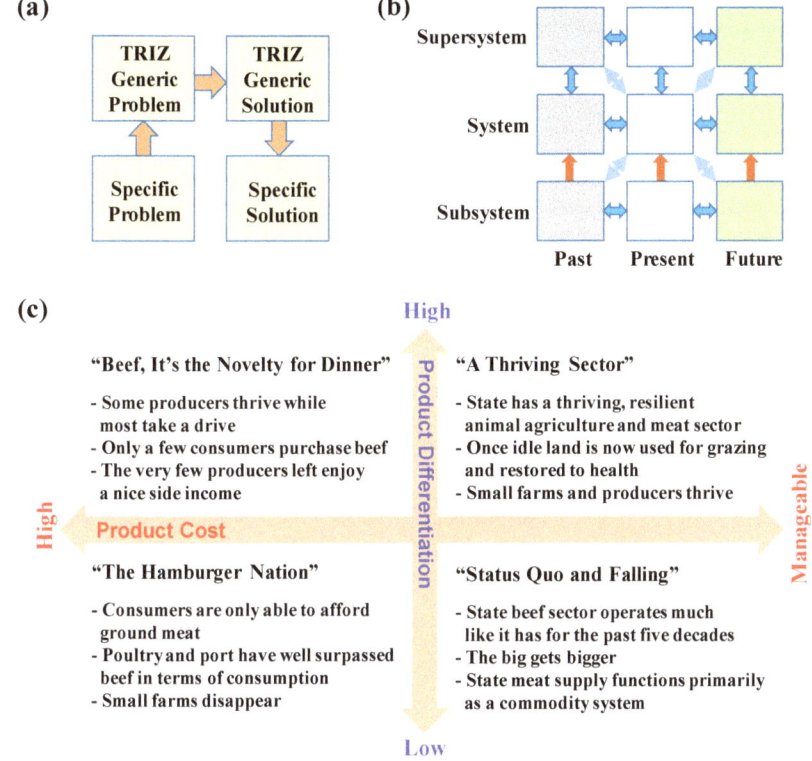

Figure 2. (**a**) Theory of Inventive Problem Solving, or Teoriya Reshniya Izobretatelskikh Zadatch (TRIZ) analysis process; (**b**) Nine Windows analysis; and, (**c**) Example of scenario planning.

While such TRIZ techniques mainly focus on solving technical and physical problems for new product/process development, the nine windows technique was selected and adopted due to its uniqueness in describing how the target domain (system) and its super and sub systems evolve over time to generate a holistic view of the smart grid industry [32]. The nine windows analysis is a multi-dimensional technique that can aid in the prediction of the future by dividing the past, present, and future into sub-system, system, and super-system [39]. By examining the sub-system, system, and super-system of the present and the past of a technology, the corresponding three levels of the future can be more accurately comprehended (Figure 2b). What differentiates this method from other future foresight methods is the relevant utilization of the sub- and super-systems to predict the future in a structural manner.

2.4. Scenario Planning

Scenario planning is a method that is used to deal with uncertainties in the future business environment [44–49]; Figure 2c illustrates a schematic example. The scenario perspective acknowledges the inability to ideally predict the future, and thus prepares for the uncertainties of the future. More specifically, scenario planning primes for the future through the attainment of planned solutions to each feasible and meaningful scenario, which is created with future uncertainty as its foundation. The advantages of scenario planning are as follows:

(i) future business opportunities are investigated via scenarios;
(ii) plans are established by modification according to various different scenarios;

(iii) a clear direction of what capabilities are required in the future is suggested appropriately;
(iv) current established plans can be tested for suitability under future circumstances; and,
(v) crises and emergencies can be dealt with beforehand.

3. Strategic R&D Planning Process

3.1. Smart Grid Strategic R&D Planning Process

In this study, a strategic R&D planning process catered toward smart grid industry has been developed and it is shown in Figure 3.

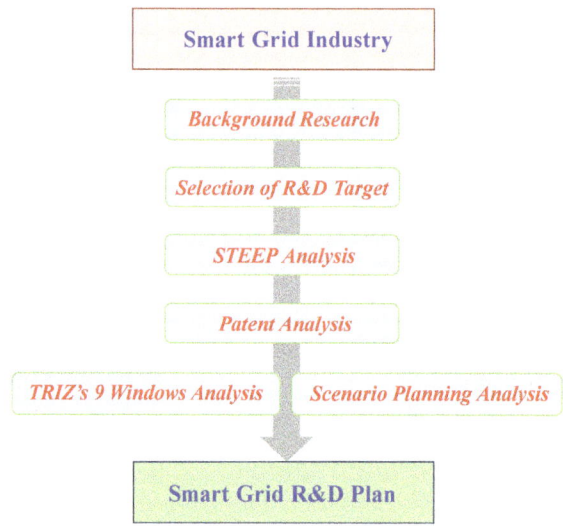

Figure 3. Strategic R&D planning process for Smart Grid.

First, thorough background research was performed to establish a strong foundation regarding the smart grid industry. To achieve this, interviews with domain experts and policy makers were conducted in parallel to obtain a concrete understanding of research trends in smart grid, in addition to surveying smart grid related text books, reports, magazines, academic literatures, and internet documents. Furthermore, the perspective target of research has been selected. The selection of such a perspective target of research is essential, because the methods and details of the analysis are distinguishable, depending on whether the perspective target is a company, industry, or nation. This study pinpoints a company as the target and its association with the smart grid industry. Since the smart grid industry is quite vast with an eclectic array of different technologies and domains, this study further focuses on one of these for an in-depth analysis, rather than a superficial overview of the entire industry containing little significance and credibility. This study also utilizes the societal, technological, economical, environmental, and political (STEEP) analysis for the development probability of smart grid industry, because the analysis is a suitable methodology that yields macro-level insights that are useful for the direction in the beginning phase of research planning. The STEEP analysis pinpoints key elements that induce significant influences on the smart grid industry (Table 2) [48].

Table 2. Societal, technological, economical, environmental, and political (STEEP) analysis results of smart grid industry.

STEEP	Smart Grid Drivers
Environmental	• Reduce green-house gas emissions • Support and lower cost of Kyoto Compliance • Increase renewable energy generation • Improve energy efficiency
Regulatory/Governmental	• Respond to governmental mandates (i.e., UK, Netherlands, Sweden, US, Australia, Brazil, etc.) • Support building renewable/distributed generation • Support performance-based rate making • Enable national security objectives • Facilitate wholesale and retail market efficiency
Economic	• Improve reliability (e.g., decreased outage duration and frequency, etc.) • Reduce labor costs (e.g., meter reading and field maintenance, etc.) • Reduce system losses (e.g., better systems planning and asset management, etc.) • Reduce non-labor costs (e.g., field vehicle, insurance and damage, etc.) • Provide revenue protection (e.g., better billing accuracy, prevention/detection of theft/fraud, etc.) • Provide new revenue sources (e.g., energy management, etc.) • Deferral of large capital projects (e.g., reduced generation requirements, etc.)
Societal	• Respond to consumer demand for sustainability and enhanced ability to manage their usage • Respond to customers increasing demand for uninterruptible power • Provide customer options for energy conservation
Technology & Infrastructure	• Invest significant capital for development of new technology (improvement of existing technologies) • Replace outdated T&D infrastructure • Decreasing workforce (e.g., 50% of technical workforce are expected to reach retirement age in 5–10 years)

T&D: transmission and distribution.

Subsequently, starting with the target company's patents performs patent analysis; them, the company's capabilities can be examined. In this case, the company's capabilities are evaluated in regard to the smart grid industry by exploring the company's patents, past developments and businesses, and changes in focus of areas and investments. With the company's capability information, patent analysis is conducted again to shed light upon the specific areas in which the company can expect business opportunities. Through the analysis of the specific areas within the smart grid industry, robust knowledge can be obtained to further investigate industry growth, development trends, core technologies and technology trends, and the insight of market competitors, thus offering a foundation for strategic R&D planning activities.

Finally, extending beyond the analyses of the target technology, future strategies can be established by using TRIZ's nine windows and scenario planning methodologies. Through the nine windows methodology, the smart grid's current system and super-system can be analyzed to chronologically draw a conclusion and prediction of the future's super-system and system. In detail, the current system's shortcomings (problems) are identified. After an evaluation of the specific problems at the current system level, the problems of the current super-system are established. The most probable solutions or portrayal of the future in accordance to the current super-system is formulated in the future super-system. Finally, under the assumptions that the future super-system holds true, specific future solutions at the system level are created. For example, while assuming that the current problem (current system) is the subpar or lack of internet access on commercial airplanes, the current super-system problem is the lack of long-range tower or satellite infrastructure and the reliability of signal receiving antenna technology. The future super-system solution may become the advancement and dominance of satellite-provided internet connection. Assuming this development in satellite-provided internet, the R&D plan would consequently suggest developing antennas that are specialized for satellite connectivity as the solution.

The scenario planning is composed of six step processes, as follows. (i) Selection of core issues: Regarding what affairs, will decisions be made? (ii) Extraction of decision-making factors: What is required to be able to make decisions? (iii) Investigation of the motive for change: What is the significant cause for change? (iv) Abstraction of the scenario: What/which scenario is significant? (v) Dictation of the scenario: How will the future unfold? (vi) Establishment of counterstrategy: How will the future be coped with? The resulting future system is utilized for R&D planning. In addition, the uncertainties of future can be dealt with the scenario planning methodology through the creation of multiple scenarios that are based on key change agents. A robust smart grid strategy can be established as a result of these processes.

3.2. The Smart Grid Industry

The future of the energy industry is rapidly expanding towards an efficient and reliable smart grid infrastructure. The smart grid industry and technologies can provide a variety of benefits, not only for developed countries, but also for developing countries. The smart grid can: (i) increase the energy efficiency; (ii) decrease the losses from the electrical grid; (iii) improve system performance and asset utilization; (iv) combine the sources of renewable energy; and, (v) actively cope with energy demands. Particularly, since the 2000s, the implementation of smart grid has received much attention [1–4], especially after immense capital losses that result from frequent major blackouts and their consequent damages [6]. From the development of local microgrids [6,10,50] to country-level case studies and R&D funding initiatives [13,17], significant strides toward smart grid planning and development have undergone thorough discussion, evaluation, and support, both nationally and internationally [14,15,51,52]. Furthermore, in support of the growing interests in the circular economy concept, the smart grid possesses prodigious capacity to accommodate efficient, renewable, and reusable energy solutions to many aspects of the energy grid, e.g., power generation, distribution, and consumption [22]. Many research endeavors have been performed to address the capabilities and opportunities of smart grid in respect to the circular economy framework [18], most notably to address the roles of digital and smart technologies [20,23] and the monitoring of energy generation [51], transmission, distribution [52], consumption [53], and management [50–54] in smart grid scenarios [24].

The smart grid can be seen as a complete package that contains electricity generation, transmission, distribution, and consumption [55], as illustrated in Figure 4. When the infrastructure for the smart grid is constructed, the current electrical grid changes from a class-like composition with distinct roles (generation, transmission, distribution, and consumption) into a network structure where multiple entities can simultaneously act as the consumer and the supplier. While the current electrical grid is the infrastructure solely existing to supply electricity, the smart grid can be perceived as an evolved version of the electrical grid, in which home appliances, telecommunications, automobiles, constructions, and energy are intertwined as a platform. The smart grid can allow a new era of innovative and pristine businesses and technologies to propagate, such as bidirectional communication between the electric supplier and consumer, real-time payment systems, reactive response to demand fluctuations, and electric automobiles. Additionally, the smart grid has great potential to act as a strong foundation for new developments that are aimed at renewable energy solutions alongside its native ability to efficiently manage energy distribution and consumption, which is in line with the circular economy concept.

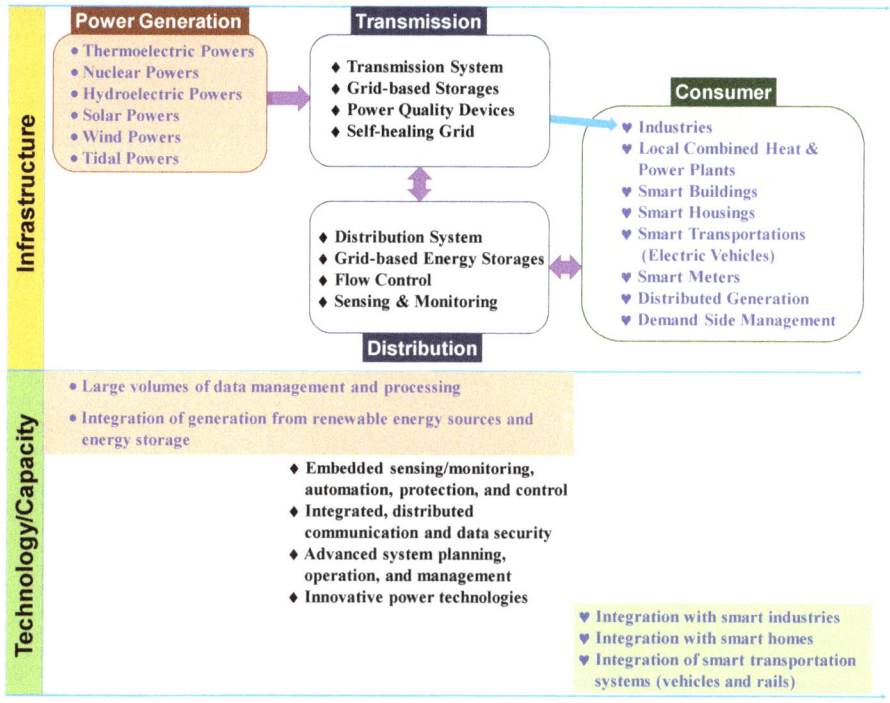

Figure 4. Conceptualization of Smart grid features.

In this study, the fourth generation R&D phase was considered, as described in Section 2.1, from the current third generation R&D processes in regard to smart grid industry. The R&D planning phase was focused on predicting future technologies, create ideas, and extract company strategies based on R&D strategies. In general, prior to strategic R&D planning, efforts have relied heavily on groups of experts from various domains. Although human efforts are ultimately needed to predict the future, establish plans, and make decisions, the significance and demand for a systematic and methodological approach to cope with today's large upsurges of information and the rapid and drastic changes in society is growing.

The smart grid industry may comprise of five industry sectors [56]; Table 3 lists their core technologies and demands. First, the smart power grid sector can be perceived as the industry that increases the reliability and management efficiency of electrical grid by integrating information and telecommunications. The smart power grid technology can be further subcategorized into smart transmission system, smart distribution system, smart power device, and smart power communication network technologies. Second, the smart consumer sector focuses on increasing energy efficiency through communications infrastructure that provides consumers with an assortment of different services. The smart consumer sector consists of automated meter infrastructure, energy management system, and bidirectional communications network technologies. Third, the smart transportation sector concentrates on the generation of new business opportunities through the unrestricted two-way connectivity of electrical grid to electric vehicles. Furthermore, this technology aids in reducing emissions of greenhouse gases, such as carbon dioxide, hydrocarbons, nitrogen monoxide, and sulfur oxide, and in increasing the efficiency of grid through its sub-sectors, such as electric vehicle parts and materials, charging infrastructure, and vehicle-to-grid technologies. Fourth, the smart renewable sector allows for the renewable energy sources to successfully and stably link with the current electric grid by managing the obstacles that hinder the supply of renewable energy. This area comprises of technologies of micro-grid, energy storage, power quality, and power exchange infrastructure. Finally, the smart electricity sector aims to increase the electric grid's efficiency through developing various electric billing schemes and consumer electric transactions system, as well as a reactive response to demand and unrestricted electricity trade between all stakeholders. This sector consists of smart real time pricing, smart demand response, and smart power exchange technologies.

Table 3. Smart grid industry sectors and core technologies.

Smart Grid Industry Sector	Core Technology
Smart Power Grid	(1) Smart transmission system: SMES, FACTS, HVDC, WAMS, WACS (2) Smart distribution system: Distributed generation, AMI, Smart switch, PCS, Converter (3) Smart power device (4) Smart power communication network
Smart Consumer	(1) Duplex transmission AMI, In home device (2) Energy management system (3) Bidirectional transmission network
Smart Transportation	(1) Electric vehicle component & material (2) Electric vehicle charging infrastructure (3) Vehicle to grid
Smart Renewable	(1) Micro-grid (2) Energy storage system (3) Power quality, Power exchange infrastructure
Smart Electricity	(1) Real time pricing (2) Demand & Response (3) Power exchange

SMES: super-conducting magnetic energy storage; FACTS: flexible AC transmission system; HVDC: high voltage direct current; WAMS: wide area monitoring systems; WACS: wide area control systems; AMI: automated metering infrastructure; PCS: power conversion system.

4. Case Study: Hyosung's Smart Grid R&D Planning

Hyosung Corporation is a Korean industrial conglomerate that was founded in 1957. It operates in various fields, including power systems, industrial machineries, chemicals, information technology, constructions, and trades. Among Hyosung's eclectic specialties, this case study has aimed to focus on strategic R&D planning to enhance the competitiveness of the smart grid business sector within the Hyosung Corporation.

4.1. Patent Analysis

4.1.1. Patent Analysis: Hyosung Corporation

In this case study, the patents that Hyosung Corporation currently possessed since its establishment were searched and analyzed; the total number of patents that are owned by Hyosung amounts to 4398 patents. The majority of these patents were registered in the period of 1998 to 2018; only 371 patents were filed for the first forty years (1957–1997) since Hyosung was founded. As shown in Figure 5a, the number of patents increased to a maximum 126 in 1998, corresponding to 34% of the total number of the patents disclosed for the first forty years. However, it turns to be reduced to 116 in 1999 and reached a minimum 47 in 2000. The number of patents slowly increased again to 69 in 2001, 63 in 2002, 98 in 2003, and 139 in 2004. Here, it is noted that the lowest numbers of patents in the period of 2000 to 2002 might be attributed to the severe economic crisis of Korea that started at the end of 1997. The number of patents drastically increased to 324 in 2005 and it reached a maximum 358 in 2006. Again, in the last quarter of 2008, the global recession is speculated to have caused the rapid decline in registered patents since its maximum in 2006. Thereafter, the number of patents annually fluctuates in an upward trend. Overall, these patent data reveal a trend that Hyosung has maintained continuous R&D growth, despite the Korean economic crisis and global recession.

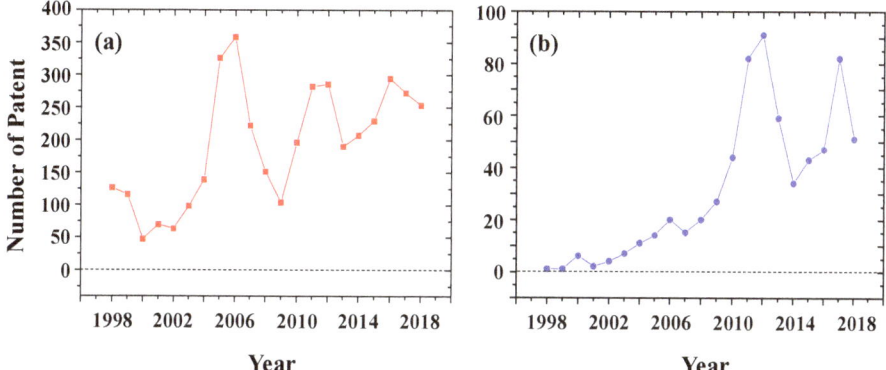

Figure 5. Hyosung's annual number of patents disclosed in the period of 1998 to 2018: (**a**) all possessed technologies (total 4027 patents); (**b**) technologies disclosed to smart grid: electric power components and systems, solar energy components and systems, wind energy components and systems, and battery components and systems (total 661 patents).

The analysis of all Hyosung patents exhibits an active surge in patent registrations across all areas. The corporation has possessed patents in diverse areas that are mainly converged around polymers and downstream products (fibers, tire cords, films, sheet, blends, and composites), chemicals, banking machines, electric power components and systems, solar energy components and systems, wind energy components and systems, and battery components and systems. Overall, the patent analysis confirmed that the company has engaged and invested in robust R&D activities since around 2003.

This study has further narrowed the scope of analysis by concentrating on patents pertaining to specific areas, such as electric power components and systems, solar energy components and systems, wind energy components and systems, and battery components and systems, which have direct associations with the smart grid. As a result, 661 patents were selected for further analysis. It is observed that the patents and technologies that are related to the smart grid begin to proliferate starting 2003 and continue to increase overall (Figure 5b). It can be noted that the decline of patents related to smart grid in 2014 may be a repercussion of the 2008 global recession. A recession can directly impact R&D related expenditures in the form of employee layoffs, project terminations, and budget cuts in forthcoming R&D proposals. Additionally, accounting for the time required for R&D to yield patentable results and the one- to two-year duration that is required for patent registration, the aftershock of the 2008 recession seems most evident in explaining the decrease in numbers of patents that were registered in 2014. The increase in patent numbers after 2014 may be an indication of recovery from the recession.

The technologies of Hyosung Corporation that were disclosed to smart grid in the period of 1998–2018 could be categorized into transformers, switchgears, electronic devices, electronic systems, wind turbine systems, solar systems, battery systems, and other related components, as illustrated in Table 4 and Figure 6. For key technologies, such as (a) transformers, (b) switchgears, (c) electronic devices, and (d) electronic systems, the numbers of patents start to increase from 2007 and steadily continued growth up to today (Figure 6a–d). Differently, the development of wind turbine systems has been intensively carried out only for the period of 2006–2014 (Figure 6e). A similar trend of patent filings is discernible for solar cells and systems, as well as for batteries and their systems (Figure 6f–g). It is noteworthy that a sharp increase of patents has been registered on the solar cells and their systems in the period of 2009–2013. The registration of patents on batteries and systems has been conducted in 2006 to 2014. While taking into consideration of the buffer time to prepare patent submissions, the concentration of research activities in these key smart grid technology areas began in around 2003 (namely, a few years after the Korean economic crisis), which coincides with when Hyosung started to translate their R&D outcomes into intellectual properties. Overall, the patent data is testimony that Hyosung has devoted significant amounts of investments and research on key smart grid technology areas. As a result, Hyosung has achieved substantial technological competitiveness within the smart grid domain.

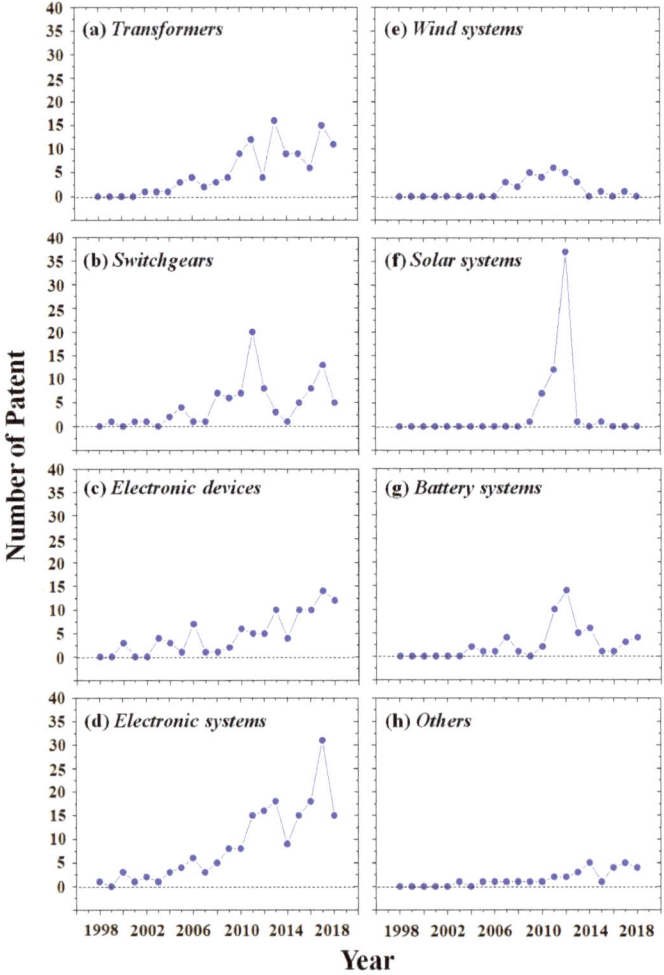

Figure 6. Hyosung's annual number of the patents registered for smart grid technology components: (a) Transformers; (b) Switchgears; (c) Electronic devices; (d) Electronic systems; (e) Wind systems; (f) Solar systems; (g) Battery systems; and, (h) Others.

Table 4. Smart grid technologies and relevant patents of Hyosung Corporation.

Smart Grid Technology	Number of Patent	Detail Technology
Transformers	110	Power transformer, Oil filled transformer, Converter
Switchgears	94	Gas insulated switch, Disconnecting switch
Electronic Devices	98	Digital protection relay, Detector
Electronic Systems	182	Generator, Control system, Power system
Wind Turbine Systems	30	Wind generator, Wind power system
Solar Cell & Systems	59	Solar cell, Solar generation system
Battery Systems	55	Fuel cell, Lithium secondary cell, NaS battery
Others	33	Motor, Welding machine

As shown in Figure 7, all of the technologies that are owned and patented by Hyosung are connected to the entirety of smart grid industry. In particular, the possessed technologies have competitiveness in the smart power grid industry, smart transportation industry, and smart renewable industry. Through the utilization of the capabilities of transformers and switchgear technologies as a basis, high potential exists regarding entering the smart transmission system business and smart distribution system business within the smart power grid industry. Additionally, the entry barrier for the electric vehicle components and charging infrastructure businesses of the smart transportation industry is considerably low when situating the capabilities of battery systems, electric motors, and electronic systems as a foundation. Aside from these business entry possibilities, Hyosung's patented technologies can provide a strong base to potentially succeed in entering the core of the future energy industry, the smart renewable energy business, and the smart grid system solution business.

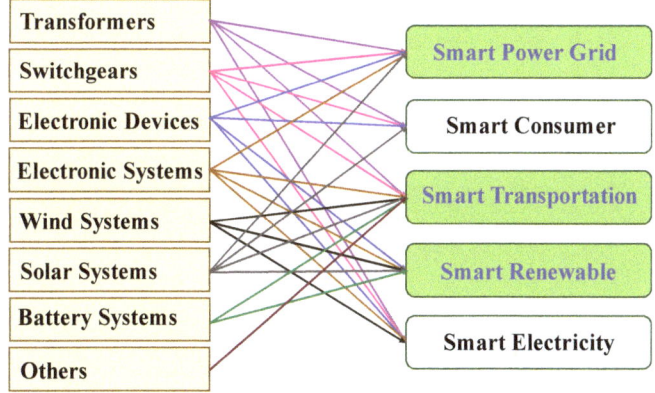

Figure 7. The map of Hyosung's technologies and smart grid industry relations.

4.1.2. Patent Analysis: Electric Vehicle Charging System

This study has selected electric vehicle charging infrastructure business as a focal point for further analysis. The smart grid allows for electric vehicles to be cost-effectively and efficiently charged through selective power usage. For example, electric vehicles can be charged during the night if the price of electricity becomes cheaper as the demand of electricity decreases. Not only does this enable electric vehicle fuel prices to decrease, but this also alleviates the problems that result from the long charging hours that are required for electric vehicle batteries. The remainder of this section analyzes patents regarding electric vehicle charging methods.

The convenient and effective charging of electric vehicles is a world-wide issue. Thus, 15,820 patents that were filed to Korea, USA, Europe, Japan, and China in the period of 2002 to 2018 for electric vehicle charging technology were retrieved (Table 5). Among these patents, 414 patents have been selected for analysis after filtering the noise and directly unrelated patents (Table 5). There is a discernible trend in the patent registration, as shown in Figure 8. The registration of such patents begins in the early 2000s. Only one to four patents are annually registered until 2007. Thereafter, the number of registered patents rapidly increases every year, reaching to a maximum (58 patents) in 2011. It turns to decrease to 44 in 2012 and 40 in 2013, reaching a minimum (28 patents) in 2014. Subsequently, it again turns back to increase to 34 in 2015, 48 in 2016, and 53 in 2017. Nevertheless, in 2018 it again drops to 24.

Table 5. Number of patents disclosed in 2002–2018: Battery charging technology.

Country	Number of Patent	Number of Patent (after filtering noise)
Korea	1407	39
USA	3234	43
Europe	1675	5
Japan	4458	147
China	5046	180

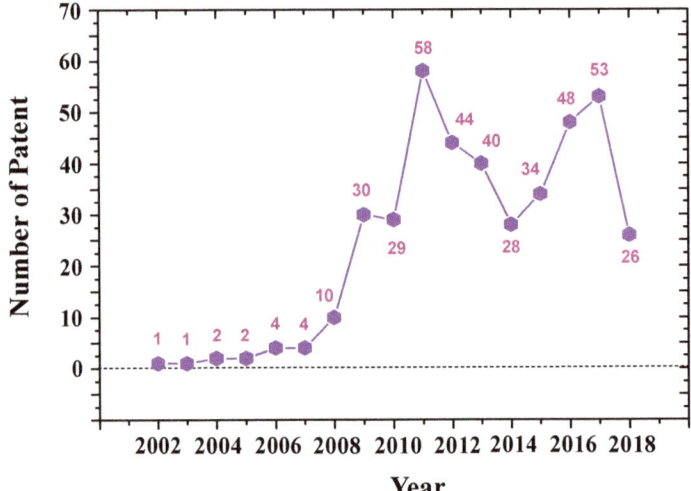

Figure 8. Annual number of patents disclosing battery charging technology.

The patents have been further analyzed in regional bases (Figure 9; Table 5). In Korea, the first patent of electric vehicle charging technology appears in 2004. Subsequently, the patent registration of this technology was actively done between 2009 and 2012, and then declined thereafter. Although similar trends of patent filing are observed in USA and Japan, the overall number of filed patents is slightly higher in USA or much higher in Japan, when compared to that of Korea. However, a slightly different trend of patent filing activity is observed in China. The first patent was filed in 2006 and two additional patents in 2008. Afterwards, the number of patent filing has drastically increased, reaching a maximum (44 patents) in 2017. Even in 2018, the number of patents is 26, which is still very high in comparison to the other countries. In contrast, the activity of patent registration is very slow and furthermore very weak in Europe. Overall, the share fraction of patents is 9.4% for Korea, 10.4% for USA, 1.2% for Europe, 35.5% for Japan, and 43.5% for China. These data confirm that Japan and China dominate the patent registrations, followed by USA, Korea, and Europe. In general, a few years of research are necessary for filing patents. Taking this fact into account, it is evident that, for the recent seventeen years, aggressive research and investments have been devoted worldwide to develop more efficient and more convenient charging technology for electric vehicles.

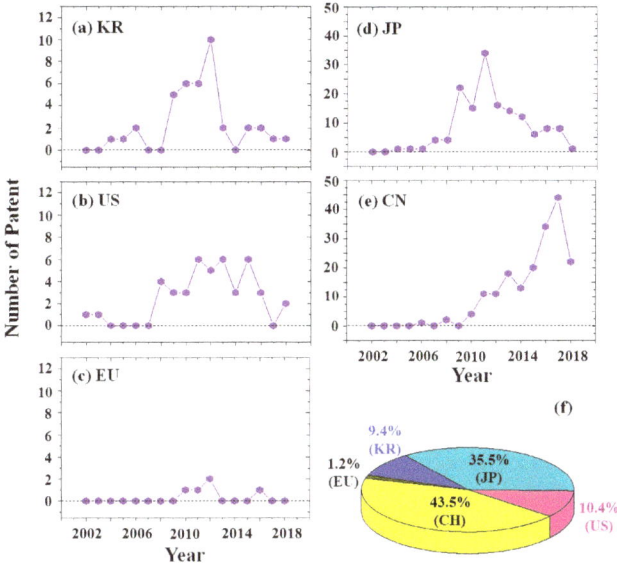

Figure 9. Annual number of regional patents: (**a**) Korea (KR); (**b**) United States of America (U.S.A.) (US); (**c**) Europe (EU); (**d**) Japan (JP); (**e**) China (CH); and, (**f**) Regional patent distributions.

As mentioned above, the R&D of electric vehicle charging technology has been carried out in the fierce competition among the Asian region (Korea, Japan, and China), the American region (USA), and the European region. These competitive R&D works have been found to be led by top 28 organizations, as shown in Figure 10. In particular, the State Grid Corporation of China, Toyota Motor, Kawamura Electric, Hyosung, Mitsubishi, Sharp, LSIS, JFE Engineering, and China Electric Power Research have played dominant leading roles by holding the majority of electric charging technologies. Hyosung and LSIS represent Korea within the top nine applicants. Among the Korean automobile companies, only Hyundai Motor and Kia Motors are barely listed within the top twenty-eight patent applicants. In addition, it is noteworthy that three universities, namely Kookmin University, KAIST (Korea Advanced Institute of Science & Technology), and Tsinghua University, are ranked within the top twenty-eight applicants. Overall, the R&D of electric vehicle charging technology has been driven mainly by automobile, electric power, heavy engineering, and electronic companies. Additionally, there exist instances where academic research groups have worked and filed patents conjointly with such leading companies.

For the electric vehicle technology patents, a thorough and meticulous examination has been performed for each patent to ultimately survey trends in the technology R&D. Furthermore, key development factors have been selected in order to categorize the patent data pool. Subsequently, the core patents of the electric vehicle charging system technology have been chosen. Figure 11 presents the analysis results. In the 2000s, research effort has been focused to develop electric vehicles, batteries, charge methods, charging devices and controls, private power station, and charge infrastructure; these R&D efforts have been continued until recently. In the late 2000s, such developments have begun to emerge in the charging methods and charge control domains. As the developments of individual parts and equipment have matured, the charge service, infrastructure, and system developments have started to emerge from the 2010s to today. Three main key R&D factors have been extracted through the patent analysis of electric vehicle charging technologies. These three factors consist of the electric vehicle battery technologies, the technologies to charge batteries, and the technologies to efficiently manage data via the smart grid. These key R&D factors are organized into specified R&D targets and directions, matched with respective relevant patents, and are summarized in Table 6.

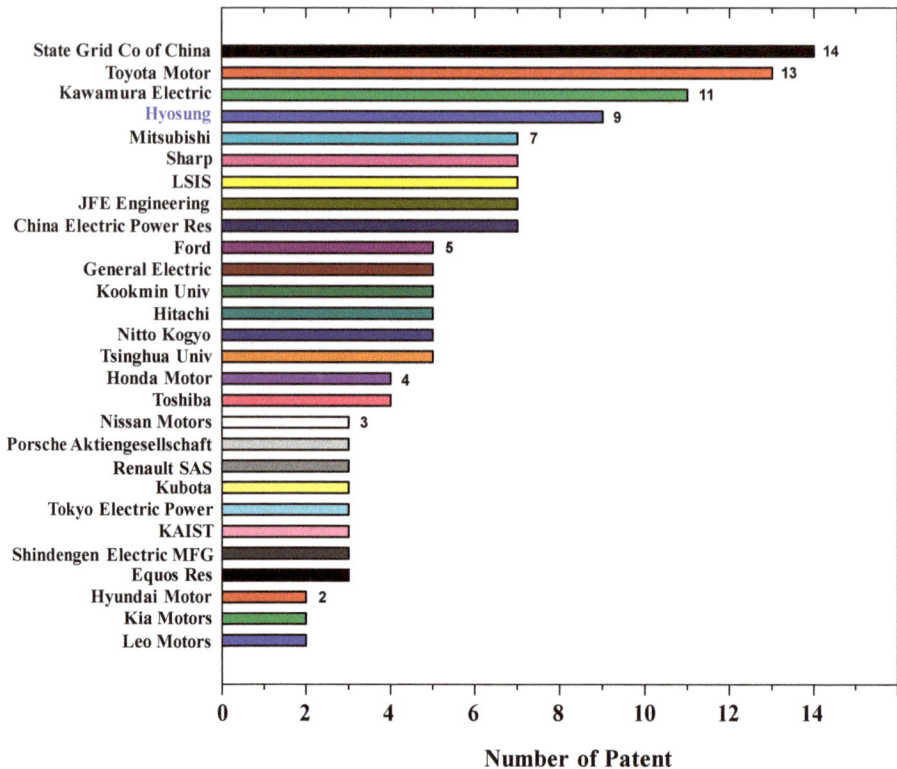

Figure 10. Top patent applicants in the electric vehicle charging technology.

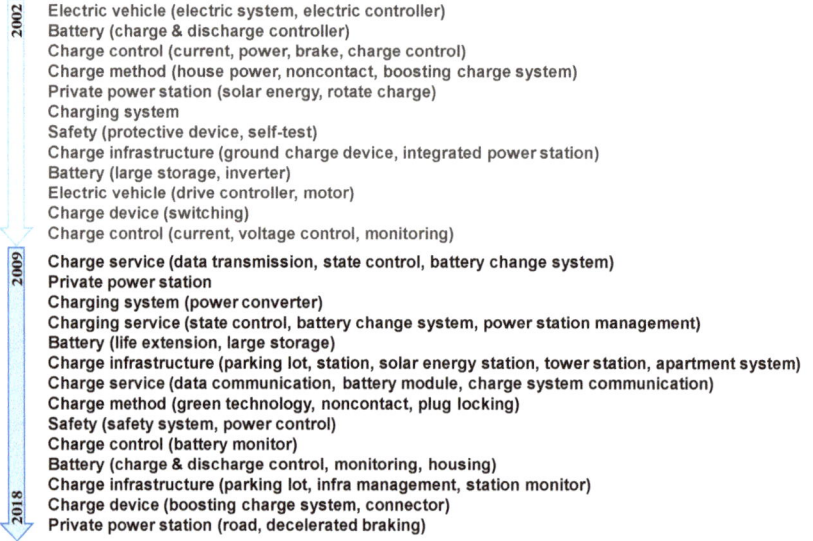

Figure 11. Results of yearly R&D trend analysis in charging technology.

Table 6. Key R&D factors, targets, and directions of electric vehicle charging technology.

Key R&D Factor	R&D Target	R&D Direction	Patents
Data management	Electric vehicle's effective data management	Charging system management	US20050199704A1P, US20090021213A1P, US20140266047A1, JP20100146569A2P, KR20100006913A_P, KR20110138564A_P, KR20100059641A_P, CN105098905A, CN102255113A, KR1020130047901A, CN103914749A
		Power data management	US20110251753A1P, US20110010281A1P, CN107462823A, CN103909910B, CN101685890A, JP20100146568A2P, JP20090227189A2P, JP20090089474A2P, JP20080043174A2P, JP20070318849A2P, JP20060117241A2P, KR20070091762A_P, KR20110101846A_P
		Information management	JP20110143923A2P, JP20110103048A2P, JP20110048430A2P, JP20070206889A2P, CN103795093A, CN102315677A, US20120253567A1
Battery	Improvement of battery performance	Charge inductor	US00006989653B2P, JP20040364481A2P, KR20100104728A_P
		Power source	US20070200532A1P, US20090212627A1P, US20100001531A1, US00008093861B2P, JP20100268576A2P, EP00001338461A3P, JP20060074870A2P
		Battery charge	US00007208914B2P, US20100114762A1P, US8330411B2, US8334676B2, US8183826B2, CN107650704A, JP20110205758A2P, JP20110151896A2P, JP20100161856A2P, JP20100158134A2P, JP20090100649A2P, JP20090095157A2P, JP20090077557A2P, JP20090060683A2P, JP20090038872A2P, EP00002157637A1P, KR20040108453A_P, JP20090027831A2P, JP20060074868A2P, JP20050210835A2P, CN101527374A, CN106696721A
		Charge device	US20100181126A1, EP00002282390A2P, EP00002279892A2P, EP00002226216A1P, CN108859816A
Charge method	Effective various charge method	On-line charge method	KR20050106313A_P, KR20070118872A_P, KR20100002934U_U, KR20110042403A_P, US00007501789B2P, US20110187321A1P, EP00002325037A1P, EP00002292877A1P, JP20100068632A2P, US20120268245A1, US20140266047A1, CN107839502A, CN107128192A
		Public place station	JP20100101082A2P, JP20100095848A2P, KR20100023908A_P, KR20110112656A_P, KR20120009923A_P, JP20120016179A2P, JP20110169043A2P, JP20110018336A2P, JP20100146564A2P, US20120229085A1, KR101245572B1, CN108859843A, CN1074591853A
		Effective charge system	KR20110047950A_P, KR20110048444A_P, US20070188126A1P, EP00001961098A2P, US20120313562A1, CN103311978A, US9517700B2, US9233618B2, KR1020180084656A
		Auto charge system	KR20050017335A_P, KR20050064899A_P, US00007479763B2P, CN103001264A, CN108407626A

4.1.3. Strategy Based on Patent Analysis

This section is dedicated towards suggesting an entry strategy that is based on R&D targets in regard to the electric vehicle charging system by examining Hyosung's smart grid capabilities. In comparison with the other areas of Hyosung, little research has been conducted in this field. Nevertheless, Hyosung has established a better position in the R&D of battery systems and charging systems, when compared to all other competitors in the world, as discussed in the earlier section. Moreover, recent increases in patents reflect the escalation of its research. When considering that the effective data management field is still an unexplored area in the world market, the battery system and charging system area is expected to be a promising opportunity, under the assumption that Hyosung concentrates its capabilities from its power management system into information and communications technology, and continues R&D efforts in this area.

The battery performance and charging technology, among all electric vehicle related areas, has been the most actively researched area. Notably, Japanese companies currently possess the technologies of the highest standards, as well as the bulk of the core technologies. Nevertheless, research of battery performance will proceed to the continual advancements in performance of electric vehicles. Although the Japanese competitors possess many of the core technologies regarding the performance of electric vehicle battery technologies, the opportunities exist, given that Hyosung allocates its capabilities of power equipment and battery technology to conduct research in reducing battery charging durations.

Various charging methods are closely related technologically and, more importantly, highly interconnected to the electric vehicle charging industry's business model. Though several business models have been developed, there has yet to be a model that can be recognized as a world-wide standard. This area is not limited to the development of charging technologies, but it must incorporate the developments of an all-inclusive mixture of the power equipment system, service, and infrastructure.

4.2. Future Strategy

This research goes beyond technological analysis by establishing a strategy from a business perspective. TRIZ's nine windows and scenario planning methodologies have been utilized to provide a strategy from a business point of view.

4.2.1. The 9 Windows Analysis

In order to stimulate the market of electric vehicles, a consumer-friendly and convenient, not to mention efficient, electric vehicle charging system is a vital element. The electric vehicle charging system must be able to use cost-reduced power during the night and to minimize the problems faced during demand data management and demand fluctuations to properly coincide with the goals of the smart grid industry. The electric vehicle charging system has been analyzed at the current system and super-system levels in a consecutive manner to achieve these. Figure 12 shows the analysis results. The current problem at the system level pertains to the problems with performance, implementation, and expansion of private charging stations. A 24 kWh Nissan Leaf charged into a normal Level 1 (1.4 kW) residential outlet for roughly twelve hours only recharges the battery to half capacity and would cost approximately $60 per month; and, it costs $1500–$2200 for users to install a Level 2 220-volt (6.6 kW) system, which is capable of fully charging the car in seven hours [45]. Essentially, long charging time, the cost of electricity, and private charging unit installation are some of the major issues in electric vehicle charging. For the super-system, system standardization is considered to be a generalized problem of the system due to the lack of a global standard in place for electric vehicle charging, which consequently affects the expansion of charging infrastructure, a vital element that is required for electric vehicle adoption [57]. For the future super-system, the current issues of industry standardization are resolved and the vitalization of smart grid technologies and its benefits have instated. In correspondence to this super-system, the future system of electric vehicle charging is expected to promise the expansion of highly available, affordable, and rapid charging stations, where the company can position its R&D planning activities.

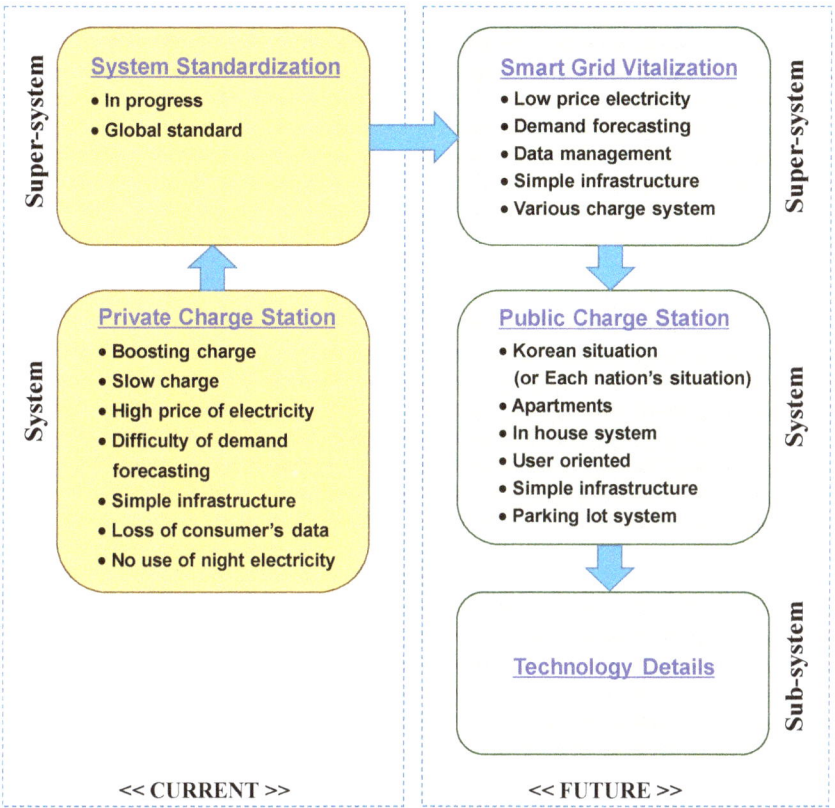

Figure 12. Results of the nine windows analysis for smart grid industry in conjuction to electric vehicle charging technology.

The contemporary plan and development of electric vehicle charging system is close to that of the current gasoline station. Although research is attempting to solve the biggest issue of the charging system, namely extensive charging durations, twenty-minute boosted charging, and five-hour slow charging stations are being implemented at present. The implementation of ultra-fast charging stations is being contemplated, but the expensive costs are acting as an unavoidable obstacle. In addition, private charging stations incur limitations, since user data management is difficult to the point where demand prediction is overly complex to satisfy the efficiency standards for the smart grid.

In its current state, it is a world-wide plan to set a universal standard for the smart grid charging system. Domestically, basic systems requirements and safety regulation standards have been arranged; and similarly, the US, Europe, and Japan have proceeded to create their own national standards. Preparation in advance is essential, because the universal standardization of the smart grid charging system will set the world-wide norm.

The progress toward the smart grid is becoming factual in the future super-system. When the smart grid becomes activated, the smart grid's advantages must be sufficiently utilized. For example, the cost-reduced power during the night is selectively used by the consumer, and, through data management, the predictions of demand and supply are accurately accomplished. Furthermore, a surplus of different charging systems will be provided to fit and satisfy individual lifestyles and preferences.

Predicting the future system based from what was foreseen in the future super-system, the charging system infrastructure must be customized to fit into all types of housing environments. For example, the charging systems are implemented within the apartment housing parking facilities for each parking spot, and to implement water-proof enclosures for outdoor charging units for other types of parking. Moreover, payment system and data management system developments are required in maintaining the charging infrastructure. The scenario planning technique is utilized to fine tune the directions of the R&D plan with the resulting future system and super-system as a reference to guide the types of electric vehicle charging innovations.

4.2.2. Scenario Planning

This research uses scenario planning to establish strategies for countering the extracted plausible future scenarios. Key change agents, namely technology and growth rates, policies and government subsidies, and system standards of the smart grid charging system, were generated and combined accordingly to produce the following four most plausible scenarios (Table 7).

Table 7. Future smart grid industry scenarios.

Scenario	Technology Rate	Growth Rate	Regulation	Standardization
1	Fast	Fast	Low (support)	Uniformity
2	Fast	Fast	Low	Diversity
3	Fast	Slow	High	Uniformity
4	Slow	Slow	Low	Diversity

In Scenario 1 (The Stabilized Settlement of the Smart Grid Industry), the future setting can be hypothesized as the following. The advancement of the technology is sufficient; the growth rate of the smart grid and electric vehicle charging system industry is accelerated; government subsidies are strongly supportive; and, the industry standard is strictly uniform. In this scenario, related markets are growing, and innumerable companies are heavily investing in R&D in the hopes of entering the smart grid industry. Although this is a prime instance for successful opportunities, the strictly uniform standardization policy plays a significant role in determining the outcomes of such opportunities. It is vital to recognize and closely monitor the standardization policies as the key change agent from a company's perspective. Regardless of the supremacy and excellence of a technology, if the technology strays from the standardization guidelines, then the risk of failure is inevitable.

In order to prepare for Scenario 1, the company must avoid investing in a new all-inclusive charging system, and focus R&D expenditures on the development of various modules or a platform that concurs with the current and predicted standards. It is suggested to capture the business opportunities in developing individual modules and related technologies that lie within the boundaries of the standards, since the future charging system will diversify into varied businesses.

Scenario 2 (The Short-lived Blue Ocean of the Smart Grid Industry) can be described as the environment, where adequate advancements of technology, rapid growth rates, hefty government aid, and the freedom from standardization are harmonized. It may seem that all of the stars have aligned in this scenario, where all of the external factors are in favor for opportunity. However, in this scenario, the lack of regulation of any form results in a highly competitive atmosphere and causes rapid and impulsive advancements of the technology.

As a result, an aggressive plan is required to cope with such competition. This plan must encompass continuous R&D activities in a portfolio of different technologies. Not only is it essential to invest in incremental improvements of existing technologies, but it is more important to prepare and further invest into the future to maintain competitive advantages. For example, while developing technologies that enhance the performance of existing charging systems, investments into disruptive innovative technologies are crucial. It is speculated that the future representative technology may become wireless charging technology and robotic charging [58]. The patent analysis results show that wireless charging related technologies have developed starting the late 2000s, though the wireless technology is still in its early stages and far from commercialization for smart grid applications. However, similar to how the wireless charging technology is becoming commercialized for small applications, like cellular phones, the wireless charging technology has great potential to solve certain problems of the electric vehicle charging system. In short, Scenario 2 illuminates a future filled with prosperity and success, but the need for concentrated R&D efforts in disruptive innovations is a difficulty that cannot be disregarded.

In Scenario 3 (The Questionable Market of the Smart Grid), the future can be represented as a situation where ample developments of technology exist and continue to be improved, but the smart grid market is showing slow growth; in addition, governmental grants are low and the standardization is strict. Thus, the situation, where companies have heavily invested in the smart grid industry, but are struggling in difficult times due to a combination of factors mentioned, is commonplace across the market.

In order to cope with Scenario 3, the company must look towards industry-convergence technologies. More specifically, in order to survive through such unfortunate events, the company must search for other industries where the developed smart grid technologies can be applied, and, once another industry is found, the existing smart grid technology must be adjusted to fit its new industry. Instead of focusing on converging technologies, like the electric vehicle charging system, technologies that have higher potential to be connected to other industries, such as data management technologies, should be emphasized. To conclude, the solution for this scenario is to be selective in choosing the technology to be developed, with one that has high amounts of possible connections with other industries.

Scenario 4 (The Stalemate of the Smart Grid Industry) describes one of the worst cases a company can face, and is as follows: Slow market growth and technology development, and incumbents have begun to cut further investments. Basically, Scenario 4 is the case where the smart grid industry fails to settle as a significant entity of the future energy industry.

The solution to Scenario 4 represses the investments into new technologies and promotes the focus on R&D efforts into technologies that maintain competitiveness. Hypothetically, Hyosung could redirect its focus on its existing power equipment and system solution technologies to overcome these difficulties. The company can place emphasis on technologies that have the potential to enter into the smart grid industry if and when the smart grid industry situation does improve due to the high risks that are involved with continuing investments in new technologies in the smart grid industry.

In summary, it is critical to note that quarterly assessments of changes in technology, market, industry standards, and policy is required to quickly adjust R&D plans accordingly. R&D planning based on scenarios 1 and 2 would result in strong thrusts for developments in the electric vehicle charging domain. From the analysis, the proper assessment of national and global governmental policies on subsidies or regulations and industry trends regarding standardization will further determine whether to direct the company's R&D plans to focus on the development of smart grid modules or platforms, or to focus on innovating disruptive technologies to obtain or maintain competitive advantages. The consideration of global policies is important, in particular, policies of allied countries that may have a conforming effect on local legislations on electric vehicle charging. When considering scenario 1, it can be suggested that the company invest into demand-side management modules targeting the enhancement of user convenience. Under the scenario where the adoption of electric vehicles is increasing, and industry standardization is uniform, R&D plans and expenditures must cater more heavily to meet the consumers' needs and approval. Alternatively, under scenario 2, a manager of R&D can propose the expansion of breakthrough developments into robotic charging stations, renewable energy sources, wireless electric vehicle charging, and energy storage solutions for electric charging infrastructure for competitive advantages over other industry contenders. Such investments would be valuable as the number of users of the electric vehicles increase and continue to desire more accessible, quick, and easy-to-use charging stations. Additionally, heavy investment into energy storage solutions can be suggested for developing charging stations that are able to balance the load of larger and higher power stations. In light of the lacking number of electric vehicle charging infrastructure available for users, it has been stated that battery storage can assist in the acceleration of the construction of charging stations, and consequently boost electric vehicle adoption.

Under the situation where scenarios 3 and 4 are imminent, the company should explore other areas where the incumbent technologies can be applied for new opportunities. Further investing directly into new smart grid technologies could be detrimental to the company. Comprehensive evaluation of existing technologies may provide solutions towards novel innovation. In accordance to scenario 3, it is suggested for the company to converge its electric vehicle charging technologies into areas, such as hydrogen charging stations and data management solutions. Although converging existing technologies into a new industry will require an appropriate assessment of the advantages and disadvantages of the particular area, the accurate selection and transfer of the electric charging technologies can act as a gateway for future developments and prevent developed technologies from becoming obsolete. Assuming scenario 4, the company is suggested to focus R&D plans on existing power equipment and system solution technologies that are closely related to the smart grid industry, such as energy storage systems, data center solutions, and power transmission and distribution infrastructure. These areas are some of the company's most developed technologies, which, with further R&D investments, are sustainable revenue sources whilst encompassing the ability to transition into electric vehicle charging infrastructure in preparations for a possible positive turn for the smart grid industry.

5. Conclusions

In this study, a strategic R&D planning process for the smart grid industry was developed. The planning process is comprised of background research in the smart grid industry, selection of perspective R&D target, STEEP analysis, patent analysis, and TRIZ's nine windows and scenario planning analyses to develop a method and process in establishing a future strategic R&D plan.

The developed planning process was applied to the Hyosung Corporation perspective in the strategic planning of smart grid R&D as a case study. The patent analysis confirmed that Hyosung has devoted significant amounts of investments and research on key smart grid technology areas (i.e., transformers, switchgears, electronic devices, electronic systems, wind turbine systems, solar systems, battery systems, and other related components) since around 2003, and, as a result, already achieved substantial technological competitiveness within the sustainable smart grid domain, including smart power grid, transportation, and renewable energy. With the patent analysis results, TRIZ's nine windows and scenario planning analyses were further performed to produce a strategic R&D plan from the business point of view. From this case study, a strategic smart grid R&D plan for Hyosung to enter the sustainable smart grid industry was suggested and further entered into the sustainable electric vehicle charging technology business. The strategic R&D plan comprises of smart grid communication system, service infrastructure, battery charging duration, public charge station system, platform and module, renewable energy sources, wireless charging, data management system, and electric storage system solution as key technology and business factors. To achieve the goals of the strategic R&D plan, a set of four scenario plans were suggested by taking into consideration technology and growth rates, policies and government subsidies, and system standards of the smart grid charging system as key change agents: Scenario 1, 'The Stabilized Settlement of the Smart Grid Industry'; Scenario 2, 'The Short-lived Blue Ocean of the Smart Grid Industry'; Scenario 3, 'The Questionable Market of the Smart Grid'; and, Scenario 4, 'The Stalemate of the Smart Grid Industry'. The R&D plans in accordance to each scenario were recommended. In scenario 1, the company R&D plan is to pursue investments in development of various modules or a platform that concurs with the current and predicted standards, such as demand-side management modules targeting the enhancement of user convenience. Under scenario 2, the suggested R&D plan is to emphasize a diverse portfolio of different technologies with incremental and disruptive innovations, such as robotic charging, wireless charging, and energy storage solutions, to gain and maintain a competitive advantage in this regulation-less, competitive environment. Scenario 3 is a high uncertainty situation with a strong technological supply-side push, but lacks the market growth and government support. In this scenario, the resulting company R&D plan suggests that R&D endeavors to concentrate on convergent technologies that have the potential to be introduced into new industries, such as data management products and services. Finally, scenario 4 restrains the investments into new technologies and it promotes the focus on R&D efforts into technologies that maintain competitiveness or further investment into incumbent technologies, such as data center solutions and power transmission and distribution infrastructure. The strategic R&D planning process of this study can be used in a variety of different ways within the smart grid industry by reselecting the target to other businesses. Moreover, the planning process can be applicable in various technologies and businesses, because of the independence of any specific subject or business area. Additionally, the planning process may gain high expectations to provide a strategy at the business level and to support an integrated and effective R&D effort.

Author Contributions: Conceptualization, J.J.R. and K.K.; data curation, J.J.R.; investigation, J.J.R.; methods, J.J.R; discussions, J.J.R. and K.K.; writing—original draft, J.J.R.; writing—review and editing, J.J.R.; supervision, K.K.

Funding: This research was supported by the National Research Foundation of Korea (NRF) grant funded by the Ministry of Science and ICT (Grant No. NRF-2016R1A2B4008381).

Acknowledgments: The authors appreciate the financial support from the NRF allowing for this research to be accomplished.

Conflicts of Interest: The authors have no conflict of interest.

References

1. World Energy Council. *World Energy Trilemma Index*; World Energy Council: London, UK, 2018.
2. U.S. Department of Commerce. *Smart Grid Top Market Report*; U.S. Department of Commerce: Washington, DC, USA, 2017.
3. U.S. Department of Energy. *Smart Grid*; U.S. Department of Energy: Washington, DC, USA, 2012.
4. Ling, A.P.A.; Kokichi, S.; Masao, M. The Japanese smart grid initiatives, investments, and collaborations. *Intern. J. Adv. Comput. Sci. Appl.* **2012**, *3*, 1–11.
5. Mesaríc, P.; Đukec, D.; Krajcar, S. Exploring the potential of energy consumers in smart grid using focus group methodology. *Sustainability* **2017**, *9*, 1463. [CrossRef]
6. Saleh, M.S.; Althaibani, A.; Esa, Y.; Mhandi, Y.; Mohamed, A.A. Impact of clustering microgrids on their stability and resilience during blackouts. In Proceedings of the 2015 International Conference on Smart Grid and Clean Energy Technologies (ICSGCE), Offenburg, Germany, 20–23 October 2015; pp. 195–200.
7. Markovic, D.S.; Branovic, I.; Popovic, R. Smart grid and nanotechnologies: A solution for clean and sustainable energy. *Energy Emiss. Control Technol.* **2015**, *3*, 1–13. [CrossRef]
8. Horta Nogueira, L.A.; Haddad, J.; Borges da Silva, L.E.; Lambert Torres, G. *Energy Efficiency and Smart Grids for Low Carbon and Green Growth in Brazil*; Inter-American Development Bank: Washington, DC, USA, 2015.
9. Zezulka, F.; Szabo, Z.; Vesely, I. Smart energo model. *IFAC-PapersOnLine* **2015**, *48*, 404–408. [CrossRef]
10. Ton, D.T.; Smith, M.A. The U.S. Department of Energy's microgrid initiative. *Electr. J.* **2012**, *25*, 84–94. [CrossRef]
11. Miskinis, V.; Galinis, A.; Konstantinaviciute, I.; Lekavicius, V.; Neniskis, E. Comparative analysis of the energy sector development trends and forecast of final energy demand in the Baltic States. *Sustainability* **2019**, *11*, 521. [CrossRef]
12. Guo, H.; Qiao, W.; Liu, J. Dynamic feedback analysis of influencing factors of existing building energy-saving renovation market based on system dynamics in China. *Sustainability* **2019**, *11*, 273. [CrossRef]
13. U.S. Department of Energy. *Distribution Automation: Results from the Smart Grid Investment Grant Program*; U.S. Department of Energy: Washington, DC, USA, 2016.
14. U.S. Department of Commerce. *2016 ITA Smart Grid Top Markets Report: Country Case Study—Japan*; U.S. Department of Commerce: Washington, DC, USA, 2016.

15. Lee, J. Comparison of smart grids demonstration projects and strategies in major leading countries. *New Renew. Energy* **2016**, *12*, 4–48. [CrossRef]
16. Ton, D.T.; Wang, W.-T.P. A more resilient grid. *IEEE Power Energy Mag.* **2015**, *13*, 26–34. [CrossRef]
17. United States Federal Energy Regulatory Commission. *Assessment of Demand Response and Advanced Metering: A Staff Report*; United States Federal Energy Regulatory Commission: Washington, DC, USA, 2008.
18. Türkeli, S.; Kemp, R.; Bleischwitz, R.; McDowall, W. Circular economy scientific knowledge in the European Union and China: A bibliometric, network and survey analysis 2006–2016. *J. Clean. Prod.* **2018**, *197*, 1244–1261. [CrossRef]
19. Macpherson, L. Amsterdam's Environmental Savior: The Circular Economy. Available online: https://towardsdatascience.com/amsterdams-environmental-saviour-the-circular-economy-c83200222e61 (accessed on 12 May 2019).
20. Pagoropoulos, A.; Pigosso, D.; McAloone, T. The emergent role of digital technologies in the circular economy: A review. *Procedia CIRP* **2017**, *64*, 19–24. [CrossRef]
21. Piergiuseppe, M.; Falcone, P.; Lopolito, A. How to promote a new and sustainable food consumption model: A fuzzy cognitive map study. *J. Clean. Prod.* **2019**, *208*, 563–574.
22. Ghorab, M. Energy hubs optimization for smart energy network system to minimize economic and environmental impact at Canadian community. *Appl. Therm. Eng.* **2019**, *151*, 214–230. [CrossRef]
23. Nižetić, S.; Djilali, N.; Papadopoulos, A.; Rodrigues, J. Smart technologies for promotion of energy efficiency, utilization of sustainable resources and waste management. *J. Clean. Prod.* **2019**. in Press.
24. Avancini, D.; Rodrigues, J.; Martins, S.; Rabelo, R.; Al-Muhtadi, J.; Solic, P. Energy meters evolution in smart grids: A review. *J. Clean. Prod.* **2019**, *217*, 702–715. [CrossRef]
25. Meyer, M.; Booker, J. *Eliciting and Analyzing Expert Judgment: A Practical Guide*; Society for Industrial and Applied Mathematics: Philadelphia, PA, USA, 2001.
26. Briand, L.; Freimut, B.; Vollei, F. Assessing the Cost-Effectiveness of Inspections by Combining Project Data and Expert Opinion. In Proceedings of the 22nd International Conference on Software Engineering ICSE, Limerick, Ireland, 4–11 June 2000.
27. Salamatov, Y.; Souchkov, V. *TRIZ: The Right Solution at the Right Time—A Guide to Innovative Problem Solving*; Insytec: Hattem, The Netherlands, 1999.
28. OECD. *OECD Factbook 2008: Economics, Environmental and Social Statistics*; OECD: Paris, France, 2008.
29. Yoon, B.; Park, Y. A text-mining-based patent network: Analytical tool for high-technology trend. *J. High Technol. Manag. Res.* **2004**, *15*, 37–50. [CrossRef]
30. Yun, J.-H.; Hyun, B.-H.; Seo, J.-H. A study on the methodology of R&D planning—Investigation on the market oriented new method of R&D planning. In Proceedings of the 2006 Conference of Korea Technology Innovation Society Conference, Jeju, Korea, 2–3 November 2006. pp. 139–154.
31. Altshuller, G.S. *Creativity as an Exact Science: The Theory of the Solution of Inventive Problems*; Gordon Breach Sci.: New York, NY, USA, 1984.
32. Kerdini, S.; Hooge, S. Can strategic foresight and creativity tools be combined? Structuring a conceptual framework for collective exploration of the unknown. In Proceedings of the 2013 International Product Development Management Conference, Paris, France, 23–25 June 2013; pp. 17–35.
33. Mann, D. The (Predictable) Evolution of Useful Things. The TRIZ Journal. Available online: http://www.triz-journal.com/archives/1999/09/b/index.htm (accessed on 17 April 2008).
34. Moehrle, M.G. How combinations of TRIZ tools are used in companies–results of a cluster analysis. *RD Manag.* **2005**, *35*, 285–296. [CrossRef]
35. Lee, K.; Lee, S. Patterns of technological innovation and evolution in the energy sector: A patent-based approach. *Energy Policy* **2013**, *59*, 415–432. [CrossRef]
36. Urraca-Ruiz, A. On the evolution of technological specialization patterns in emerging countries: Comparing Asia and Latin America. *Econ. Innov. New Technol.* **2019**, *28*, 100–117. [CrossRef]
37. Yang, C.J.; Chen, J.L. Forecasting the design of eco-products by integrating TRIZ evolution patterns with CBR and simple LCA methods. *Expert Syst. Appl.* **2012**, *39*, 2884–2892. [CrossRef]
38. Cempel, C. The ideal final result and contradiction matrix for machine condition monitoring with TRIZ. *Key Eng. Mater.* **2014**, *588*, 276–280. [CrossRef]
39. Mann, D.L. *Hands-On Systematic Innovation*; CREAX Press: Kuala Lumpur, Malaysia, 2002; pp. 63–85.

40. Barry, K.; Domb, E.; Slocum, M. TRIZ—What Is TRIZ? Available online: http://www.trizjournal.com/archives/what_is_triz (accessed on 27 April 2019).
41. Moehrle, M. TRIZ-based technology roadmapping. *Int. J. Technol. Intell. Plan.* **2004**, *1*, 87–99. [CrossRef]
42. Ilevbare, I.; Probert, D.; Phaal, R. A review of TRIZ, and its benefits and challenges in practice. *Technovation* **2013**, *33*, 30–37. [CrossRef]
43. Marconi, J. ARIZ: The Algorithm for Inventive Problem Solving. Available online: https://triz-journal.com/ariz-algorithm-inventive-problem-solving/ (accessed on 27 April 2019).
44. Schoemaker, P.J.H. Scenario planning: A tool for strategic thinking. *Sloan Manag. Rev.* **1995**, *36*, 25–40.
45. Lindgren, M.; Bandhold, H. *Scenario Planning: The Link between Future and Strategy*; Palgrave MacMillan: New York, NY, USA, 2009.
46. Allwood, J.; Laursen, S.; Russell, S.; Malvido de Rodríguez, C.; Bocken, N. An approach to scenario analysis of the sustainability of an industrial sector applied to clothing and textiles in the UK. *J. Clean. Prod.* **2008**, *16*, 1234–1246. [CrossRef]
47. Bishop, P.; Hines, A.; Collins, T. The current state of scenario development: An overview of techniques. *Foresight* **2007**, *9*, 5–25. [CrossRef]
48. Hiltunen, E. *Foresight and Innovation: How Companies and Coping with the Future*; Palgrave MacMillan: New York, NY, USA, 2003.
49. Ramirez, R.; Selin, C.L. Plausibility and probability in scenario planning. *Foresight* **2014**, *16*, 54–74. [CrossRef]
50. Porumb, R.; Seritan, G. Chapter 5—Integration of advanced technologies for efficient operation of smart grids. In *Green Energy Advances*; Enescu, D., Ed.; IntechOpen: London, UK, 2019; pp. 105–180.
51. Momoh, J. Smart grid design for efficient and flexible power networks operation and control. In Proceedings of the Power Systems Conference and Exposition, Seattle, WA, USA, 15–18 March 2009; pp. 1–8.
52. Li, F.; Qiao, W.; Sun, H.; Wan, H.; Wang, J.; Xia, Y.; Xu, Z.; Zhang, P. Smart transmission grid: Vision and framework. *IEEE Trans. Smart Grid* **2010**, *1*, 168–177. [CrossRef]
53. Wang, Y.; Huang, Y.; Wang, Y.; Zeng, M.; Li, F.; Wang, Y.; Zhang, Y. Energy management of smart micro-grid with response loads and distributed generation considering demand response. *J. Clean Prod.* **2018**, *197*, 1069–1083. [CrossRef]
54. Gungor, V.; Sahin, D.; Kocak, T.; Ergut, S.; Buccella, C.; Cecati, C.; Hancke, G. Smart grid technologies: communication technologies and standards. *IEEE Trans. Ind. Inform.* **2011**, *7*, 529–539. [CrossRef]
55. Major Economies Forum on Energy and Climate. *Technology Action Plan: Smart Grids*; Major Economies Forum on Energy and Climate: Washington, DC, USA, 2009.
56. Ministry of Knowledge Economy of Korea. *The National Roadmap: Smart Grids*; Ministry of Knowledge Economy of Korea: Sejong, Korea, 2010.
57. Lee, H.; Clark, A. *Charging the Future: Challenges and Opportunities for Electric Vehicle Adoption*; HKS Working Paper No. RWP18-026; Elsevier: Amsterdam, The Netherlands, 2018; pp. 1–77.
58. IDTechEx: Electric Vehicle Charging Infrastructure 2019–2029: Forecast, Technologies and Players. Available online: https://www.idtechex.com/research/reports/electric-vehicle-charging-infrastructure-2019-2029-forecast-technologies-and-players-000633.asp?viewopt=showall (accessed on 27 April 2019).

© 2019 by the authors. Licensee MDPI, Basel, Switzerland. This article is an open access article distributed under the terms and conditions of the Creative Commons Attribution (CC BY) license (http://creativecommons.org/licenses/by/4.0/).

Article

A Model for Predicting Energy Usage Pattern Types with Energy Consumption Information According to the Behaviors of Single-Person Households in South Korea

Sol Kim [1], Sungwon Jung [1,*] and Seung-Man Baek [2]

1 Department of Architecture, Sejong University, Seoul 05006, Korea; kkimssor@gmail.com
2 School of Architecture, Yeungnam University, Gyeongsan 38541, Korea; smbaek@yu.ac.kr
* Correspondence: swjung@sejong.ac.kr; Tel.: +82-02-3408-3289

Received: 13 December 2018; Accepted: 27 December 2018; Published: 7 January 2019

Abstract: Residential energy consumption accounts for the majority of building energy consumption. Physical factors and technological developments to address this problem have been researched continuously. However, physical improvements have limitations, and there is a paradigm shift towards energy research based on occupant behavior. Furthermore, the rapid increase in the number of single-person households around the world is decreasing residential energy efficiency, which is an urgent problem that needs to be solved. This study prepared a large dataset for analysis based on the Korean Time Use Survey (KTUS), which provides behavioral data for actual occupants of single-person households, and energy usage pattern (EUP) types that were derived through K-modes clustering. The characteristics and energy consumption of each type of household were analyzed, and their relationships were examined. Finally, an EUP-type predictive model, with a prediction rate of 95.0%, was implemented by training a support vector machine, and an energy consumption information model based on a Gaussian process regression was provided. The results of this study provide useful basic data for future research on energy consumption based on the behaviors of occupants, and the method proposed in this study will also be applicable to other regions.

Keywords: occupant behavior; single-person household; energy consumption; Korean Time Use Survey; EnergyPlus; data mining; K-modes clustering; support vector machine; Gaussian process regression

1. Introduction

Approximately 40% of global energy is consumed in buildings, and residences account for approximately 3/4 of the total energy consumption in buildings [1]. Energy consumption in residences is expected to continuously increase until 2040 [2]; consequently, active research is being conducted on energy savings in residences [3–5]. Since energy consumption in the residential sector accounts for a large part of total energy consumption, energy saved in this sector through continuous research and technological development will have a positive impact on the reduction of total energy consumption and greenhouse gases.

To examine the research trends related to residential energy, the paradigm is shifting toward an emphasis on the behaviors of occupants, as well as the physical and environmental factors that affect energy consumption [6–9]. The behaviors of occupants are considered an important factor that affects residential energy consumption since, given identical physical and environmental situations, energy consumption may still differ depending on the behaviors of occupants. Previous studies have indicated that occupants behaviors are a major factor in energy consumption, and have therefore predicted energy consumption based on these behaviors. However, researchers are facing difficulties in data collection due to privacy issues [10].

The recent increase in single-person households due to demographic changes is becoming an issue in various fields of research, and residential energy researchers are also paying attention to this. Yu et al. [11] investigated future scenarios of energy consumption through demographic changes, and anticipated that the increase in single-person households would increase energy consumption and carbon emissions, arguing that energy consumption measures must consider these factors. In South Korea, which is the target region of this study, single-person families have steadily increased since 1975. According to the Population and Housing Census [12], which was conducted in 2015, single-person households account for 27.2% of all households, emerging as the main household type in South Korea.

According to a study on single-person households and residential energy, the increase in single-person households has decreased the total energy consumption per household but doubled the power consumption of residential energy. This is due to the fact that even with only one occupant, energy consumption efficiency decreases with the use of basic household appliances [13]. Thus, energy experts are pointing out the problems with the increasing number of single-person households, and the need for countermeasures. Previous studies found that energy research, based on the behaviors of occupants, was necessary to effectively reduce energy consumption in residences. In particular, they observed demographic changes from the global trend of increasing single-person households and recognized the importance of, and the need for, research into energy consumption in single-person households.

Therefore, this study derives types of "energy usage patterns" (EUPs) using K-modes clustering for single-person households in South Korea. In addition, energy consumption data were extracted from EnergyPlus based on the behaviors of occupants, and three types of energy consumption data (total energy, cooling, and heating) were provided for each EUP type using a Gaussian process regression, in order to improve the use of this studies results. Finally, based on the results of a support vector machine (SVM), a model for predicting EUP types was implemented via household characteristics and living patterns.

2. Materials and Methods

In Section 2, we describe the overall research process, the data, and methods used for the purposes of the research.

2.1. Research Process

Figure 1 shows the research process for creating a model to predict EUP types with energy consumption information, according to the behaviors of single-person households in South Korea.

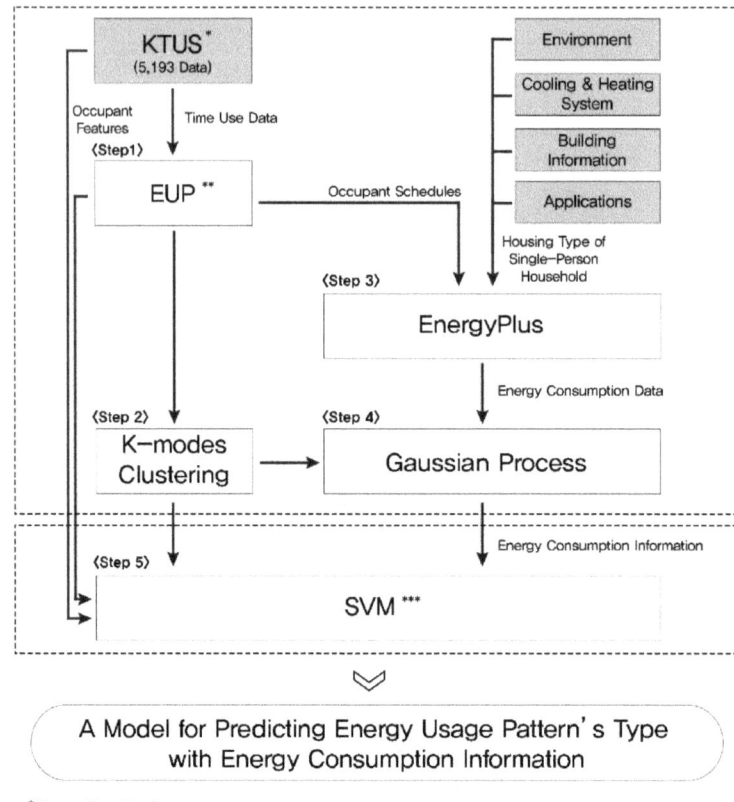

Figure 1. Research process for predicting energy usage pattern types with energy consumption information.

- Step 1: Build 5193 EUP datasets by redefining 5193 "Korean Time Use Survey (KTUS)" data points [14] through codes related to energy consumption.
- Step 2: Derive EUP types by performing K-modes clustering analysis with the EUP dataset built in Step 1.
- Step 3: Create 5193 "occupant schedules" to become inputs into EnergyPlus using the EUP dataset built in Step 1. Implement representative residential environments of Korean single-person households, and extract 5193 points of "energy consumption data" based on occupant behaviors through energy simulation.
- Step 4: Provide an "Energy Consumption Information" model for three items (total, cooling, and heating) for each type through a Gaussian process with the EUP types and energy consumption data, which are the results of Steps 2 and 3, as input.
- Step 5: Train the SVM model to predict EUP types through 5193 occupant features and EUPs, and verify the suitability of the model through the test process of the trained predictive model.

2.2. Generating EUP Datasets from the Korean Time Use Survey

The purpose of the KTUS is to determine the lifestyles of Koreans for 24 h and provide basic data for policy formulation and academic research in the sectors of labor, welfare, culture, transportation, etc. [15]. The KTUS is performed every five years in 800 regions in South Korea. The survey data largely consist of three parts: Residences, residents, and time logs. This allows us to understand basic information and living patterns of the residents. The time logs are created for 144 times slots that divide 24 h into 10-min units based on the actual living pattern of residents. The behaviors and places of the residents in each time slot are recorded as codes. The behavior classification codes consist of nine major categories, 42 middle categories, and 138 subcategories. These data allowed us to find out the detailed behaviors of residents. Furthermore, it includes the residents' location information, which allowed us to find out what the residents did at specific locations in their residences. We obtained the KTUS data through the Microdata Integrated Service (MDIS) [14].

In the KTUS data for 2014, which was obtained through the MDIS in this study, 5240 pieces of data were collected that satisfied the conditions of urban residents, with household members 10 years or older in single-person households. Among them, 5193 KTUS data points were used for this study, excluding 47 data points in which the residents were not in their residences for 24 h.

First, to perform the K-modes clustering and EnergyPlus analysis based on the KTUS data, the 138 pieces of behavior classification data were redefined as codes related to energy consumption, as shown in Figure 2.

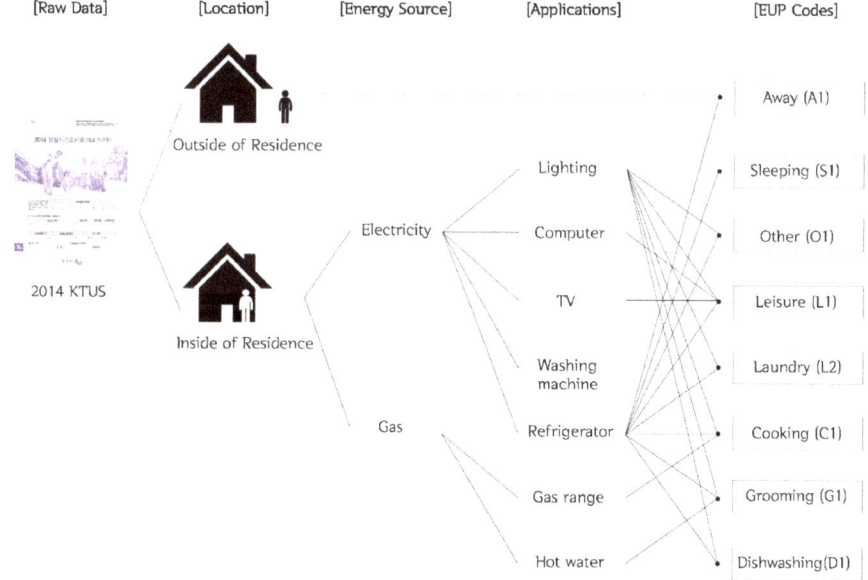

Figure 2. EUP data code definitions according to 5193 resident behaviors in the KTUS.

Applications related to energy consumption in the residences were assumed to be related to lighting, computer, TV, washing machine, refrigerator, gas range, and hot water usage. Occupant behaviors were classified into eight energy consumption behaviors by identifying which application was used [16].

The location of the occupants was divided into inside and outside the residence through the "behavior location" information of the KTUS; when the occupant was outside the residence, it was set as Away (A1). Refrigerators are assumed to operate continuously regardless of the presence or behavior of occupants. By identifying which application was used according to the occupants' behavior, the behaviors were classified into the behavior codes of: Leisure (L1), Laundry (L2), Cooking (C1), Grooming (G1), and Dishwashing (D1). In addition, when the occupant slept, it was assumed that the lights were out, and the behavior was classified as Sleeping (S1). Behaviors that were not directly related to energy consumption were defined as Other (O1). Table 1 lists the energy application and energy consumption behavior codes defined through this process.

Table 1. EUP codes related to energy-using applications.

	Electricity				Gas		Energy Consumption Behavior
Lighting	Computer	TV	Washing Machine	Refrigerator	Gas Range	Hot Water	EUP Codes
✓	✓			✓			A1, S1
✓				✓			O1
✓		✓		✓			L1
✓			✓	✓			L1
✓				✓		✓	L2
✓				✓	✓		C1
✓				✓		✓	G1, D1

2.3. Deriving EUP Types and an Energy Consumption Information Model for Each Type

2.3.1. K-Modes Clustering

Clustering is a machine learning analysis method used for various purposes, such as pattern analysis, grouping, and classification. Among the clustering methods, K-means clustering, which was first proposed by MacQueen [17], finds appropriate center values by clustering data that have the most similar shapes from the centers of k items that have been randomly selected, and combines them into a group of items that have similar characteristics. This is the most popular clustering method, it is however limited in that it can only be applied to data consisting of numerical variables [18]. K-means clustering is not appropriate for this study because clustering must be performed with categorical variables consisting of behavior classification codes rather than numerical variables.

To complement the limitations of the existing K-means clustering, Huang [19] proposed K-modes clustering. K-modes clustering is applicable to data formats consisting of categorical variables rather than numerical variables. Categorical variables refer to data that cannot be expressed by numbers, such as men and women. These data only provide information but cannot express quantities such as numbers.

K-modes clustering is composed of a distance function for measuring the distance between objects, and a cost function for optimized analysis. These functions are explained in detail below.

D is a set of objects X, which consist of nominal variables. The number of X's is n, and $D = [X_1, X_2, \ldots, X_n](1 \le i \le n)$. If object X_i has m nominal variables, $X_i = [x_{i,1}, x_{i,2}, \ldots, x_{i,m}](1 \le j \le m)$. Z is a set of k cluster centers, and $Z = [Z_1, Z_2, \ldots, Z_k](1 \le l \le k)$. If Z_l, which is the center of cluster C_l, has m nominal variables, $Z_l = [z_{l,1}, z_{l,2}, \ldots, z_{l,m}] \, (1 \le j \le m)$.

The equation for obtaining the distance between the center of cluster C_l, $Z_l = [z_{l,1}, z_{l,2}, \ldots, z_{l,m}]$, and the object $X_i = [x_{i,1}, x_{i,2}, \ldots, x_{i,m}]$ through the distance function is as follows:

$$\text{Distance function}: d\left(x_{i,j}, z_{l,j}\right) = \begin{cases} 0, & x_{i,j} = z_{l,j} \\ 1, & x_{i,j} \ne z_{l,j} \end{cases}, \tag{1}$$

where $d\left(x_{i,j}, z_{l,j}\right)$ is a function for measuring the distance between Z_l and X_i, and calculates the distance between two objects by comparing the jth variables. The distance between each object X and the cluster center Z is calculated through Equation (1), and the model is optimized in such a way that the result value of Equation (2) is minimized:

$$\text{Cost function}: P(U, Z) = \sum_{l=1}^{k} \sum_{i=1}^{n} \sum_{j=1}^{m} u_{i,l} d\left(x_{i,j}, z_{l,j}\right) \tag{2}$$

$U = [u_{i,l}]$ is an $n \times k$ matrix consisting of 0 and 1. $u_{i,l} = 1$ indicates that object X_i is assigned to the closest cluster C_l. To optimize the model, the cluster center is reset, and this is repeated until the minimum value of the cost function $P(U, Z)$ is obtained. Then, the analysis is stopped, and the result of the cluster analysis is derived.

This study aims to derive EUP types using K-modes clustering, which consists of the following process. First, a dataset for 144 time slots for 5193 objects was prepared through the EUP codes defined in Section 2.2 (Table 2). The analysis environment was built through the K-modes clustering algorithm suggested by Huang [19] using Python 3.6.

Table 2. EUP dataset format for K-modes clustering.

Household No.	Time Step 1 (00:00–00:10)	Time Step 2 (00:10–00:20)	Time Step 3 (00:20–00:30)	...	Time Step 143 (23:40–23:50)	Time Step 144 (23:50–24:00)
1	S1	S1	O1	...	G1	A1
2	G1	G1	S1	...	L1	L1
3	A1	A1	A1	...	S1	S1
...	...	...	...	...	...	...
5192	A1	O1	O1	...	S1	S1
5193	C1	C1	L2	...	A1	A1

Next, the analysis for each k was performed 1000 times to derive the minimum of the cost function, and find the optimized k. The K-modes clustering analysis and the method of finding the optimized k is as follows:

1. Select k random cluster centers.
2. Calculate the distance between each object and cluster center, and allocate each object to the closest cluster center. Allocate every object to clusters, and move the cluster center in the direction for which the distance between each object and cluster center becomes minimized.
3. Compare the center of the moved cluster with the previous center. If the result is different, return to Step 2 and repeat. The analysis is stopped if the same result value, when compared with the previous center, is generated.

Finally, to select the optimized k, k is found by drawing an elbow curve using the errors of the model for each k (Figure 3). In this study, $k = 7$ was found to be the most optimized model through visualization of the analysis results 6, 7, and 8, which had small variations in the error reduction in the elbow curve.

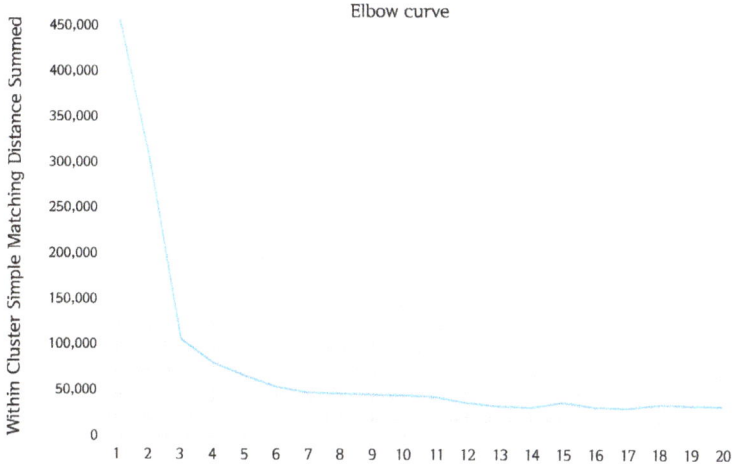

Figure 3. Elbow curve for error values of the model for each k.

2.3.2. EnergyPlus

EnergyPlus is a dynamic energy simulation based on the strengths of Building Loads Analysis and System Thermodynamics (BLAST) and DOE-2 in the U.S. in 1996 [20]. EnergyPlus can implement more concrete building materials and facility systems, and makes it easier to control facility systems than do other energy simulations [21]. Furthermore, EnergyPlus can consider the behaviors of occupants in the simulation. It is widely used in energy research that considers the behaviors of occupants. The use of applications, cooling and heating, and natural ventilation can be controlled in 1-min units according to the behavior of occupants, resulting in a simulation environment that is more similar to real-world situations. Therefore, in this study, 5193 energy consumption amounts were derived based on the occupant behavior through EnergyPlus.

The target building of this study is located in a residential area, and the inside of the building has a floor plan that is typical in South Korea. Hence, it can be regarded as a representative residence type for single-person households in South Korea (Figure 4). Happy houses are for households consisting of one or two persons, such as college students, newlyweds, and career starters. The construction of happy houses started in 2014, and the goal was to build 150,000 happy houses by 2017 [22]. For this study, we obtained the design drawings and energy savings plan for the "S Happy House" from the Korea Land and Housing Corporation (LH), which we used as the basic data for accurate energy simulation. S Happy House was a 41 m^2 unit on the fourth floor of a building in Seoul and was certified as Class 3 in the energy efficiency rating system, and can be used as a reference for energy consumption in future research. The fourth floor, which is the middle floor of the residential floors, was selected because the top and bottom floors are heavily influenced by the external environment [23].

(a)

Figure 4. *Cont.*

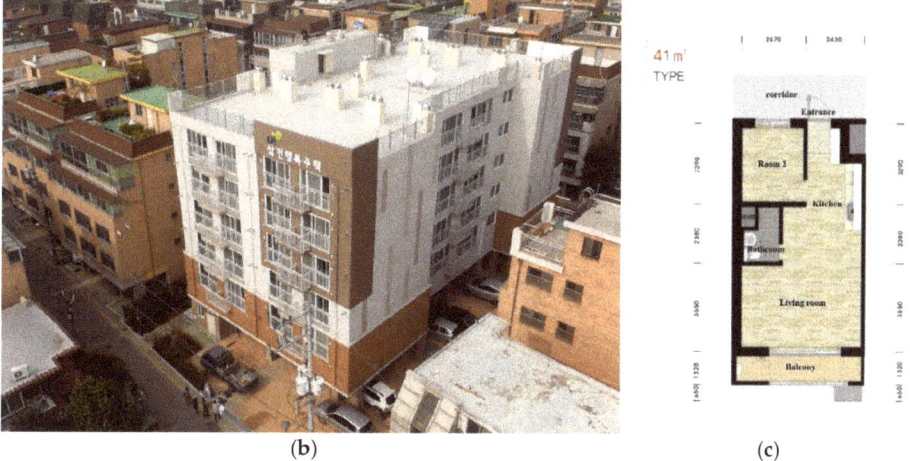

Figure 4. The "S Happy House" located in Seoul, which is the energy simulation target area: (a) The target site located in a residential area; (b) a bird's-eye view of the S Happy House; (c) a floor plan for the 41 m² module.

The main inputs and settings for the EnergyPlus simulation are listed in Table 3. The five main input items of EnergyPlus are the environment, building information, cooling and heating systems, applications, and occupant schedules.

1. The input data for "Environment" included the location, which was based on the latitude and longitude of the target site, Greenwich Mean Time, surrounding city environment, and weather data. This was the external environment information that affected the building energy. For weather data, which was not provided by EnergyPlus, the "Korea standard weather data" from Seoul were used [24].
2. "Building Information" included a 3D model, material, and the composition of the simulation object. To build a simulation environment that was similar to the actual environment, we received the energy savings plan and drawings of the S Happy House from LH.
3. For the "Cooling and Heating System", a wall-mounted air conditioner and floor heating system were applied [25]. The cooling and heating temperatures for a comfortable indoor environment were set as 20 °C and 26 °C, respectively, by referring to the Certification Standard for Building Energy Efficiency Rating and Zero Energy Building [26]. Water was delivered from an external source, but hot water was heated through an individual boiler.
4. "Applications" include electric lights, refrigerator, TV, computer, washing machine, and gas range. For the power standards of these applications, the "Survey Of Household Appliance Penetration And Household Power Consumption" was referenced [16].
5. "Occupant Schedules" corresponds to every schedule used for simulations based on the occupant's behaviors. In this study, 5193 data points obtained through the KTUS were used.

The physical elements such as the S Happy House and surrounding environment were constructed using SketchUp 2017, through the drawings received from LH. Furthermore, the framework of the cooling and heating systems structure was constructed by converting the three dimensional (3D) model with OpenStudio2.5.1. The Korean standard weather data for Seoul were converted to an epw file for weather data to input into EnergyPlus. Based on the EUPs derived in Section 2.2, 5193 occupant schedules were created.

As shown in Figure 5, 5193 IDF files is the extension name for the EnergyPlus were created through the collected information, from which 5193 energy consumption data points were finally obtained through the EnergyPlus simulation.

Table 3. Energy simulation settings through EnergyPlus.

EnergyPlus Input Items		Details	Used Tools
Environment	Location	Seoul Latitude: 37.49 Longitude: 127.09 Time zone: GMT + 9	Elements1.0.6
	Weather	Standard weather data in Seoul	
Building Information	3D Modeling	S Happy House	SketchUp2017 OpenStudio2.5.1 IDF Editor
	Material	Glass, Concrete, EPS, etc.	
	Construction	External · Internal wall External · Internal window Floor, Roof, Door etc.	
Cooling and Heating Systems	Ductless air-conditioning	Set point: 26 °C	OpenStudio2.5.1 IDF Editor
	Low temperature radiant system	Set point: 20 °C	
Applications	Lighting	25 W * 4 ea	IDF Editor
	Refrigerator	40.6 W * 1 ea	
	TV	130.6 W * 1 ea	
	Computer	255.9 W * 1 ea	
	Washing machine	242.8 W * 1 ea	
	Gas range	2100 W * 1 ea	
	Hot water	3 ea	
Occupant Schedule	Activity level	5193 schedules based on EUP data (time step: 10 min)	Excel 2013 IDF Editor
	Applications Operation		

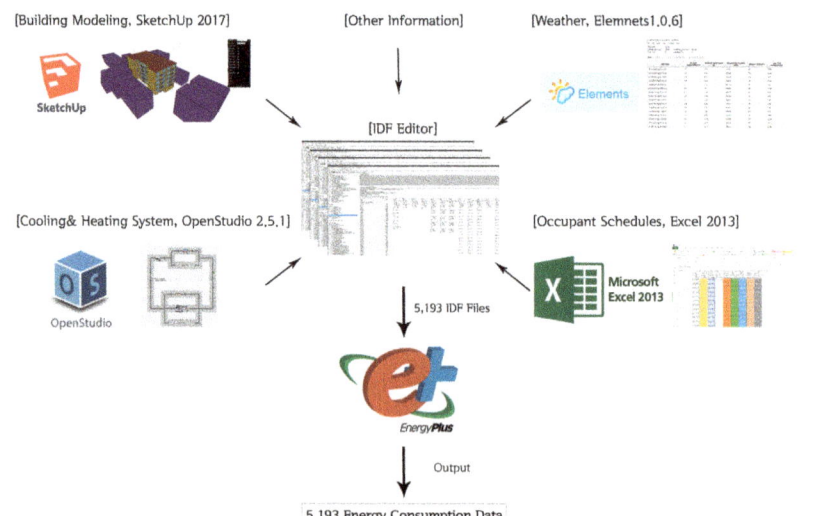

Figure 5. Process of deriving the 5193 energy consumption data points based on occupants' behaviors using EnergyPlus.

2.3.3. Gaussian Process Regression

Gaussian process regression (GPR) [27] is a black box model that can flexibly cope with nonlinear data using a Bayesian approach. It is appropriate for highly uncertain energy research because it can perform probabilistic prediction (Figure 6) [28]. Research is being conducted to build energy-related models using GPR [29–32]. In this study, an energy consumption information model was provided through GPR using the daily energy consumption data for 365 days according to each EUP type.

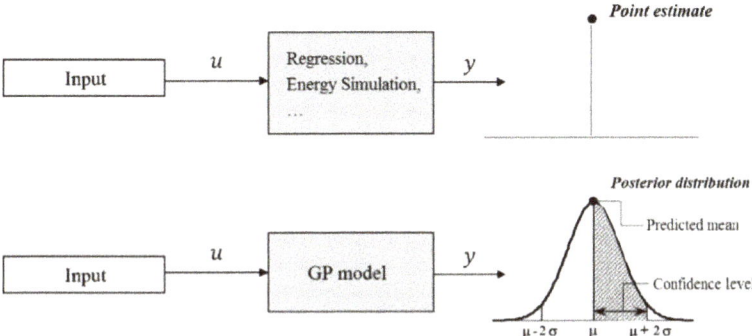

Figure 6. Two different Gaussian process models: A general predictive model and probabilistic predictive model. Source Reference [28].

GPR is a linear regression model estimated from n training data $\{(x_i, y_i); i = 1, 2, \ldots n\}$, as shown in Equation (3). When the mean is zero and the variance is σ^2, ε is Gaussian noise, as shown in Equation (4). β is a coefficient estimated from the data, and x^T is expressed as an input vector:

$$y = x^T \beta + \varepsilon, \tag{3}$$

$$\varepsilon \sim N(0, \sigma^2). \tag{4}$$

When a training dataset is given, the post distribution of β can be estimated through the Bayesian approach, which consists of a pre-distribution and a likelihood functions. The equation of the GPR model is composed of a mean function shown in Equation (5), and a covariance function shown in Equation (6). Because the mean function has a value of zero, it can be expressed as Equation (7):

$$E(f(x)) = m(x), \tag{5}$$

$$k(x, x') = E[\{f(x) - m(x)\}\{(f(x') - m(x')\}], \tag{6}$$

$$f(x) \sim GP(0, k(x, x')). \tag{7}$$

Therefore, $f(x)$ has a zero mean and follows the covariance function $k(x, x')$. The covariance function can be defined as various kernel functions, such as squared exponential, exponential, matern, rational quadratic, and automatic relevance determination. The most universal function is the squared exponential kernel (SE) function [33], which is widely used in building energy prediction research [28,34–37]. In this study, the SE method was adopted, and the kernel function is expressed as Equation (8). $k(x, x')$ is parameterized by the kernel parameter or hyper parameter θ, and $k(x, x')$ is dependent on the θ value. Therefore, it can be expressed as $k(x, x'|\theta)$. Here, σ_f denotes the signal standard deviation, and σ_l denotes the characteristic length scale:

$$k(x_i, x_j|\theta) = \sigma_f^2 \exp[-\frac{1}{2} \frac{(x_i - x_j)^T (x_i - x_j)}{\sigma_l^2}]. \tag{8}$$

For the GPR tool, the statistics and machine learning toolbox of MATLAB R2018b were used. For the algorithm, Williams et al. [33] was referenced. For a detailed explanation of the equation, the MATLAB User's Guide [38] can be referenced. To predict the energy consumption by EUP type for 365 days in this study, three energy consumption information models were created for total energy, cooling, and heating for each type, with x as the date and y as the daily energy consumption data.

2.4. Methodology for Predicting EUP Types Used by the SVM

SVM is a model that can solve the classification and regression problems through supervised learning. The SVM model can effectively respond to various problems because it can be applied to linear and nonlinear problems [39]. The idea behind SVM is to create lines or hyperplanes that classify data into classes [40]. Therefore, it creates lines or hyperplanes that can distinguish different outputs (i.e., classes) by entering input x and output y data. SVM models include linear, quadratic, cubic, fine Gaussian, medium Gaussian, and coarse Gaussian SVMs, depending on the kernel function for implementing the lines or hyperplanes that distinguish classes (Table 4). In this study, six SVM models were trained through the Classification Learner App of MATLAB R2018b, and the most appropriate SVM model was selected [38].

Table 4. Six SVM models put through the Classification Learner App of MATLAB R2018b.

SVM Type	Model No.	Kernel Function	Kernel Scale Mode	Multiclass Method
Linear SVM	1.1	Linear kernel	Auto *	
Quadratic SVM	1.2	Quadratic kernel	Auto *	
Cubic SVM	1.3	Cubic kernel	Auto *	One-vs-One **
Fine Gaussian SVM	1.4	Gaussian kernel	3.1	
Medium Gaussian SVM	1.5	Gaussian kernel	12	
Coarse Gaussian SVM	1.6	Gaussian kernel	49	

* When you set the Kernel scale mode to Auto, the software uses a heuristic procedure to select the scale value; ** only for data with three or more classes. This method reduces the multiclass classification problem to a set of binary classification subproblems, with one SVM learner for each subproblem. One-vs-One trains one learner for each pair of classes.

To train the SVM model, input data x and output data y are required. The x data provide information for predicting y, and consist of the factors affecting y. The y data are the results obtained from x, and y is the answer that we want to eventually obtain through the predictive model. In this study, x consisted of five occupant features and 144 EUP data points. Furthermore, the EUP types defined by K-modes clustering in Section 2.3.1 correspond to y.

Table 5. Dataset format for building a predictive model for EUP types through SVM.

Household No.	Occupant Features					EUP Data			Type
	Age	Gender	Income	Working	Care Needs	Time1	...	Time144	
	x_1	x_2	x_3	x_4	x_5	x_6	...	x_{149}	y
1	42	woman	3	yes	no	S1	...	A1	6
2	67	woman	1	no	no	G1	...	L1	7
3	70	man	1	no	yes	A1	...	S1	2
...	...	...	...	...	...	...	...	...	...
5192	34	man	2	yes	no	A1	...	S1	3
5193	25	man	1	yes	no	C1	...	A1	4

Occupant behavior varies by occupant features such as age, gender, income, working status, and care needs, which can lead to differences in EUP. Many studies have been conducted on occupant behavior and energy consumption according to occupant features [41–43]. In this study, a predictive model for the SVM EUP type was constructed using occupant features and EUP, and the dataset for this is shown in Table 5.

3. Results and Discussion

As a result of the research, Section 3 has the following contents. Seven EUP types of single-person households have been derived in South Korea. In addition, we analyzed the differences in household characteristics and energy consumption of each type and constructed an energy information model through Gaussian process regression. Finally, we developed an EUP type predictive model with 95% accuracy through SVM.

Deriving the EUP Types and Analysis of Occupant Features through K-Modes Clustering

As a result of K-modes clustering with 5193 EUP data points using the method outlined in Section 2.3.1, seven EUP types of single-person household occupants were derived. The household characteristics of seven EUP types based on the analysis results can be seen in Figure 7, which visualizes the energy consumption behavior probability of occupants by time slot for the seven EUP types, and in Table 6, which outlines the five household characteristics used in the SVM.

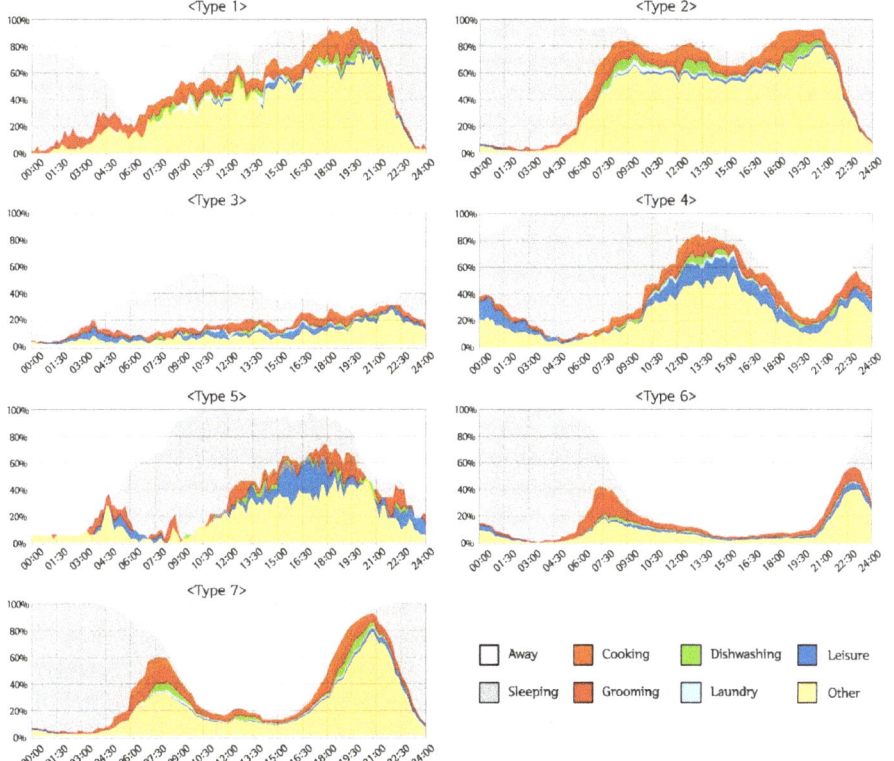

Figure 7. Energy consumption behavior probability by time slot for the seven EUP types derived through K-modes clustering.

- Type 1 was a cluster that mainly consists of people aged 65 or older (51.7%) with low incomes. They were mainly involved in outside activities between 04:30 and 09:00, and in indoor activities in the residence at other times.
- Type 2 had the highest ratios of household members who are 65 or older (66.0%), female (68.8%), had monthly incomes less than 1 million won (75.8%), and had care needs (65.78%). They had almost no outside activities or hobbies. Their main space of activity was their residence.

- Type 3 had the highest ratio of economic activities (81.4%), and their daily work started in the afternoon, after 13:00. After their economic activities were finished in the late evening, they mainly slept in their residence between 04:00 and 13:00. Besides sleeping, their activities related to energy consumption were few, and their living patterns were irregular.
- Type 4 spent a similar amount of time in the residence as Type 1. Their genders, monthly incomes, and jobs were similar, but their age was different. The time slot for the main outside activities of Type 4 was 18:00 to 22:30. They spent a high percentage of time in hobby activities in their residence.
- Type 5 had a sleeping time period similar to that of Type 3, but after sleeping they mainly spent time in their residence enjoying hobbies. They went out in the evening between 22:00 and 08:00.
- Type 6 had a high percentage of youth who were active in economic and outside activities, with the lowest ratio of people aged 65 or older (9.0%). Furthermore, they spent the shortest amount of time in their residences. The main activity in the residence was personal hygiene, and they had the least amount of time spent on other activities.
- Type 7 comprised 2074 out of the 5193 data points. This seemed to be the living pattern type of general office workers. However, based on the percentage of those 65 years or older (45.4%) and without work (51.2%), this type also included people who spent their personal time mainly outside their residence, besides office workers.

Table 6. Percentages of five household characteristics by EUP type.

Type	Age		Gender		Income (Unit: 10,000 w)		Working		Care Needs
Type 1 (60 persons)	~30 31~50 51~64 65~	7 (11.7) 10 (16.7) 12 (20.7) 31 (51.7)	M W	28 (46.7) 32 (53.3)	~100 100~200 200~300 300~400 400~500 500~	31 (51.7) 14 (23.3) 9 (15.0) 3 (5.0) 3 (5.0) 0 (0.0)	Y N	36 (60.0) 24 (40.0)	2 (1.07)
Type 2 (1422 persons)	~30 31~50 51~64 65~	71 (5.0) 143 (10.1) 270 (19.0) 938 (66.0)	M W	444 (31.2) 978 (68.8)	~100 100~200 200~300 300~400 400~500 500~	1078 (75.8) 219 (15.4) 60 (4.2) 32 (2.3) 13 (0.9) 20 (1.4)	Y N	283 (19.9) 1139 (80.1)	123 (65.78)
Type 3 (161 persons)	~30 31~50 51~64 65~	43 (26.7) 66 (41.0) 37 (23.0) 15 (9.3)	M W	94 (58.4) 67 (41.6)	~100 100~200 200~300 300~400 400~500 500~	42 (26.1) 53 (32.9) 50 (31.1) 8 (5.0) 5 (3.1) 3 (1.9)	Y N	131 (81.4) 30 (18.6)	4 (2.14)
Type 4 (176 persons)	~30 31~50 51~64 65~	79 (44.9) 62 (35.2) 14 (8.0) 21 (11.9)	M W	87 (49.4) 89 (50.6)	~100 100~200 200~300 300~400 400~500 500~	85 (48.3) 47 (26.7) 33 (18.8) 7 (4.0) 2 (1.1) 2 (1.1)	Y N	101 (57.4) 75 (42.6)	2 (1.07)
Type 5 (33 persons)	~30 31~50 51~64 65~	15 (45.5) 12 (36.4) 2 (6.1) 4 (12.1)	M W	14 (42.4) 19 (57.6)	~100 100~200 200~300 300~400 400~500 500~	7 (21.2) 6 (18.2) 12 (36.4) 2 (6.1) 2 (6.1) 4 (12.1)	Y N	31 (93.9) 2 (6.1)	2 (1.07)
Type 6 (1267 persons)	~30 31~50 51~64 65~	384 (30.3) 500 (39.5) 269 (21.2) 114 (9.0)	M W	673 (53.1) 594 (46.9)	~100 100~200 200~300 300~400 400~500 500~	325 (25.7) 425 (33.5) 309 (24.4) 104 (8.2) 50 (3.9) 54 (4.3)	Y N	1004 (79.2) 263 (20.8)	7 (3.74)

Table 6. Cont.

Type	Age		Gender		Income (Unit: 10,000 w)		Working		Care Needs
Type 7 (2074 persons)	~30 31~50 51~64 65~	206 (9.9) 384 (18.5) 543 (26.2) 941 (45.4)	M W	731 (35.2) 1343 (64.8)	~100 100~200 200~300 300~400 400~500 500~	1180 (56.9) 476 (23.0) 237 (11.4) 71 (3.4) 47 (2.3) 63 (3.0)	Y N	1012 (48.8) 1062 (51.2)	47 (25.13)

a. *Comparison of Energy Consumption by EUP Type According to the Energy Simulation Results*

A total of 5193 energy simulations were performed based on occupants through EnergyPlus. To compare the energy consumption among the seven EUP types, the data were expressed as boxplots, as shown in Figure 8. The analysis was performed for five categories: Electric light, applications, cooling, heating, and total energy consumption for one year. Furthermore, to facilitate the comparison of energy consumption, a ranking list was prepared for each item, as shown in Table 7.

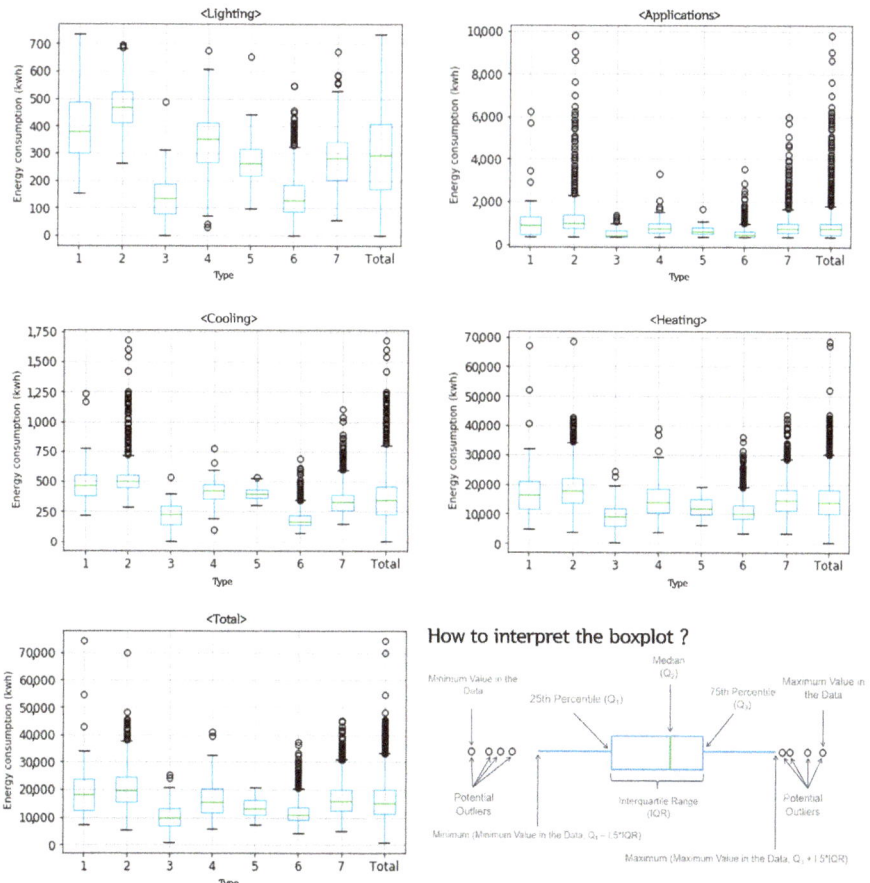

Figure 8. Comparison of the total annual energy consumption for lighting, applications, cooling, heating, and the overall total for each EUP type.

As shown in Table 7, the annual total energy consumption was the highest for Type 2, followed by Type 1, Type 7, Type 4, Type 5, Type 6, and Type 3, respectively. Type 1 and Type 2 ranked highest for all five items. They seem to be largely affected by the occupancy time, need for a comfortable indoor environment, and personal life. For Type 2, the percentage of households that had care needs was 65.78%, and, since they had a high percentage of people staying at home, they had the highest energy consumption. Furthermore, Type 2 had the highest energy consumption, which was twice as high as that of Type 3, which had the lowest energy consumption in the same conditions (except for daily living pattern).

Table 7. Energy consumption ranking by type.

	Type 1	Type 2	Type 3	Type 4	Type 5	Type 6	Type 7
Lighting	2	1	6	3	5	7	4
Equipment	2	1	7	4	5	6	3
Cooling	2	1	6	3	4	7	5
Heating	2	1	7	4	5	6	3
Total	2	1	7	4	5	6	3

Type 4, which had a similar occupancy time as Type 1, showed lower energy consumption in every item. The largest differences in household characteristics between the two types were age and occupancy time slot. Type 4, which had a high percentage of youth, mainly saw occupants spending time for sleeping, leisure, and personal management. Type 1, which had a high percentage of elders, had large energy consumption amounts for residential environment maintenance, such as food preparation and dishwashing. Even though their occupancy times were similar, the energy consumption patterns of the two types were different due to the differences in household characteristics and living patterns.

Furthermore, there were some cases in which the occupancy time was long, but the energy consumption was low. Type 4 had a longer occupancy time than did Type 7, but Type 7 had greater energy consumption in applications and heating. Type 7, which had similar household characteristics as Type 1, had high energy consumption for applications because their living pattern focused on maintaining their residential environment. In the case of heating energy, unlike Type 7, Type 4 household members stayed at home during the day when the outside temperature was higher in the winter, and because their time outside was short, they consumed less energy in returning the indoor temperature to a proper, comfortable temperature. However, in the summer, Type 4 members consumed higher cooling energy than did Type 7 due to the high outdoor temperatures during the day. It appears that Type 7 reduced energy consumption due to cooling energy by maintaining the indoor temperature at 26 °C or lower through natural ventilation during the evening. This was clearly seen through a comparison of Type 5 and Type 6, which had opposite time slots for staying at home, although their occupancy times were similar.

According to Fong et al. [44], age and sex influence energy consumption, and the higher the percentages of elders and women, the higher the energy consumption was. The results of this study also suggest that age and sex have a large effect on energy consumption. Out of the seven types, Type 1, Type 2, and Type 7 had the highest percentages of people 65 years or older, and they correspond to 1st, 2nd, and 3rd for their annual total energy consumption, and all three types had the highest percentages of women. Next, Type 4 and Type 5, which ranked 4th and 5th, respectively, also had a higher percentage of women than men. Type 6 and Type 3, which ranked 6th and 7th, respectively, had a higher percentage of men. Therefore, these results suggest that age and sex can cause differences in lifestyles and energy consumption.

Income and work had an effect on occupancy time, and caused differences in energy consumption. Type 3, Type 5, and Type 6 had jobs, and the higher the income, the lower the energy consumption was. For Type 3 and Type 5, the percentage of people with jobs was 81.4% and 93.9%, respectively. According to a previous study on energy consumption based on income for Koreans, the higher the economic level, the lower the constraints for energy consumption and the higher the energy consumption to maintain the environment of residence [45,46]. However, in this study, the higher the income, the lower the energy consumption was. This result seems to be because of the nature of single-person households; the more actively they are involved in economic activities, the higher their income, and the lower their occupancy time in the residence are.

b. *Energy Consumption Information Model for Total Energy, Cooling, and Heating for Each Type Through Gaussian Process Regression*

To provide energy consumption data for each EUP type, GPR was performed, as described in Section 2.3.3, and the information model for total energy, cooling, and heating was created. The information model is composed of three elements. The total model (Figure S1) can identify the overall energy consumption pattern. The cooling (Figure S2) and heating (Figure S3) models are most heavily influenced by the climate. To examine the energy consumption by period, the three information models are presented in the Appendix A [47].

In the figures in the Appendix A, the blue dots indicate the actual data, and the red dots comprise the boundary representing a 99% confidence interval. A larger space between the top and bottom lines means greater diversity in energy consumption. The black solid line is the mean value where the largest amount of data is located. The top and bottom 99% prediction interval lines are drawn above and below this line.

c. *Predictive Model of EUP Types Through Occupant Features and EUPs*

The predictive model was evaluated through six SVMs, according to the kernel function types and options presented in Section 2.4. Each model was trained using 80% of the total data as the training data, and the model was evaluated with 20% data. As a result, the prediction rate of each model was as shown in Table 8. Thus, Model 1.2 was most appropriate for predicting the EUP type.

Table 8. Comparison of the prediction rates of six SVM models.

SVM Type	Model No.	Prediction Speed (No. of Observations Per s)	Training Time (s)	Accuracy (%)
Linear SVM	1.1	~1900	7.6457	94.0%
Quadratic SVM	1.2	~1300	8.1307	95.0%
Cubic SVM	1.3	~1400	8.4946	94.5%
Fine Gaussian SVM	1.4	~170	57.533	51.6%
Medium Gaussian SVM	1.5	~1300	9.383	93.8%
Coarse Gaussian SVM	1.6	~1000	12.078	91.0%

To examine the prediction performance of Model 1.2 in detail, a confusion matrix was drawn, as shown in Figure 9. The probabilities that the EUP-type prediction model would accurately predict the type were generally high: 42%, 93%, 96%, 90%, 75%, 98%, and 97%, respectively, according to type. However, when we examined Type 1 and Type 5, which showed relatively low prediction rates, it seemed that they were not sufficiently trained because they had a smaller number of data points compared to other types. This issue could be improved if we were to acquire more data in the future. Therefore, the EUP-type prediction model with a prediction rate of 95.0% was implemented through SVM. The results of this study showed applicability of EUP prediction in other countries and regions.

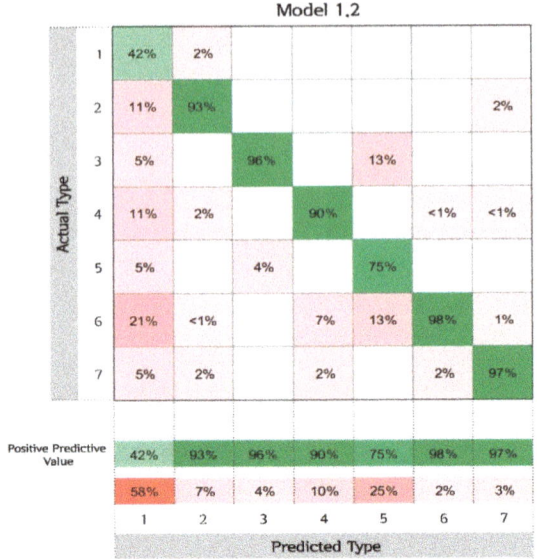

Figure 9. Confusion matrix for verifying the performance of the EUP-type prediction model.

4. Conclusions

The residential sector accounts for 3/4 of the total building energy consumption. Research and development of technologies to reduce energy consumption in residences is being actively conducted. However, the improvement of physical elements in buildings has limitations for energy reduction purposes. The paradigm is shifting towards considering the behaviors of occupants, as a means of overcoming these limitations and to create sustainable energy savings in the residential sector. Although studies considering the behaviors of occupants have been conducted, there have been difficulties due to limitations in data collection and privacy issues. Furthermore, it is difficult to analyze the behaviors of occupants because of their inherent complexity. This study attempted to overcome this limitation through data mining.

This study prepared large datasets for analysis based on the KTUS, which provides behavior data of actual residential occupants. Furthermore, EUP types were derived through K-modes clustering, and the household characteristics and energy consumption of each type were analyzed. As a result of K-modes clustering, seven EUP types were derived. Comparisons of the five household characteristics and energy consumption among the types revealed that people aged 65 or older and females had higher amounts of energy consumption than other groups. Unlike the results of previous studies, we found that the higher the economic level of a resident, the lower the energy consumption was, because of their occupancy time and occupancy time slots. This can be considered a characteristic of single-person households. The energy consumption showed a two-fold difference depending on the EUP type. Finally, an EUP-type prediction model with a prediction rate of 95.0% was implemented by training an SVM, and an energy consumption information model provided through GPR.

The processes of deriving the EUP types based on actual behavior data, energy simulation, and implementation of the EUP-type prediction model and the energy consumption information model, can be used as basic research data in future studies based on the behaviors of occupants, and they can be applied to other regions. In addition, EnergyPlus, Openstudio, and Python used in this study are open-source software and can increase the usefulness of this study in the future.

The limitation of this study is that the energy consumption data were created through energy simulations. If this limitation can be overcome, the outputs of the obtained research, through the process used in this study, can help prepare the framework for a building occupant energy management system in the coming era of smart buildings, which will lead to energy savings in the residential sector. There is also a limitation that the subject of this study is limited to single-person households. Recently, however, technology development and research on smart cities and buildings, enabling data collection, is in progress and a smart monitoring system for occupants will be available. Once the monitoring system is established, energy management will be possible considering the energy consumption patterns of occupants. The limitations could be overcome if later studies were conducted from data collection of Internet of Things (IoT) devices in multi-person households.

Supplementary Materials: The following are available online at http://www.mdpi.com/2071-1050/11/1/245/s1, Figure S1. Energy Consumption Information Model for total energy consumption over 365 days. Figure S2. Energy Consumption Information Model for cooling for 365 days. Figure S3. Energy Consumption Information Model for heating for 365 days.

Author Contributions: Conceptualization, S.K., and S.J.; data curation, S.K. and S.-M.B.; formal analysis, S.K.; methodology, S.K. and S.J.; project administration, S.K.; software, S.K.; supervision, S.J. and S.-M.B.; visualization, S.K.; writing—original draft, S.K.; writing—review and editing, S.K., S.J., and S.-M.B.

Funding: This work was supported by the National Research Foundation of Korea (NRF) grant funded by the Korean government (MSIT) (No. NRF-2018R1A2B2005528).

Acknowledgments: In this section you can acknowledge any support given which is not covered by the author contribution or funding sections. This may include administrative and technical support, or donations in kind (e.g., materials used for experiments).

Conflicts of Interest: The authors declare no conflict of interest.

Appendix A

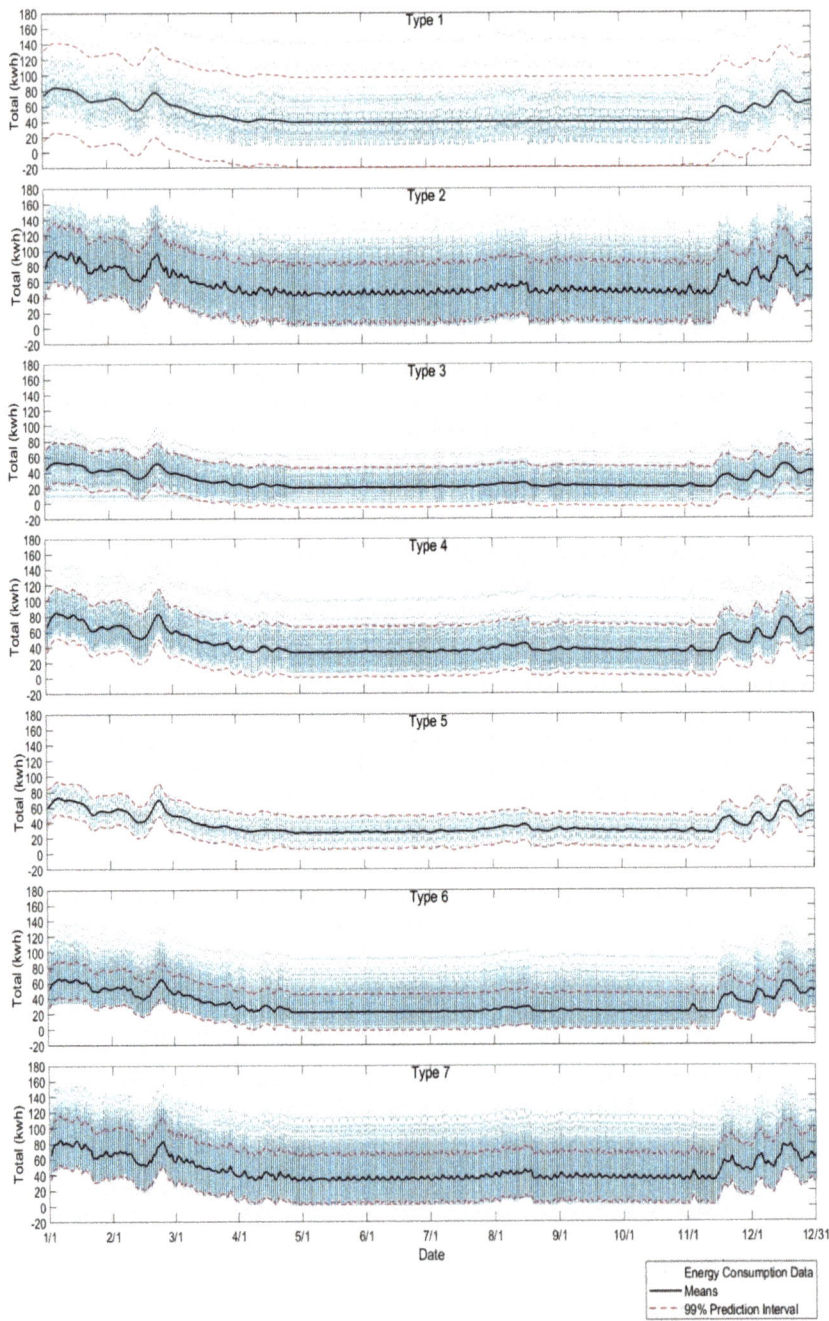

Figure A1. Energy Consumption Information Model for total energy consumption over 365 days.

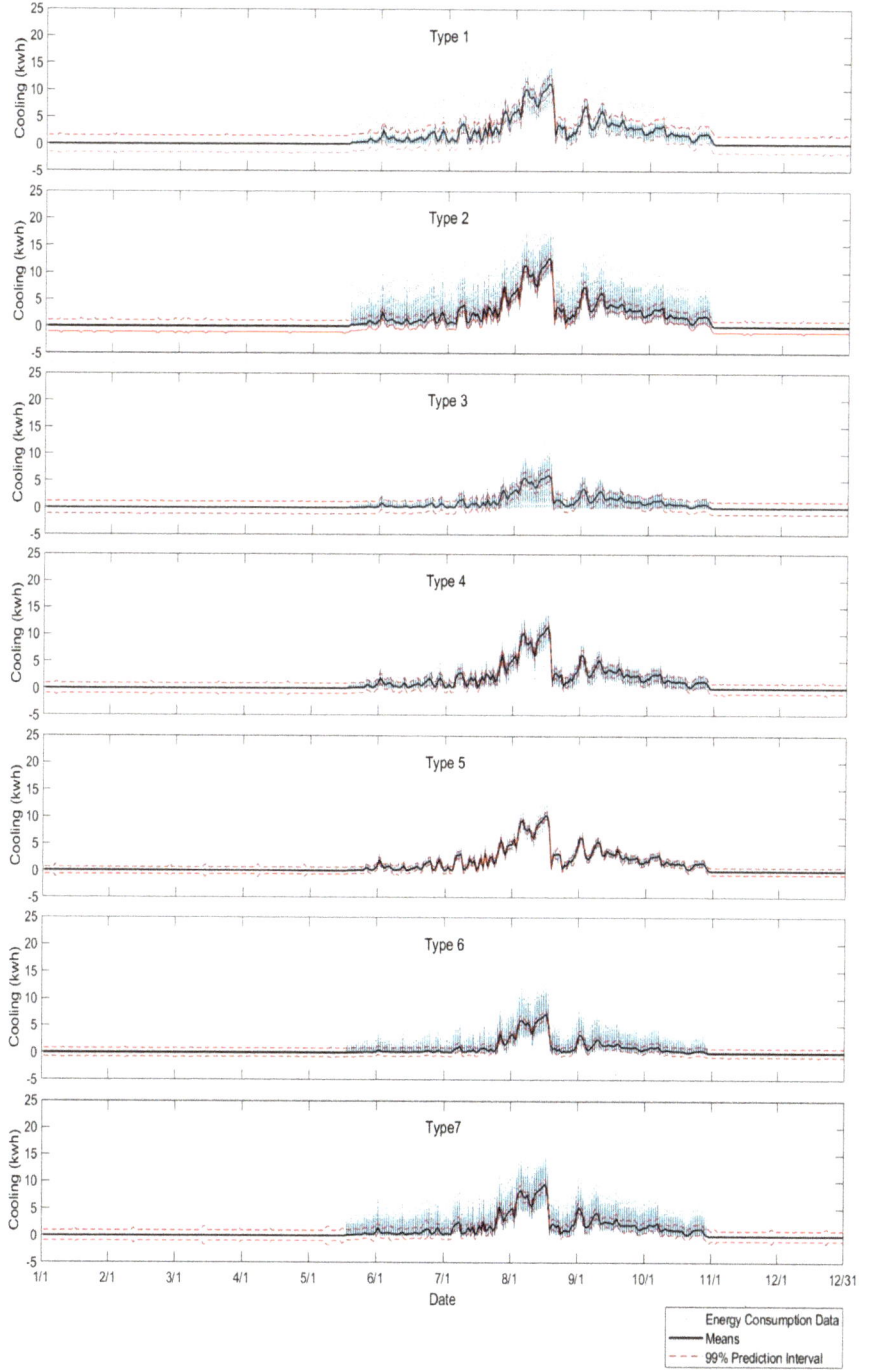

Figure A2. Energy Consumption Information Model for cooling for 365 days.

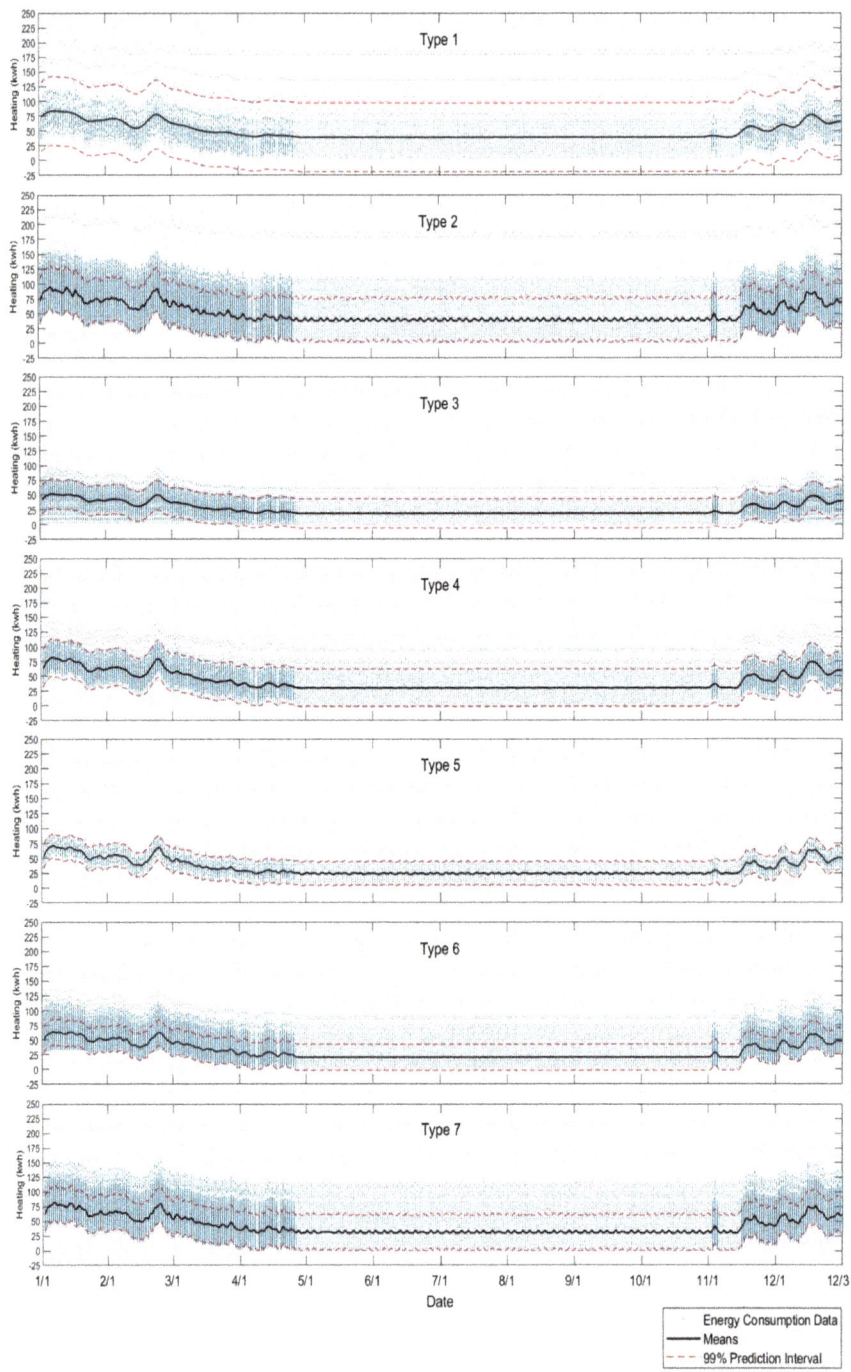

Figure A3. Energy Consumption Information Model for heating for 365 days.

References

1. Nejat, P.; Jomehzadeh, F.; Taheri, M.M.; Gohari, M.; Majid, M.Z.A.J.R. A global review of energy consumption, CO_2 emissions and policy in the residential sector (with an overview of the top ten CO_2 emitting countries). *Renew. Sustain. Energy Rev.* **2015**, *43*, 843–862. [CrossRef]
2. Conti, J.; Holtberg, P.; Diefenderfer, J.; LaRose, A.; Turnure, J.T.; Westfall, L. *International Energy Outlook 2016 with Projections to 2040*; USDOE Energy Information Administration (EIA), Office of Energy Analysis: Washington, DC, USA, 2016.
3. Guo, F.; Akenji, L.; Schroeder, P.; Bengtsson, M.J.E. Static analysis of technical and economic energy-saving potential in the residential sector of Xiamen city. *Energy* **2018**, *142*, 373–383. [CrossRef]
4. Hoicka, C.E.; Parker, P.J.E.E. Assessing the adoption of the house as a system approach to residential energy efficiency programs. *Energy Effic.* **2018**, *11*, 295–313. [CrossRef]
5. Gonzalez, J.M.; Pouresmaeil, E.; Canizares, C.A.; Bhattacharya, K.; Mosaddegh, A.; Solanki, B.V. Smart Residential Load Simulator for Energy Management in Smart Grids. *IEEE Trans. Ind. Electron.* **2018**, *66*, 1443–1452.
6. Menezes, A.C.; Cripps, A.; Bouchlaghem, D.; Buswell, R.J.A.E. Predicted vs. actual energy performance of non-domestic buildings: Using post-occupancy evaluation data to reduce the performance gap. *Appl. Energy* **2012**, *97*, 355–364. [CrossRef]
7. Ouyang, J.; Hokao, K.J.E. Energy-saving potential by improving occupants' behavior in urban residential sector in Hangzhou City, China. *Energy Build.* **2009**, *41*, 711–720. [CrossRef]
8. Guerra-Santin, O.; Bosch, H.; Budde, P.; Konstantinou, T.; Boess, S.; Klein, T.; Silvester, S.J.E.E. Considering user profiles and occupants' behaviour on a zero energy renovation strategy for multi-family housing in the Netherlands. *Energy Effic.* **2018**, 1–24. [CrossRef]
9. Brounen, D.; Kok, N.; Quigley, J.M.J.E.E.R. Residential energy use and conservation: Economics and demographics. *Eur. Econ. Rev.* **2012**, *56*, 931–945. [CrossRef]
10. Amasyali, K.; El-Gohary, N.M.J.R.; Reviews, S.E. A review of data-driven building energy consumption prediction studies. *Renew. Sustain. Energy Rev.* **2018**, *81*, 1192–1205. [CrossRef]
11. Yu, B.; Wei, Y.-M.; Kei, G.; Matsuoka, Y.J.N.E. Future scenarios for energy consumption and carbon emissions due to demographic transitions in Chinese households. *Nat. Energy* **2018**, *3*, 109–118. [CrossRef]
12. Statistics Korea. *Population and Housing Census*; Statistics Korea, Ed.; Statistics Korea: Seoul, Korea, 2015.
13. Lee, S.G.; Lee, S.I.; Na, I.G. *A Study on Improvement Mesures for the National Energy Saving and Efficiency Improvement System*; Korea Energy Economics Institute; KEEI: Ulsan, Korea, 31 December 2008.
14. Statistics Korea. *2014 Korea Time Use Survey*; Statistics Korea, Ed.; Micro Intergreted Service: Seoul, Korea, 2015.
15. Statistics Korea. *Guidelines for Users of 2014 Korean Time Use Survey*; Statistics Korea, Ed.; Statistics Korea: Seoul, Korea, 2016.
16. Exchange, K.P. *A Survey of Hosehold Appliance Penetration and Household Power Consumption Patterns*; Statistics Korea, Ed.; Statistics Korea: Seoul, Korea, 2013.
17. MacQueen, J. Some methods for classification and analysis of multivariate observations. In Proceedings of the Fifth Berkeley Symposium on Mathematical Statistics and Probability, Berkeley, CA, USA, 21 June–18 July 1965; pp. 281–297.
18. Huang, Z.; Ng, M.K. A note on k-modes clustering. *J. Classif.* **2003**, *20*, 257–261. [CrossRef]
19. Huang, Z. Extensions to the k-means algorithm for clustering large data sets with categorical values. *Data Min. Knowl. Discov.* **1998**, *2*, 283–304. [CrossRef]
20. Crawley, D.B.; Lawrie, L.K.; Winkelmann, F.C.; Buhl, W.F.; Huang, Y.J.; Pedersen, C.O.; Strand, R.K.; Liesen, R.J.; Fisher, D.E.; Witte, M.J.J.E.; et al. EnergyPlus: Creating a new-generation building energy simulation program. *Energy Build.* **2001**, *33*, 319–331. [CrossRef]
21. Zhou, X.; Hong, T.; Yan, D. Comparison of HVAC system modeling in EnergyPlus, DeST and DOE-2.1 E. In *Building Simulation*; Springer: Berlin/Heidelberg, Germany, 2014; pp. 21–33.
22. Ministry of Land, Infrastructure and Transport. Introduction of Happy House. Available online: www.molit.go.kr/happyhouse/info.jsp (accessed on 30 January 2018).

23. Jang, H.; Kang, J.J.E. An energy model of high-rise apartment buildings integrating variation in energy consumption between individual units. *Energy Build.* **2018**, *158*, 656–667. [CrossRef]
24. The Korean Solar Energy Society; KIAEBS. *Korean Standard Weather Data*; The Korean Solar Energy Society, KIAEBS, Eds.; The Korean Solar Energy Society; KIAEBS: Seoul, Korea, 2015.
25. US DOE. *Getting Started with Energy Plus*; US Department of Energy: Washington, DC, USA, 2009.
26. Ministry of Land, Infrastructure and Transport. *Standards for Building Energy Efficiency Rating and Zero Energy Building Certification*; Ministry of Land, Infrastructure and Transport, Ed.; Ministry of Land, Infrastructure and Transport: Sejong, Korea, 2017.
27. Williams, C.K.; Rasmussen, C.E. Gaussian processes for regression. In *Advances in Neural Information Processing Systems*; NIPS: Montreal, QC, Canada, 1996; pp. 514–520.
28. Yoon, Y.R.; Moon, H.J.J.E. Energy consumption model with energy use factors of tenants in commercial buildings using Gaussian process regression. *Energy Build.* **2018**, *168*, 215–224. [CrossRef]
29. Heo, Y.; Choudhary, R.; Augenbroe, G.J.E. Calibration of building energy models for retrofit analysis under uncertainty. *Energy Build.* **2012**, *47*, 550–560. [CrossRef]
30. Kim, Y.-J.; Park, C.-S. Gaussian process model for real-time optimal control of chiller system. *J. Archit. Inst. Korea Plan. Des.* **2014**, *30*, 211–220. [CrossRef]
31. Nghiem, T.X.; Jones, C.N. Data-driven demand response modeling and control of buildings with gaussian processes. In Proceedings of the 2017 American Control Conference (ACC), Seattle, WA, USA, 24–26 May 2017; pp. 2919–2924.
32. Sheng, H.; Xiao, J.; Cheng, Y.; Ni, Q.; Wang, S. Short-term solar power forecasting based on weighted Gaussian process regression. *IEEE Trans. Ind. Electron.* **2018**, *65*, 300–308. [CrossRef]
33. Rasmussen, C.E. Gaussian processes in machine learning. In *Advanced Lectures on Machine Learning*; Springer: Berlin, Germany, 2004; pp. 63–71.
34. Heo, Y.; Zavala, V.M.J.E. Gaussian process modeling for measurement and verification of building energy savings. *Energy Build.* **2012**, *53*, 7–18. [CrossRef]
35. Bhinge, R.; Biswas, N.; Dornfeld, D.; Park, J.; Law, K.H.; Helu, M.; Rachuri, S. An intelligent machine monitoring system for energy prediction using a Gaussian Process regression. In Proceedings of the 2014 IEEE International Conference on Big Data (Big Data), Washington, DC, USA, 27–30 October 2014; pp. 978–986.
36. Blum, M.; Riedmiller, M.J.P. Electricity demand forecasting using Gaussian processes. *Power* **2013**, *10*, 104.
37. Manfren, M.; Aste, N.; Moshksar, R. Calibration and uncertainty analysis for computer models—A meta-model based approach for integrated building energy simulation. *Appl. Energy* **2013**, *103*, 627–641. [CrossRef]
38. The MathWorks, Inc. *Statistics and Machine Learning Toolbox™ User's Guide*; Matlab: Natick, MA, USA, 2018.
39. Huang, J.; Lu, J.; Ling, C.X. Comparing naive Bayes, decision trees, and SVM with AUC and accuracy. In Proceedings of the Third IEEE International Conference on Data Mining, Melbourne, FL, USA, 22 November 2003; p. 553.
40. Noble, W.S. What is a support vector machine? *Nat. Biotechnol.* **2006**, *24*, 1565–1567. [CrossRef] [PubMed]
41. Al-Mumin, A.; Khattab, O.; Sridhar, G.J.E. Occupants' behavior and activity patterns influencing the energy consumption in the Kuwaiti residences. *Energy Build.* **2003**, *35*, 549–559. [CrossRef]
42. Huebner, G.; Shipworth, D.; Hamilton, I.; Chalabi, Z.; Oreszczyn, T. Understanding electricity consumption: A comparative contribution of building factors, socio-demographics, appliances, behaviours and attitudes. *Appl. Energy* **2016**, *177*, 692–702. [CrossRef]
43. Kavousian, A.; Rajagopal, R.; Fischer, M.J.E. Determinants of residential electricity consumption: Using smart meter data to examine the effect of climate, building characteristics, appliance stock, and occupants' behavior. *Energy* **2013**, *55*, 184–194. [CrossRef]
44. Fong, W.-K.; Matsumoto, H.; Lun, Y.-F.; Kimura, R. Influences of indirect lifestyle aspects and climate on household energy consumption. *J. Asian Archit. Build. Eng.* **2007**, *6*, 395–402. [CrossRef]
45. Jung, J.; Yi, C.; Lee, S. An Integrative Analysis of the Factors Affecting the Household Energy Consumption in Seoul. *J. Korea Plan. Assoc.* **2015**, *50*, 75–94. [CrossRef]

46. Noh, S.C.; Lee, H.Y. An Analysis of the Factor Affecting the Energy Consumption of the Household in Korea. *J. Korea Plan. Assoc.* **2013**, *48*, 295–312.
47. Gi, K.; Sano, F.; Hayashi, A.; Tomoda, T.; Akimoto, K. A global analysis of residential heating and cooling service demand and cost-effective energy consumption under different climate change scenarios up to 2050. *Mitig. Adapt. Strateg. Glob. Chang.* **2018**, *23*, 51–79. [CrossRef]

© 2019 by the authors. Licensee MDPI, Basel, Switzerland. This article is an open access article distributed under the terms and conditions of the Creative Commons Attribution (CC BY) license (http://creativecommons.org/licenses/by/4.0/).

Review

Framing Smart Meter Feedback in Relation to Practice Theory

Hanna Mela [1,*], Juha Peltomaa [1], Marja Salo [1,2], Kirsi Mäkinen [1] and Mikael Hildén [1]

1. Finnish Environment Institute, P.O. Box 140, FI-00251 Helsinki, Finland; juha.peltomaa@ymparisto.fi (J.P.); marja.salo@ymparisto.fi (M.S.); kirsi.makinen@ymparisto.fi (K.M.); mikael.hilden@ymparisto.fi (M.H.)
2. Faculty of Biological and Environmental Sciences, University of Helsinki, P.O. Box 65, FI-00014 Helsinki, Finland
* Correspondence: hanna.mela@ymparisto.fi

Received: 29 August 2018; Accepted: 28 September 2018; Published: 3 October 2018

Abstract: Smart metering is advancing rapidly and consumption feedback from smart meters is expected to help residents to reduce their energy and water consumption. In recent years, more critical views have been expressed based on theories of social practice, arguing that smart meter feedback ignores the role of various mundane practices where energy and water are consumed and instead targets individuals as active decision-makers. We present a review of qualitative studies on smart meter feedback and results of a survey to European smart metering projects. We argue that theories of social practice can be used to reframe the challenges and potentials of smart meter feedback that have been identified in the literature and our survey. This presents challenges of smart meter feedback as resulting from normalised resource intensive practices rather than from uninterested and comfort-loving individuals. Potentials of improving the effectiveness of smart meter feedback relate to supporting communities and peer-learning and combining smart meter feedback with micro-generation of renewable energy. This has implications for how domestic energy and water consumption is targeted by policy.

Keywords: smart metering; feedback; households; energy and water consumption; theories of social practice

1. Introduction

Energy consumption is a key contributor to climate change, and a continuously increasing level of energy consumption is unsustainable with the current energy mix that is heavily based on fossil fuels [1]. Water consumption in many parts of the world has already exceeded sustainability limits, leading to declining groundwater levels and river flows in densely populated areas [2]. While water scarcity itself is a problem, household water usage is also closely linked to energy consumption due to water heating.

Demand management is increasingly seen as a way to decrease both overall consumption of resources such as energy and water and to deal with peak loads that are particularly problematic for system management. Demand management can involve different types of energy users but in this paper, we focus exclusively on households. Smart metering has been identified as an important element of demand side management and the development of smart grids [3,4] that would lead to a transition towards low carbon energy systems.

Smart metering can be linked to the concept of energy transition. There is no standard definition of energy transition, but it usually refers to a change in an energy system, involving for example, changes of fuel sources, technologies and a whole set of actors including suppliers, distributors and end users [5,6] or structural changes in the way energy services are delivered [6]. In this paper we use the term energy transition to refer to a systemic change from (the current), fossil-fuel based energy system where residents are mere consumers of energy, towards a system based on intermittent renewable energy sources and also involving power generation in smaller units, including the household and community level. The role of residents is expected to change in future smart energy systems and concepts of 'energy citizen' or 'citizen-consumer' and 'co-provider' have been used to describe residents who are actively engaged in the management of energy [7,8]. There are already examples of communities of front-runners in which learning from peers has contributed to increasing energy efficiency [9–11]. In a more critical context, the concept of 'Resource Man' has been introduced [12], referring to the ideal energy consumer that smart metering projects are targeted to.

The objectives of smart electricity metering range from peak load management and reducing total demand to fraud detection and accurate billing [13] (p. 447). In the case of water metering, improved leak detection is also an important factor motivating smart meter roll-outs, in addition to the aim of decreasing the overall consumption of water e.g., [14]. There is variation in what residents see of the meters, from no direct information to in-home displays, websites or mobile applications visualising consumption in real-time [15].

Large smart metering roll-out programs are under way or being planned throughout the developed countries. It is expected that by 2020 nearly 72% of European consumers will have a smart meter for electricity and 40% for gas [16]. The Energy Efficiency Directive of the EU also encourages the use of in-home displays [17], and the EU's 2015 Strategic Energy Technology Plan emphasises the need to provide smart solutions for energy consumers to enable them to optimise their energy consumption (and production) [18].

Despite the expectations for smart metering, empirical analyses of the effects of smart meter feedback show mixed results. Some review studies have indicated energy or water savings within the range of 5 to 15%, at best nearly 30% [14,19–21]. A meta-analysis looking at the effect of different information and feedback strategies on energy consumption found an average reduction of 7.4%, with the most effective interventions including relatively high involvement such as home audits and individual energy consultation [22]. Some studies report modest (2–4%) reductions [23] or no statistically significant reductions [24,25]. Studies looking at the effect of smart meter feedback have been criticised for not being always able to differentiate between the effects of feedback, self-selection bias of the participants and/or the Hawthorne effect (participants behaving differently because they know they are taking part in a study). The long-term sustainability of achieved savings has also been questioned [26–28]. As a result, a growing body of literature questioning the optimism related to smart meter feedback has emerged [29]. This gap between expectations and actual experiences calls for investigations of the underlying reasons for energy and water consumption. Practice theories can help to formulate and address relevant research questions, as they place smart meters in the context of everyday practices [12].

Both energy and water consumption are embedded in everyday practices such as cooking, laundering, cleaning or showering [30]. Energy and water consumption are not practices in themselves but rather a by-product, 'just what happens while going about our daily lives' [31]. Following this line of thinking, a focus on practices helps to extend the scope from flows of resources to the meaningful activities they enable. In this paper we explore the challenges and potentials related to the interaction between residents and smart meters through the lens of social practice theories. This leads to a discussion on what smart metering can and cannot do in contributing to energy transitions. The questions guiding our analysis are:

- What challenges and potentials related to smart meter feedback have been identified?
- How can the identified challenges and potentials be analysed and reframed through theories of social practice?
- Can smart meter feedback support energy transitions?

Our conceptual approach is based on practice theories, e.g., Reference [32], implying that patterns of energy and water consumption change when links between the elements of energy and water consuming practices are broken or created. We review key literature on smart metering trials where attention has been paid to the interaction between residents and smart meters, in order to explore the challenges and potentials related to smart meter feedback.

The paper is organised as follows. In Section 2 we first introduce our conceptual approach to smart metering, after which our data and methods are described in Section 3. In Section 4 we present results from the reviewed literature and a survey we have conducted, identifying challenges and potentials for effective smart metering and interpreting these based on practice theories. Section 5 reflects on the results in the light of practice theories and discusses whether smart metering could contribute to energy transitions. Section 6 summarises key findings and gives recommendations for the development of smart meter feedback.

2. Conceptual Approach

Smart meter feedback is assumed to make rational consumers alter their behaviour in response to information about their consumption (and the related costs). This assumption has in recent years been challenged by some researchers, and more critical approaches to smart metering have been expressed. Yolande Strengers [12] (p. 2) argues that the global vision for smart energy technologies is part of a 'Smart Utopia' that is underpinned by a smart ontology, in which human action and social change are understood as being mediated by information and communication technology (ICT) and data. In a Smart Utopia, environmental problems are solved with the help of new technology without compromising current lifestyle. In this vision, technologies, such as smart meters, act in various roles: They promise rational and efficient control of residents, they assign responsibility for complex environmental problems to individuals rather than states or companies and they 'seamlessly manage the home environment while maintaining or enhancing current lifestyle expectations' [12] (p. 8, 23). This vision is based on viewing residents as highly informed and rational micro-resource managers of energy (and water), who are interested in their own consumption and manage it with the help of smart meter feedback. The concept of 'Resource Man' has been introduced to describe this type of an ideal energy consumer that smart metering projects are targeted at. 'Resource Man' is described as a 'gendered, technologically minded, information-oriented and economically rational consumer of the Smart Utopia' [12] (pp. 36, 54).

In contrast to the smart ontology, Strengers [12] uses the ontology of everyday practice, in which 'human action and social change is mediated by and through participation in routinely performed practices'. Energy and water are not valuable in themselves, but they are merely 'ingredients' [30] or 'resources' [33] that enable domestic practices, but operate in the background without the resident paying attention to them. When smart meter feedback concentrates solely on energy or water as commodities, measured in kWhs or litres, it ignores the role of practices in which they are consumed [12].

These ideas belong to an increasing body of literature on theories of social practice. Practice theories build on the works of e.g., Giddens [34] and have been developed further by e.g., Schatzki [35], Reckwitz [36] and Shove et al. [32]. As interpretations of the practice approach vary between scholars, the theory is often referred to in plural [32], as we also do in our paper. In practice theories, the unit of analysis and enquiry is social practice, which has been defined as a 'routinised type of behaviour' [36] (p. 249), which 'endures between and across specific moments of enactment' [37]. Practices consist of three elements: Materials, competences and meanings. Materials include all tangible objects, infrastructure and technologies, competences include skills, know-how and technique, and meanings encompass ideas, aspirations and symbolic meanings [32]. A practice exists simultaneously as a performance and as an entity [36] (p. 249), [32]. Practice as a performance is the observable behaviour of individuals, such as heating or cooling spaces, cooking or showering. However, when practices are understood as entities, it becomes clear that the observable behaviour (practice as performance) is not a result of individual choices but by nature social, embedded in socially shared ideas and meanings. Seen in this way, the observable behaviour of individuals is only the 'tip of the iceberg' [38], underpinned by socially shared elements of practice: Competences, materials and meanings. Examples of this analytical distinction are described in Table 1.

Table 1. Elements of practice as an entity. Adapted from [38].

Observable Behaviour of Individuals (Practice as Performance): e.g., Heating or Cooling Spaces, Cooking, Showering, Watching TV etc.		
Skills and Competences	Materials and Infrastructure	Socially Shared Meanings and Conventions
(Embodied) knowledge and skills needed to perform practices, knowing how to use appliances, timing and ordering of activities etc.	Constant availability of heat, electricity and hot water, showers installed in bathrooms, ovens in kitchens, availability of a TV	Ideas of 'normality', ideas of comfort, socially accepted ways of e.g., eating and levels of cleanliness

Each practice-as-entity has its history and a trajectory along which it develops. A process of co-evolution happens between technologies and infrastructures, competences and meanings that together constitute a practice. The development paths of practices illustrate how normality is constructed and how unsustainability can become standardised. When certain expectations of comfort, cleanliness and convenience have become standardised, in terms of what is regarded as 'normal', patterns of consumption may escalate as what previously was regarded as 'luxury' gradually becomes 'normal'. These developments have led to more energy-intensive lifestyles [39]. For example, air-conditioning has become normalised in many countries, making passive thermal design in buildings more rare [39]. An example related to water consumption is how the frequency of washing bodies and clothes has increased, rendering daily showering and frequent laundering normal practices and thus increasing water consumption [40,41].

Practices are, on the one hand, relatively stable and to an extent resist change, which emphasises their path-dependent character. On the other hand, people, when performing practices, also create possibilities for change by experimenting, adapting and improvising. Even though there is individual variation between performing a practice, practices are social by character and individuals participating in them seek recognition as competent practitioners by performing practices well, according to the shared standards and understandings of normality, which vary between social groups and geographical contexts [42]. The importance of communities for developing low-carbon ways of living and challenging conventions has been recognised in previous research. Communities hold potential that individuals alone do not have [43].

Practice theoretical approaches differ from research focusing on individuals as active decision-makers, which is often presumed in consumption feedback studies. Approaches focusing on individuals see behaviour as a set of choices determined by identifiable factors, including attitudes and values and have been described as the 'ABC paradigm', A standing for attitude, B for behaviour and C for choice [44]. This paradigm draws on economic theories and psychology, for example on the theory of planned behaviour [45]. In the language of the ABC, there are barriers and motivators for individuals to change their behaviour, and the role of policy is to remove those barriers and persuade individuals to make sustainable choices [32]. The distinction between the two approaches is illustrated in Figure 1a,b. In Figure 1a, residents are presented as rational managers of resources. In Figure 1b, residents are practitioners taking part in social practices, each of which consists of the interlinked elements of competences, materials and meanings. Practices are sequenced and linked with each other in time and space. They may also overlap in timing, location and elements. Consumption of resources such as energy and water is an integrated part of practices but is essentially instrumental rather than an end in itself.

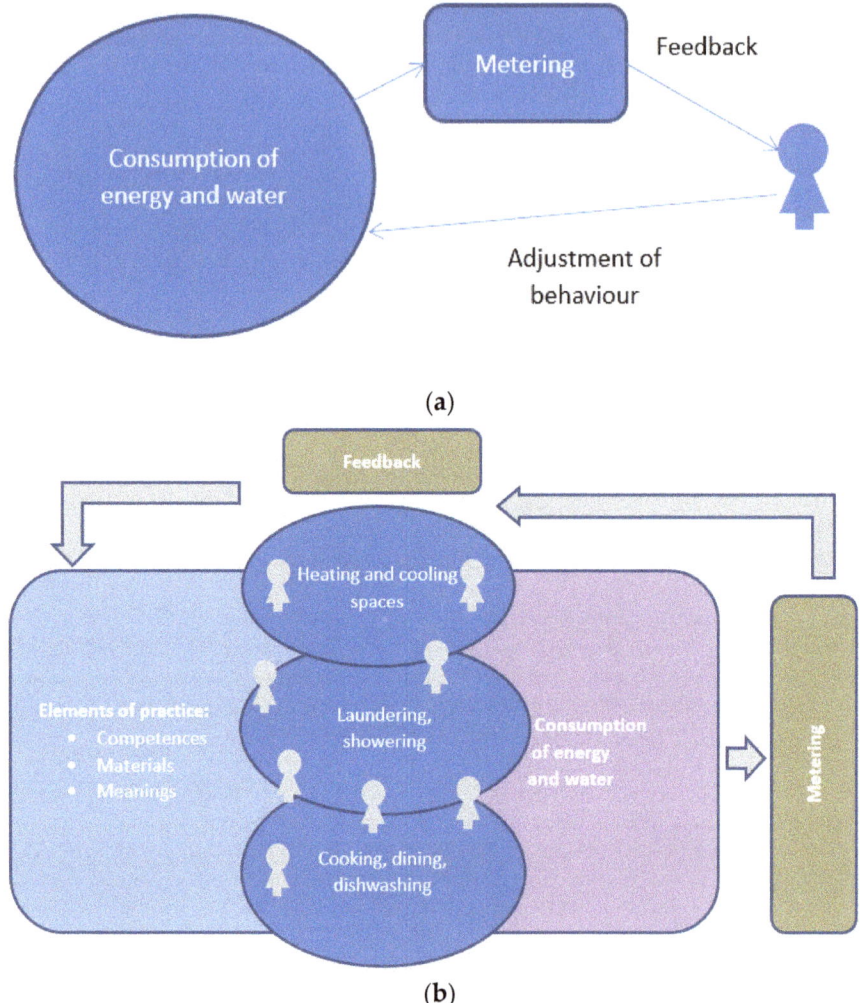

Figure 1. (**a**). Smart meter feedback according to the ABC paradigm; (**b**). Smart meter feedback in the context of everyday social practices.

Despite the criticism of smart metering and rational optimisation of energy consumption, there have also been attempts to see if and how smart metering could become a more integrated element of different practices contributing to energy transition. It has been suggested that smart systems, including smart metering, could be an element in potentially emerging home energy management (HEM) practices involving energy monitoring, microgeneration, energy storage, energy trading/sharing, timing of demand and energy conservation, which are being promoted by smart grid development [46].

HEM-practices could support energy transitions, and three orders of sociotechnical innovation related to energy transitions have been suggested. The orders differ in terms of the level of integration between domestic consumption practices and HEM-practices [46]. The first order innovations do not change the basic set-up of the household and its connection with the wider energy system, and do not connect domestic consumption practices with HEM-practices. Examples include using an in-home energy display to monitor consumption or substitution of old appliances with more energy efficient ones. In second order innovations there is a stronger connection between HEM and domestic consumption practices, as well as the wider energy systems, e.g., timing of laundry or vacuuming practices to fit local energy production in order to de-link consumption from carbon intensive production and distribution practices [46]. Third order innovations present a more radical departure from the current system, involving increasingly sharing the tasks and responsibilities for energy production and distribution practices between households, existing service providers and new intermediaries (ibid.). Third order innovations can also involve energy self-sufficient districts or energy cooperatives, transferring the power from centralised to de-centralised actors. In one scenario households belong to a personal carbon allowance trading scheme, and some domestic consumption practices are no longer private but become shared, e.g., through shared cars and shared laundry and cooking facilities [46] (pp. 130–133). For our study, the three orders of innovations provide a benchmark to look at the level of integration of smart meter feedback into practices.

3. Data and Methods

We carried out a literature review and conducted a survey within a sample of European smart metering projects in order to analyse what issues are seen as limiting the effectiveness of smart meter feedback and what issues, on the other hand, are seen as potentials for effective smart meter feedback, within a sample of European smart metering projects. The type of the review is integrative. An integrative literature review has been defined as 'a form of research that reviews, critiques and synthesises representative literature on a topic in an integrated way such that new frameworks and perspectives on the topic are generated' [47]. Articles for the literature review were searched in the Web of Science in January 2017, where we restricted the analysis to papers published in the last 10 years (between 2007 and 2017). The search was repeated in September 2018 to cover articles published in 2017 and 2018. Several combinations of keywords were tested, and a search was done with the following set of keywords: TS = (Domestic OR household* OR residen* OR consumer*) AND TS = (electric* OR energ* OR heat* OR water) AND TS = (meter* OR feedback OR smart) AND TS = (behav* OR awareness) AND TS = (conserva* OR sav* OR reduc*).

We focused on studies where the main aim has been to affect overall consumption and in some cases also load shifting by providing consumption feedback information in the form of a smart meter display or similar medium. Our focus is on the challenges and potential of smart meter feedback to affect overall energy and water consumption. Thus, we have excluded papers from our review where the focus has been only on load shifting. Most of the available research addresses smart electricity metering, while smart water, heat or gas metering have been studied less. Therefore, most of the articles we have reviewed concentrate on smart electricity metering. However, several domestic practices use both energy and water simultaneously (e.g., laundering, dishwashing, showering). The background assumption of smart meter feedback (i.e., provision of consumption data to residents makes them reduce their consumption) is the same, regardless of whether the resource is electricity, water, gas or heat.

In order to better understand the potentials and challenges of smart metering feedback in relation to practices, we have limited our analysis to studies that have qualitatively addressed the interaction between residents and smart meter feedback. Thus, studies that have concentrated on quantitatively measuring the achieved savings, customer satisfaction of smart meter trials or comparing different features of in-home displays and the residents' preferences regarding the display features but not addressing the residents' experiences in more detail have been excluded, along with studies looking at smart metering from the point of view of utility companies, for example its economic profitability. Another requirement was that the trials should enable residents to monitor their consumption and access their consumption data via an in-home display, mobile application or a website. The final sample included 20 articles, which are listed and briefly described in Appendix A. These articles form our core material. In addition, we use a larger body of literature on smart metering and energy and water feedback to support our analysis. The reported trial settings vary according to their size, duration, whether the participants volunteered or not, or whether other interventions e.g., energy efficiency measures were introduced at the same time. In some studies, changes in consumption were measured or a quantitative survey accompanied the qualitative analysis, but our focus is on the qualitative data, most of which consists of interviews. Only articles containing new empirical material were included in our sample, thus excluding review articles. The analysis is qualitative. We have read through the final sample of articles, collected and grouped all the identified challenges and potentials and organised them under the three elements of practice theories (materials, competences and meanings).

In addition to the literature review, we use data from an online survey. The survey questions were formulated based on an analysis of documents describing metering initiatives in various European countries. Information about these metering projects was collected on the ICLEI (International Council for Local Environmental Initiatives) website in the form of documents and descriptions on metering projects' websites, as well as with a snowball method from these websites and documents and through the researchers' contacts. The survey was sent in December 2016 to 39 organisations that have implemented energy and/or water metering trials in Europe. Of these, 11 responded. The responses covered 15 smart metering projects in 10 European countries. The aim of the survey was to gather information on the barriers and enablers of European metering projects and the role of residents in them. The survey contained multiple choice and open-ended questions related to these themes. The results of the survey are used to complement the results of the literature review.

From the literature we have identified challenges and on the other hand, potentials of smart metering in the spheres of competences and skills, materials and infrastructures, and socially shared meanings. We have applied practice theories to reframe these three spheres to assess the possibilities to tackle the challenges and harness the potentials.

4. Results

In this section we present our findings on the challenges and potentials identified in the analysed articles and the survey. We summarise our findings as described in the literature and in parallel interpret them in the light of the theories of practice presented in Section 2.

Through the interpretation we shift perspective from the meters, management and data to challenges and potentials in the light of the everyday practices that people already carry out in order to complete meaningful tasks. We link the observations from the survey with those reported in the reviewed literature to identify supporting or conflicting findings.

4.1. Challenges Related to Smart Meter Feedback

In Tables 2–4, we summarise the challenges of smart metering as identified in the literature and interpret the observations from a practice theories perspective. The challenges are grouped into the three categories of practice elements: Competences and skills, materials and infrastructures and socially shared meanings.

4.1.1. Challenges Related to Competences and Skills

First, we look at what in the practice terminology is referred to as competences and skills (Table 2). In the literature, insufficient knowledge and inactivity of residents is often referred to as hindering energy or water saving.

Table 2. Challenges of smart meter feedback related to competences and skills as identified in the literature and as interpreted through theories of social practice.

Challenges of Smart Metering Feedback Related to Competences and Skills	
Identified in the literature	Interpreted through theories of practice
Residents' knowledge on energy or water consumption is insufficient. Smart meter feedback is perceived as too complicated and abstract by residents in general [27,48–53], and by the elderly population in particular [48]. Residents are not interested in monitoring their consumption [27,48–51].	Competence and skills that save energy or water are currently not integral to the practices where energy and water are used. Understanding energy or water use at an abstract level is not relevant for the performance of domestic practices.
After an initial interest in the new 'gadget' and playing with it, smart meters tend to get 'backgrounded' and forgotten [31,51,54–57].	Smart meters do not manage to successfully connect to existing practices where energy and water are used and remain isolated from them. They do not support or strengthen competences that would use less energy or water. Thus, they are rarely able to change the practices.
Residents do not know how to further save energy or water. Smart meters do not provide any assistance with how to actually organise daily activities in a way that would save energy and water. [47,50,51]	From a practice perspective, the observation on the lack of assistance to organise daily activities is valuable. Competences related to energy or water saving cannot be built simply by providing consumption feedback.

Lack of knowledge and interest among residents was also seen as a challenge to the effectiveness of smart meter feedback by nearly all the respondents of our survey, and residents were characterised as apathetic and uninterested. Within this kind of problem-framing (residents are not interested, they quickly lose interest) attempts to improve the effectiveness of the intervention are typically reduced to making the feedback more salient and diverse by adding more features to it in order to prevent the residents from losing their interest, e.g., [56]. This is in line with Figure 1a, a simple feedback loop where information directly affects the way resources are used. From a point of view of practice theories, the impact of information provision efforts on the way energy and water are used can be questioned. The challenge identified in the literature is that residents do not (know how to) use the meter and data in their daily lives. Practice framing would turn the problem around: People already have knowledge and skills to go on with their everyday activities but lack knowledge and skills to embed and translate feedback from the meters to transform the way they perform practices. The elderly has been recognised as a group that faces special challenges in engaging with smart meter feedback [48].

4.1.2. Challenges Related to Materials and Infrastructures

Access to unlimited volume of energy and (hot) water, and appliances to assist in housekeeping and cleanliness form an important material element of daily energy and water consuming practices. The material elements have evolved over time, at the same time shaping domestic practices and changing our perceptions about standards of cleanliness and comfort. Smart meter feedback always takes place in a material context, which may sustain energy and water intensive practices and thus pose a challenge for the effectiveness of smart meter feedback (Table 3).

Table 3. Challenges of smart meter feedback related to materials and infrastructures as identified in the literature and as interpreted through theories of social practice.

Challenges of Smart Meter Feedback Related to Materials and Infrastructures	
Identified in the literature	Interpreted through theories of practice
Technical problems related to the installation and functioning of smart meters, e.g., reception issues or sensor faults [27,52,56,58,59].	These observations do not relate to any material elements of existing practices but rather highlight the material essence of the meter itself. Technical problems are associated with reliability and can form an obstacle for smart meters to become a material element of practices as intended.
Housing developers, local authorities and manufacturers of appliances create hindrances to making use of smart meter feedback and thus energy saving, for example by making the installation of heat pumps or micro-generation of renewable energy difficult or not being proactive about energy efficiency in the design of new buildings and appliances [55,60].	The material elements underlying current energy and water consuming practices are often complex and expensive and therefore slowly replaced. Although energy efficiency has become a selling point, the material elements do not address the rebound effect related to increased use.

The results of our survey are to a large extent in line with the results of the literature review regarding material elements. However, various technical problems were represented in our survey as the most important factor limiting the effectiveness of smart metering, while in the reviewed literature technical issues were rather rarely mentioned. This underlines that a functioning technology is crucial for smart metering [61], which in itself is not surprising, but may obscure the important interactions between technologies and practice. Although a functioning technology is a prerequisite for the smart system to be integrated into practices, technically viable solutions do not guarantee that the technology will become part of relevant practices. The metering projects aimed at providing material elements for consumers with an objective to encourage them to come up with practices to manage resource consumption. However, what seems to be missing is a link between the new material component and other elements of existing practices. Our findings from the review and the survey support the notion that the elements of practices and links between them need to change in a consistent way in order to increase the likelihood of achieving the final goal: more sustainable patterns of resources use (Figure 1b).

Other, nearly as important limiting factors recognised in the survey were the residents' low interest in their energy and water consumption and habits and culture regarding energy and water consumption. From a practice perspective, it is interesting that one respondent highlighted how customers were only interested in data when it was related to specific events. A practice interpretation of the low interest emphasises the instrumental and invisible role of the flows of energy and water in conducting meaningful practices.

A third finding from our survey regarding the materiality of the smart meter is that in the trials that responded to it, the most common means to access the information was via a website. This is rather disconnected from the actual practices that consume water and energy. Based on the objectives of the projects and the focus on metering, many projects aimed at introducing new practices of managing energy, but these did not often realise as intended. The survey responses indicate that parallel measures to support users and provide financial incentives would be needed to support the residents' engagement with consumption data more actively. Our interpretation is that if the objective is to introduce new practices of managing energy in households, the practice approach would be useful in mapping the elements of practice that need to be reconfigured instead of simply focusing on introducing a specific technology (Figure 1b).

4.1.3. Challenges Related to Socially Shared Meanings

The third category of challenges is related to socially shared meanings (Table 4). The literature highlights numerous observations related to the taken-for-granted levels of comfort that are difficult to challenge.

Table 4. Challenges of smart meter feedback related to socially shared meanings as identified in the literature and as interpreted through theories of social practice.

Challenges of Smart Meter Feedback Related to Socially Shared Meanings	
Identified in the Literature	Interpreted through Theories of Practice
Residents are not willing to compromise certain living standards and to reduce their consumption. Certain appliances and perceived levels of comfort are not negotiable and residents receiving smart meter feedback become defensive about them [27,51,55,56,58,60,62,63].	Meanings of comfort and normality are integral to many practices where energy and water are consumed. Rather than focusing on individuals and their feelings, practice theories try to understand the trajectory of socially shared meanings related to those mundane domestic practices where energy and water are consumed. By examining how these meanings have developed over time, how they are currently sustained and how they are changing, insights can be gained that help to identify paths to more sustainable practices.
Smart meter feedback only appeals to a small group of 'monitor enthusiasts' who are already interested in following their energy consumption, leaving the large audience unaffected [27,56,58].	These 'monitor enthusiasts' can be seen as the early adopters of a practice of energy monitoring. Whether this practice manages to recruit larger groups of practitioners depends on the broader context where domestic energy and water consuming practices take place.
Unequal participation of household members. Smart metering projects have often appealed to the technically minded, typically male members of households, leaving other family members more passive [27,31,52–55,57,60,63–65].	Patterns of inequality and power are present in practices, as well as shared conventions, roles and responsibilities related to e.g., gender and age within households. These are reflected in the way smart meter feedback is received and used in households.
Smart meter feedback can serve as a 'normaliser' of consumption, as residents receiving smart meter feedback quickly develop a sense of their usual level of consumption. Smart meter feedback helps to identify abnormal, excess use and 'waste', at the same time legitimising certain basic level of consumption [27,52,57,59,60,64]. Residents see themselves as energy or water efficient and do not see how they could further save energy or water [27,31,50,52–54,58,59].	Meanings of normality are important in understanding how certain practices become established as part of everyday life. Smart meter feedback is not capable of challenging meanings of normality. Rather, it may serve to confirm what is already regarded as normal. Purposeful 'wasting' of energy and water is not part of practices, but the 'normality' of practices hides resource intensive ways of living that residents do not easily question.
Residents refer to the importance of health and well-being, especially of children or the elderly, and maintaining harmony in the home. The goal of energy saving is perceived as conflicting with these aims [27,58,60,63].	Saving energy is not currently involved as an element in domestic practices, while other meanings such as care and well-being especially of children and the elderly, belong to the energy-intensive practices of heating and cooling.
Residents have feelings of distrust towards utilities and smart meters and are concerned about loss of privacy or autonomy. Smart meters blur the boundaries between the private and the external and are sometimes hidden from sight by residents who feel they are too visible in the home environment. Smart meters cause conflicts between family members [27,49,53,56,65].	Meanings of privacy and autonomy that are integral to domestic practices are threatened when smart meters presume disclosure of consumption information vertically (between household and utility) or horizontally (between or within households, e.g., between parents and children).
There is not enough economic incentive for residents to engage with smart meter feedback as the average achieved economic savings are too small to motivate changes in daily routines [27,52,55,58,60].	These observations reflect a view of residents as rational individuals who respond to economic incentives by weighing the costs and making decisions based on the smart meter feedback. This assumption is questioned by practice theories. However, a wider economic context is important for how energy and water consuming practices have developed over time and how they will develop in the future.

As seen in the results Tables 2–4, the analysed literature has its main focus on individuals, their attitudes and feelings, preferences and behaviour. These challenges look different when the focus is on the social practices where energy and water are consumed rather than on the individuals that consume. As highlighted in Section 4.1.2. on challenges related to material elements, availability of resources and appliances have normalised increasing standards of comfort and cleanliness and as a result also practices have changed. Pure consumption-based feedback may not be powerful enough to affect practices if material elements and socially shared meanings remain unchanged (Figure 1b).

When discussing the absence of a supportive policy and market context, theories of social practice would pay attention to the ways policies have shaped and sustained the unsustainable ways of life with high energy and water consumption. The residents' experience that it is not easy to save money when trying to save electricity [60] may be well justified. However, the conclusion according to practice theories would not straightforwardly claim that residents need more economic incentives to make better choices. Pricing of energy and water does play a role in creating meanings of abundancy or scarcity, but above a certain minimum level of income the pricing may be less important than the ideas of normal levels of comfort, cleanliness and convenience, which have co-evolved with infrastructure, policy and technical development of appliances.

The importance of a practice-based perspective is illustrated by the reported observation that residents find the shifting of activities to low-peak periods of the day inconvenient [31,52,63]. This does not clearly relate to only one element of practices, but rather emphasises the role of the organisation and sequences of daily activities and the connections between practices. The temporal flexibility of practices varies considerably, dining being among the least flexible domestic practices as it is tied to bodily rhythms as well as timing of work, study and leisure activities [66].

4.2. Potentials of Smart Meter Feedback

In Tables 5–7, we summarise the potentials of smart metering and suggested ways of improving it. Similarly as the challenges in the previous section, the potentials are grouped into the three categories of competences and skills, materials and infrastructures and socially shared meanings.

4.2.1. Potentials Related to Competences and Skills

Potentials related to competences and skills (Table 5) show that the role of improving the provision of information is central in the literature.

Table 5. Potentials of smart meter feedback related to competences and skills as identified in the literature and as interpreted through theories of social practice.

Potentials of Smart Meter Feedback Related to Competences and Skills	
Identified in the literature	**Interpreted through theories of social practice**
Smart meter feedback functions as a learning tool, increasing the residents' awareness and understanding of their energy and water consumption [27,31,50–55,57,59,60,63–65].	Smart meter feedback evidently increases knowledge of energy and water consumption among (some) residents. However, this increased knowledge does not automatically actualise as reduced consumption, because consumption results from practices that do not change simply by providing information.
Tailoring smart meter feedback to various user groups and contexts, in order to make consumption data more relevant and interesting [50,56,64,67].	Paying attention to the variation among individuals and social contexts where domestic practices are performed makes feedback less general and can create a stronger link to specific practices.
Using smart meter feedback as an educational aid in school projects among children and young people [52,60,67]	In these examples smart meter feedback is adopted as an element in teaching practices.

In our survey, the importance of choosing simple enough smart metering technology that suits the residents' needs was emphasised as the most important success factor for smart metering projects. This may reflect the fact that the meter and monitoring activities as such are not important but how these can be embedded in other meaningful practices. In addition, the importance of recognising windows of opportunity to disrupt current patterns of consumption (see also Reference [11]) was mentioned in the survey.

Providing comparative consumption data within a peer group was considered important by most respondents of our survey, while most of the reviewed studies did not address comparative feedback. Some researchers have paid attention to the potential 'boomerang effects' of social comparisons, if residents who are defined as 'low users' increase their consumption [28] as the "normal" level justifies higher consumption. Others [67] note that contextualising feedback by providing social comparisons with the community was experienced positively by most residents participating in the trial but a negative sense of competition was felt by some. There is some indication that co-designing smart meter feedback together with residents can yield higher engagement with the feedback [68]. Smart meter—based gamification of energy and water consumption has also successfully managed to engage residents and resulted in reductions of water consumption in a small trial setting [69]. From a practice theories approach this could be seen as recruiting individuals to a practice of playing a game, which, when played by its rules, results in reduced water consumption. Practice theories would also invite us to focus on how smart meter feedback is embedded in the current arrangement of elements of practice, not only on interactions of metering feedback and the user.

From a practice perspective tailoring feedback makes sense in that the contexts in which practices take place vary in terms of materials, competences and meanings, and there is no 'one size fits all' solution that would work in all contexts [70]. Also, if tailoring involves interaction with intermediaries, such as energy advisers, it is more likely to build competences that can engender concrete changes in energy-using practices. The potential of intermediaries for building competences in low-carbon ways of living has been recognised in previous studies, e.g., Reference [71]. There is also some indication that targeting particular practices such as showering, (via separate smart meter displays next to the shower) [72] or heating (measuring gas consumption with a smart meter and providing tailored, user- and building specific advice to lower room temperatures through an in-home display) [73], may be successful in decreasing the resource consumption in these practices. From a practice theories point of view, this creates a stronger link to certain practices than providing information on (inconspicuous) energy or water consumption in general, which results from various domestic practices that are not always easily separable from each other.

4.2.2. Potentials Related to Materials and Infrastructures

Smart meter feedback happens in a material context, which consists of the built infrastructure, resource flows and appliances of the household. Currently, these are often unsupportive of low-carbon ways of living. Thus, only few potentials related to the material context were identified in the literature (Table 6).

Combining smart meter feedback with microgeneration can be seen not only as creating new material elements to practices but also to the competences and meanings associated with domestic energy-using practices. Only four of the studies we analysed included households that were micro-producers of solar power. Microgeneration provides potentially an entry point to a radically different relationship to energy [46] (p. 90), as confirmed by two of the studies in our sample [52,62], in which some households with smart metering and photovoltaic panel installed adapted their domestic practices to match the solar peak periods. An analogy with farming was presented by a participant, referring to not wanting to waste the 'crops' one produces [62]. Meanings of self-sufficiency have in some trials also been found to be an important part of prosumerism [74]. It has been observed that becoming a prosumer requires certain competences that are evolving over time and requires integrating 'home grown' energy in everyday practices [70]. However, in two of the reviewed studies [31,63], similar impacts were not observed.

Table 6. Potentials of smart meter feedback related to materials and infrastructures as identified in the literature and as interpreted through theories of social practice.

Potentials of Smart Meter Feedback Related to Materials and Infrastructures	
Identified in the Literature	Interpreted through Theories of Social Practice
Linking engagement with smart meter feedback to an existing activity such as watching TV [65].	Practice theories may help to explain why in one study, where smart meter feedback could be viewed on TV, it became the most popular medium of engaging with it. It was connected to an existing practice and its material element (TV) and associated with commercial breaks.
Smart meter feedback combined with microgeneration of solar energy can engender changes in the way energy is used [31,52,62].	Participating in the production of intermittent solar energy can change the configuration and rhythm of certain practices by reconfiguring the meaning of energy in domestic practices as something that is temporally limited.

The above mentioned examples indicate that micro-generation, when used as a complementary source of energy and still having a constant and reliable access to centralised power grid, does not necessarily challenge existing practices and energy intensive ways of life, while the situation is completely different for practitioners 'off-the-grid', who have to organise their practices and daily rhythms based on the availability of energy [12] (pp. 135–154)). The different responses by households in the reviewed literature are of interest as all households in the studies we reviewed were connected to the centralised grid. The reasons for successes (or failures) are beyond our study but would be a topic to address in future research.

4.2.3. Potentials Related to Socially Shared Meanings

Socially and culturally shared meanings and conventions are deeply embedded in and shape the practices using energy and water. Potentials related to shared meanings that have been identified in our sample of reviewed studies are presented in Table 7.

Communities hold potential for shaping practices towards a more sustainable direction through experimentation and peer learning. The way knowledge is shared and becomes part of practices for example among frontrunner households that experiment with sustainable energy technologies, includes interaction with peers, home visits and conversations face to face or in online communities [9–11]. This is very different from most of the reviewed literature where individuals or families receive abstract consumption information that is not shared with others, or sharing of it is not meaningful as it is not linked to any other activities of the community (if there even is one). However, some of the analysed studies recognise the potential of supporting communities in feedback experiments [31,67]. It can also be argued that 'frontrunner' households experimenting with smart meter feedback and presenting the monitoring systems to their peers serve to promote a potentially emerging practice of energy monitoring [46].

Several of the reviewed studies argue that smart meter feedback should look beyond individuals and instead focus on and try to understand household dynamics and the community context in which feedback is received [49,51,55,60,63–65,67]. One of the studies [60] also suggest that providing smart meters to households as part of broader collective efforts and partnerships on energy conservation could make a difference regarding how actively residents engage in using them. In our survey, one of the key issues identified for the effectiveness of smart metering was linking smart metering to other energy or water saving actions such as energy coaching or awareness raising campaigns.

Table 7. Potentials of smart meter feedback related to socially shared meanings as identified in the literature and as interpreted through theories of social practice.

Potentials of Smart Meter Feedback Related to Socially Shared Meanings	
Identified in the Literature	Interpreted through Theories of Practice
Combining smart meter feedback with community-level action on energy or water conservation, such as workshops and group discussions [51,67]. Residents taking part in smart meter feedback trials feeling proud of the system and presenting it to their peers, which can have potential spillover effects outside the household [52,53,60,65].	Social networks such as communities are important for the diffusion of practices. Thus, community-level activities together with smart meter trials may help to create and support a community of practice, in which social learning takes place, ideas are shared, and utilising smart meter feedback could take root as a practice.
Smart meter feedback playing other roles and attaining other meanings than the management of energy or water consumption: source of reassurance when being able to predict the size of the next bill, a tool for controlling family members, providing security and maintaining sense of togetherness between family members who are located separately or a tool for detecting failures, leaks, incorrect billing or underperformance of PV-panels [52,59,60,63,64].	Meanings attached to certain elements of practices are always on the move and their development is to some extent unpredictable. Smart meters, when entering household contexts, are likely to attain other meanings than those originally intended. This may affect consumption negatively or positively. For example, using smart meter feedback to detect system failures and thus improving the performance of e.g., PV-panels decreases the climate impacts of a household even if did not affect the ways energy is actually consumed.

Social networks such as communities are important for the diffusion of practices, and it is also within these that practices can change [32]. Building trust and establishing a community partnership based on social, face-to-face interaction in the community e.g., in the form of home energy assessments have been recognised as key success factors in some community-based energy interventions [75,76]. Personal contacts with intermediaries have been linked to better effectiveness of feedback interventions [77]. In the light of practice theories, community-level activities could also be integrated into smart meter feedback initiatives, as they would help to create and support a community of practice in which social learning takes place, ideas are shared and utilising smart meter feedback could take hold as a practice, which would provide peer-support and help to build new competences. However, this does not happen automatically and belonging to a certain geographic community or neighbourhood is not necessarily enough for a community of practice to emerge [78].

Some of the reviewed studies [55,57] call for policies and smart metering strategies that would challenge what is seen as 'normal' everyday consumption, by using existing policy techniques to support low-carbon social practices and even experimenting with radical policy measures such as personal carbon allowances [55]. It has been suggested that smart meter feedback could, instead of providing only consumption data, concentrate on the meanings and competences associated with certain practices and give advice on how they could be performed differently [57]. These suggestions resonate with practice theories, as they would radically shape the meanings and competences related to energy and water consuming practices. However, these suggestions should be further developed and empirically studied in order to find out their potential in a real-life setting.

5. Discussion

This paper is an attempt to look at smart meter feedback and the challenges and potentials associated with it from the point of view of practice theories. One guiding question in practice theories when looking at energy use is 'What is energy for?' [30]. As the behaviour change paradigm 'ABC' and practice theories ask different questions, interpreting the challenges and potentials of smart metering from a point of view of practice theories is not a straightforward task. Based on our analysis it seems that it is especially fruitful to use practice theories to reframe challenges that are usually seen as individuals' lack of interest or knowledge or their love of comfort. When these are not framed as characteristics of the individuals but understood as being integral parts of normalised, socially shared practices that are shaped by the material arrangements of our everyday environments, new opportunities for intervention can emerge.

Attempts to affect the 'tip of the iceberg', the observable behaviour of individuals (see Section 1) without recognising the material features, competences and shared understandings that underpin visible behaviour, miss an opportunity to shape the elements of which practices are made and thus affect their sustainability in the long term. Smart meter feedback attempts to improve the efficiency of the 'performance' of various practices and in this it has sometimes succeeded. Some residents do reduce 'excess' consumption of energy and water at least in the initial phase of smart metering trials. Residents taking part in smart meter feedback trials typically describe undertaking various 'curtailment' behaviours such as turning off lights or appliances that are not in use (e.g., [52,53,57]). However, it seems that this does not contribute to the sustainability of energy and water consuming practices as a whole, the practices as entities. On the contrary, several developments within homes are pointing towards more energy-intensive ways of life [12,39], and various smart technologies may even reinforce this trend [79,80].

Smart metering and other smart home appliances are often marketed to consumers as a way of managing resource use without compromising certain lifestyle or level of comfort [12] (p. 26). Smart meter feedback and its sustainability should be viewed and evaluated as part of a wider development where ICT has an increasing role in various domestic practices [81]. Homes are becoming equipped with smart appliances and home-automation. Smart meter feedback systems are also becoming more developed and complex, integrating many kinds of data, e.g., Reference [82]. Attention has been paid to the possible drawbacks of smart home technologies, as they could create new energy-demanding practices, and new expectations of normality related to comfort, convenience, entertainment, security and health may rise as a result of increasing smart features in homes [80,83,84]. The escalating character of such expectations is also of concern [39]. Many smart meter feedback devices are marketed and used together with other smart home features which have several functions, energy management being only one of them, and not necessarily the most important one for residents or marketers of these systems [79].

As argued earlier in this paper, the assumption of residents being rational individuals that actively make decisions based on smart meter feedback is strongly built in smart meter feedback initiatives. The examples described in the reviewed literature can be interpreted in a way that the idealised energy consumer is to some extent present in the smart meter trials, appealing to few, perhaps technologically enthusiastic members of a household. The 'Resource Man' is very similar to the description of an active energy citizen, whose emergence seems to be a precondition but not a guarantee for the effectiveness of smart meter feedback and for participating in smart grids. This is very far from the everyday reality of how energy and water consumption is an unconscious and embedded part of our everyday living.

If the unit of change is social practice rather than individual choice as theories of social practice suggest, interventions based on the emergence of active energy citizens or 'Resource Men' cannot achieve a true transition. However, it has been recognised that community energy schemes involving microgeneration may be able to change the residents' relationship to energy by making it more salient. Thus, for residents who produce energy through microgeneration and consume it, energy can become an active component in their practices, instead of something that is taken for granted. It has been argued that microgeneration or community energy schemes should be an essential part of introducing smart meters and that without them rollouts of smart meters are 'likely to be a missed opportunity' [8]. On the other hand, engagement does not happen automatically, as observed in Section 4.2.2, and conventional smart meter feedback may not even serve the needs of prosumers [85]. Members of energy cooperatives have been reported to quickly lose interest in engaging with the feedback and finding it difficult to shift activities to the solar peak periods of the day [31]. This brings us back to the importance of looking at practices rather than individuals. If the interlinked and sequenced practices where energy and water are used are not temporally flexible enough and if their competences do not relate to the conservation of energy and water, simply introducing consumption feedback is not enough.

An analysis of the ongoing implementation of the UK smart meter roll-out programme has identified barriers related to technology, vulnerability and resistance and given policy recommendations to overcome these barriers. The recommendations include, among others, early engagement with residents, recognising lack of trust and concerns over security and privacy, assessing how smart meters connect with prosumption, assessing how smart metering affects vulnerable groups and conducting lifecycle analysis on smart meters [61].

The policy recommendations [61] are likely to improve the technical reliability, transparency and acceptability of the smart meter roll-out as a project, as well as its environmental sustainability. However, they do not question the effectiveness of smart meter feedback as such. The third recommendation (assessing how smart meters connect with prosumption) may be crucial for the future effectiveness of smart metering, as microgeneration has potential for making energy a more important element in the practices where it is used.

A shift towards low-carbon ways of living does not necessarily require feedback but more fundamental changes in the socio-technical system of energy provision and consumption. This transition is not only a technical matter but is likely to deeply affect our society and culture [86]. Examples in the literature we have analysed belong almost solely to the category of early first order innovations [46] (see Section 2), while second and third order innovations are almost totally missing, with the exception of some examples where the timing of domestic practices had been adapted as a result of microgeneration of solar energy [52,62]. It could be argued that in order for smart meter feedback to be effective, it should be implemented only in a context where second and third order innovations are supported, advancing a wider transition towards low-carbon energy systems. There is not one straightforward and easy way to make practice-based policy. Rather, policy based on an understanding of practice as a unit of change instead of an individual, can take many forms and utilise many currently existing policy instruments. However, the fundamental difference is that practice-based policymaking is not targeted at persuading individuals to change their behaviour as a result of incentives or information but rather creating conditions for more sustainable practices to take hold and recruit practitioners from less sustainable ones [32]. An important first step is to build an understanding of practices as a unit of societal change.

6. Conclusions

Smart metering technologies are politically popular and there is confidence in their role in reducing consumption. However, there is also a growing body of literature challenging the assumption that making energy or water consumption data visible or accessible will make residents use energy and water differently.

Challenges of smart meter feedback are to a great extent related to the resource-intensity of current domestic practices, including all their elements: infrastructures and appliances, competences and skills as well as shared meanings. Practices do change, but they cannot be changed simply by providing consumption feedback to residents. Energy and water are generally not featuring as elements in domestic practices but operating in the background and taken for granted in current provision systems. This is why residents are generally not interested in monitoring their consumption and why feedback often seems abstract and irrelevant to them and makes them defensive. To change this will require a transition where saving energy starts to matter, e.g. through increasing competences related to microgeneration and energy communities. At the same time, expectations of normality when it comes to standards of living, comfort and cleanliness, should change towards directions that are less energy intensive. This requires policies that address practices rather than individuals.

Smart meter feedback has managed to successfully engage some residents who are experimenting with energy or water management. They can be seen as pioneers of an emerging 'home energy management (HEM) practice'. Whether this practice will successfully recruit more practitioners and eventually have an effect on energy consumption will depend on several developments within energy provision and the role of energy and water in domestic practices. By itself, monitoring is not enough if it does not lead to any or more than minor decreases in consumption.

Monitoring and data collection are increasingly becoming an element of various practices as a result of digitalisation and development of monitoring technologies. For example, monitoring body functions and physical performance has become more popular as part of exercise routines and fostering well-being. It could be argued that as monitoring gains ground in different spheres, it can also become part of domestic energy and water using practices. However, there is a risk that efficiency gains from home energy management will not keep up with the increasing levels of consumption due to rising standards and expectations that come along with the increasing smartness of home environments.

Researchers have an important role to play in contributing to the discussion around smart metering and how it is framed. Is it a project to activate individuals to make informed choices when using energy and water? Or could it be a project to transform the meanings related to energy and water consuming practices, together with an ongoing, (though still in its initial phase) transition towards a de-centralised renewable energy supply system?

We argue that in order to increase the impact of smart meter feedback, it should be developed as an embedded part of the practice arrangements, not as a stand-alone technological solution to measure resource consumption in households. Its relevance for energy and water using practices could be improved by combining it with microgeneration, thus creating an opportunity for energy-monitoring practices to emerge and become more established. This would require simultaneously supporting a wider energy transition towards a low-carbon society, involving second and third order innovations that would create a link between current domestic practices and currently non-existent HEM-practices. Participatory, household or community-level approaches with active involvement of intermediaries can also support the building of new competences and meanings related to the use of smart meter feedback in performing domestic practices. At the same time, material trends in home environments and domestic practices that have implications on what is regarded as normal standard of living should be carefully monitored and their links to policy identified.

Author Contributions: Conceptualisation, M.H.; H.M.; K.M.; J.P.; M.S. methodology, M.H.; H.M.; J.P.; M.S. formal analysis, H.M.; writing—original draft preparation, H.M.; writing—review and editing, M.H.; H.M.; K.M.; J.P.; M.S. supervision, M.H. funding acquisition, M.H.; J.P.

Funding: This research was funded by the Academy of Finland Strategic Research Council, project Dwellers in Agile Cities (DAC), grant number 303481. The contribution of M.S. was supported by Kone Foundation.

Conflicts of Interest: The authors declare no conflict of interest.

Appendix A

List of reviewed studies. Full references to the studies are available in References. When describing type of feedback we use terminology that has been used in the articles themselves, e.g., energy monitor or in-home display.

Table A1. List of reviewed core literature.

Quantitative Changes in Consumption (if any)	Qualitative Study Setting	Duration of Experiment (if applicable)	Type of Feedback	Authors
Not measured.	19 households interviewed	9 months	In-home display for electricity consumption.	Barnicoat & Danson 2015
Not measured	25 households interviewed	Between 1 and 4 years	In-home display for electricity, water and gas consumption.	Berry et al., 2017
Not measured.	5 households interviewed	5 months	Home energy monitor, solar panels and other energy efficiency devices.	Bickerstaff et al., 2016
Not measured.	125 user reviews of in-home displays	Not applicable	In-home display for electricity consumption.	Buchanan et al., 2014
Not measured.	30 households interviewed	2 years	In-home display for electricity. Weekly e-mails and community activities.	Burchell et al., 2016
6.8% decrease after one year, 6.4% after 5 years	82 households included in the study, does not specify how many were interviewed.	1 year + a longitudinal study of 5 years	In-home display for water consumption	Davies et al. 2014
8.1% decrease	20 households interviewed	5 months	In-home display for electricity consumption	Grønhøj & Thøgersen 2011
Not measured	112 blogposts on smart meters	Over one year	Smart electricity meter with real-time consumption data	Guerreiro et al., 2015
Not measured	15 households interviewed	1–12 months, in average 5.8 months.	In-home display for electricity consumption, some models also for heating and hot water.	Hargreaves et al., 2010
Not measured	11 households interviewed	Over one year	In-home display for electricity, some models also for heating and hot water.	Hargreaves et al., 2013
Not measured	21 households interviewed	Over 6 months	In-home display for electricity consumption	Murtagh et al., 2014
No changes in consumption	9 households interviewed	Approximately 4 months	In-home display for electricity consumption	Nilsson et al., 2014
Not measured	17 individuals interviewed	3 weeks	In-home display for electricity consumption	Oltra et al., 2013
Not measured	7 households interviewed	18 months	Smart electricity meter with monitoring software. Consumption data could be accessed via TV, computer, tablet or mobile phone.	Schwartz et al., 2014
Not measured	28 households interviewed	4–6 months	In-home display for electricity, gas and water consumption	Snow et al., 2015
Not measured	7 individuals interviewed	From 2 months to 2 years	In-home display for electricity. One model also measuring water and gas consumption, room temperature and solar power production.	Strengers 2011
Not measured	21 households interviewed	6 months	Smart meter data (electricity and gas) on a website and a smart phone app	Verkade & Höffken 2017
Not measured	33 households interviewed	3–6 weeks	In-home display for electricity consumption	Wallenborn et al., 2011
12.3% reduction in one pilot study, not measured in another pilot.	24 individuals interviewed,	From 1 month to 1.5 years	In-home display for electricity consumption	Westskog et al. 2015
Not measured	5 focus group discussions with 21 participants	Approximately 3 months	In-home display for electricity consumption.	Winther & Bell 2018

References

1. Girod, B.; van Vuuren, D.P.; Hertwich, E.G. Climate policy through changing consumption choices: Options and obstacles for reducing greenhouse gas emissions. *Glob. Environ. Chang.* **2014**, *25*, 5–15. [CrossRef]
2. Mekonnen, M.M.; Hoekstra, A.Y. Four billion people facing severe water scarcity. *Sci. Adv.* **2016**, *2*, 1–6. [CrossRef] [PubMed]
3. Torriti, J.; Hassan, M.; Leach, M. Demand response experiences in Europe: Policies, programmes and implementation. *Energy* **2010**, *35*, 1575–1583. [CrossRef]
4. Siano, P. Demand response and smart grids—A survey. *Renew. Sustain. Energy Rev.* **2014**, *30*, 461–478. [CrossRef]
5. Araújo, K. The Emerging Field of Energy Transitions: Progress, Challenges, and Opportunities. *Energy Res. Soc. Sci.* **2014**, *1*, 112–121. [CrossRef]
6. Sovacool, B. How long will it take? Conceptualizing the temporal dynamics of energy transitions. *Energy Res. Soc. Sci.* **2016**, *13*, 202–215. [CrossRef]
7. van Vliet, B.J.M. Sustainable innovation in network-bound systems: Implications for the consumption of water, waste water and electricity services. *J. Environ. Policy Plan.* **2012**, *14*, 263–278. [CrossRef]
8. Goulden, M.; Bedwell, B.; Rennick-Egglestone, S.; Rodden, T.; Spence, A. Smart grids, smart users? The role of the user in demand side management. *Energy Res. Soc. Sci.* **2014**, *2*, 21–29. [CrossRef]
9. Hyysalo, S.; Juntunen, J.; Freeman, S. User innovation in sustainable home energy technologies. *Energy Policy* **2013**, *55*, 490–500. [CrossRef]
10. Hyysalo, S.; Juntunen, J.; Freeman, S. Internet forums and the rise of the inventive energy user. *Sci. Technol. Stud.* **2013**, *26*, 25–51.
11. Jalas, M.; Hyysalo, S.; Heiskanen, E.; Lovio, R.; Nissinen, A.; Mattinen, M.; Rinkinen, J.; Juntunen, J.K.; Tainio, P.; Nissilä, H. Everyday experimentation in energy transition: A practice theoretical view. *J. Clean. Prod.* **2017**, *169*, 77–84. [CrossRef]
12. Strengers, Y. *Smart Energy Technologies in Everyday Life: Smart Utopia?* Palgrave Macmillan: Hampshire, UK, 2013; ISBN 978-1-349-44325-3.
13. Darby, S. Smart metering: What potential for householder engagement? *Build. Res. Inf.* **2010**, *38*, 442–457. [CrossRef]
14. Sonderlund, A.L.; Smith, J.R.; Hutton, C.; Kapelan, Z.; Savic, D. Effectiveness of smart-meter based consumption feedback in curbing household water use: Knowns and unknowns. *J. Water Resour. Plan. Manag.* **2016**, *142*. [CrossRef]
15. Darby, S. *Literature Review for the Energy Demand Research Project*; Environmental Change Institute: Oxford, UK, 2010; pp. 17–24.
16. USmartConsumer 2016. European Smart Metering Landscape Report "Utilities and Consumers". USmartConsumer Project. Available online: http://www.escansa.es/usmartconsumer/documentos/USmartConsumer_European_Landscape_Report_2016_web.pdf (accessed on 7 February 2017).
17. European Commission. Energy Efficiency Directive 2012/27/EU. Available online: https://eur-lex.europa.eu/legal-content/EN/TXT/?uri=celex:32012L0027 (accessed on 21 August 2018).
18. European Commission. Towards an Integrated Strategic Energy Technology (SET) Plan: Accelerating the European Energy System Transformation. Available online: https://ec.europa.eu/energy/sites/ener/files/documents/1_EN_ACT_part1_v8_0.pdf (accessed on 21 August 2018).
19. Darby, S. *The Effectiveness of Feedback on Energy Consumption. A Review for Defra of the Literature on Metering, Billing and Direct Displays*; Environmental Change Institute: Oxford, UK, 2003; p. 3.
20. Fischer, C. Feedback on household electricity consumption: A tool for saving energy? *Energy Effic.* **2008**, *1*, 79–104. [CrossRef]
21. Faruqui, A.; Sergici, S.; Sharif, A. The impact of information feedback on energy consumption—A survey of the experimental evidence. *Energy* **2009**, *35*, 1598–1608. [CrossRef]
22. Delmas, M.A.; Fischlein, M.; Asensio, O.I. Information strategies and energy conservation behavior: A meta-analysis of experimental studies from 1975 to 2012. *Energy Policy* **2013**, *61*, 729–739. [CrossRef]
23. Klopfert, F.; Wallenborn, G. *Empowering Consumers through Smart Metering, a Report for the BEUC*; Bureau Europeen des Unions des Consommateurs (BEUC): Brussels, Belgium, 2011; p. 2.

24. Allen, D.; Janda, K. The Effects of Household Characteristics and Energy Use Consciousness on the Effectiveness of Real-Time Energy Use Feedback: A Pilot Study. In *Proceedings of the ACEEE 2006 Summer Study on Energy Efficiency in Buildings*; American Council for an Energy-Efficient Economy: Washington, DC, USA, 2006; pp. 1–12.
25. Alahmad, M.A.; Wheeler, P.G.; Schwer, A.; Eiden, J.; Brumbaugh, A. A Comparative Study of Three Feedback Devices for Residential Real-Time Energy Monitoring. *IEEE Trans. Ind. Electron.* **2012**, *59*, 2002–2013. [CrossRef]
26. Abrahamse, W.; Steg, L.; Vlek, C.; Rothengatter, T. A Review of Intervention Studies Aimed at Household Energy Conservation. *J. Environ. Psychol.* **2005**, *25*, 273–291. [CrossRef]
27. Wallenborn, G.; Orsini, M.; Vanhaverbeke, J. Household appropriation of electricity monitors. *Int. J. Consum. Stud.* **2011**, *35*, 146–152. [CrossRef]
28. Buchanan, K.; Russo, R.; Anderson, B. The question of energy reduction: The problem(s) with feedback. *Energy Policy* **2015**, *77*, 89–96. [CrossRef]
29. Hargreaves, T. Beyond energy feedback. *Build. Res. Inf.* **2018**, *46*, 332–342. [CrossRef]
30. Shove, E.; Walker, G. What Is Energy For? Social Practice and Energy Demand. *Theory Cult. Soc.* **2014**, *31*, 41–58. [CrossRef]
31. Verkade, N.; Höffken, J. Is the Resource Man coming home? Engaging with an energy monitoring platform to foster flexible energy consumption in the Netherlands. *Energy Res. Soc. Sci.* **2017**, *27*, 36–44. [CrossRef]
32. Shove, E.; Pantzar, M.; Watson, M. *The Dynamics of Social Practice: Everyday Life and How It Changes*; Sage: London, UK, 2012; ISBN 978-0-85702-043-7.
33. Spaargaren, G. Theories of practices: Agency, technology, and culture: Exploring the relevance of practice theories for the governance of sustainable consumption practices in the new world-order. *Glob. Environ. Chang.* **2011**, *21*, 813–822. [CrossRef]
34. Giddens, A. *The Constitution of Society*; Polity Press: Cambridge, UK, 1984; ISBN 0745600069.
35. Schatzki, T. *The Site of the Social: A Philosophical Account of the Constitution of Social Life and Change*; Penn State Press: University Park, PA, USA, 2002; ISBN 978-0-271-02292-5.
36. Reckwitz, A. Toward a theory of social practices: A development in culturalist theorizing. *Eur. J. Soc. Theory* **2002**, *5*, 243–263. [CrossRef]
37. Shove, E.; Watson, M.; Hand, M.; Ingram, J. *The Design of Everyday Life*; Berg: Oxford, UK, 2007; ISBN 1845206835.
38. Spurling, N.; McMeekin, A.; Shove, E.; Southerton, D.; Welch, D. Interventions in Practice: Re-Framing Policy Approaches to Consumer Behavior. Available online: http://www.sprg.ac.uk/uploads/sprg-report-sept-2013.pdf (accessed on 21 September 2018).
39. Shove, E. *Comfort, Cleanliness and Convenience: The Social Organization of Normality*; Berg: Oxford, UK, 2003; ISBN 9781859736302.
40. Hand, M.; Shove, E.; Southerton, D. Explaining Showering: A Discussion of the Material, Conventional, and Temporal Dimensions of Practice. *Sociol. Res. Online* **2005**. [CrossRef]
41. Yates, L.; Evans, D. Dirtying Linen: Re-evaluating the sustainability of domestic laundry. *Environ. Pol. Gov.* **2016**, *26*, 101–115. [CrossRef]
42. Warde, A. Consumption and Theories of Practice. *J. Consum. Cult.* **2005**, *5*, 131–153. [CrossRef]
43. Heiskanen, E.; Johnson, M.; Robinson, S.; Vadovics, E.; Saastamoinen, M. Low-carbon communities as a context for individual behavioural change. *Energy Policy* **2010**, *38*, 7586–7595. [CrossRef]
44. Shove, E. Beyond the ABC: Climate Change Policy and Theories of Social Change. *Environ. Plan. A Econ. Space* **2010**, *42*, 1273–1285. [CrossRef]
45. Ajzen, I. The theory of planned behaviour. *Organ. Behav. Hum. Decis. Process.* **1991**, *50*, 179–211. [CrossRef]
46. Naus, J. The Social Dynamics of Smart Grids: On Households, Information Flows & Sustainable Energy Transitions. Ph.D. Thesis, Wageningen University, Wageningen, The Netherlands, 2017. [CrossRef]
47. Torraco, R.J. Writing Integrative Literature Reviews: Guidelines and examples. *Hum. Resour. Dev. Rev.* **2005**, *4*, 356–367. [CrossRef]
48. Barnicoat, G.; Danson, M. The Ageing Population and Smart Metering: A Field Study of Householders' Attitudes and Behaviours towards Energy Use in Scotland. *Energy Res. Soc. Sci.* **2015**, *9*, 107–115. [CrossRef]
49. Guerreiro, S.; Batel, S.; Lima, M.L.; Moreira, S. Making energy visible: Sociopsychological aspects associated with the use of smart meters. *Energy Effic.* **2015**, *8*, 1149–1167. [CrossRef]

50. Nilsson, A.; Jakobsson Bergstad, C.; Thuvander, L.; Andersson, D.; Andersson, K.; Meiling, P. Effects of continuous feedback on households' electricity consumption: Potentials and barriers. *Appl. Energy* **2014**, *122*, 17–23. [CrossRef]
51. Oltra, C.; Boso, A.; Espluga, J.; Prades, A. A qualitative study of users' engagement with real-time feedback from in-house energy consumption displays. *Energy Policy* **2013**, *61*, 788–792. [CrossRef]
52. Berry, S.; Whaley, D.; Saman, W.; Davidson, K. Finding faults and influencing consumption: The role of in-home energy feedback displays in managing high-tech homes. *Energy Effic.* **2017**, *10*, 787–807. [CrossRef]
53. Winther, T.; Bell, S. Domesticating in home displays in selected British and Norwegian households. *Sci. Technol. Stud.* **2018**, *31*, 19–38. [CrossRef]
54. Davies, K.; Doolan, C.; van den Honert, R.; Shi, R. Water-saving impacts of Smart Meter technology: An empirical 5 year, whole-of-community study in Sydney, Australia. *Water Resour. Res.* **2014**, *50*, 7348–7358. [CrossRef]
55. Hargreaves, T.; Nye, M.; Burgess, J. Keeping energy visible? Exploring how householders interact with feedback from smart energy monitors in the longer term. *Energy Policy* **2013**, *52*, 126–134. [CrossRef]
56. Buchanan, K.; Russo, R.; Anderson, B. Feeding back about eco-feedback: How do consumers use and respond to energy monitors? *Energy Policy* **2014**, *73*, 138–146. [CrossRef]
57. Strengers, Y. Negotiating everyday life: The role of energy and water consumption feedback. *J. Consum. Cult.* **2011**, *11*, 319–338. [CrossRef]
58. Murtagh, N.; Gatersleben, B.; Uzzell, D. 20:60:20—Differences in Energy Behaviour and Conservation between and within Households with Electricity Monitors. *PLoS ONE* **2014**, *9*, 1–12. [CrossRef] [PubMed]
59. Westskog, H.; Winther, T.; Sæle, H. The Effects of In-Home Displays—Revisiting the Context. *Sustainability* **2015**, *7*, 5431–5451. [CrossRef]
60. Hargreaves, T.; Nye, M.; Burgess, J. Making energy visible: A qualitative field study of how householders interact with feedback from smart energy monitors. *Energy Policy* **2010**, *38*, 6111–6119. [CrossRef]
61. Sovacool, B.; Kivimaa, P.; Hielscher, S.; Jenkins, K. Vulnerability and resistance in the United Kingdom's smart meter transition. *Energy Policy* **2017**, *109*, 767–781. [CrossRef]
62. Bickerstaff, K.; Hinton, E.; Bulkeley, H. Decarbonisation at home: The contingent politics of experimental domestic energy technologies. *Environ. Plan. A* **2016**, *48*, 2006–2025. [CrossRef]
63. Snow, S.; Vyas, D.; Brereton, M. When an eco-feedback system joins the family. *Pers. Ubiquit. Comput.* **2015**, *19*, 929–940. [CrossRef]
64. Grønhøj, A.; Thøgersen, J. Feedback on household electricity consumption: Learning and social influence processes. *Int. J. Consum. Stud.* **2011**, *35*, 138–145. [CrossRef]
65. Schwartz, T.; Stevens, G.; Jakobi, T.; Denef, S.; Ramirez, L.; Wulf, V.; Randall, D. What People Do with Consumption Feedback: A Long-Term Living Lab Study of a Home Energy Management System. *Interact. Comput.* **2015**, *27*, 551–576. [CrossRef]
66. Powells, G.; Bulkeley, H.; Bell, S.; Judson, E. Peak electricity demand and the flexibility of everyday life. *Geoforum* **2014**, *55*, 43–52. [CrossRef]
67. Burchell, K.; Rettie, R.; Roberts, T.C. Householder engagement with energy consumption feedback: The role of community action and communications. *Energy Policy* **2016**, *88*, 178–186. [CrossRef]
68. Peacock, A.D.; Chaney, J.; Goldbach, K.; Walker, G.; Tuohy, P.; Santonja, S.; Todoli, D.; Owens, E.H. Co-designing the next generation of home energy management systems with lead-users. *Appl. Ergon.* **2017**, *60*, 194–206. [CrossRef] [PubMed]
69. Novak, J.; Melenhorst, M.; Micheel, I.; Pasini, C.; Fraternali, P.; Rizzoli, A.E. Integrating behavioural change and gamified incentive modelling for stimulating water saving. *Environ. Model. Softw.* **2018**, *102*, 120–137. [CrossRef]
70. Hansen, M.; Hauge, B. Prosumers and smart grid technologies in Denmark: Developing user competences in smart grid households. *Energy Effic.* **2017**, *10*, 1215–1234. [CrossRef]
71. Salo, M.; Nissinen, A.; Lilja, R.; Olkanen, E.; O'Neill, M.; Uotinen, M. Tailored advice and services to enhance sustainable household consumption in Finland. *J. Clean. Prod.* **2016**, *121*, 200–207. [CrossRef]
72. Tiefenbeck, V.; Goette, L.; Degen, K.; Tasic, V.; Fleisch, V.; Lalive, R.; Staake, T. Overcoming Salience Bias: How Real-Time Feedback Fosters Resource Conservation. *Manag. Sci.* **2018**, *64*, 1458–1476. [CrossRef]

73. Mogles, N.; Walker, I.; Ramallo-Gonzáles, A.P.; Lee, J.; Natarajan, S.; Padget, J.; Gabe-Thomas, E.; Lovett, T.; Ren, G.; Hyniewska, S.; et al. How smart do smart meters need to be? *Build. Environ.* **2017**, *125*, 439–450. [CrossRef]
74. Wittenberg, I.; Matthies, E. How do PV households use their PV system and how is this related to their energy use? *Renew. Energy* **2018**, *122*, 291–300. [CrossRef]
75. Morris, P.; Vine, D.; Buys, L. Critical success factors for peak electricity demand reduction: Insights from a successful intervention in a small island community. *J. Consum. Policy* **2018**, *41*, 33–54. [CrossRef]
76. Heiskanen, E.; Johnson, M.; Vadovics, E. Learning about and involving users in energy saving on the local level. *J. Clean. Prod.* **2013**, *48*, 241–249. [CrossRef]
77. Gupta, R.; Barnfield, L.; Gregg, M. Exploring innovative community and household energy feedback approaches. *Build. Res. Inf.* **2018**, *46*, 284–299. [CrossRef]
78. Hitchings, R. Sharing conventions. Communities of practice and thermal comfort. In *Sustainable Practices. Social Theory and Climate Change*; Shove, E., Spurling, N., Eds.; Routledge Advances in Sociology: London, UK, 2013; pp. 103–115, ISBN 978-0-415-54065-0.
79. Strengers, Y.; Nicholls, L. Convenience and energy consumption in the smart home of the future: Industry visions from Australia and beyond. *Energy Res. Soc. Sci.* **2017**, *32*, 86–93. [CrossRef]
80. Tirado Herrero, S.; Nicholls, L.; Strengers, Y. Smart home technologies in everyday life: Do they address key energy challenges in households? *Curr. Opin. Environ. Sustain.* **2018**, *31*, 65–70. [CrossRef]
81. Røpke, I.; Christensen, T.H.; Jensen, J.O. Information and communication technologies—A new round of household electrification. *Energy Policy* **2010**, *38*, 1764–1773. [CrossRef]
82. Francisco, A.; Truong, H.; Khosrowpour, A.; Taylor, J.E.; Mohammadi, N. Occupant perceptions of building information model-based energy visualizations in eco-feedback systems. *Appl. Energy* **2018**, *221*, 220–228. [CrossRef]
83. Nyborg, S.; Røpke, I. Energy impacts of the smart home—Conflicting visions. Energy Efficiency First: The Foundation of a Low-Carbon Society. In Proceedings of the ECEEE 2011 Summer Study, Belambra Presquile de Giens, France, 6–11 June 2011; European Council for an Energy Efficient Economy: Stockholm, Sweden, 2011; pp. 1849–1860.
84. Darby, S. Smart technology in the home: Time for more clarity. *Build. Res. Inf.* **2018**, *46*, 140–147. [CrossRef]
85. Miller, W.; Senadeera, M. Social transition from energy consumers to prosumers: Rethinking the purpose and functionality of eco-feedback technologies. *Sustain. Cities Soc.* **2017**, *35*, 615–625. [CrossRef]
86. Järvensivu, P. A post-fossil fuel transition experiment: Exploring cultural dimensions from a practice-theoretical perspective. *J. Clean. Prod.* **2017**, *169*, 143–151. [CrossRef]

© 2018 by the authors. Licensee MDPI, Basel, Switzerland. This article is an open access article distributed under the terms and conditions of the Creative Commons Attribution (CC BY) license (http://creativecommons.org/licenses/by/4.0/).

Article

Game Theoretical Energy Management with Storage Capacity Optimization and Photo-Voltaic Cell Generated Power Forecasting in Micro Grid †

Aqdas Naz [1], Nadeem Javaid [1,*], Muhammad Babar Rasheed [2], Abdul Haseeb [3], Musaed Alhussein [4] and Khursheed Aurangzeb [4,*]

1. Department of Computer Science, COMSATS University Islamabad, Islamabad 44000, Pakistan; aqdasmalik17@gmail.com
2. Department of Electronics and Electrical Systems, The University of Lahore, Lahore 54000, Pakistan; babarmeher@gmail.com
3. Department of Electrical Engineering, Institute of Space Technology (IST), Islamabad 44000, Pakistan; haseeb_karak@yahoo.com
4. Computer Engineering Department, College of Computer and Information Sciences, King Saud University, Riyadh 11543, Saudi Arabia; musaed@ksu.edu.sa
* Correspondence: nadeemjavaidqau@gmail.com (N.J.); kaurangzeb@ksu.edu.sa (K.A.)
† This manuscript is an extended version of paper published in the proceedings of 33rd International Conference on Advanced Information Networking and Applications (AINA), Matsue, Japan, 27–29 March 2019.

Received: 18 March 2019; Accepted: 25 April 2019; Published: 14 May 2019

Abstract: In order to ensure optimal and secure functionality of Micro Grid (MG), energy management system plays vital role in managing multiple electrical load and distributed energy technologies. With the evolution of Smart Grids (SG), energy generation system that includes renewable resources is introduced in MG. This work focuses on coordinated energy management of traditional and renewable resources. Users and MG with storage capacity is taken into account to perform energy management efficiently. First of all, two stage Stackelberg game is formulated. Every player in game theory tries to increase its payoff and also ensures user comfort and system reliability. In the next step, two forecasting techniques are proposed in order to forecast Photo Voltaic Cell (PVC) generation for announcing optimal prices. Furthermore, existence and uniqueness of Nash Equilibrium (NE) of energy management algorithm are also proved. In simulation, results clearly show that proposed game theoretic approach along with storage capacity optimization and forecasting techniques give benefit to both players, i.e., users and MG. The proposed technique Gray wolf optimized Auto Regressive Integrated Moving Average (GARIMA) gives 40% better result and Cuckoo Search Auto Regressive Integrated Moving Average (CARIMA) gives 30% better results as compared to existing techniques.

Keywords: forecasting; solar generation; storage capacity; game theory; nash equilibrium; distributed energy management algorithm; micro grid; meta heuristic techniques

1. Introduction

Despite the ever increasing economic development attained by the world, many challenges are being faced in context of environmental inefficiency, environmental pollution, etc. With the passage of time, energy demand rises and infrastructure needs to be upgraded. Therefore, new power grid is required, that enhances power supply as well as it integrates renewable energy resources. In order to overcome such challenges, Smart Grid (SG) is brought to the light [1–4]. Demand Response (DR) is a crucial component of SG technology that tends to maintain the balance among electricity supply and demand using peak load shaving [5]. Real Time Pricing (RTP) is an efficient mechanism among all the schemes that are included in DR [6]. Monopoly of energy generation companies that are owned by the state causes pricing schemes to be ineffective for enabling the user to be actively involved in trading process of energy [7]. In order to focus on these issues, energy Internet is considered as one of the key enablers of third industrial evolution [8]. Energy Internet is considered as new paradigm shift for user and generation system [9]. By analogy with Internet properties, complete framework is offered by energy Internet for integration of each equipment that performs energy production, issuance, transformation, storage and usage with basic Information and Communication Technologies (ICT) [10]. Standard and modular energy units, i.e., solar panel, wind turbine, hydrogen, fuel cell, biomass and storage system can be operated by plug and play modules [11]. Open standard based communication protocol are added in plug and play paradigm to enhance the capability and interoperability for many products, technologies, systems and solutions, which construct energy Internet. In this new evolving paradigm, the supplier and consumer are connected very closely and promptly because of implementing distributed and flexible systems [12]. Moreover, energy consumer with co-located energy provisioner formulates local Internet of energy. Where, MG relieves the stress at reasonable degree, that is caused by increasing energy demands. MG is considered as one of the reliable networks for establishing connection between renewable resources and consumers along with managing storage units [13]. It can either be treated as controllable load or production system and can work in connection with grid. Nevertheless, owing to intermittent and changing nature of renewable energy resources, restricted energy generation capacity and greater dependency of MG on uncontrolled renewable energy resources lead to high level of fluctuation and disturbance of the system. For example, the state of unreliability that is brought by renewable energy resources will cause significant difference between production and demand, which rises several issues regarding power imbalance, voltage instability and frequency instability [14]. Thus, energy management techniques are needed to harness in order to reduce energy supply demand imbalance. To achieve ideal economic performance by MG while ensuring reliability, various factors involve in energy Internet. It includes conventional fossil fuel based dispatch able generators and renewable energy based distributed producers. It is clearly non feasible to take each detail into account as it increases computational complexity dramatically. However, small uncertainty that rises from implementation or estimation in real world energy management system makes the system completely incomprehensible in practical point of view [15]. For instance, considering MG with Photo-Voltaic Cell (PVC), solar radiations suddenly become intense at day time. That will increase generation of MG at certain time [16]. Similarly in case of wind turbine, speed of wind goes up any time and become stronger, which will cause grid frequency goes up [14]. Therefore, dynamics of energy generation behavior of renewable resources can not be ignored [17–19]. Literature work is restricted to limited application of MG, where real world data is not considered while managing energy distribution. However, comprehensive framework is required to improve energy management of real world data. Moreover, the prior statistical knowledge of uncertain renewable resources energy production was considered to be precisely known and power trade among various market players is completely neglected This is the motivation behind proposed algorithm that performs distributed energy management along with integration of linear forecasting techniques, which makes proposed system more effective and reliable.

To utilize renewable energy resources effectively, Distributed Energy Management (DEM) algorithm has been proposed that optimizes payoff of each player, i.e., users and MG. In order to overcome the uncertainties that are caused by renewable resources, forecasting techniques have been proposed, i.e., Gray wolf optimized Auto Regressive Integrated Moving Average (GARIMA) and Cuckoo search Auto Regressive Integrated Moving Average (CARIMA). In DEM, GARIMA forecasting data has been used for energy management as it performs more efficiently as compared to CARIMA in current scenario. List of acronyms and list of symbols used in this paper are shown in table at the end of this paper.

This work is an extension of [20]. Whereas, remaining paper has following organization. Section 2 discusses related work regarding load forecasting and game theory. Section 3 gives details of problem statement and contribution. Section 4 explains method and material regarding proposed forecasting techniques and game theory. It includes DEM algorithm for optimization of MG and users cost. Section 5 explains simulations results of proposed techniques. Finally, Section 6 provides conclusion of presented work.

2. Related Work

Proposed work focuses to solve distributed energy management at Micro Grid (MG) level in energy Internet, using both game theory and data analytics techniques. There is recent surge in data analytics that introduces mathematical tool, which addresses uncertainties [21]. While managing energy, two main methodologies are used, i.e., stochastic optimization and robust optimization. These are widely used to handle data uncertainties [22]. Stochastic optimization in energy management solutions is considered as one of the effective techniques for optimization of statistical objective function. In case, the undefined numerical data has to be assumed in order to follow well-known probability distribution. Real time energy management techniques that are having stochastic optimization are proposed to reduce operational cost [23]. In [24], multistage framework is proposed in order to reduce the cost of total energy system that is based on stochastic optimization. In order to cater multi dimensional energy management stochastic dynamic programming methods are used [25]. Nonetheless, the precise calculation of probability distribution of the data that is not certain may get tremendous challenge in practical applications, which considers complex operation details and various complex constraints. Optimality performance is affected by impact of data uncertainties. However, it is not cater for in energy management approaches that contain stochastic optimization.

Robust optimization based energy management approaches merely rely on limited information and enable distribution free model for uncertainties [26]. In energy management system, worst case operation scenario is considered in optimization process. Consequently, energy management approaches can remove the negative impact of uncertainty in optimality performance. Thus, it outperforms the stochastic optimization. Novel pricing strategy is proposed, which promotes robustness against uncertainties of power input [27]. Nonetheless, the robust version of controllable electric load management issues are not ensured to be tractable that relies on appropriate design for objective function modeling and building of uncertainty set.

Ever since the growing advancement in the field of advance ICT, large amount of data is collected regarding consumer behavior, states of battery, substations, customer devices, distributed energy resources, renewable output, weather conditions, video surveillance, etc. [28]. The energy generation using PVC is considered to be one of the important components of electricity sector. It is also attaining attention of government because of increasing environmental issue and being cost effective. In [29], it is stated that PVC energy will be responsible to fulfill 16% of total energy consumption. PVC energy relies on solar radiation does not remain constant always. The intermittent nature of PVC energy is experienced because of position of sun and movement of clouds at particular position. It causes variation at any point in time within a day. Subsequently, it can be predicted accurately in a year. Information of solar energy is greatest concern for operator and planners of electrical system. Hence, there is sheer need of forecasting PVC power generation [30] using machine learning techniques. Forecasting models that are mainly used for prediction of electrical load and the renewable energy resources are defined under three categories, i.e., statistical model: Auto Regressive (AR), Exponential Smoothing (ES) models [31], Artificial Intelligence (AI) Model: Neural Network (NN), Convolution Neural Network (CNN), Hybrid Models: neuro fuzzy models [32,33]. As a case study, electricity demand of 10 countries are taken into account in order to analyze these methods. In [34] 5 different forecasting techniques are used, i.e., multi model, iterative, single model multivariate forecasting are analyzed in detail. It also covers issues such as NN designing, implementation and validation. In [35], a combination of NN and Enhanced Particle Swarm Optimization (PSO) are used in order to perform power forecasting. It focuses on feature selection. PVC generation is affected by many factor apart of solar radiations. It may be affected by cloud movement, location of sun, etc. Nevertheless, time series of power generation by PVC contains many well defined patterns such as there are peaks at afternoon and off peak in morning and evening. Subsequently, there is no generation at night timings. Therefore time series patterns can not be ignored [36]. Auto Regressive Integrated Moving Average (ARIMA) is used for univariate time series forecasting because solar generation tends to follow specific pattern and it is proven to be efficient on account of flexibility. It also performs its orderly searching at every level. Thus it determines best fit model for particular time series. Optimization of parameters has significant impact on the performance of forecasting algorithms. Meta heuristic techniques are applied on many forecasting techniques, i.e., ARIMA [37] Neural Network (NN) [38], Support Vector Regression (SVR) [39] in literature for improving performance. Game theory concept is widely applied on MG energy management studies. It provides distributed self organizing and self optimizing solution of the problem having conflicting objective function.In broader aspect, game theory is characterized in two categories in context of players. One category includes players with binding agreement among them. Second category includes those players who are not having binding agreement among them [40]. In non cooperative game theory, main focus is on predicting individual strategies and it also asses players that make decisions to find Nash Equilibrium (NE). It provides framework for performing analytical framework, i.e., DEM that is devised for characterizing the interaction among players and decision making process to achieve NE. The strategic outcome among players can be improved under mutual commitment.

In order to handle non cooperative game theory based energy management, multi user based Stackelberg game is used to optimize the payoff of each player [41]. In [42], multi stage market model is proposed, that is based on cooperative game to reduce the cost of utility, whereas it maximizes the total profit of the market. In order to cater dispatch problem in integration of renewable resources generation and energy storage, a cooperative distributed energy scheduling algorithm was proposed in [43].

3. Problem Statement and Contributions

To make efficient use of renewable resources, this paper focuses on distributed energy management problem. It aims to maximize objective function of each player and satisfying user demand of electricity and guaranteeing reliable system operations. Due to uncontrollable and uncertain nature of energy generation by renewable resources [44], we used electricity generation forecasting techniques to get short term prediction value. Afterwards, distributed MG energy management problem has been addressed using non cooperative game theory as it ignores common commitment of players and it contains low communication overhead [40]. Optimization of storage capacity in MG has vital role in context of efficient management of electric load, which has also been addressed in this paper. Following are the contributions of our work:

1. Game theory and data-centric approaches are adapted in order to address MG electric load management problem. In order to overcome uncertainties caused by PVC generation, linear forecasting technique ARIMA has been used for forecasting. Parameters of ARIMA, i.e., AR and Moving Average (MA) are optimized through GWO and named as GARIMA,
2. Energy management problem has been solved using two stage Stackelberg game theory to capture the dynamic interaction and interconnection among users and MG. Where, MG acts as a leader and users act like followers. Besides, if there exist a scenario where energy demand of users increases as compare to MG capacity of energy generation, MG purchases energy from utility. Furthermore, energy cost of MG has also been reduced by using energy storage mechanism,
3. Two proposed techniques, i.e., GARIMA and CARIMA have been used for forecasting purpose. Parameters optimization of ARIMA has been performed using Gray Wolf Optimizer (GWO) and Cuckoo Search (CS) algorithm, where GARIMA gives better result as compare to CARIMA

and other conventional techniques. Forecasting results of GARIMA technique are used in DEM algorithm in order to reduce uncertainties that are caused by renewable resources historic data and

4. For non cooperative game of MG and users, existence of NE is proved using Stackelberg game theory. Furthermore, iterative DEM algorithm is proposed for MG to prove NE.

4. Material and Method

This section elaborates material and method regarding proposed system model. Two stage Stackelberg game theory in SG is presented in system model. Multiple residential users $\mathbb{N} = \{1, 2, ...n\}$ along with single MG are taken into account. MG is taken as supplier of power in order to provide power stability to users. MG contains smart meter in order to help users to schedule their energy usage. It is also equipped with PVC storage system. PVC provides power to fulfill requirements of residential users and to charge the battery. Furthermore, surplus energy is sent to the utility as shown in Figure 1. After receiving price policy of utility from information network, users send demand to MG. In the presented system model, users have both shift able and non shift able loads. In order to perform energy usage scheduling at users side, single day is considered. $\mathcal{K}$ represents each time slot in a day. In this scenario, the utility receives electricity demand from users for each time slots in a single day and real time price are communicated to users regarding each time slot in a day. $\{P^k = p_1^k, ...p_j^k, ...p_m^k\}$ where dataset shows time slots in a single day. MG sets its prices, to optimize the payoff, according to the demand of users in real time.

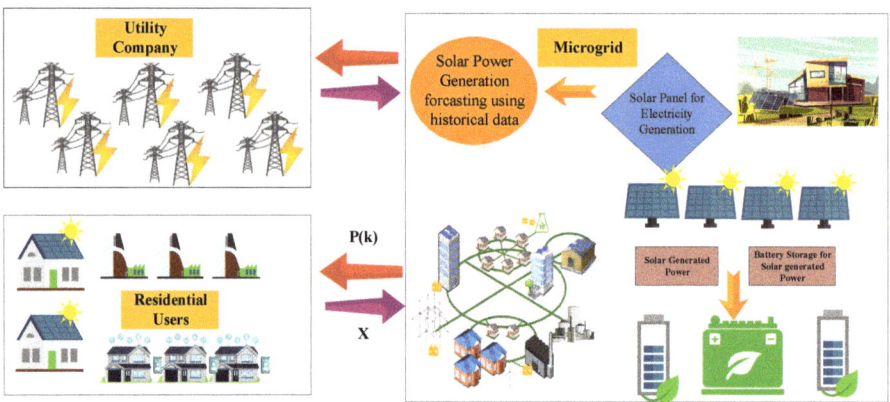

Figure 1. Interaction among User and Micro grid.

4.1. Cost Model of Users

Let us assume, multiple users are considered in the proposed system model with a set of $\mathbb{N} = \{1, 2, ...N\}$. Whole day is divided into $\mathcal{K}$ time slots. $l_n(k)$ represents energy utilization of n users that contains both shift able and non shift able load. Equations (1)–(6) have been taken from [45]. Daily energy consumption of load is explained with the help of equation below:

$$l_n = [l_n(1), ..., l_n(k), ...l_n(\mathcal{K})] \quad (1)$$

MG offers energy for daily load consumption. Lets assume $x_n(k)$ is energy demand that is sent to MG by users in k time slots, where $k \varepsilon \mathcal{K}$. So that energy that is demanded by users from MG is written as:

$$x_n = [x_n(1), ..., x_n(k), ...x_n(\mathcal{K})] \quad (2)$$

Sum of the total amount of energy used by users is represented as:

$$X_k = \sum_{n=1}^{N} x_n^k \tag{3}$$

In order to calculate Peak Average Ratio (PAR) of energy that is demanded by users. Peak consumption is represented as:

$$X(peak) = \max_{k \epsilon \mathcal{K}} X_k \tag{4}$$

Average user consumption is calculated as:

$$X(average) = \frac{1}{K} \sum_{k \epsilon \mathcal{K}} X_k \tag{5}$$

Thus, PAR of total energy consumption by users is calculated as:

$$PAR = \frac{X(peak)}{X(average)} \tag{6}$$

Cost function of user $\mathcal{U}_n$ relies on type of pricing mechanism that has been formulated by MG. Effective pricing scheme should be employed by the utility and MG to encourage users to actively become a part of energy consumption scheduling planning. Which helps to make effective plan to charge electricity price from users. Following assumptions are made:

1. MG is responsible for providing energy to user at any time. Hence, the cost function of user regarding energy consumption by user at any time slot $k \; \epsilon \; \mathcal{K}$ is function of energy consumption X_k by N users,
2. In daily life, energy consumption by user at certain time slot is smooth function or at least it is piecewise smooth function and always increasing. Likewise, cost function of user follows the demand X_k,
3. Cost function also depends on timings of energy consumption, apart of energy consumption by user.

Owing to the assumptions discussed above, quadratic cost function is used because it is non linear and strictly convex in nature. As it is mostly utilized in literature [40]. Thus, cost function of users is as follows [45]:

$$C_k X_k = a_k X_k^2 + b_k X_k \tag{7}$$

where, $a_k > 0$ and $b_k > 0$ are fixed parameters that have maximum value at peak demand hours. $C_k X_k$ represents total cost of energy consumption by k^{th} user. Whereas, total cost of N user is calculated as follows [45]:

$$\mathcal{U}_{n,k} = (a_k X_k + b_k) x_n^k = p_k(X_k) x_n^k \tag{8}$$

where, $p_m^k(X_k)$ represents energy price by MG at time slot k. Price model confirms that energy price increases as increase in energy consumption X_k happens. Consequently, users are convinced to shift their daily load from on peak hour to off peak hours. Moreover, behavior of the user in different time slots regarding consumption of energy is also affected via modifying a_k and b_k. Total cost of user n in a day is calculated as:

$$\mathcal{U}_n(x_n, x_{-n}, p_m^k) = \sum_{k=1}^{K} [P_k(X_k) x_n^k] \tag{9}$$

where, $x_{-n} = [x_1, ..., x_{n-1}, x_{n+1}, ...x_N]$ represents energy demand of N users except n user. Equation (9) clearly shows that cost function of user is directly dependent on consumption of energy. Consumption of energy can be managed by shifting demand of on peak demand hours to off peak hours. Hence, cost function of users can be reduced as:

$$\min_{x_n^k} \mathcal{U}_n(x_n, x_{-n}, p_m^k) \tag{10}$$

4.2. MG Cost Model with Storage Capacity

For efficient energy management, MG and $\mathbb{N}$ users are agreed upon the energy parameter in order to avoid any kind of conflict. Total amount of energy that is needed by all users is x_m^k and the price p_m^k is decided by MG in such way to maximize its cost function, i.e., $\mathcal{B}$. Likewise, y_m is maximum storage required by MG to store the energy generated by PVC. Optimal solution must satisfy objective of both players, i.e., users and MG. Moreover, energy that is generated using PVC will be forecast, in order to decide energy price p_m^k per time slot k effectively [35]. Total electricity that is demanded from the users has to be less or equal than electricity which is produced by MG as represented in equation below:

$$X_m^k \leq G_m^k \ \forall k \in \mathcal{K} \tag{11}$$

where, G_m^k is the total capacity of the MG in time slot k. In order to complete energy trading successfully, both players of game, i.e., users and MG exchange messages among each other and both of them agree on the trading parameters.

However, energy x_m^k demanded by total no. of users from MG m must satisfy following constraint:

$$\sum_{k=1}^{K} G_m^k + D_m \geq \sum_{n=1}^{N}\sum_{k=1}^{K} x_m^k, \text{ and } e_m \leq \sum_{k=1}^{K} G_m^k \tag{12}$$

where, D_m represents amount of energy that is requested to utility, if renewable resources do not generate enough energy for fulfilling demand of $\mathbb{N}$ users. e_m is predicted energy that is generated by PVC explained in Section 4.3.1. We consider amount of energy, generated by PVC, in a single day. Moreover, price per unit energy P_m^k is decided and announced by MG. In current scenario, P_m^k is price that is charged by MG. Therefore, energy that is required by users should satisfy constraint given in Equations (11) and (12). The main challenge that is being faced by MG is to decide optimum energy prices P_m^k and optimum battery storage y_m to increase MG revenue.

Most of the hybrid systems, which contain both generation plant and renewable resources, are difficult to manage; specifically if storage system is placed with every user and resultantly surplus energy is supplied to the utility [45]. Therefore, it is required to have centralized system to formalize the mechanism which decides the distribution of energy among all stake holders. Distribution of energy is purely on the basis of generation and requirement basis. Assumption is made that generation of PVC is cheaper than conventional energy generation methods. MG prioritize the demand of $\mathbb{N}$ users and to charge the batteries. While at second priority, surplus energy is to be transmitted back for generating revenue by trading. For MG, it is assumed that PVC generation is $e_m(k) \geq 0$ in time slot k. PVC power generation provides energy, $e_m(k) - e_m^l(k) - e^b(k) = 0$ that means surplus energy does not exist that can be sold back to utility. Where, $e_m^l(k)$ represents PVC energy to be distributed among users to fulfill their demand. Whereas, $e^b(k)$ depicts power that is needed to charge the battery. In case $e_m(k) - e_m^l(k) - e_m^b(k) \geq 0$, MG is having extra energy which has been generated by PVC for selling. Profit of MG, that it generates when it sells surplus energy to the utility is represented as:

$$\mathcal{UC} = \sum_{k=1}^{K} \lambda_s (e_m(k) - e_m^l(k) - e_m^b(k)) \tag{13}$$

where, λ_s depicts the PVC power selling price in cents/kWh. Currently, profit generated by selling electricity to utility is in form of price subsidies that is provided by the government to MG. These days, subsidy standard is λ_s is 6.3 cents/kWh for distribution of energy [46]. Total cost of PVC generation is as follows:

$$C_m^{PVC} = a_m(\hat{L} + \Delta) + b_m(\hat{L} + \Delta)^2 + c_m + F|\Delta| \tag{14}$$

where a_m, b_m and c_m are cost parameters of MG with PVC power generation and $((\hat{L}) + \Delta)$ represents the prediction of PVC power [47]. Whereas, $\hat{L}$ shows prediction of power generated by PVC in next year and Δ shows prediction error. F is chosen as penalty factor of prediction. Where, $F < 0$ shows that payoff of MG is decreased if prediction result is not accurate. Hence, it causes restriction of power agreement in market.

PVC power often needs storage system because it has to store surplus energy once users requirements are fulfilled. Storage system is irreplaceable need for PVC energy generation. Then, MG will provide energy to users at demand peak hours. Where, utility charges more price from consumer, in order to reduce cost of MG. These days, number of types of batteries that are available in market, i.e., Sodium/Sulfur batteries, Zinc/Bromine batteries and lithium-ion batteries [48]. Each battery type has certain charging and discharging cycles depending on their material, technology and size. Afterwards, battery life is expired. Hence, depreciation cost needs to be taken under consideration. Battery cost function is either defined as quadratic function that is based on charging and discharging capacity or it is taken as linear function. In the proposed scenario, battery cost function is taken as linear function. Equations (15)–(23) are taken from [45]. It is assumed that y_m is the battery capacity that is needed by MG and it may ranges within certain limits:

$$y_m \varepsilon [y_m^l, y_m^u] \tag{15}$$

where y_m^l and y_m^u represent lower and upper limits of the battery capacity. Equation (16) gives daily depreciation cost function:

$$C_m^{bat}(y_m) = \lambda^{bat} y_m \tag{16}$$

where $C_m^{bat}(y_m)$ represents cost of battery depreciation. Its unit is cents/kWh and it also maintains correlation with the material and type of the battery. $C_m^{bat}(y_m)$ shows linear increasing function that is based on total capacity y_m of the storage. However, if storage of the battery is not enough to store generated PVC power, it will be waisted and consequently, the payment will be increased. Therefore, it is required to decide optimal capacity of the battery y_m. Other parameters that have important role in storage optimization other than battery capacity parameters. Charging and discharging efficiency of battery are parameters that are required to be taken under consideration other than battery capacity parameters. It is assumed, $0 < \eta_{ch} < 1$ and $0 < \eta_{disch} < 1$ show battery charging and discharging efficiency. $s = [s_1, ..., s_k, ..., s_K]$ gives state of the battery for complete day. Battery capacity has also been defined. Hence, Equation (17) gives state of battery and capacity of battery inequality constraints:

$$0 \leq s_k \leq y_m \tag{17}$$

h_{ch}^k and h_{disch}^k are variables that represent pattern of battery charging and discharging in each time slot $\mathcal{K}$ in binary form. At one time, charging or discharging of the battery can be taken place that is given in Equation (18):

$$h_{ch}^k + h_{disch}^k \leq 1 \tag{18}$$

state of the energy of battery at any time slot k is given as under:

$$s_{k+1} = s_k + \eta_{ch} e_k^b - \frac{1}{\eta_{disch}} \cdot b_k^l \tag{19}$$

$$e^b(k) \leq h_{ch}(k) B_{ch} \tag{20}$$

$e^b(k)$ is amount of energy which is required to charge the battery from PVC generation. Besides, $b^l(k)$ shows total energy to discharge the battery that is consumed while satisfying users requirement. Upper and lower limit of the battery must be satisfied while charging and discharging the battery. Hence, Equations (21) and (22) must be satisfied by the values of $e^b(k)$ and $b^l(k)$.

$$e^b(k) \leq h^k_{disch} B_{ch} \tag{21}$$

$$b^l(k) \leq h^k_{disch} B_{disch} \tag{22}$$

e^k_m shows total energy generated by PVC. It is on the basis of balancing principle, and $b^l(k)$ represents energy stored by battery and D_m is amount of energy received from utility or vice versa. It is shown as:

$$x^k_m = e^l_m(k) + b^l(k) + D_m \tag{23}$$

Energy management becomes complicated because of existence of PVC-battery storage. MG has to pay cost for PVC generation and depreciation cost of battery. Besides, MG generates revenue by selling energy to users and surplus energy to utility. Thus, the total cost of MG is calculated as:

$$\mathcal{B}^k_m = x^k_m + C^{bat}_m(y) + C^{PVC}_m + D_m \tag{24}$$

where, x^k_m represents energy that is demanded by $\mathbb{N}$ from MG in time slot k. C^{bat}_m and C^{PVC}_m are explained in Equations (13) and (15). Equations (13), (15) and (23) show total load of MG. In proposed scenario, if PVC power generation plant uses large battery that cannot be fully charged, consequently large amount of battery storage will be wasted, which affects daily cost of MG. Therefore, it is required to optimize battery storage capacity, so that it may be used optimally. Solar power generation fully charge it in working time. Solar generation provides $e^l_m(k)$ energy for charging the battery in time slot k. Thus, battery obtains $\eta_{ch} e^b(k)$ considering efficiency of charging. Which is shown as follows:

$$\lambda_s \sum_{k=1}^{K} e^b(k) = \lambda_s y / \eta_{ch} \tag{25}$$

Furthermore,

$$\sum_{k=1}^{K} e^l_m = \sum_{k=1}^{K} x^k_m - \sum_{k=1}^{K} x^k_n - \sum_{k=1}^{K} b^l(k) \tag{26}$$

where $\sum_{k=1}^{K} b^l(k) = \eta_{disch} y_m$

Now Equation (21) is represented as:

$$\mathcal{B}(X^k_m(p^k_m), y_m) = (\sum_{k=1}^{K} p_k(X^k) - \lambda_s) + x^k \lambda^{bat}_m + y_m + \varphi \tag{27}$$

$$\lambda^{bat} = \lambda^{bat} + \lambda_s / \eta_{ch} - \eta_{disch} \lambda_s \tag{28}$$

$$\varphi = \lambda_s \sum_{k=1}^{K} l^k + (\lambda^S P - \lambda_s) \sum_{k=1}^{K} e_k \tag{29}$$

The sole objective behind this work is to find an optimal battery capacity and best strategy for pricing of energy for the $\mathbb{N}$ users so that daily cost of MG will be reduced:

$$\min_{p^k_m} \mathcal{B}(x^k_m(p^k), y_m) \tag{30}$$

where x_m^k represents total amount of energy demanded by $\mathbb{N}$ users and y_m shows total storage capacity.

4.3. Game Formulation and Analysis

Single leader and multi follower Stackelberg game is proposed in [49] that studies communication between MG and users. Basically, multi player game is used in which MG, being leader decides price p_m^k of power that has to be charged from users and optimal storage capacity of the battery. Whereas, users as a followers decide amount of energy x_m^k to be demanded from MG. Proposed work is extension of the work related to game theory presented in [45]. Proposed game theory is shown in strategic form ς in equation below.

$$\varsigma = (\mathbb{N} \cup \mathbb{M}), (x_n^k, \mathscr{U}_n)_{n \in N}, (p_m^k, \mathscr{B})_{k \in \mathcal{K}}, (\vec{\mathbb{P}}^k)_{k \in \mathcal{K}} \quad (31)$$

Strategic form of proposed game has following components:

1. Users cost function $\mathscr{U}$ shows cost of energy consumption that is received by $\mathbb{N}$ users in time slot k,
2. Whereas, $\mathscr{B}$ captures the benefit that is gained by MG after supplying energy x_m^k to $\mathbb{N}$ set of users,
3. P_m^k defines price of energy that is defined by MG against each time slot k,
4. y_m define optimal storage capacity that is required to minimize the cost function of MG,
5. Cost function of user: $\mathscr{U}_n(x_n, x_{-n}, p_m^k)$,
6. Cost function of MG: $\mathscr{B}(x_m^k(p^k), y_m)$.

Nonetheless, MG tends to opt optimal price per unit energy. Thus, proposed algorithm reaches the NE. Where, leader, i.e., MG will utilize optimal amount of energy on the basis of given strategy ς and $\mathbb{N}$ users will demand optimal amount of energy x_m^k from MG. In this paper, proposed strategic form ς of proposed algorithm has been used to achieve NE for non cooperative game theory, if and only if cost function of MG, i.e., leader and cost function of users, i.e., follower $\mathbb{N}$; must satisfy following inequalities shown as:

$$\mathscr{U}_n(x_n^{k*}, x_{-n}^{k*}, p_m^{k*}) \leq \mathscr{U}_n(x_n^k, x_{-n}^{k*}, p_m^k)$$
$$\mathscr{B}_m(p_m^{k*}, x_n^k(p_m^{k*}), y_m*) \leq \mathscr{B}_m(p_m^{k*}, x_n^k(p_m^k), y_m) \quad (32)$$

where, $p_m^{k*} \epsilon p^{k*}$, p_m^{k*} are represented as price per unit time energy at NE that is finalized by MG for time slot k. Whereas, y_m^* shows total storage capacity of battery after achieving NE. Similarly, x_m^{k*} is total energy that is requested by users $\mathbb{N}$ for time period k.

In non cooperative game where multi level hierarchy is involved, pure solution for equilibrium is not ensured in every case. Therefore, DEM algorithm has been proposed to determine the presence of NE. Primarily, variational equality is considered to be more socially stable as compare to other equilibrium methods [41]. Each user is considered while determining variational equality as mentioned in [42].

Proposition 1. *In case of users $n \epsilon \mathbb{N}$, daily cost function $\mathscr{U}_n$ is continuously differentiable in x_n for price p_m^k and electricity consumption by users x_n. Therefore, strategy space of cost function of users $\mathscr{U}_n$ is a non empty convex compact subset of a Euclidean space.*

Proof. It is continuously differentiable in x_n because of consistent characteristics of the daily cost function $\mathscr{U}(x_n, x_{-n}, p^k)$. The Hessian of $\mathscr{U}(x_n, x_{-n}, p_k)$ is positive semi definite. Resultantly, cost function of user n is convex in x_n. Proposition 1 shows daily cost function $\mathscr{U}(x_n, x_{-n}, p_k)$ is continuously differentiable. It is also convex in x_n. Owing the fact, energy cost $\mathscr{U}_n(x_n, x_{-n})$ has continuous quadratic form in context of x_n. Preposition 1 is prerequisite of Proposition 2. □

Proposition 2. *For $\forall n \epsilon \mathbb{N}$ and time slot $k \epsilon \mathcal{K}$, the NE of the non cooperative game exists and it is also unique.*

Proof. According to the proof already mentioned in [45], because cost function $\mathcal{U}_N(x_n, x_{-n}, p^k)$ is convex in x_n, the NE is proved to be present in non cooperative game as well as it is proven to be distinct. □

Proposition 3. *The distinct NE (x_n^*, x_{-n}^*) that is proven in Preposition 2 is Pareto optimality.*

Proof. Pareto optimality is defined as the opted strategy state where no player changes its payoff by updating the user's strategies without affecting remaining user's payoff [45]. As per proof of Proposition 2, non cooperative game has achieved NE among all users. Where, each user increases his payoff, which depends on others users strategy. Afterwards, no user can change its payoff without taking consent to modify other users strategies [48]. Subsequently, it is said that NE (x_n^*, x_{-n}^*) is achieved, i.e., Pareto optimality. Preposition 2 expresses that, there exist distinct NE of non cooperative game for fixed price of MG. Nonetheless, if price will be changed then the NE will be different. On the basis of Preposition 3, it depicts the strategy against energy consumption by users will reach to pareto optimality and each user cannot change its payoff without affecting payoff of others. □

Proposition 4. *For battery storage capacity $y_m \in [y_m^l, y_m^u]$ and the energy consumption vector of MG. There is unique battery capacity y_m^*. It is understood that there is specific value of cost $\mathcal{B}(x_m, y_m)$ with a certain battery capacity. It is also assumed that there is only specific battery capacity that can minimize MG cost $\mathcal{B}$. Proof of proposition is mentioned in [45].*

From Propositions 1–3, it is clearly seen that non cooperative game is based on payoff function. It encourages residential user and MG to choose optimal strategy to minimize cost function of MG and user. Thus, payoff of follower and leader may be increased.

4.3.1. PVC Power Forecasting Algorithm

This section explains forecasting techniques to be further utilized in DEM.

ARIMA: Forecasting is performed using this technique by utilizing historic values. *AR* represents lags of differenced series that is given in Equation (33) [50]; *MA* is known as lag of time series and prediction error, which is required to be subtracted in order to make it static that is termed as "integrated". Non seasonal ARIMA model that is represented as $ARIMA(p,d,q)$ [51], is given as under:

$$Y_t = \phi_1 Y_{t-1} + \phi_2 Y_{t-2} + \ldots + \phi_p Y_{t-p} + \epsilon_t - \\ \theta_1 \epsilon_{t-1} - \theta_2 \epsilon_{t-2} - \ldots - \theta_2 \epsilon_{t-2} - \ldots - \theta_q \epsilon_{t-q} \tag{33}$$

where Y_t represents actual values of time series and ϵ_t shows error at certain time t; ϕ_i and θ_i represent vectors based on values of AR and MA; p and q are integers in vector $\phi_i (i = 1, 2, 3, \ldots p)$ and $\theta_i (i = 1, 2, \ldots q)$. Random errors ϵ_t is distributed with mean zero and constant variance σ_ϵ^2. In order to find optimal results, three stage model was proposed: (i) Recognition of the model, (ii) Estimation and (iii) Diagnostic Checking of the proposed model [52].

Recognition: No. of potential AR and MA orders to be selected using Auto Correlation Function (ACF) and Partial ACF (PACF). In order to analyze stationarity of time series, famous method is used that includes Augmented Dickey Fuller (ADF) and Phillips-Perron unit root test. It considers null hypothesis. These tests ensure that time series is not stationary. Details related to these methods are already mentioned in literature.

Estimation: In this stage, all the parameters that are identified in stage 1 are estimated for the ARIMA model by using iterative least square. Akaike Information Criterion (AIC) and Bayesian Information Criterion (BIC) values are used to find best values in model as mentioned in [51] is given as under:

$$AIC = Mlog(\sigma^2) + 2(p + q + 1) \tag{34}$$

and
$$BIC = Mlog(\sigma^2) + 2(p + q + 1)logT \quad (35)$$

where M shows total no. of observation that are used for estimation of parameters and σ^2 shows mean square error.

DiagnosticChecking: Depending on ACF and PACF of the residual, in dependency of residual can be analyzed. If residual is according to white noise, residual of the model are proven to be random in nature. The sample space-time ACFs is required to be effectively zero.

In [50], Ljung-Box test is used on actual time series or to the residual. Null hypothesis is considered for fitting the model which shows that series is representing the noise, and the alternative hypothesis states that one or more autocorrelation till lag m are not zero. The test static is shown as under mentioned in [51]:

$$Q^* = M(M+2) \sum_{K=1}^{m} \frac{r_k^2}{M-K} \quad (36)$$

where M shows total no. of observations that are used to finalize the model and m represent total lags. The statistics Q^* go along with $\mathcal{X}^2$ distribution having $(M-K)$ level of freedom. Where, K shows number of parameters that are estimated in model and r_k shows ACF of residual at lag k. In case it not appropriate, stage 1 is again used for recognition of another model.

Cuckoo Search (CS): Idea was given by Yang and Deb [53]. It is one of the latest meta heuristic techniques. It is also proven in recent studies that it works better than Genetic Algorithm (GA) and Particle Swarm Optimization (PSO). No. of AR and MA parameters are pre assumed ($pmax+qmax$) previously. Each parameter is individual in population, which contains respective solution from solution set. Functionality of CS is shown in Figure 2. First of all, the corresponding cost function of weight optimization should be taken out [38]. where, p and q are row of population matrix, which contain their best solution after optimization.

Gray Wolf Optimizer (GWO): Mirjalili et al. proposed GWO in 2014. It is a meta heuristic technique, which mimics the leadership and hunting behavior of GWO. Grey wolves population consist of 4 types, i.e., alpha (α), beta (β), delta (δ) and omega (ω) [54,55]. Flow diagram of GWO algorithm is shown in Figure 3.

Two proposed techniques are proposed, i.e., CARIMA and GARIMA, both of them perform better due to optimization of parameters. Simulation section proves that proposed forecasting techniques outperformed conventional techniques in predicting energy generation by PVC. However, GARIMA forecasting results outperformed CARIMA algorithm also. Therefore, GARIMA forecasting technique has been used in DEM algorithm to ensure accuracy of work.

Evaluation parameter: Objective function is defined in order to evaluate parameters. It can either be mathematical or experimental function that will give desired output. Basically, function is based on subtraction of actual values with predicted value. In our proposed scenario, Root Mean Absolute Error (RMSE) and MAPE is termed as evaluation parameter and that is shown as:

$$MAPE = \frac{1}{T} \sum_{tm=1}^{TM} |\frac{A_v}{F_v}| * 100 \quad (37)$$

$$RMSE = \sqrt{\frac{1}{T} \sum_{tm=1}^{TM} (A_v - F_v)^2} \quad (38)$$

where, A_v is actual time series and $\hat{F}_v$ is predicted time series.

Distributed Energy Management (DEM) Algorithm: After analyzing energy generation e_m by PVC using prediction method, i.e., GARIMA and energy demand x_m^k from $\mathbb{N}$ users, MG tries to find optimum price p_m^k and storage capacity of battery y_m that is used to store the energy generated by PVC using DEM Algorithm. Each of them will be reached to NE as per different battery capacities and price. After comparing the daily cost corresponding to battery capacity, DEM selects best response strategy that will choose minimal cost p_m^k. Further details are mentioned in Algorithm 1.

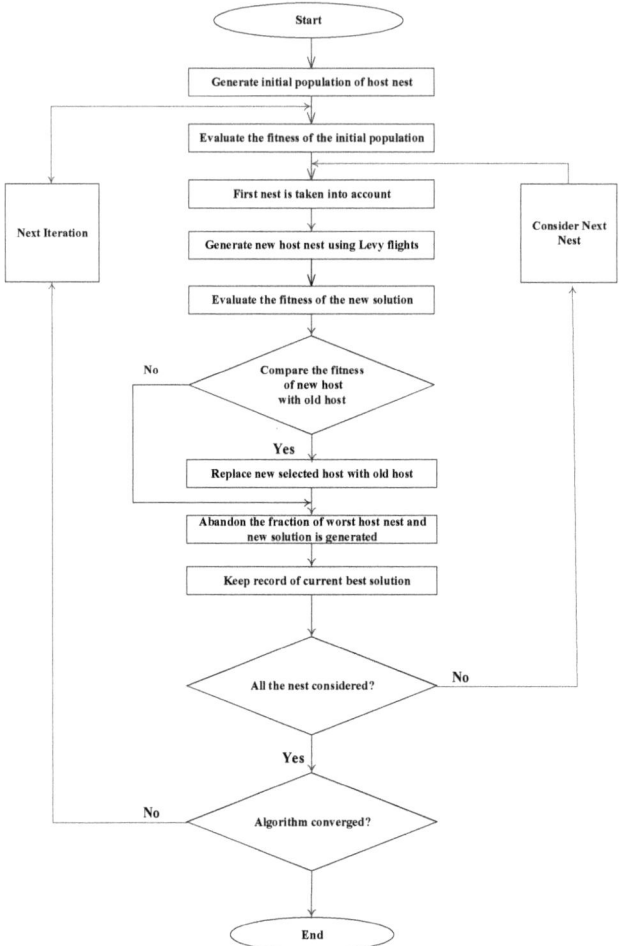

Figure 2. Flow Diagram of CS Algorithm.

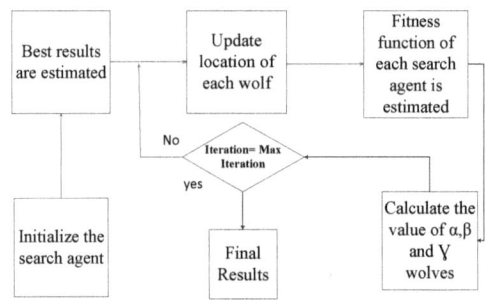

Figure 3. Flow Diagram of GWO Algorithm.

Algorithm 1: Executed by MG

```
 1  y_m = y_m^l
 2  while y_m^l ≤ y_m^u do
 3    │ Initialize x_m^k(p_m) repeat
 4    │   │ solve problem (29) ;
 5    │   │ if p_m changes then
 6    │   │   │ update and broadcast p_m;
 7    │   │ end
 8    │ until;
 9    │ Price at each slot k remains same
10    │ y_m = y_m + Δy_m
11  end
12  Select minimal daily cost of MG B_m(x_m^k(p_m), y_m) ;
13  Return optimal strategy for price scheme p_m^k and battery storage capacity y_m
```

5. Simulation and Discussion

5.1. Data Description

In forecasting techniques, PVC power production data is used. Dataset is taken from Elia, Belgium's electricity transmission system operator site [56]. Data contains 15 min interval power production of 5 years, January 2013 to December 2018.

5.2. Experimental Results

In this section, simulations are used in order to verify the effectiveness of game theoretic energy management using Stackelberg game theory. Apart of game theory, accuracy of PVC production forecasting has also been analyzed. Assumption is made that 6 users are connected to MG for getting supply of electricity. $\mathbb{N}$ users have both shift able load and non shift able load, furthermore power consumption of users are shown in Figure 4. Single day is represented in form of 24 h time slots, i.e., k. Usage of electricity varies on the basis of peak demand hours, off peak demand hours and mid peak hours at each slot.

Here, effectiveness of proposed game theory is analyzed between users and MG. n^{th} user can increase or decrease his payoff by adjustment of energy usage. Furthermore, MG focuses on optimization of payoff by announcing optimal energy price p_m^k and size of the storage battery, i.e., y_m.

In proposed work, pricing scheme that has been used is ToU. The whole day is split into 24 h time slots k. Moreover, single day is composed of three main chunks. Peak demand hours, off peak demand hours and mid peak demand hours make basis of chunks creation. Different pricing schemes are followed by pricing parameters.

User consumption is based on building and homes energy consumption. Each user is having 15 to 20 non shift able home appliances that cannot be shifted in any other hour of the day. Demand Side Management (DSM) has already been discussed in [45]. It has not been discussed for each user in our work. Unanimously, energy consumption by all users $\mathbb{N}$ is considered. Nevertheless, energy distribution by MG, energy pricing p_m^k and capacity of the battery y_m have been discussed in detail. MG contains renewable resources, i.e., PVC in current scenario.

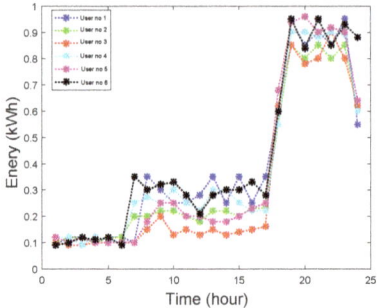

Figure 4. Consumption of Users in 24 h.

Figure 5 represents the energy distribution that is done by MG in $\mathcal{K}$ time slots without game optimization. However, Figure 6 shows evenly energy distribution is being carried out among users, battery and utility, after equilibrium achieved by NE. Energy is received from utility, if solar radiations are not enough for production of PVC. Besides, in remaining hours power is optimally used that is generated by solar radiations and stored in battery. Figure 7 shows energy consumption in a single day by MG without incorporation of game theory and applying DEM using Tables 1 and 2. Figure 8 shows solar power energy consumption by users, battery and company with game, which shows the evenly distribution of energy generated by PVC.

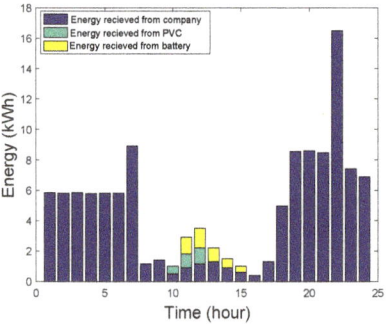

Figure 5. Energy Distribution without Game.

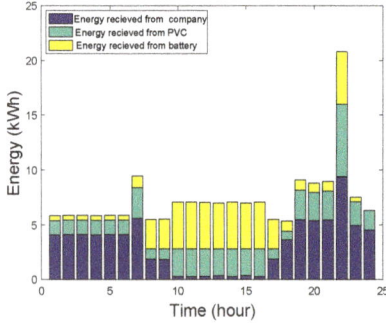

Figure 6. Energy Distribution with Game.

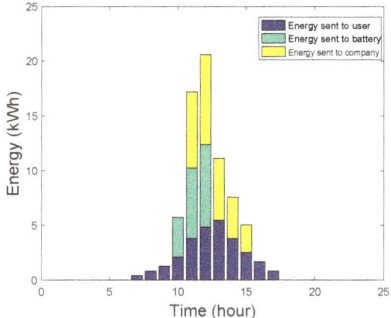

Figure 7. PVC Energy Distribution without Game.

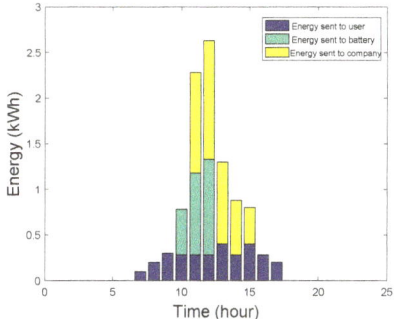

Figure 8. PVC Power Generation using Game.

Table 1. Pricing Model Parameter.

Time Chunks	a_h (cents/kWh)	b_h (cents/kWh)
0.00–6:00	0.03	4.9
6:00–1400	0.06	12.1
1400–2000	0.07	18.1
2100–2400	0.06	12.5

Table 2. PVC Generation Storage Battery Parameters.

Parameter	Value
y	[1.0, 5.0]
B_{disch}	2.0
B_{ch}	2.0
η_{disch}	92%
η_{ch}	92%
$\lambda^{bat}(y)$	7.2
λ_s	7.1
λ^{PVC}	4.3
ChargeTime	6.00–1700
DischargeTime	1700–2200

MG achieves maximum benefit from PVC power when sun light is present with full intensity. Once MG ensures energy supply to users, it also affects the battery capacity optimization. Nevertheless, proposed game theoretic energy management will not be affected by timings of day light and intensity. MG contains 2 kW PVC production capacity. Solar cells energy production per day is represented in Figure 9. PVC power is generated during 0600 to 1700 h. Parameters related to battery storage and charging and discharging are given in Table 1 and battery can be discharged at any time except charging hours. Cost of users depends on discharging time. It has minimal impact on the cost except battery discharges at peak demand hours, i.e., 17:00 to 22:00 h. In single hour, charging and discharging of the battery is 1.5 kWh [9]. Besides, 7.2 cents/kWh is battery charging and discharging efficiency [20]. In a current scenario, MG finds optimal solution for energy management using NE. To facilitate optimization and making error free prediction, MAPE is also brought under light. Figure 10 represents MAPE values of 5 forecasting algorithms that includes Back Propagation (BP), SVM, Genetic Stacked Auto Encoder (GSAE), CARIMA and GARIMA techniques. Step 1 shows that historical data is used for further forecasting PVC production. Data after forecasting is added in historical data and used for forecasting future PVC production is called as step 2 and so on. Figure 10 clearly shows that value of MAPE increases as forecasting steps increases. Simulation represents existing techniques as well as proposed techniques, i.e., CARIMA and GARIMA in Figure 10. Results are clearly shown in Figure 11a,b that GARIMA technique not only outperformed existing technique as well as second proposed technique, i.e., CARIMA in terms of RMSE in hourly and daily load forecasting. Results show actual and predicted results of proposed techniques. Hence, for achieving NE in game theory, GARIMA predicted results are used in DEM algorithm for performing energy management more efficiently.

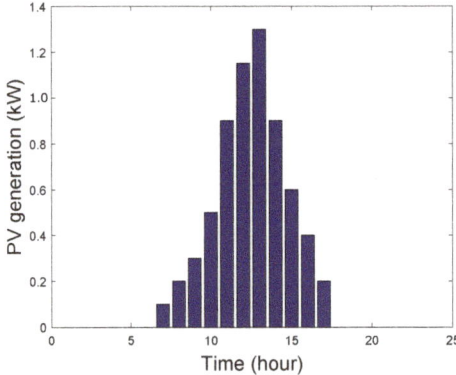

Figure 9. PVC Power Generation in a Certain Day.

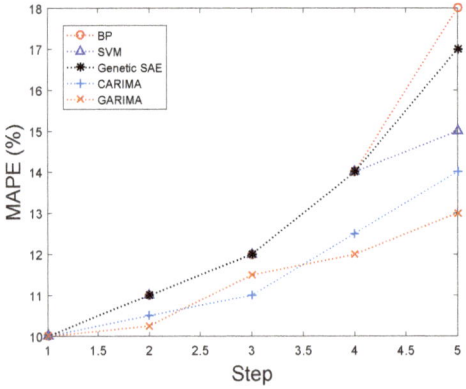

Figure 10. MAPE of Five Different Models with PVC Power Forecasting Steps.

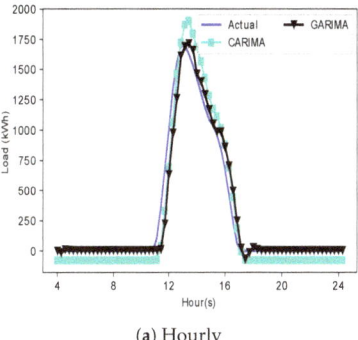
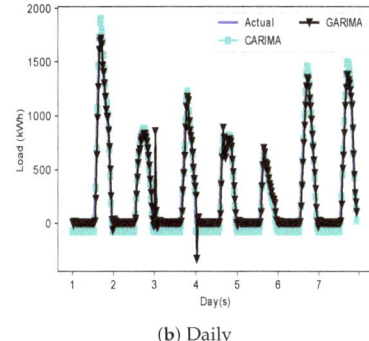

(a) Hourly (b) Daily

Figure 11. PVC Generation Forecasting.

6. Conclusions and Future Work

In our work, energy management system has been presented that contains single MG and multiple users. In order to analyze game theory using proposed distributed algorithm, i.e., DEM, which represents the complete mechanism that how MG optimally decides price that has to be charged from users. Moreover, battery storage capacity is also optimized. The continuous non cooperative game theory is used to model interaction among MG and consumers. To ensure efficient use of PVC generated energ, short term PVC power production forecasting is performed. Two forecasting techniques are proposed, i.e., GARIMA and CARIMA. GARIMA forecasting results turned out to be better as compared to CARIMA and other conventional techniques in MAPE. The forecasting results are then used to analyze the impact on payoff of both players in game theory. In non cooperative scenario, Stackelberg game theory is used to prove NE among two players, i.e., users and MG. DEM algorithm has been proposed to prove NE among two players. Energy distribution among users after applying game theory gives better results as compared to results produced without applying game theory. Forecasting results of hourly and daily generated energy data also validate that GARIMA forecasting algorithm performs best as compared to CARIMA forecasting algorithm. Consequently, GARIMA forecasting data has been used in DEM in order to ensure precision in energy management. The future directions of our work will be focused on non cooperative energy management among multiple MGs. Apart of inclusion of multiple MG, no. of players in game will also be increased, i.e., utility in order to improve flexibility in the proposed system model. Multiple renewable resources will be considered for forecasting purpose in order to make the system dynamic in context of distributed energy management. In order to prove efficiency of DEM algorithm, comparison with existing energy management algorithm will be performed.

Author Contributions: A.N., N.J. and M.B.R. proposed, implemented and wrote heuristic schemes. A.H., M.A. and K.A. wrote rest of the paper. All authors together organized and refined the paper.

Acknowledgments: The authors extend their appreciation to the Deanship of Scientific Research at King Saud University for funding this work through research group NO (RG-1438-034).

Conflicts of Interest: The authors declare no conflict of interest.

Abbreviations

Acronyms

Abbreviations	Full Form
AR	Auto regressive
AI	Artificial intelligence
ARIMA	Auto Regressive Integrated Moving Average
ACF	Auto Correlation Function
ADF	Augmented Dickey Fuller
AIC	Akaike Information Criteria

BIC	Bayesian Information Criteria
BP	Back Propagation
CARIMA	Cuckoo Search Optimized ARIMA
CS	Cuckoo Search
CNN	Convolution Neural Network
DEM	Distributed Energy Management
DR	Demand Response
DSM	Demand Side Management
ES	Exponential Smoothing
GARIMA	Gray Wolf Optimized ARIMA
GSAE	Genetic Stacked Auto encoder
GA	Genetic Algorithm
GWO	Grey Wolf Optimization
ICT	Information and Communication Technology
MG	Micro Grid
MAPE	Mean Absolute Percentage Error
RMSE	Root Mean Square Error
NN	Neural Network
NE	Nash Equilibrium
PSO	Particle Swarm Optimization
PVC	Photo Voltaic Cell
PAR	Peak Average Ratio
PACF	Partial Auto Correlation Function
RTP	Real Time Pricing
RMSE	Root Mean Square Error
SG	Smart Grid
SVM	Support Vector Machine
ToU	Time of Use

Nomenclature

Symbols	Descriptions
A_v	Actual Time Series
$X_{average}$	Average User Consumption
η_{ch}	Charging Efficiency of Battery
σ_ϵ^2	Constant Variance
$\mathcal{U}$	Cost Function of Users
$\mathcal{B}$	Cost Function of Micro Grid
l_n	Daily Energy Usage of User
η_{disch}	Discharging Efficiency of Battery
$C_m^{bat}(y)$	Daily Depreciation Cost Function
$C_k x_k$	Cost Function of User
ϵ_t	Error at Time t
$x_n(k)$	Energy Demand in Particular Slot
e_n	Energy Demanded from Users
$e_m^b(k)$	Energy Required to Charge Battery
a_k, b_k	Fixed Parameters
y_l^m	Lower Limit of Battery Capacity
$\mathbb{N}$	Multiple Residential Users
Q^*	Micro Grid
σ^2	Mean Square Error
X_{peak}	Peak Consumption
U_n	Profit of Micro grid
F	Penalty Factor
δ	Prediction Error
F_v	Predicted Time Series

h_{ch}^k	Pattern of Battery Charging
h_{disch}^k	Pattern of Battery Discharging
p^k	Real Time Price of Each Slot
X_k	Sum of Total Energy
ς	Strategy Form
λ_s	Solar Power Selling Price
s	State of Battery
$\mathcal{K}$	Time Slots in a Period
$b^l(k)$	Total Energy Required to Discharge Battery
$C_{n,k}$	Total Cost of $\mathbb{N}$ Users
T	Total no. of Observations
C_n^{sp}	Total Cost of Solar Power Generation
Q^*	Test Static
y_m^u	Upper Limit of Battery Capacity
ϕ_i	Vector Based on Rules of AR
θ_i	Vector Based on Rules of MA

References

1. Huang, B.; Li, Y.; Zhang, H.; Sun, Q. Distributed optimal co-multi-microgrids energy management for energy internet. *IEEE/CAA J. Autom. Sin.* **2016**, *3*, 357–364.
2. Wollenberg, B.F. Toward a smart grid: Power delivery for the 21st century. *Power Energy Mag.* **2005**, *3*, 34–41.
3. Ahmad, A.; Javaid, N.; Mateen, A.; Awais, M.; Khan, Z. Short-Term Load Forecasting in Smart Grids: An Intelligent Modular Approach. *Energies* **2019**, *12*, 164. [CrossRef]
4. Zahid, M.; Ahmed, F.; Javaid, N.; Abbasi, R.A.; Kazmi, Z.; Syeda, H.; Javaid, A.; Bilal, M.; Akbar, M.; Ilahi, M. Electricity Price and Load Forecasting using Enhanced Convolutional Neural Network and Enhanced Support Vector Regression in Smart Grids. *Electronics* **2019**, *8*, 122. [CrossRef]
5. Wang, Y.; Mao, S.; Nelms, R.M. Distributed online algorithm for optimal real-time energy distribution in the smart grid. *IEEE Internet Things J.* **2014**, *1*, 70–80. [CrossRef]
6. Muralitharan, K.; Sakthivel, R.; Shi, Y. Multiobjective optimization technique for demand side management with load balancing approach in smart grid. *Neurocomputing* **2016**, *177*, 110–119. [CrossRef]
7. Muratori, M.; Rizzoni, G. Residential demand response: Dynamic energy management and time-varying electricity pricing. *IEEE Trans. Power Syst.* **2016**, *31*, 1108–1117. [CrossRef]
8. Rifkin, J. *The Third Industrial Revolution: How Lateral Power Is Transforming Energy, the Economy, and the World*; Macmillan: New York, NY, USA, 2011.
9. Yi, P.; Zhu, T.; Jiang, B.; Jin, R.; Wang, B. Deploying Energy Routers in an Energy Internet Based on Electric Vehicles. *IEEE Trans. Veh. Technol.* **2016**, *65*, 4714–4725. [CrossRef]
10. Xia, Z.; Chen, L.; Sun, X.; Liu, J. A multi-keyword ranked search over encrypted cloud data supporting semantic extension. *Int. J. Multimed. Ubiquitous Eng.* **2016**, *11*, 107–120. [CrossRef]
11. Ellabban, O.; Abu-Rub, H.; Blaabjerg, F. Renewable energy resources: Current status, future prospects and their enabling technology. *Renew. Sustain. Energy Rev.* **2014**, *39*, 748–764. [CrossRef]
12. Sato, T.; Kammen, D.M.; Duan, B.; Macuha, M.; Zhou, Z.; Wu, J.; Tariq, M.; Asfaw, S.A. *Smart Grid Standards: Specifications, Requirements, and Technologies*; John Wiley and Sons: Hoboken, NJ, USA, 2015.
13. Uski, S.; Forssén, K.; Shemeikka, J. Sensitivity Assessment of Microgrid Investment Options to Guarantee Reliability of Power Supply in Rural Networks as an Alternative to Underground Cabling. *Energies* **2018**, *11*, 2831. [CrossRef]
14. Zhang, Y.; Fu, L.; Zhu, W.; Bao, X.; Liu, C. Robust model predictive control for optimal energy management of island microgrids with uncertainties. *Energy* **2018**, *164*, 1229–1241. [CrossRef]
15. Khalid, A.; Javaid, N.; Mateen, A.; Ilahi, M.; Saba, T.; Rehman, A. Enhanced Time-of-Use Electricity Price Rate Using Game Theory. *Electronics* **2019**, *8*, 48. [CrossRef]
16. Nuño, E.; Maule, P.; Hahmann, A.; Cutululis, N.; Sørensen, P.; Karagali, I. Simulation of transcontinental wind and solar PV generation time series. *Renew. Energy* **2018**, *118*, 425–436. [CrossRef]
17. Zhou, Z.; Bai, J.; Dong, M.; Ota, K.; Zhou, S. Game-theoretical energy management design for smart cyber-physical power systems. *Cyber-Phys. Syst.* **2015**, *1*, 24–45. [CrossRef]

18. Ahmad, F.; Rasool, A.; Ozsoy, E.; Sekar, R.; Sabanovic, A.; Elitaş, M. Distribution system state estimation—A step towards smart grid. *Renew. Sustain. Energy Rev.* **2018**, *81*, 2659–2671. [CrossRef]
19. Hussain, B.; Khan, A.; Javaid, N.; Hasan, Q.U.; Malik, S.A.; Dar, A.H.; Kazmi, A.; Ahmad, O. A Weighted-Sum PSO Algorithm for HEMS: A New Approach for the Design and Diversified Performance Analysis. *Electronics* **2019**, *8*, 180. [CrossRef]
20. Naz, A.; Javaid, N.; Khan, A.B.M.; Iqbal, M.M.; ur Rehman Hashmi, M.A.; Abbasi, R.A. Game-Theoretical energy management for residential user and micro grid for optimum sizing of photo voltaic battery systems and energy prices. In Proceedings of the 33rd International Conference on Advanced Information Networking and Applications (AINA), Matsue, Japan, 27–29 March 2019; pp. 1097–1106.
21. Farhath, Z.A.; Arputhamary, B.; Arockiam, L. A Survey on ARIMA Forecasting Using Time Series Model. *Int. J. Comput. Sci. Mob. Comput.* **2016**, *5*, 104–109.
22. Wei, W.; Liu, F.; Mei, S.; Hou, Y. Robust energy and reserve dispatch under variable renewable generation. *IEEE Trans. Smart Grid* **2015**, *6*, 369–380. [CrossRef]
23. Iqbal, Z.; Javaid, N.; Iqbal, S.; Aslam, S.; Khan, Z.; Abdul, W.; Almogren, A.; Alamri, A. A Domestic Microgrid with Optimized Home Energy Management System. *Energies* **2018**, *11*, 1002. [CrossRef]
24. Chen, Z.; Wu, L.; Fu, Y. Real-time price-based demand response management for residential appliances via stochastic optimization and robust optimization. *IEEE Trans. Smart Grid* **2012**, *3*, 1822–1831. [CrossRef]
25. Reka, S.S.; Ramesh, V. Demand side management scheme in smart grid with cloud computing approach using stochastic dynamic programming. *Perspect. Sci.* **2016**, *8*, 169–171. [CrossRef]
26. Hu, W.; Wang, P.; Gooi, H.B. Toward optimal energy management of microgrids via robust two-stage optimization. *IEEE Trans. Smart Grid* **2018**, *9*, 1161–1174. [CrossRef]
27. Cui, S.; Wang, Y.W.; Xiao, J.; Liu, N. A Two-Stage Robust Energy Sharing Management for Presumer Microgrid. *IEEE Trans. Ind. Inform.* **2018**. [CrossRef]
28. Lew, D.; Bird, L.; Milligan, M.; Speer, B.; Carlini, E.M.; Estanqueiro, A.; Flynn, D.; Gómez-Lázaro, E.; Holttinen, H.; Menemenlis, N.; et al. Wind and solar curtailment: International experience and practices. In Proceedings of the 12th International Workshop on Large-Scale Integration of Wind Power into Power Systems as Well as on Transmission Networks for Offshore Wind Power Plants, London, UK, 22–24 October 2013.
29. Ibanez, E.; Brinkman, G.; Hummon, M.; Lew, D. A solar reserve methodology for renewable energy integration studies based on sub-hourly variability analysis. In Proceedings of the 2nd International Workshop on Integration of Solar Power in Power Systems Proceedings, Lisbon, Portugal, 12–13 November 2012.
30. Mills, A.; Ahlstrom, M.; Brower, M.; Ellis, A.; George, R.; Hoff, T.; Kroposki, B.; Lenox, C.; Miller, N.; Stein, J.; et al. *Understanding Variability and Uncertainty of Photovoltaics for Integration with the Electric Power System*; No. LBNL-2855E; Lawrence Berkeley National Lab. (LBNL): Berkeley, CA, USA, 2009.
31. Taylor, J.; McSharry, P. Short-term load forecasting methods: An evaluation based on European data. *IEEE Trans. Power Syst.* **2007**, *22*, 2213–2219. [CrossRef]
32. Melin, P.; Soto, J.; Castillo, O.; Soria, J. A new approach for time series prediction using ensembles of ANFIS models. *Expert Syst. Appl.* **2012**, *39*, 3494–3506. [CrossRef]
33. Potter, C.; Negnevitsky, M. Very short-term wind forecasting for Tasmanian power generation. *IEEE Trans. Power Syst.* **2006**, *21*, 965–972. [CrossRef]
34. Hippert, H.; Pedreira, C.; Souza, R. Neural networks for short-term load forecasting: A review and evaluation. *IEEE Trans. Power Syst.* **2001**, *16*, 44–55. [CrossRef]
35. Amjady, N.; Keynia, F.; Zareipour, H. Wind power prediction by a new forecast engine composed of modified hybrid neural network and enhanced particle swarm optimization. *IEEE Trans. Sustain. Energy* **2011**, *2*, 265–276. [CrossRef]
36. Adamowski, K.; Dalezios, N.R.; Mohamed, F.B. Space-time ARIMA modeling for regional precipitation forecasting. *J. Comput. Math.* **1987**, *5*, 249–263.
37. Ong, C.-S.; Huang, J.-J.; Tzeng, G.-H. Model identification of ARIMA family using genetic algorithms. *Appl. Math. Comput.* **2005**, *164*, 885–912.
38. Yeh, W.-C.; Yeh, Y.-M.; Chang, P.-C.; Ke, Y.-C.; Chung, V. Forecasting wind power in the Mai Liao wind farm based on the multi-layer perceptron artificial neural network model with improved simplified swarm optimization. *Int. J. Electr. Power Energy Syst.* **2014**, *55*, 741–748. [CrossRef]

39. Chen, R.; Liang, C.-Y.; Hong, W.-C.; Gu, D.-X. Forecasting holiday daily touristflow based on seasonal support vector regression with adaptive geneticalgorithm. *Appl. Soft Comput.* **2015**, *26*, 435–443. [CrossRef]
40. Zhou, D.; An, Y.; Zha, D.; Wu, F.; Wang, Q. Would an increasing block carbon tax be better? A comparative study within the Stackelberg Game framework. *J. Environ. Manag.* **2019**, *235*, 328–341. [CrossRef]
41. Ji, X.; Yang, K.; Na, X.; Lv, C.; Liu, Y. Shared Steering Torque Control for Lane Change Assistance: A Stochastic Game-Theoretic Approach. *IEEE Trans. Ind. Electron.* **2019**, *66*, 3093–3105. [CrossRef]
42. Han, L.; Morstyn, T.; McCulloch, M. Incentivizing Prosumer Coalitions With Energy Management Using Cooperative Game Theory. *IEEE Trans. Power Syst.* **2019**, *34*, 303–313. [CrossRef]
43. Li, C.; Cai, W.; Luo, H.; Zhang, Q. Power utilization strategy in smart residential community using non-cooperative game considering customer satisfaction and interaction. *Electr. Power Syst. Res.* **2019**, *166*, 178–189. [CrossRef]
44. Haupt, S.E.; Kosović, B. Variable generation power forecasting as a big data problem. *IEEE Trans. Sustain. Energy* **2017**, *8*, 725–732. [CrossRef]
45. Gao, B.; Liu, X.; Yu, C.; Tang, Y. Game-theoretic energy management with storage capacity optimization in the smart grids. *J. Mod. Power Syst. Clean Energy* **2018**, *6*, 656–667. [CrossRef]
46. Su, J.; Zhou, L.; Li, R. Cost-benefit analysis of distributed grid-connected photovoltaic power generation. *Proc. CSEE* **2013**, *33*, 50–56.
47. Liang, X.; Li, X.; Lu, R.; Lin, X.; Shen, X. UDP: Usage-based dynamic pricing with privacy preservation for smart grid. *IEEE Trans. Smart Grid* **2013**, *4*, 141–150. [CrossRef]
48. Stephens, E.R.; Smith, D.B.; Mahanti, A. Game theoretic model predictive control for distributed energy demand-side management. *IEEE Trans. Smart Grid* **2015**, *6*, 1394–1402. [CrossRef]
49. Sinha, A.; Malo, P.; Frantsev, A.; Deb, K. Finding optimal strategies in a multi-period multi-leader–follower Stackelberg game using an evolutionary algorithm. *Comput. Oper. Res.* **2014**, *41*, 374–385. [CrossRef]
50. Iwueze, I.S.; Nwogu, E.C.; Nlebedim, V.U.; Imoh, J.C. Comparison of Two Time Series Decomposition Methods: Least Squares and Buys-Ballot Methods. *Open J. Stat.* **2016**, *6*, 1123. [CrossRef]
51. Cadenas, E.; Rivera, W.; Campos-Amezcua, R.; Heard, C. Wind speed prediction using a univariate ARIMA model and a multivariate NARX model. *Energies* **2016**, *9*, 109. [CrossRef]
52. Dickey, D.A.; Fuller, W.A. Distribution of the estimators for autoregressive time series with a unit root. *J. Am. Stat. Assoc.* **1979**, *74*, 427–431.
53. Yang, X.S.; Deb, S. Cuckoo search via Lévy flights. In Proceedings of the 2009 World Congress on Nature & Biologically Inspired Computing (NaBIC), Coimbatore, India, 9–11 December 2009.
54. Naz, M.; Iqbal, Z.; Javaid, N.; Khan, Z.A.; Abdul, W.; Almogren, A.; Alamri, A. Efficient Power Scheduling in Smart Homes Using Hybrid Grey Wolf Differential Evolution Optimization Technique with Real Time and Critical Peak Pricing Schemes. *Energies* **2018**, *11*, 384. [CrossRef]
55. Mirjalili, S.; Mirjalili, S.M.; Lewis, A. Grey wolf optimizer. *Adv. Eng. Softw.* **2014**, *69*, 46–61. [CrossRef]
56. Available online: http://www.elia.be/en/grid-data/power-generation/Solar-power-generation-data/Graph (accessed on 1 January 2019).

© 2019 by the authors. Licensee MDPI, Basel, Switzerland. This article is an open access article distributed under the terms and conditions of the Creative Commons Attribution (CC BY) license (http://creativecommons.org/licenses/by/4.0/).

Article

Two-Stage Coordinate Optimal Scheduling of Seawater Pumped Storage in Active Distribution Networks

Ning Liang [1], Changhong Deng [1,*], Yahong Chen [1], Weiwei Yao [1], Dinglin Li [2], Man Chen [2] and Peng Peng [2]

1 School of Electrical Engineering, Wuhan University, Wuhan 430072, China; liangning2688@163.com (N.L.); yahongchen@foxmail.com (Y.C.); Yaoww@whu.edu.cn (W.Y.)
2 Power Generation Company of China Southern Power Grid, Guangzhou 510630, China; lidinglin@126.com (D.L.); 13926159826@139.com (M.C.); 13926169785@139.com (P.P.)
* Correspondence: dengch-whu@163.com

Received: 19 May 2018; Accepted: 10 June 2018; Published: 12 June 2018

Abstract: The percentage of penetration in renewable energy generation (REG) in distribution networks has dramatically increased. Variable speed seawater pumped storage, which has a large power controllable range and flexible modes of operation, is an important tool to be applied in distribution networks to realize peak shaving and valley filling, and to mitigate the negative effects of REG. This paper presents a two-stage coordinated optimal scheduling model for the day-ahead and real-time operation of active distribution networks containing seawater pumped storage, REG, and flexible loads. In the model, seawater pumped storage and flexible loads are dispatched in the first day-ahead stage based on short-term forecast information of REG and load demands to minimize total operational costs. Then in the second real-time stage, the operation schedule of seawater pumped storage is adjusted to mitigate the negative effects of forecast errors of REG on the operation of active distribution networks. Network nodes power quality is improved and power loss is reduced. Applying the model, disadvantages of low accuracy short-term forecast are minimized whereas advantages of high accuracy ultra-short term forecast are fully taken. This model is tested using a modified Institute of Electrical and Electronics Engineers 33-bus system. Numerical results demonstrate the effectiveness of the proposed approach.

Keywords: seawater pumped storage; renewable energy; active distribution networks; two-stage; scheduling

1. Introduction

There are rich renewable energy resources in China offshore. In recent years, the percentage of penetration in renewable energy in distribution networks has increased dramatically [1–3]. Power output of renewable energy generation (REG) like wind turbine (WT) generators and photo voltaic (PV) arrays have high randomness and fluctuation attributes. The controllability of WT and PV output is low [4,5], which rese new challenges to the operation of distribution networks. To increase its operation reliability and economy, energy storage provides an attractive solution because it has fast power response speed and strong energy shifting capabilities [6–8]. Compared with traditional energy storage technologies such as lead acid battery banks, hydrogen energy storage, and flywheel energy storage, etc., variable speed seawater pumped storage has many advantages, such as larger storage capacity, higher energy storage efficiency, lower per kilowatt capital and maintenance costs, longer service life, and it also can provide additional inertia energy for networks [9,10]. Meanwhile, compared with traditional pumped storage, offshore ocean is used as the lower reservoir for variable speed seawater pumped storage. The amount of water resources are inexhaustible and total construction costs can be reduced. But how to dispatch the distribution networks that contain variable speed seawater pumped storage and REG, reaching a high operation performance and having high economic benefits, is still a problem for the network operators.

Current distribution networks were operated in a passive fashion. The traditional operating method is no longer applicable to new distribution networks which contain REG, energy storage, and flexible loads, etc. In recent years, the concept of active distribution networks was proposed [11–13]. Compared with traditional distribution networks, this network is able to manage the power of REG, energy storage, and flexible loads actively, having the advantages of increasing the operation reliability and minimizing the total operation cost and power losses of the distribution networks [14,15]. Besides, the active distribution networks can provide additional spinning reserve capacities to alleviate the negative impacts of the stochastic fluctuations of REG and load demands on itself at the distribution network level. Problems of declining power quality are induced by stochastic, uncertain fluctuations of power of REG and load demands. These problems are solved locally by appropriate dispatch of distributed generators and energy storage [16]. In the dispatch of active distribution networks, an optimization model that can both achieve a global optimal solution and provide good performance is demanded [17]. Dynamic economic dispatch method [18] and consensus based dispatch method [19] can be applied to generate a dispatch schedule. However, the performance cannot meet this demand. The power flexibility and energy storability characteristics of energy storage in active distribution networks are not fully utilized in the present market management systems (MMS) and energy management systems (EMS) [20]. Operation schedules are generally determined at day-ahead time stages, ignoring the effect of real-time adjustment which can compensate for uncertainties and forecast errors of REG [21,22]. Day-ahead scheduling is performed, which is based on short-term forecasts of REG and load demands. However, because of the stochastic fluctuation attributes in wind strength, illumination intensity, etc., it is difficult to make accurate short-term predictions for PV and WT power and load demands.

For reducing the impact of renewable energy sources which are connected to power systems, the two-stage optimal scheduling method has been presented by some scholars in recent years. For example, there is the disturbance of power systems caused by the randomness and volatility of wind power, which can be reduced by a two-stage dynamic scheduling method, as noted in Reference [23]. In addition, a two-stage optimal scheduling method for micro-grid economy is presented in Reference [24]. In the first stage, based on the real-time forecast data for future power, optimal management for power source in a micro-grid is able to be calculated. In the second stage, according to the limitation of diesel generator output which can be figured out before, the output of diesel generators is adjusted for improving the robustness of the system. The dispatch schedule is made in day-ahead for seawater pumped storage and renewable energy has not yet utilized their potential capabilities to increase the performances and economics of active distribution networks adequately. Compared to short-term forecasts, ultra-short forecasts for PV and WT power is quite accurate. Based on the ultra-short forecasts information, real-time scheduling for seawater pumped storage can solve or mitigate the negative impacts caused by forecast errors effectively. In this paper, a two-stage scheduling method is proposed, because it has high computation efficiency and takes advantages of the flexibility of seawater pumped storage and flexible loads to dispatch the active distribution networks. In a day-ahead scheduling stage, optimal dispatch schedule of active distribution networks is determined. In a real-time scheduling stage, negative impacts of forecast errors of REG on the operation of active distribution networks is compensated and mitigated.

Main contributions are as follows.

i. Variable speed seawater pumped storage is first utilized for dispatch in offshore local active distribution networks in China;
ii. A two-stage scheduling method considering variable speed seawater pumped storage, flexible loads, and REG in active distribution networks is presented. Both advantages of day-ahead and real-time scheduling are fully utilized and exploited.

Remaining sections are organized as follows: Section 2 introduces the dispatch model for variable speed seawater pumped storage. Section 3 presents the problem formulation of the two-stage scheduling in active distribution networks and the solving algorithm. Section 4 gives the test results of the proposed approach on a modified Institute of Electrical and Electronics Engineers (IEEE) 33-bus distribution system. Finally, conclusions are drawn in Section 5.

2. Variable Speed Seawater Pumped Storage Model

The structure of a traditional pumped storage station consists of a generator unit, pumping station, and upper and lower reservoirs. Capacities of the upper and lower reservoirs are mainly determined by the conditions of natural water resources. However, for a variable speed seawater pumped storage station, one of the prominent features is that the lower reservoir is the sea. Its water resources are unlimited and its capacity can be considered infinitely large. Only the water amount of the upper reservoir should be considered in the optimal scheduling. Another prominent feature is that the speed of the motor rotor in generating or pumping mode can vary within a large range. Therefore, the efficiency of the overall system can be improved in the generating mode. In the pumping mode, the range of input power is expanded. From the point of view of scheduling, the above features can gain great benefits for the operation of the power system. Operation models of a variable speed seawater pumped storage station in generating and pumping modes are expressed as follows.

2.1. Generating and Pumping Modes

In the generating mode, output power is formulated by:

$$P_g(t) = k_g \rho g h \times \int_{t-1}^{t} q_g(\tau) d\tau \tag{1}$$

In the pumping mode, energy for the pump is directly input from local active distribution networks. The flow rate of seawater sucked from the sea is stated as follows.

$$\int_{t-1}^{t} q_p(\tau) d\tau = \frac{k_p P_p(t)}{\rho g h} \tag{2}$$

2.2. Operation and Maintenance Cost of Seawater Pumped Storage

Seawater contains high salinity and corrosiveness to pipelines, which increases the maintenance cost of pipelines. Therefore, the cost of seawater pumped storage contains the installation fee, start-up cost, pipelines, and generator maintenance cost. Equation (3) is the operation and maintenance cost of seawater pumped storage; Equations (4) and (5) are the start-up cost of a turbine generator and pump-motor unit in period t, respectively; Equation (6) is the maintenance cost of turbine generator and pump-motor unit in period t; Equation (7) is the piping maintenance cost of the seawater pumped storage.

$$C_{sea}(t) = C_g^{cr}(t) + C_p^{cr}(t) + C_{sea}^{in}(t) + C_{sea}^{run}(t) \tag{3}$$

$$C_g^{cr}(t) = C_g \times \mu_g(t) \times [1 - \mu_g(t-1)] \tag{4}$$

$$C_p^{cr}(t) = C_p \times \mu_p(t) \times [1 - \mu_p(t-1)] \tag{5}$$

$$C_{sea}^{in}(t) = [\frac{C_{sea}^{az}}{8760\tau_{sea}} \frac{r_{sea}(1+r_{sea})^{n_{sea}}}{(1+r_{sea})^{n_{sea}}-1} + \lambda_{sea}^{co}][P_g(t) + P_p(t)] \tag{6}$$

$$C_{sea}^{run}(t) = \lambda_g^{run} P_g(t) + \lambda_p^{run} P_p(t) \tag{7}$$

2.3. Operation Constraints of Variable Speed Seawater Pumped Storage Station

The proposed coordinate optimal scheduling models for active distribution networks containing seawater pumped storage have two stages. The first stage is the day-ahead, and the second stage is the real-time.

2.3.1. Day-Ahead Operation Constraints

In period t, the quantity of seawater stored in upper reservoir is given as follows.

$$Q_{up}(t) = (1-l) \times Q_{up}(t-1) + \int_{t-1}^{t} q_p(\tau)d\tau - \int_{t-1}^{t} q_g(\tau)d\tau \quad t = 1, 2, \ldots, T \tag{8}$$

Seawater pumped storage stations can operate only in one mode in a given time interval, on-off operation mode constraint:

$$\mu_g(t) \times \mu_p(t) = 0 \quad t = 1, 2, \ldots, T \tag{9}$$

Power range constraint in generating mode:

$$0 \leq P_g(t) \leq \mu_g(t) P_g^{\max} \tag{10}$$

Power range constraint in pumping mode:

$$\mu_p(t) P_p^{\min} \leq P_p(t) \leq \mu_p(t) P_p^{\max} \tag{11}$$

Water amount constraint for upper reservoir:

$$Q_{sea}^{\min} \leq Q_{up}(t) \leq Q_{sea}^{\max} \tag{12}$$

$$\sum_{t=1}^{T} [\int_{t-1}^{t} q_p(\tau)d\tau] = \sum_{t=1}^{T} [\int_{t-1}^{t} q_g(\tau)d\tau] \tag{13}$$

2.3.2. Real-Time Operation Constraints

The quantity of seawater stored in upper reservoir:

$$Q_{up}(t) = (1-l) \times Q_{up}(t-1) + \int_{t-1}^{t} q_p(\tau)d\tau - \int_{t-1}^{t} q_g(\tau)d\tau \quad t = 1, 2, \ldots, T \tag{14}$$

On-off operation mode constraint:

$$\mu_g'(t) \times \mu_p'(t) = 0 \quad t = 1, 2, \ldots, T' \tag{15}$$

Power range constraint in generating mode:

$$0 \leq P_g'(t) \leq \mu_g'(t) P_g^{\max\prime} \tag{16}$$

Power range constraint in pumping mode:

$$\mu_g'(t) P_p^{\min\prime} \leq P_p'(t) \leq \mu_p'(t) P_p^{\max\prime} \tag{17}$$

Water amount constraint for upper reservoir:

$$Q_{sea}^{\min} \leq Q_{up}'(t) \leq Q_{sea}^{\max} \tag{18}$$

Restricted by technical conditions, the frequency of switching between generating and pumping modes must be limited. Minimum switching time is 30 min, (i.e., a time of $2\Delta t$ is required). Formulation below must be satisfied in real-time scheduling.

$$\begin{cases} \mu_g'(t) \times \mu_p'(t+1) = 0 & t = 1, 2, \ldots, (T'-1) \\ \mu_g'(t) \times \mu_p'(t+2) = 0 & t = 1, 2, \ldots, (T'-2) \\ \mu_g'(t+1) \times \mu_p'(t) = 0 & t = 1, 2, \ldots, (T'-1) \\ \mu_g'(t+2) \times \mu_p'(t) = 0 & t = 1, 2, \ldots, (T'-2) \end{cases} \quad (19)$$

2.3.3. Problem Formulation

Due to the stochastic fluctuation attributes in wind strength and illumination intensity, people's ability to predict REG power output accurately has yet to be improved. Ultra-short forecasting of REG are far more accurate than short-term's. Day-ahead scheduling of variable speed seawater pumped storage is performed based on the short-term forecast information of REG power and load demands. The objective of this stage is to minimize total system operation cost. Then real-time scheduling of variable speed seawater pumped storage is carried out based on ultra-short forecast information to mitigate the negative effects of the forecast errors on the operation of active distribution networks [25].

3. Optimal Scheduling Model

3.1. Day-Ahead Scheduling

3.1.1. Objective Function

The objective of day-ahead scheduling is to make the most cost-effective operation schedule for active distribution networks. Costs of power obtained from main grid, penalty for the transition between generating and pumping modes of seawater pump storage, interruption, and incentive to the flexible loads are all taken into consideration. We assume that all renewable energy sources are absorbed. Therefore the costs of wind and solar energy power generation are neglected. These costs are induced by operation, maintenance, and depreciation.

Objective function is as follows.

$$\min F_{Day} = \sum_{t=1}^{T} [P_{Grid}(t) \times \Omega(t) + C_{sea}(t) + C_{load}(t)] \quad (20)$$

$$C_{load}(t) = fc(t) + fe(t) \quad (21)$$

$$f_c(t) = \sum_{i=1}^{M} u_c(i,t)[\alpha_1 \times P_c^2(i,t) + \alpha_2 \times P_c(i,t) - \alpha_2 \times P_c(i,t) \times k_c(i)] \quad (22)$$

$$f_e(t) = \sum_{j=1}^{N} u_e(j,t)[\beta_1 \times P_e^2(j,t) + \beta_2 \times P_e(j,t) - \beta_2 \times P_e(j,t) \times k_e(j)] \quad (23)$$

3.1.2. Constraints

Power balance constraint is formulated by:

$$\sum_{m=1}^{N_{WT}} P_{WT}(m,t) + \sum_{n=1}^{N_{PV}} P_{PV}(n,t) + P_{Grid}(t) + \mu_g(t)P_g(t) = \sum_{i=1}^{N_L} P_L(i,t) + P_f(t) + \mu_p(t)P_p(t) \quad (24)$$

Reserve constraints are given below.

$$P_g^{\max} - \mu_g(t)P_g(t) \geq P_g^{Rup}(t) \quad (25)$$

$$P_p^{max} - \mu_p(t)P_p(t) \geq P_p^{Rup}(t) \tag{26}$$

$$P_g^{Rup}(t) = p_g^{Rup} + \eta_1 \sum_{m=1}^{N_{WT}} P_{WT}(m,t) + \eta_2 \sum_{n=1}^{N_{PV}} P_{PV}(n,t) \tag{27}$$

$$P_p^{Rup}(t) = p_p^{Rup} + \lambda_1 \sum_{m=1}^{N_{WT}} P_{WT}(m,t) + \lambda_2 \sum_{n=1}^{N_{PV}} P_{PV}(n,t) \tag{28}$$

Constraint of power obtained from main grid is:

$$0 \leq P_{Grid}(t) \leq P_{grid}^{max} \tag{29}$$

3.2. Real-Time Scheduling

3.2.1. Objective Function

The day-ahead scheduling is seen as the foundation for real-time scheduling. However, the forecast data is gotten in day-ahead, which cause a large forecast error. As a result, the reliability of distribution network operation could be influenced if the forecast error cannot be compensated in the real-time stage. So we assume that the forecast error of renewable energy is suppressed by seawater pumped storage in priority at real-time stage. The detail objective function for real-time optimal scheduling is shown as Equation (30).

$$\min F_{Real} = \sum_{t=1}^{T'} \left| \Delta P_g(t) + \Delta P_p(t) + \sum_{i=1}^{N_{WT}} \Delta P_{WT}(t) + \sum_{i=1}^{N_{PV}} \Delta P_{PV}(t) \right| \tag{30}$$

3.2.2. Operation Constraints

Power balance constraint is as below.

$$\sum_{m=1}^{N_{WT}} P_{WT}'(m,t) + \sum_{n=1}^{N_{PV}} P_{PV}'(n,t) + \mu_g'(t)P_g'(t) + P_{grid}'(t) = \sum_{i=1}^{N_L} P_L(i,t) + P_f(t) + \mu_p'(t)P_p'(t) \tag{31}$$

Constraint of power obtained from main grid is:

$$0 \leq P_{grid}'(t) \leq P_{grid}^{max\prime} \tag{32}$$

3.3. Approach to Solve This Model

A flowchart of the two-stage optimal scheduling of the seawater pumped storage and flexible loads in active distribution networks is shown in Figure 1. ILOG Cplex 12.7 was called by MATLAB and is used to solve the proposed model, and case studies were carried out on a 3.5 GHz ASUS PC. Day-ahead scheduling was updated each day. Real-time scheduling was carried out every 15 min according to the rolling update ultra-short forecasting data of REG.

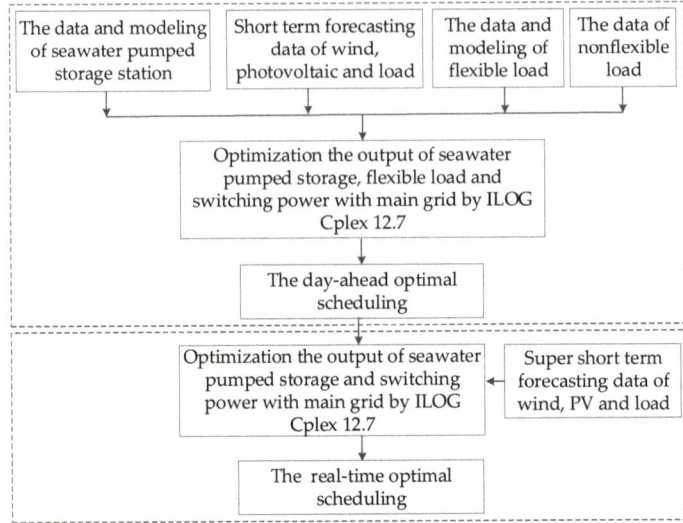

Figure 1. Flowchart of the proposed approach.

4. Case Study

4.1. System Description

A modified IEEE 33-bus distribution system was used to validate the proposed approach, which is shown in Figure 2. Node 1 was connected to the main grid. The seawater pumped storage station, PV, WT, and flexible loads were connected to bus 6, bus 8, bus 13, and bus 17, respectively. Forecasted power outputs of REG were based on the information collected on the spot in the south of China. Typical forecast data is shown in Figure 3. Operation parameters of the seawater pumped storage and flexible loads are given in Tables 1 and 2, respectively. In Table 2, Pf is the rated power of flexible load. Time-of-use price of general industry is given in Table 3, and these data were taken from a distribution network in China.

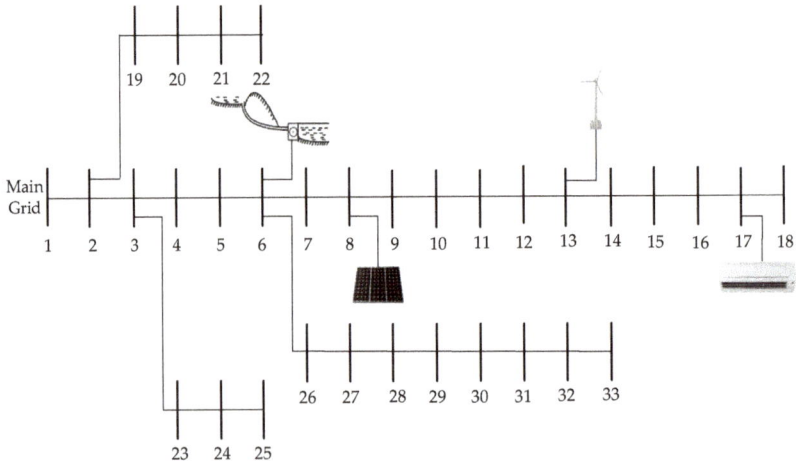

Figure 2. Modified IEEE 33-bus distribution system.

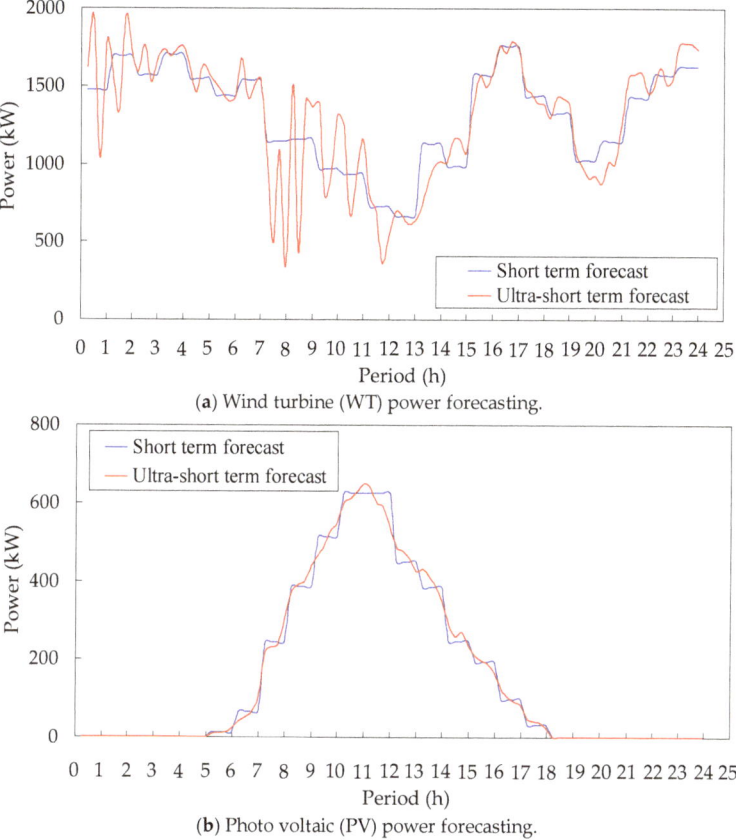

(a) Wind turbine (WT) power forecasting.

(b) Photo voltaic (PV) power forecasting.

Figure 3. Renewable energy forecasting data.

Table 1. Parameters of seawater pumped storage.

	Maximum Power (kW)	Life Time (y)	k_g/k_p
Generating mode	3000	40	0.91
Pumping mode	3000	40	0.82

Table 2. Parameters of flexible loads.

Variation Range	α_1 ¥/h	α_2 ¥/kW·h	β_1 ¥/(kW·h)	β_2 ¥/(kW·h)
[0.6Pf, 1.4Pf]	0.0002	0.25	0.00025	0.1

Table 3. Time-of-use price of general industry.

	Period (h)	Price ¥/kW·h
Peak	[11:00, 16:00); [19:00, 22:00)	1.2
Off-peak	[08:00, 11:00); [16:00, 19:00); [22:00, 24:00)	0.7
Valley	[24:00, 08:00)	0.3

4.2. Day-Ahead Scheduling Results and Analysis

In order to prove the algorithm applicability of the proposed model, improved particle swarm optimizer (IPSO) [26], genetic algorithms (GA), and Cplex are used to solve the proposed model. As can be seen in Table 4, by the achievement of a better total operation cost, the ILOG Cplex over the other approaches is substantiated.

Table 4. Comparison of total operation.

Method	Total Operation Cost (¥)	Running Time (min)
IPSO	136,247	6.4
GA	135,774	8.8
Cplex	131,886	3.6

IPSO = improved particle swarm optimizer; GA = genetic algorithms.

Day-ahead scheduling was performed based on short-term forecast data of WT, PV, and loads. Two cases were simulated and compared to analyze the proposed methodology. Case 1 did not contain seawater pumped storage and flexible loads. Case 2 contained seawater pumped storage and flexible loads, and the proposed methodology was applied in it. Base reserve capacities of p_g^{Rup} and p_p^{Rup} were 500 kW. Optimization results are shown in Figures 4–6.

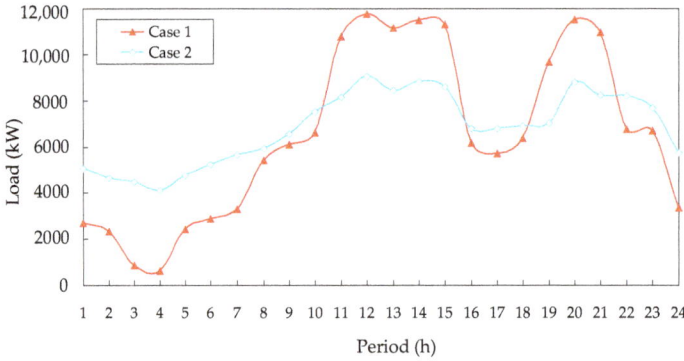

Figure 4. Power exchange with main grid of two cases in day-ahead scheduling.

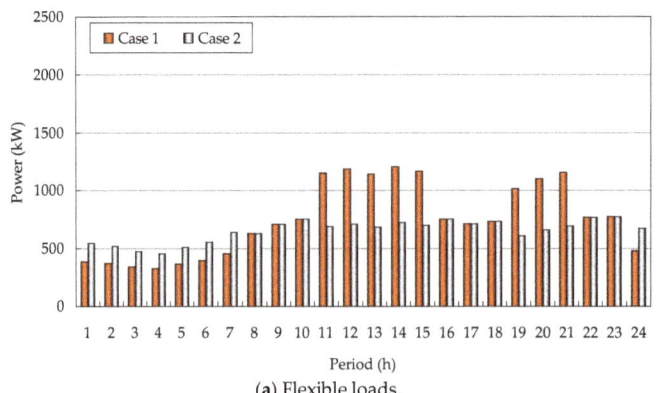

(a) Flexible loads.

Figure 5. Cont.

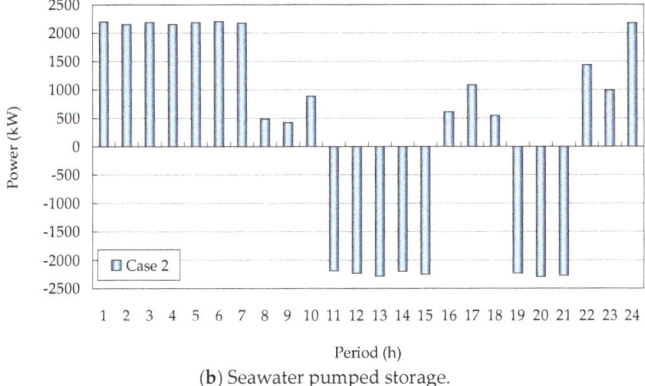

(b) Seawater pumped storage.

Figure 5. Optimization results of flexible loads and seawater pumped storage.

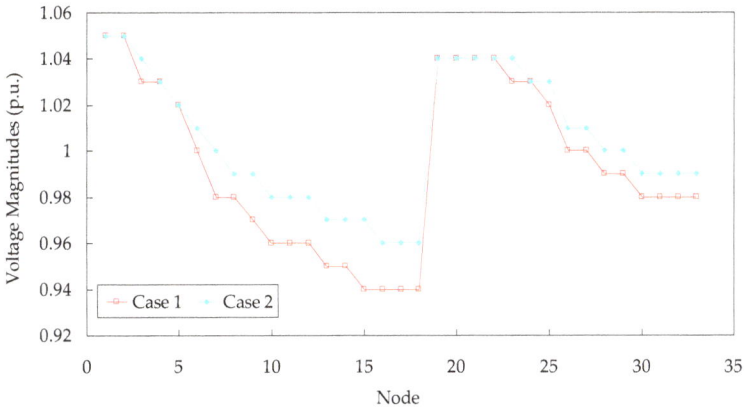

Figure 6. Power flow results of voltage magnitudes in peak load period.

As shown in Figure 4, results showed that by applying the proposed model, not only did the maximum exchange power between local active distribution networks and main grid decrease about 25%, but also the maximum rate of power change was effectively reduced. Peak loads from 11:00 to 15:00 and from 19:00 to 21:00 were successfully shifted and maximum power was reduced. Valley loads from 1:00 to 8:00 were effectively filled, and minimum valley loads increased by about eight times. In Table 3, the electricity price on the open market was high from 11:00 to 15:00 and from 19:00 to 21:00, and was low from 24:00 to 7:00. As shown in Figure 5, flexible loads and seawater pumped storage could respond to the price and coordinate their power to reduce total operation cost. Power demands of flexible loads increased about 30% at low electricity price from 24:00 to 7:00, and decreased about 40% at high electricity price from 11:00 to 15:00 and from 19:00 to 21:00.

Based on the specific power system structure, power flow calculation was implemented. Periods during peak loads were taken as an example. As shown in Figure 6, utilizing the proposed methods, except for voltage magnitudes at nodes 1 to 2 and 19 to 22 which did not increase, voltage magnitudes at nodes 3 to 18 and 23 to 33 increased. This proves that the power quality of the active distribution networks were effectively improved by using the proposed method. In Table 5, cost of purchase power from the main grid and network energy loss are compared. Cost of purchase power from the main grid decreased from ¥146,915 to ¥131,886 by 10.2%. Network loss decreased from 4481 kW·h to 3862 kW·h by 13.8%.

Table 5. Cost of purchase power and network loss.

	The Cost of Purchase Power from Main Grid (¥)	Network Loss (kW·h)
Case 1	146,915	4481
Case 2	131,886	3862

4.3. Real-Time Scheduling Results and Analysis

Real-time scheduling was carried out every 15 min based on ultra-short forecast data of WT and PV. Power exchange results between active distribution networks and the main grid are shown in Figure 7. Optimization results of the seawater pumped storage is shown in Figure 8. After correction, the power exchange curve is much more flat and smooth, especially in the periods with large forecast errors. Higher forecast errors needed greater correction amounts. The results demonstrated that real-time scheduling can effectively compensate for the forecast errors of REG and reduce its negative effects on the operation of local active distribution networks.

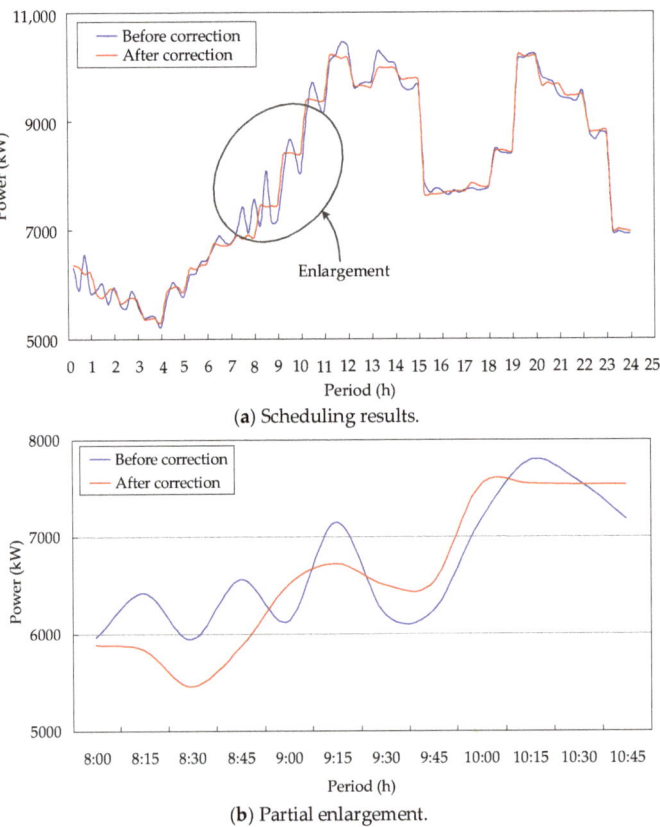

(a) Scheduling results.

(b) Partial enlargement.

Figure 7. Power exchange with main grid in real-time scheduling.

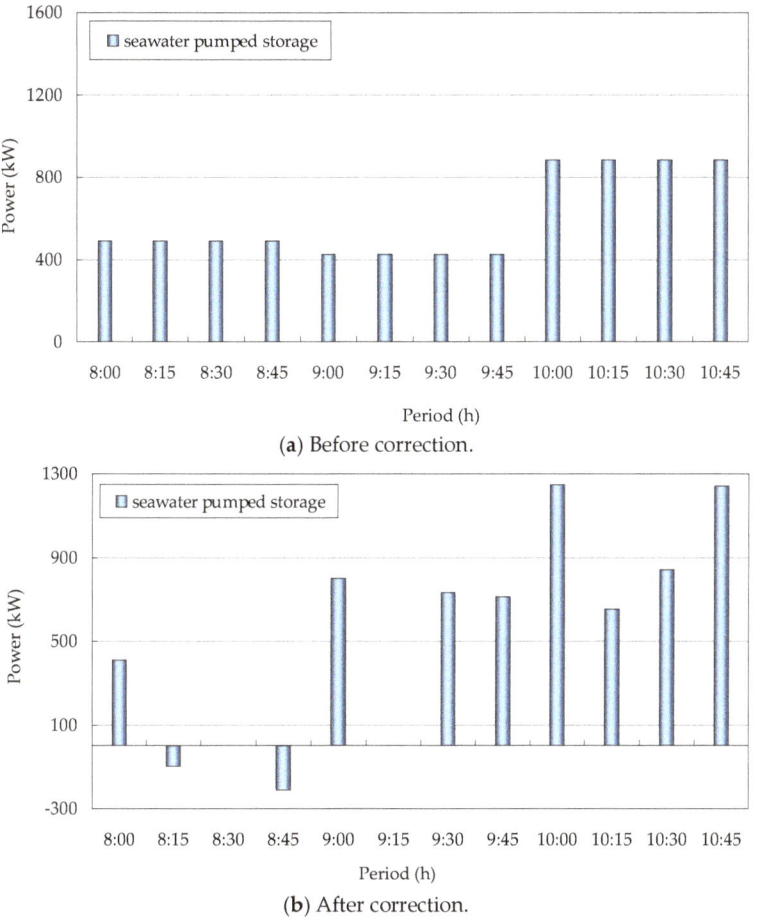

Figure 8. Real-time optimization results of seawater pumped storage.

In the simulation, average execution time for day-ahead scheduling was 3.6 min and for real-time scheduling it was 19.6 s. As a result, the proposed model can be applied online for active distribution networks. Applying the proposed strategy, the overall economic benefits for active distribution networks can be promoted in the day-ahead scheduling stage. Negative influence of the forecast errors on the operation of active distribution networks can be minimized in the real-time scheduling stage. As a consequence, reserve capacity for synchronous generators in the main grid and regulation burden for automatic generation control (AGC) can be reduced.

5. Conclusions

This paper has presented a two-stage optimal scheduling model for active distribution network integration of seawater pumped storage stations, flexible loads, and renewable energy. ILOG Cplex 12.7 was called to solve the proposed model. In addition, the model was successfully tested in a modified IEEE 33-bus. Numerical results demonstrated that by applying the proposed methods in the day-ahead scheduling stage, total operation cost and the network power loss of active distribution networks could be reduced. At the same time, peak loads were shifted. In a real-time scheduling stage, the negative effects of forecast errors of renewable energy on the operation of active distribution networks were compensated and alleviated by proper coordination of a seawater pumped storage station. Utilizing the two-stage scheduling model, the performances and power quality of active distribution networks were improved. In addition, reserve capacity and regulation burden for the main grid were also effectively reduced.

Author Contributions: Conceptualization, N.L. and C.D.; Formal analysis, N.L., C.D., Y.C., W.Y., D.L., M.C., and P.P.; Funding acquisition, C.D., D.L., M.C., and P.P. All of the authors revised the manuscript.

Funding: This work was supported by National Key Research and Development Program of China (No. 2017YFB0903700, 2017YFB0903705).

Conflicts of Interest: The authors declare no conflict of interest.

Nomenclature

Abbreviations

REG	renewable energy generation
WT	wind turbine
PV	photo voltaic
MMS	market management systems
EMS	energy management systems
IPSO	improved particle swarm optimizer
GA	genetic algorithms

Sets, parameters and constants

T, T', t	set of indexes of the dispatch time periods
k_g	efficiency of the turbine generator
ρ	density of seawater (1050 kg/m³)
g	gravity acceleration (9.8 m/s²)
h	water head height
$q_g(\tau)$	flow rate of water volumetric input into the turbine at the time of τ
$q_p(\tau)$	flow rate of water volumetric pumped from the sea at the time of τ
k_p	efficiency of the pump-motor unit
C_g, C_p	start-up cost coefficient of turbine generator and pump-motor unit
c_{sea}^{az}	average annual installation fee of the seawater pumped storage
r_{sea}	depreciation rate of the seawater pumped storage
n_{sea}	life time of the seawater pumped storage
λ_{sea}^{co}	coefficient of pipelines corrosiveness cost
$\lambda_g^{run}, \lambda_p^{run}$	piping maintenance cost coefficient of turbine generator and pump-motor unit
l	leakage loss and evaporation
P_g^{max}, P_p^{max}	maximum output in generating and pumping modes
P_p^{min}	minimum power of seawater pumped storage in pumping modes
$Q_{sea}^{min}, Q_{sea}^{max}$	bottom and top limit of water amount in upper reservoir
$P_g^{max\prime}, P_p^{max\prime}$	maximum output in generating and pumping modes
$P_p^{min\prime}$	minimum output in pumping modes
$\Omega(t)$	price of purchase electricity
M, N	number of interruptible loads and incentive loads
α, β	coefficients of interruptible loads and incentive loads
$k_c(t), k_e(t)$	willingness factors of interruptible loads and incentive loads
N_{WT}, N_{PV}, N_L	number of WT, PV and nodes
$P_{WT}(m,t)$	output of m th WT
$P_{PV}(n,t)$	output of n th PV
$P_L(i,t)$	rigid loads at node i
$P_g^{Rup}(t)$	minimum reserve constraints in generating modes
$P_p^{Rup}(t)$	minimum reserve constraints in pumping modes
p_g^{Rup}, p_p^{Rup}	base reserve capacities in generating and pumping modes
η, λ	proportional coefficient, values of both were taken as 0.2
P_{grid}^{max}	maximum purchase power from main grid in day-ahead

$\Delta P_{WT}(t)$	prediction errors of WT
$\Delta P_{PV}(t)$	prediction errors of PV
$P_{WT}{}'(m,t)$	output of m th WT
$P_{PV}{}'(n,t)$	output of n th PV
$P_{grid}^{max\prime}$	maximum power purchase from the main grid

Variables

$C_g^{cr}(t)$, $C_p^{cr}(t)$	start-up cost of turbine generator and pump-motor unit
$C_{sea}^{in}(t)$	maintenance cost of the seawater pumped storage
$C_{sea}^{run}(t)$	piping maintenance cost of the seawater pumped storage
$\mu_g(t)$, $\mu_p(t)$	binary variables of turbine generator and pump-motor unit
$P_g(t)$, $P_p(t)$	output in generating and pumping modes
$\mu_g{}'(t)$, $\mu_p{}'(t)$	binary variables of turbine generator and pump-motor unit
$P_g{}'(t)$, $P_p{}'(t)$	output in generating and input in pumping modes
$C_{sea}(t)$, $C_{load}(t)$	seawater pumped storage and the response of flexible loads cost
$Q_{up}(t)$, $Q_{up}{}'(t)$	seawater quantity of upper reservoir in day-ahead and real-time
$f_c(t)$, $f_e(t)$	interruptible loads and incentive loads cost
$u_c(i,t)$, $u_e(i,t)$	status of i th interruptible loads and i th incentive loads
$P_f(t)$	flexible loads
$P_{Grid}(t)$	purchase power from main grid
$P_c(t)$, $P_e(t)$	input of interruptible loads and incentive loads
$P_L(i,t)$, $P_f(t)$	rigid loads at node i and flexible loads
$\Delta P_g(t)$, $\Delta P_p(t)$	power adjustment in generating and pumping modes
$P_g{}'(t)$, $P_{grid}{}'(t)$	power from seawater pumped storage and main grid

References

1. Bogdanov, D.; Breyer, C. North-East Asian Super Grid for 100% renewable energy supply: Optimal mix of energy technologies for electricity, gas and heat supply options. *Energy Convers. Manag.* **2016**, *112*, 176–190. [CrossRef]
2. Liu, W.; Wang, S.; Chen, Y.; Chen, X.; Niu, S.; Liu, Z. Coordinate optimization of the distribution network electricity price, energy storage operation strategy, and capacity under a shared mechanism. *Sustainability* **2017**, *9*, 1080. [CrossRef]
3. Zhao, H.; Guo, S. External benefit evaluation of renewable energy power in China for sustainability. *Sustainability* **2015**, *7*, 4783–4805. [CrossRef]
4. Sirjani, R. Optimal Capacitor Placement in Wind Farms by Considering Harmonics Using Discrete Lightning Search Algorithm. *Sustainability* **2017**, *9*, 1669. [CrossRef]
5. Zhang, G.; Jiang, C.; Wang, X.; Li, B.; Zhu, H. Bidding strategy analysis of virtual power plant considering demand response and uncertainty of renewable energy. *IET Gener. Trans. Distrib.* **2017**, *11*, 3268–3277. [CrossRef]
6. Harsha, P.; Dahleh, M. Optimal management and sizing of energy storage under dynamic pricing for the efficient integration of renewable energy. *IEEE Trans. Power Syst.* **2015**, *30*, 1164–1181. [CrossRef]
7. Shi, N.; Luo, Y. Energy Storage System Sizing Based on a Reliability Assessment of Power Systems Integrated with Wind Power. *Sustainability* **2017**, *9*, 395. [CrossRef]
8. Ho, W.S.; Macchietto, S.; Lim, J.S.; Hashim, H.; Muis, Z.A.; Liu, W.H. Optimal scheduling of energy storage for renewable energy distributed energy generation system. *Renew. Sust. Energy Rev.* **2016**, *58*, 1100–1107. [CrossRef]
9. Katsaprakakis, D.A.; Christakis, D.G. Seawater pumped storage systems and offshore wind parks in islands with low onshore wind potential. A fundamental case study. *Energy* **2014**, *66*, 470–486. [CrossRef]
10. Pannatier, Y.; Kawkabani, B.; Nicolet, C.; Simond, J.J.; Schwery, A.; Allenbach, P. Investigation of control strategies for variable-speed pump-turbine units by using a simplified model of the converters. *IEEE Trans. Ind. Electron.* **2010**, *57*, 3039–3049. [CrossRef]
11. Liu, H.; Tang, C.; Han, J.; Li, T.; Li, J.; Zhang, K. Probabilistic load flow analysis of active distribution network adopting improved sequence operation methodology. *IET Gener. Trans. Distrib.* **2017**, *11*, 2147–2153. [CrossRef]

12. Zheng, W.; Wu, W.; Zhang, B.; Wang, Y. Robust reactive power optimisation and voltage control method for active distribution networks via dual time-scale coordination. *IET Gener. Trans. Distrib.* **2017**, *11*, 1461–1471. [CrossRef]
13. Ahmadian, A.; Sedghi, M.; Elkamel, A.; Aliakbar-Golkar, M.; Fowler, M. Optimal WDG planning in active distribution networks based on possibilistic–probabilistic PEVs load modelling. *IET Gener. Trans. Distrib.* **2017**, *11*, 865–875. [CrossRef]
14. Ochoa, L.F.; Dent, C.J.; Harrison, G.P. Distribution network capacity assessment: Variable DG and active networks. *IEEE Trans. Power Syst.* **2010**, *25*, 87–95. [CrossRef]
15. Cecati, C.; Buccella, C.; Siano, P.; Piccolo, A. Optimal operation of Smart Grids with demand side management. In Proceedings of the IEEE International Conference on Industrial Technology, Cape Town, South Africa, 25–28 February 2013; pp. 2010–2015.
16. Tan, J.; Wang, L. Enabling reliability-differentiated service in residential distribution networks with PHEVs: A hierarchical game approach. *IEEE Trans. Smart Grid* **2016**, *7*, 684–694. [CrossRef]
17. Gill, S.; Kockar, I.; Ault, G.W. Dynamic optimal power flow for active distribution networks. *IEEE Trans. Power Syst.* **2014**, *29*, 121–131. [CrossRef]
18. Lin, C.; Wu, W.; Chen, X.; Zheng, W. Decentralized dynamic economic dispatch for integrated transmission and active distribution networks using multi-parametric programming. *IEEE Trans. Smart Grid* **2017**, 1. [CrossRef]
19. Wang, D.; Meng, K.; Luo, F.; Coates, C.; Gao, X.; Dong, Z.Y. Coordinated dispatch of networked energy storage systems for loading management in active distribution networks. *IET Renew. Power Gener.* **2016**, *10*, 1374–1381. [CrossRef]
20. Westermann, D.; Nicolai, S.; Bretschneider, P. Energy management for distribution networks with storage systems—A hierarchical approach. In Proceedings of the IEEE PES General Meeting, Conversion and Delivery of Electric Energy in the 21st Century, Pittsburgh, PA, USA, 20–24 July 2008.
21. Doostizadeh, M.; Ghasemi, H. Day-ahead scheduling of an active distribution network considering energy and reserve markets. *Int. Trans. Electr. Energy Syst.* **2013**, *23*, 930–945. [CrossRef]
22. Zhang, L.; Tang, W.; Liang, J.; Cong, P.; Cai, Y. Coordinated day-ahead reactive power dispatch in distribution network based on real power forecast errors. *IEEE Trans. Power Syst.* **2016**, *31*, 2472–2480. [CrossRef]
23. Liu, Y.; Nair, N.K.C. A Two-stage Stochastic dynamic economic dispatch model considering wind uncertainty. *IEEE Trans. Sustain. Energy* **2016**, *7*, 819–829. [CrossRef]
24. Sachs, J.; Sawodny, O. A two-stage model predictive control strategy for economic diesel-PV-battery island microgrid operation in rural areas. *IEEE Trans. Sustain. Energy* **2016**, *7*, 903–913. [CrossRef]
25. Tan, Y.; Cao, Y.; Li, C.; Li, Y.; Zhou, J.; Song, Y. A two-stage stochastic programming approach considering risk level for distribution networks operation with wind power. *IEEE Syst. J.* **2016**, *10*, 117–126. [CrossRef]
26. Zhao, J.; Wen, F.; Dong, Z.Y.; Xue, Y.; Wong, K.P. Optimal dispatch of electric vehicles and wind power using enhanced particle swarm optimization. *IEEE Trans. Ind. Inform.* **2012**, *8*, 889–899. [CrossRef]

© 2018 by the authors. Licensee MDPI, Basel, Switzerland. This article is an open access article distributed under the terms and conditions of the Creative Commons Attribution (CC BY) license (http://creativecommons.org/licenses/by/4.0/).

Article

Analyses of Distributed Generation and Storage Effect on the Electricity Consumption Curve in the Smart Grid Context

Simona-Vasilica Oprea [1,*], Adela Bâra [1], Adina Ileana Uță [1], Alexandru Pîrjan [2] and George Căruțașu [2]

1   Department of Economic Informatics and Cybernetics, Bucharest Academy of Economic Studies, Romana Square 6, Bucharest 010374, Romania; bara.adela@ie.ase.ro (A.B.); adina.uta@ie.ase.ro (A.I.U.)
2   Department of Informatics, Statistics and Mathematics, Romanian-American University, Expoziției 1B, Bucharest 012101, Romania; alex@pirjan.com (A.P.); georgecarutasu@yahoo.com (G.C.)
*   Correspondence: simona.oprea@csie.ase.ro; Tel.: +40-752-294-422

Received: 9 May 2018; Accepted: 29 June 2018; Published: 1 July 2018

Abstract: The householders' electricity consumption is about 20–30% of the total consumption that is a significant space for demand response. Mainly, the householders are becoming more and more active and interested in diminishing their expenses related to the electricity consumption, considering different rates of the advanced tariffs. Therefore, in the smart grid context, especially for prosumers with energy sources and storage devices (SD), the electricity consumption optimization becomes attractive since they obtain significant benefits. On the other hand, the electricity suppliers design appropriate tariffs in order to reduce the consumption peaks and avoid the occurrence of new peaks. Based on the effect of these tariffs on consumers' behavior, the stress on generators decreases and the electricity suppliers improve the demand forecast and adjust their strategies on the market. In addition, the grid operators are interested in the minimization of the consumption peak that leads to loss reduction and avoidance of congestions that would ensure at least the delay of the onerous investment in grid capacities. In this paper, we will run several scenarios for electricity consumption optimization in the context of smart grid that includes: sensors, actuators, smart meters, advanced tariff schemes, smart appliances and electricity home control applications. Our goal is to analyze the effect of the Renewable Energy Systems (RES) distributed generation (such as photovoltaic panels—PV) and storage on the consumption curve. The results show that consumption optimization with RES distributed generation and SD brings sustainable development of the power systems and significant benefits from the consumption peak and savings point of view.

Keywords: distributed generation; storage device; MILP; ToU tariff; optimization; daily consumption curve; peak/off-peak; programmable appliances

1. Introduction

Lately, more and more authorities of the countries from the European Union (EU) are concerned about issues regarding the requirements of moving away from an economy driven by fossil fuels, based on a centralized, supply-side approach, relying on old technologies and outdated business models, as it has been emphasized by the European Community Energy Union Package 2015 [1]. In order to fulfill this requirement, one of the most important steps that need to be taken consists in implementing the advancements in smart grid technologies that enable various options and strategies for consumption optimization in electricity markets. However, there are still barriers for the widespread adoption of smart grid technologies at the EU level, as emphasized by the European Technology and Innovation Platform—Smart Networks for Energy Transition, in the document Final 10-year ETIP SNET R&I

Roadmap (2017–26) [2]. Embracing the same approach, the European Network of Transmission System Operators for Electricity highlights the need to activate the demand as a new source/tool for system operation and to integrate it in the planning and operation stages and market design in ENTSO/E R&I Roadmap 2017–2026 [3]. In Ireland, the Commission for Energy Regulation (CER), the regulator for the natural gas and electricity sector, has initiated in 2007 the Smart Metering Project within which it has performed rigorous benchmark tests in order to evaluate the performance characteristics of the smart metering devices, analyzing their influence in regard to consumers, energy consumption, tariff schemes and economic aspects in view of implementing these devices at a national level [4].

Starting from the interested parties in the consumption optimization, different approaches can be envisioned. On one hand, the optimization can lead to minimization of the electricity payment (the consumers' approach), by shifting their appliances to the cheapest time intervals. On the other hand, it can lead to minimization of the consumption peak (the grid operators' and suppliers' approach) that means the peaks are shaved and valleys are filled by uniformly distributing the programmable appliances on top of the fixed consumption. Both approaches require careful analyses and comparisons in order to reach a sustainable power systems development.

From these perspectives, the consumers can be stimulated to change their behavior that leads to higher demand response and electricity consumption reduction. In this respect, the advanced tariff scheme provides incentives for optimal planning of the appliances. In this way, according to the literature, the consumers can save up to 50% of electricity payment due to the fact that off-peak tariff rate can be one third of the peak tariff rate [5]. Moreover, indirect savings are coming from less grid investments and reduction of losses, therefore achieving a smoother operation of generators that have the possibility to operate closer to their optimum capacity.

From the demand point of view, the electricity supply does not allow interruptions or massive storage, therefore knowing the consumers' behavior is significant for grid operators, electricity producers and suppliers. The system operator will dispatch the generators so that the demand is optimally supplied by the available generators based on the existing market mechanism. The grid operators will plan the grid capacity based on the consumption requirements, the suppliers will estimate the consumption level at any time and buy or sell on the various electricity markets while the generators will be prepared to produce the necessary energy as to satisfy the demand. Consequently, the predictability of electricity consumption is one of the key factors that influence the sustainable development of power systems.

Moreover, the consumption curve flattening creates the premises for important improvements, by reducing the stress on the generation units and on the entire grid, by diminishing the electricity price, by increasing the potential to achieve an accurate forecast and even by improving the market strategies. The consumption peak has direct impact on grid loading, therefore by shaving the peaks and filling the valleys, the grid will load uniformly during the 24 h of the day and onerous investments in grid infrastructure can be eliminated or at least diminished. Furthermore, by using smart grid technologies, especially smart metering systems and home control applications, the volume of investments in grid facilities can be diminished by reducing the consumption peak and integrating a higher volume of RES distributed generation. Therefore, if the consumption peak is reduced by means of optimization and the electricity is locally generated, then less grid infrastructure will be needed. Consequently, the analyzed issues have been addressed extensively in the scientific literature due to their significant and ever-increasing importance.

For example, the purpose of the study conducted in [6] consists in developing a solution that is able to manage optimally the battery storage at a large-scale level using a three-layer battery hierarchical control structure by taking into consideration specific control circuits. The authors propose models for the photovoltaics, storage batteries and supercapacitors for developing a distribution network structure that offers the possibility of hierarchical storage. In order to regulate the high and low frequency photovoltaic fluctuations the authors implement a low pass filter. After having performed a series of simulation tests, the authors state that the obtained results confirm the supercapacitor's and the battery's control effect. The authors acknowledge the fact that supplementary experiments must be performed in order to check the efficiency of the control when there are considered several distributed storage systems.

In the paper [7], the authors propose an Energy Scheduling and Distributed Storage (ESDS) algorithm in view of installing it into the smart meter devices that have the capacity to record the detailed Time of Use (TOU) consumption. The proposed algorithm aims to minimize the electricity consumption of the consumer while increasing his satisfaction towards his electricity needs in the same time. In this purpose, the authors have used the ESDS algorithm in three scenarios: in the first scenario they have analyzed the case in which there was not considered the Demand Side Management (DSM), in the second scenario the authors have taken into consideration the DSM without ESDS, while in the third scenario they have considered DSM with ESDS. The authors' obtained results confirmed that scenario 3 registered the highest level of performance with regard to the energy and financial savings, minimizing the dissatisfaction of consumers during peak periods, reducing the Peak-to-Average-Ratio demand, the sustainability of the energy grid, a series of socio-economic and environmental benefits. In their future work, the authors intend to analyze the possibility to adapt their algorithm so that it can be implemented in the case of large non-household consumers.

Related to the above-mentioned case of large non-household consumers, a study that takes them into account is [8], a paper in which a part of our research team has developed solutions regarding the hourly forecasting of the energy consumption, situation in which the accurate electricity consumption is extremely important due to the implied costs. The authors have developed several forecasting solutions consisting in a series of artificial neural networks based on the non-linear autoregressive (NAR) model and the non-linear autoregressive with exogenous inputs (NARX) one, trained using the Levenberg-Marquardt, Bayesian Regularization and Scaled Conjugate Gradient algorithms. In the study, the authors have used large datasets recorded by the smart metering devices of the consumer, along with exogenous variables in the case of the NARX model, case in which the authors have built meteorological and time stamp datasets. By means of extensive experimental tests, the authors have identified in each case the best mix between the training algorithm, the number of neurons in the hidden layer and the delay parameter. Based on a series of forecasting scenarios, using specific performance metrics, the developed solutions for the hourly consumption energy prediction have offered a high level of forecasting accuracy, having the potential to become useful tools for both the electricity producers and consumers. The authors consider that the further step of their research consists in refining the study regarding the consumption forecasting up to the individual appliances or devices level.

Paper [9] describes a method based on a hybrid energy storage system designed to smooth the fluctuating output of a photovoltaic power plant. The Renewable Energy Systems distributed generation (PV) and the hybrid energy storage systems have been modeled in Matlab/Simulink and Piecewise Linear Electrical Circuit Simulation software environment. Using a series of extensive simulations, the authors confirm the effectiveness of their proposed approach for the power control strategy. In [10], the authors propose a three-layer management system designed to handle the issues raised by large capacity energy storage batteries. The authors discuss the hardware and software implementations of the Battery Management Unit, Battery Cluster Management System and Battery Array Management System, considered as bottom, middle and respectively top layers of the proposed hierarchical management system that is useful in performing a variety of tasks such as: measurement and computation, balance of cells, the safe management of high voltage, data management, the management of the charging and discharging operations, issuing warnings.

The authors of [11] propose a model for optimizing the operation of a micro-grid, taking into account the distributed generation, the environmental factors as well as the demand response. According to the authors, the main advantages of the devised model consists in cutting down the operational costs at the micro-grid level, without affecting the consumers, therefore creating the premises of consuming clean energy. The authors' approach consists in employing a genetic algorithm for implementing the objective function and a demand response scheduling strategy. The validation of the model was done on a smart micro-grid from Tianjin. The authors state that the obtained results highlight the strong effect that the distributed generation and the demand response have on the micro-grid operation from the economical point of view. In the paper, there are also analyzed the main factors that impact the optimization results, concluding that the price of the natural gas has the most important influence on the micro-grid's operation costs and the demand response. The authors conclude that their model has the potential to be implemented in a real production working environment and that they intend to do so in their future researches.

The volatility of the PV output influences to a great extent the stability and the power quality of the PV systems, causing difficulties regarding the grid integration of PV systems. In order to address these issues, the authors of [12] have developed a control strategy based on energy storage systems in order to attenuate the generation fluctuations of the PV system and the impact of the PV systems to the grid. The active energy flow between the grid and the storage systems is ensured through the usage of bidirectional converters, a battery and ultracapacitors, the direct current constant voltage is maintained using an inverter control strategy, the photovoltaic system is managed as to provide the maximum power. In order to confirm the efficiency of their proposed model, the authors devise a dynamic modeling approach and a simulation study developed using Matlab/Simulink.

In the paper [13], the authors investigate the feasibility of replacing the autonomous thermal power plants that exist on a small Greek island with distributed energy resources that derive from RES, using also modern storage technologies. This replacement is very important in order to solve many of the problems that electricity autonomous islands have to deal with on a daily basis, for example energy and water shortages. In their study, the authors have chosen as main objectives the minimization of costs, of carbon dioxide emissions and the reliability maximization of the distributed energy resources. In order to conduct the multi-objective analysis, the authors have used a decision support tool, the Distributed Energy Resources Customer Adoption Model, that implements mathematical software designed to find the optimal solutions. After analyzing the obtained results, the authors propose a solution for all the investors that are interested in implementing the proposed approach in the case of other autonomous islands that deal with similar issues. The authors conclude that the development of the distributed energy resources has the potential to bring numerous advantages to the consumers: Energy independence, increased quality of the delivered electricity, stability of the electricity dispatch system along with economic benefits resulting from the local business.

The paper [14] proposes a reclosing method for distribution systems in the case when the distributed generation and the battery storage system are employed. The system also contains a circuit breaker in the distribution line. The method proposed in this paper is based on a model that uses an electromagnetic transient program. The authors have performed a series of simulations, which reflect the capacities of the distributed generation and the clearance time of the faults. The authors acknowledge that their proposed method has two limitations, the first one being related to the cost and the second one to the type of the distributed generation. As regarding the costs, they represent a limitation of the proposed method because of the circuit breakers and the communication's electrical installation that both must be installed. The limitation regarding the distributed generation's type is also worth to be considered, as in their study the authors have taken into account a photovoltaic system, connected through an inverter. If a system that employs a synchronous generator is used instead of this type of distributed generation, then the devised method does not provide satisfactory results. Therefore, the authors state that in a future work they intend to address the above-mentioned limitations of the method by taking into consideration the economic aspects and also the type of the distributed generation.

In the paper [15], the authors devise a simulation model of a hybrid renewable energy system generation that comprises also an energy storage system, targeting several households that have an annual electricity demand of 30 MWh. The authors state that the obtained results reveal that an energy storage system comprising a 9 kW PV array, an 8 kW wind turbine, a 2 kW water turbine as well as a

256 kWh storage capacity for the energy storage system can be a sustainable and reliable energy source, covering the needs of entire families. After having performed the simulations, the authors acknowledge that the households still have to be connected to the electricity grid taking into consideration the irregular intervals of occurrences that characterize the energy sources of the system, otherwise there is a risk of energy surpluses if the system is oversized or a risk of having an energy deficit. The authors also acknowledge the fact that achieving a proper sizing of the energy storage system as to cover appropriately the seasonal and diurnal patterns of electricity generation and demand is a complex task and emphasize that the model can be improved in the future if one takes into consideration the economic and environmental aspects of the hybrid energy sources.

The authors of [16] develop a multi-sensor system for a smart home environment by incorporating technology from wearables along with artificial intelligence positioning algorithms and multiple sensors for data fusion. The developed system allows one to operate household appliances remotely, allows the smart home environment to be aware of the residents' indoor position and the environment of each living space to be monitored in order to assure automated household appliance control, enhance home security and ultimately achieve a smart energy management. The authors state that the experimental results depicted an accuracy for the 3D gesture recognition algorithm in view of operating remotely the appliances ranging from 87.7% to 92.0% while for the positioning within the building and the smart energy management, the accuracy concerning the distance and the positioning was around 0.22% and 3.36% of the total distance that was traveled indoors. The authors stated that they intend in future studies to reduce the size of the wearable sensor as to increase the comfort level for the residents and make use of multiple positioning methods, like WiFi and Beacon systems to refine the accuracy of positioning along with developing a more practical and user-friendly monitoring interface on a smartphone device.

Another study that tackles issues regarding the smart homes and the household appliances is [17], where a part of our research team has developed and applied a method for the accurate forecasting of the residential consumption of electricity up to the appliance level. The authors have used data recorded by the sensors of the smart homes from residential area, characterized by the fact that they are equipped with solar panels able to sustain a part of the consumed energy. The main difficulties and limitations in developing this research consisted in the fact that historical meteorological data were not available and even if they had been, they would have imposed supplementary costs, as in the future the beneficiary of the forecast had to acquire from time to time accurate forecasts for a short-term horizon from a meteorological institute. In order to overcome all these limitations, the authors have developed an approach based on a mix of non-linear autoregressive with exogenous inputs artificial neural networks and function fitting neural networks. The method represents a useful tool to the contractors and it also has the potential to be incorporated in a cloud solution, in order to be offered as a service to both the operators and the consumers, for a monthly fee. As regarding future work, the authors consider that their method could be refined by taking into consideration the energy class of the appliances, the number, structure and behavioral patterns of the households' inhabitants, by using future versions of the development environment and by improving the retraining process.

In [18], the authors develop an approach based on game theory in order to achieve an appropriate balancing electricity consumption within clustered wireless sensor networks. In the approach, the authors have modified the payoff function by including a penalty system within it in order for the nodes that contain a higher amount of energy to be able to strive more vigorously for the cluster head. The authors state that they have employed convex optimization techniques in order to achieve for the clustering game the Nash equilibrium strategy. Due to the devised strategy, the authors made sure that every sensor node has the possibility to attain the highest possible payoff when the decision is taken. After having performed numerous simulations, the authors conclude that their developed approach achieves a good performance concerning the energy balance, therefore extending the lifetime of the network to a great extent. The authors state that in the future they intend to approach issues regarding the multi-hop clustered wireless sensor networks, the cluster head selection in energy harvesting wireless sensor networks and finally to tackle the security issues that arise in clusters that are based on wireless sensor networks.

The authors of [19] conduct a study that has as a main goal to assess the benefits that can be obtained after having peak shaved the demand response considering in their study different storage devices such as capacitors, magnetic storage, different types of batteries, flywheel, compressed air and pumped hydro. In order to generate randomly the electricity demand profiles at the appliance level for a typical household from the United States, the authors employed an agent-based stochastic model. The authors used the total-energy-throughput lifetime in order to introduce a storage cost model on several levels and afterwards they have tried to optimize both the electricity storage capacity and the limit concerning the demand by devising a strategy that dispatches the stored energy appropriately. The authors found out that the storage devices bring benefits, the annual profit of the demand response mechanism varying from 1 to 39% when compared to the previous period. The authors acknowledge that in the future an even higher profit rate can be obtained if one incorporates in the dispatch strategy weather forecasting aspects.

In the paper [20], the authors analyze a solution based on a battery energy storage system designed to shave the energy peaks and smooth the load curve, therefore improving the operation of a power system located in Hawaii, in the island of Maui. As the distribution circuit is mainly based on the photovoltaic generation, a series of issues regarding the renewable energy integration are posed to the operator of the power system, principally due to the transmission system. In order to approach this issue, the authors have proposed two methods for obtaining the load forecasting that is mandatory in order to employ safely at its maximum capacity the battery energy storage system and to achieve the shaving of the energy peaks and the smoothing of the load curve. The first method is based on a nonlinear programming technique, while the second one employs a control strategy in real time that offers the possibility to achieve simultaneously the two proposed targets (shaving of the energy peaks and smoothing of the load curve). In order to confirm the usefulness and accuracy of their forecasting methods, the authors benchmark their methods using a real case study, comprising data collected for 108 days.

In [21], the authors propose an energy management system for residential consumers, taking into account in the same time the energy that has been consumed and the one that has been generated. The system is based on the electricity price that motivates the consumers to reschedule their appliances so that to benefit from the price variation. The approach is analyzed considering a smart home with modern appliances, micro wind power plant, PV and batteries as storage device for a period covering 24 h. The authors have used a series of scenarios in order to evaluate their proposed model and afterwards they have analyzed and compared the obtained results, highlighting situations in which a consumer has to buy, use, store or even sell the available amount of electricity, therefore obtaining an improved efficiency and economic benefits. Based on the obtained results, the authors state that the scheduling of the electrically controllable appliances, thermostatically controllable appliances and optically controllable appliances can be done in the same time, therefore obtaining a viable solution for the management of residential electricity usage.

The authors of [22] propose an optimization model for residential consumption by shaving the peaks and filling the gaps of the load curve. The authors state that the developed method is useful in reducing the production cost of the energy at peak hours and in increasing the income at off-peak hours. For this purpose, the demand response program is based on rewards for load control. Therefore, the authors implemented a credit function within their model to influence the consumption level and provide incentives to consumers as to attain the desired load curve. After having performed simulations, the authors state that the best approach to reduce the electricity costs is to use the appliances that consume the highest amount of electricity only during the dawn, the midday or the night. The authors consider that if their method would be accepted and enforced by policy makers at the global level, one would achieve clean energy generation. In their future work, the authors intend to adjust the model as to use a multi-objective non-linear approach, to optimize the parameters of the utility function as to be related to the focus of the objective one.

An extensive literature review related to consumption peak shaving is proposed in [23] analyzing the effect of the following strategies for peak load shaving: Integration of energy storage systems, integration of electric vehicles to grid, demand side management. The paper analyzes in each of the three cases, the possible benefits brought by the peak shaving to both the supplier and the consumer, the main challenges that must be overcome, the future research directions that have to be approached in view of achieving a proper peak load shaving, the main technologies used for energy storage in the purpose of peak shaving, the operation of the energy storage systems, their management and appropriate sizing, the methods for peak shaving in the case of electric vehicles. As a limitation of their paper, the authors mention the fact that they have not analyzed in detail the aspect of using a mix of multiple RES when performing the peak shaving.

Taking into account the ever-growing importance of the consumption optimization process, our paper aims to reveal the effect of the PV and SD on the load curve, emphasizing the impact on consumption peak and electricity payment. Consequently, we develop a set of sensors based on Arduino for monitoring and controlling household appliances and analyze the simulation results of two optimization approaches with different objective functions that provide significant insights on PV and SD benefits.

This paper is structured in four sections. In the current section, the importance of consumption curve flattening, a series of related works and the structure of the paper are presented. In the second section, we propose an optimization method and design the required architecture for its implementation, describe the appliances considered in the optimization process and describe the objective functions implemented in Matlab R2016a by means of *intlinprog* and *fminimax* functions. The consumer preferences are configured in terms of hourly and daily consumption of each appliance. Also, we formalize de optimization problem, by setting the variables, constraints and objective functions: consumption peak minimization and electricity payment minimization. In the third section several simulations have been performed underlying the two optimization approaches, using six three-level ToU tariffs. Also, *intlinprog* and *fminimax* functions are also compared in terms of optimization results and we analyze the effect of the PV and SD on the consumption curve; both the consumption peak and electricity payment decreased. In section four, the main conclusions based on the simulation findings are presented. However, our approach has several limitations and policy implications that are also presented in the conclusion section.

2. Method and Objective Functions

Our optimization method aims to provide an efficient solution for minimization the consumption peak and payment by integrating the PV and SD. The method consists in several steps that allow us to configure the appliances, to set up the sensors for monitoring and controlling their operation, perform consumption peak minimization and payment minimization with two optimization functions (*intlinprog* and *fminimax* implemented in Matlab), compare the results and choose the best option in terms of objective function. Then, peak minimization is performed with *intlinprog*, adding PV and SD to analyze their effect on the consumption curve. These steps are briefly illustrated in Figure 1.

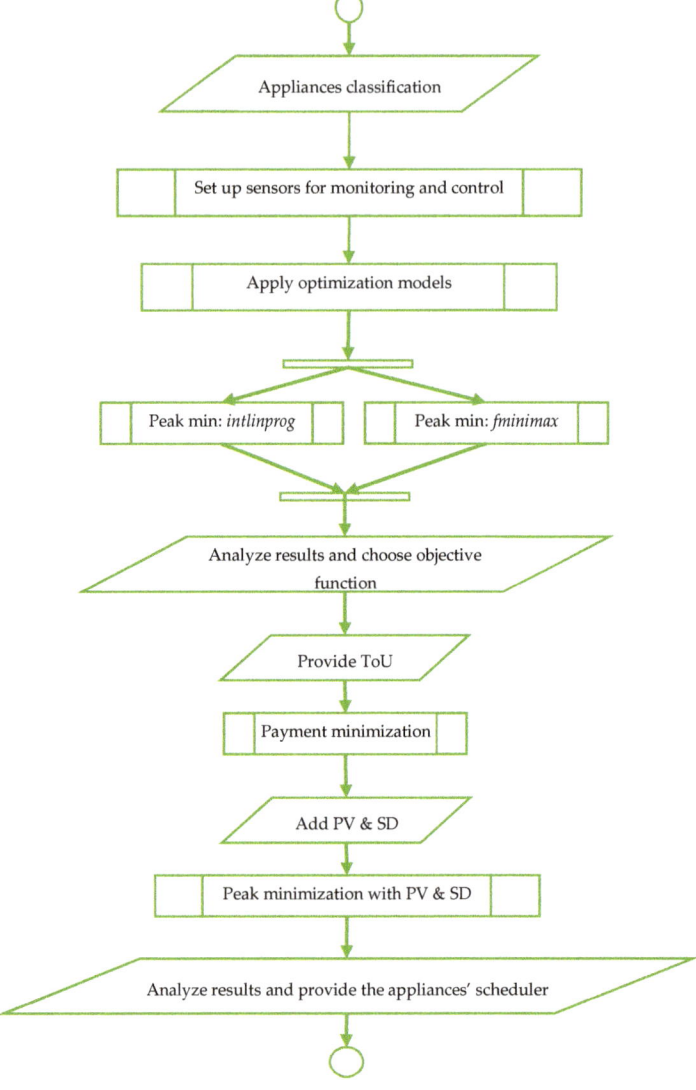

Figure 1. Flowchart of the proposed method.

The method is implemented on an architecture that consists in: Arduino-based sensors for monitoring and controlling the appliances; a relational database running Oracle Database 12c for storing the appliances consumption data, electricity tariffs and the scheduler for each programmable appliance that provides the start and operation interval based on the optimization results; a set of optimization models developed in Matlab R2016a based on two functions: *intlinprog* and *fminimax*. The electricity supplier provides ToU tariff, while the consumers provide appliances constraints and desired operational schedule via portal developed in JDeveloper. The architecture is presented in Figure 2.

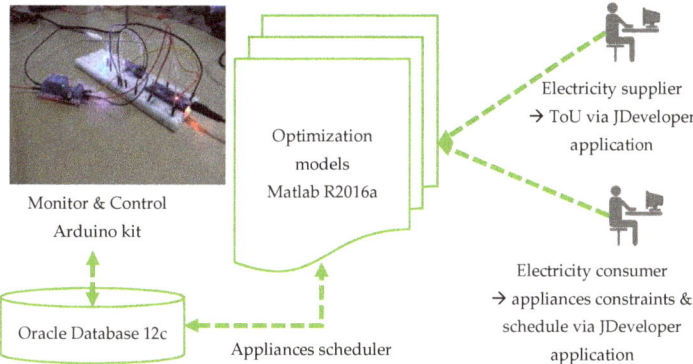

Figure 2. The architecture for implementing the optimization method.

The controlling and surveillance solution of the appliances has components for collecting and receiving data among appliances and the gateway with two parts: Transmitter and receiver. For transmitter, Micro Arduino with ATmega32U4 microcontroller with 20 digital pins was used for I/O. It has an automatic voltage regulator, an oscillator, a micro USB connector, a serial circuit programming/an In-Circuit Serial Programming.

An YHDC SCT-013 current transformer based on AC current sensor of 100A max was connected to Micro Arduino for measuring the electricity consumption of the appliance (for instance a desk lamp). Against overloading, 1 × 220 μF capacitor and 2 × 330 Ohm resistors are also connected. In the emergency cases, the Micro Arduino is energized by a battery with a capacity of 3000 mAh that is connected to the micro USB charger. For controlling the consumption of the appliances, a relay with optocoupler is also connected to Micro Arduino.

A transmitter-controlling device with relay is set up for each programmable appliance and a transmitter-monitoring device for each non-programmable appliance. The receiver consists in Raspberry Pi that communicates with the transmitter component via a RF 433 MHz transmitter-receiver module. The transmitter-receiver components shown in Figure 3 are managed through a Node.js application that passes the data inserting the records into an Oracle Database.

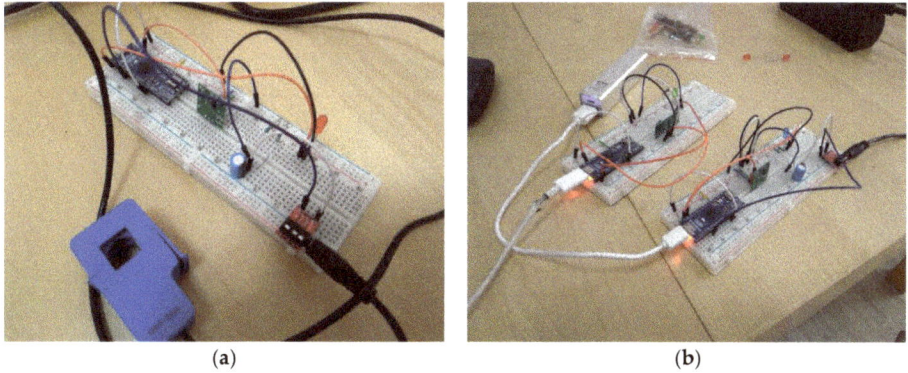

Figure 3. AC current sensor (**a**), transmitter-receiver components (**b**).

An ETL process is used for extracting (E), loading (L) and transforming (T) the input into valid data. The process is implemented in Oracle PLSQL language and consists is several procedures and functions that allow data validation and transformation. Data measured at one-minute intervals by AC current sensors is validated first using ETL subroutines to check data consistency and replace missing values with average neighboring values. Then data is aggregated at hourly interval and loaded into the Oracle Database tables for further evaluation and processing.

In Section 2.1 the appliances are classified and in Section 2.2 the objective functions are described in terms of variables and models.

2.1. Appliances Description and Classification

The appliances can be classified based on their operation mode in relation with the electricity consumers' interaction. From this point of view, three appliances categories are identified:

- Background appliances are always or periodically in operation without notably intervention from consumers side (i.e., refrigerator, house monitoring system);
- Active appliances require the presence of the consumers (i.e., vacuum, electric grill);
- Passive appliances are started by consumers, but operate without assistance (i.e., washing machine, bread oven).

Starting from this classification, some appliances such as: refrigerator, lighting cannot be programmed due to the fact that the comfort of the consumer would be prejudiced. However, the operation of other (programmable) appliances can be shifted to operate at lower tariff rate without any comfort prejudice. Some programmable appliances can be interrupted, while others are uninterruptible due to their intrinsic characteristics. Thus, for simulations, all types of appliances that belong to regular residential consumers are taken into account. Therefore, adding, modifying or eliminating an appliance should be an easy task. Describing each appliance in terms of operation mode is essential in order to define the optimization problem.

For simulations performed in Matlab R2016a, we considered twelve modern appliances such as: Oven, heating, refrigerator, water heating, car battery, TV and speakers, washing machine, dish washer, electric hob, lighting, vacuum and bread oven.

The operation period, hourly and daily consumption are required for setting the optimization problem in terms of operation constraints, limits also known as lower/upper bounds and linear equations. In Table 1, operational mode of the twelve appliances in terms of operation time interval, hourly and daily consumption requirements is described.

Table 1. Description of consumption requirements.

No.	Appliance Name	Consumption Requirements
1	Oven	Operation period: hour 20 Hourly consumption: 1 kWh
2	TV and speakers	Operation period: hours 19–23 Hourly consumption: 0.3 kWh
3	Heating	Operation period: hours 4–5 and 21–23 Hourly consumption: 1 kWh
4	Electric hob	Operation period: hour 7 Hourly consumption: 1 kWh
5	Refrigerator	Operation period: 24 h Hourly consumption: 0.1 kWh
6	Lighting	Operation period: hours 6–8 and 19–23 Hourly consumption: 0.11 kWh
7	Water heater	Hourly consumption: 0–1 kWh Daily consumption: 2.5 kWh

Table 1. Cont.

No.	Appliance Name	Consumption Requirements
8	Car battery	Loading period: hours 20–8 Power: 0.05 kW–2 kW Daily consumption: 5 kWh
9	Vacuum	Operation between hours 6–18 Hourly consumption: 0.5 kWh Daily consumption: 0.5 kWh
10	Washing machine	Uninterruptible operation twice a day 1 kWh for first hour 0.5 kWh for second hour
11	Bread oven	Operate 1 h/day, without interruptions Daily consumption: 1 kWh
12	Dish washer	Hourly consumption: 1 kWhDaily consumption: 1 kWh

The electricity consumers/prosumers become more and more involved in consumption optimization increasing the usage of the PV and SD due to their benefits in cutting-off the electricity bills. At off-peak/peak hours, SD can storage/supply the energy generated by PV, aiming to reduce both the consumption peak and electricity payment.

The usage of batteries as SD are the most appropriate energy storage form at the consumers' level (also known as distributed storage), but the main problem consists in the efficiency of the successive charging/discharging due to the fact that it may vary between 50% and 90%.

Also, the price for 1 kWh storage varies between 100 and 1050 Euro; the lifetime of the SD varies from 10^3 to 10^4 charging/discharging cycles. The efficiency of SD may vary between 72% and 100% based on the SD type; the lifetime may also vary between 200 and 10^4 charging/discharging cycles.

Therefore, it is difficult to estimate the feasibility investment in SD. However, this goal is beyond our scope that is to analyze the influence of PV and SD on the consumption curve.

Modeling PV takes into account the operating characteristics that are specific for solar-irradiance based technology. Therefore, since PV generate depending of solar irradiance intensity, they have been considered to generate gradually between 6 a.m. and 7 p.m., the correlation coefficient between the solar irradiance and PV output varying between 96 and 99%.

Starting from the consumption requirements of the appliances, characteristics of PV, SD and electricity consumers' preferences, in Section 4 we performed several simulations in Matlab R2016a.

2.2. Objective Functions

Our method consists in two optimization approaches: minimization of the consumption peak (suppliers, grid operators approach) and minimization of the electricity payment (consumers approach). For these approaches, we provide the variables of the model and use two optimization functions implemented in Matlab R2016a: *intlinprog* and *fminimax*.

The consumption of all home appliances (HA) for an hour h, $\forall h \in \{1,\ldots,24\}$ should be less or equal to the objective function $f(x_h)$ that is first to minimize the peak accumulated hourly consumption (suppliers and grid operators approach).

$$x_h = \sum_{a \in HA} x_{ah} \leq f(x_h), \forall h \in \{1,\ldots,24\} \tag{1}$$

where: x_h—consumption of all home appliances (HA) for an hour h;
x_{ah}—consumption of appliance a at hour h;
$f(x_h)$—objective function.

The sum of hourly consumption of an appliance *a* represents the total daily consumption for *a*. Regardless the hourly operation schedule of an appliance *a*, its daily consumption remains the same.

$$\sum_{h=1}^{24} x_{ah} = D_a, \forall a \in HA \qquad (2)$$

where D_a—total daily consumption for appliance *a*.

The total consumer's daily consumption T_c is the sum of the daily consumption of each appliance *a*. It remains constant after consumption optimization is implemented.

$$\sum_{a \in HA} \sum_{h=1}^{24} x_{ah} = T_c \qquad (3)$$

Considering the different types of appliances (nonprogrammable *np*, programmable with interruption *pi* and programmable without interruption *pwi*, $a \in \{np, pi, pwi\}$), we need to define the requirements for individual appliances and user preferences as supplementary constraints that are part of the optimization problem.

For a non-programmable appliance *np* with fixed operation requirements given by the consumer's preferences or intrinsic characteristics of the appliance, the consumption scheduling is fixed to a certain time interval considering that hourly consumption cannot be shifted:

$$x_a = [C_{a1}, C_{a2}, \ldots, C_{a24}]^T, \forall a \in np \qquad (4)$$

Fixing the operation limits of *np* appliances can be done by setting the upper and lower bounds. In this case, the consumer indicates the hourly consumption usually by setting the upper and lower bounds to the same value, imposing fixed consumption.

For a programmable appliance with interruption *pi*, the consumer predefines a certain consumption schedule intervals that can be written as a vector, where only the starting time can be optimized.

$$x_a = [C_{a1}, C_{a2}, \ldots, C_{a24}]^T, \forall a \in pi \qquad (5)$$

Therefore, all possible operating combinations form a matrix that shows the programmable options of appliance *pi*.

$$Xa = \begin{bmatrix} C_{a1} & \cdots & C_{a24} \\ \vdots & \ddots & \vdots \\ C_{a24} & \cdots & C_{a1} \end{bmatrix}, \forall a \in pi \qquad (6)$$

Another vector is also defined by consumer in order to indicate possible starting time *ST* (in or not in operation) for appliance *pi*.

$$ST_a = [ST_{a1}, ST_{a2}, \ldots, ST_{a24}]^T \qquad (7)$$

where the sum of starting time for 24 h is equal to 1 multiplied by the number of operation times *t* per day, considering that the appliance is in operation at least once a day. Thus, in case the appliance operates only once a day, then *t* = 1.

$$\sum_{h=1}^{24} ST_{ah} = t \qquad (8)$$

The consumption of appliance *pi* can be obtained by multiplying the matrix of all possible operating combinations *Xa* by vector ST_a.

$$x_a = Xa \times ST_a, \forall a \in pi \qquad (9)$$

For a programmable appliance without interruption pwi, we split their operation into distinct parts or tasks, such as: $p1, p2, \ldots, pj$. For instance, in case of washing machine, we may assume that first operation part is related to washing, while second part to rinsing and third part to drying. Therefore, the daily consumption D_a of each appliance pwi is the sum of its parts:

$$D_a = x_{p1} + x_{p2} + \ldots + x_{pj}, \ \forall a \in pwi \tag{10}$$

After the division of pwi operation into distinct parts, each part of programmable appliance without interruption becomes a programmable appliance with interruption pi with its hourly consumption requirement. Therefore, the consumption of part pj of an appliance pwi is:

$$x_{pj} = X_{pj} \times ST_{pj}, \ \forall a \in pwi, \ \forall pj \in a \tag{11}$$

where X_{pj}—matrix of all possible operation combination of part pj;
ST_{pj}—starting time of part pj.
Similar to (8), sum of starting time of each part pj for a certain hour h is equal to t.

$$\sum_{h=1}^{24} ST_{pjh} = t, \ \forall a \in pwi, \ \forall pj \in a \tag{12}$$

However, the starting time of all parts pj should be less than or equal to 1 in order to ensure consecutive operation of the different parts of the same appliance that cannot be interrupted.

$$\sum_{pj} ST_{pj} \leq 1, \ \forall pj \tag{13}$$

Second, the objective function is to minimize the electricity payment (consumers' approach). In this case, the objective function $p(x)$ is:

$$p(x) = \sum_{h=1}^{24} (x_{ah} \times t_h) \tag{14}$$

where t_h—time of use tariff vector.
Storage device is considered either as consumer or generator; its operation constraints are described in the following equations:

$$E_{SD}^{t+1} = E_{SD}^t \times (1 - \sigma) + \left(P_{SD,C} \times \eta_{SD,C} - \frac{P_{SD,D}}{\eta_{SD,D}} \right) \times \Delta t \tag{15}$$

$$0 \leq P_{SD,C} \leq P_{SD,C}^{max} \tag{16}$$

$$0 \leq P_{SD,D} \leq P_{SD,D}^{max} \tag{17}$$

$$E_{SD}^{min} \leq E_{SD}^t \leq E_{SD}^{max} \tag{18}$$

where E_{SD}^t and E_{SD}^{t+1} represent the output of SD at t and $t + 1$;
σ is the loss ratio;
$P_{SD,C}$ is charging power of SD, $P_{SD,D}$ is discharging power of SD;
$P_{SD,C}^{max}$ is the maximum charging capacity, $P_{SD,D}^{max}$ is the maximum discharging capacity;
E_{SD}^{min} and E_{SD}^{max} represent the minimum and maximum output of SD;
$\eta_{SD,C}$ and $\eta_{SD,D}$ represent the charging and discharging efficiency of the SD.
To avoid simultaneous charging/discharging, the following constraints have been added:

$$\text{IF } P_{SD,C} > 0 \text{ THEN } P_{SD,D} = 0 \tag{19}$$

$$\text{IF } P_{SD,D} > 0 \text{ THEN } P_{SD,C} = 0 \qquad (20)$$

When considering PV and SD into the objective function, Equations (1) and (14) become:

$$\sum_{a \in HA} x_{ah} - E_{PV}^h + E_{SD}^h \leq f(x_h), \ \forall h \in \{1, \ldots, 24\} \qquad (21)$$

$$p(x) = \sum_{h=1}^{24} \left(x_{ah} - E_{PV}^h + E_{SD}^h \right) \times t_h \qquad (22)$$

where E_{PV}^h represents the output of PV at hour h;
E_{SD}^h represents the output of SD at hour h;
t_h—time of use tariff vector.

The electricity consumption optimization is a *Mixed Integer Linear Programming* (MILP) problem being defined by: variables, objective function, lower/upper bounds and other constraints in terms of equations. MILP restricts the relaxed solution of the linear programming problem due to the fact that part of variables should be integers. To solve the linear programming problem, we use *intlinprog* optimization function that can be modeled as:

$$\min f(x_h) \text{ or } p(x_h), \begin{cases} A \times x_h \leq b \\ Aeq \times x_h = beq \\ lb \leq x_h \leq ub \end{cases} \qquad (23)$$

where:

- $f(x_h)$ is hourly consumption of all appliances;
- $p(x_h)$ is the hourly consumption multiplied by ToU tariff hourly rates;
- $A \times x_h \leq b$ linear inequalities system. A—$k \times n$ matrix; k—number of inequalities; n—number of variables; b—k length vector;
- $Aeq \times x_h = beq$ linear equations system. Aeq—$m \times n$ matrix; m—number of equations; n—number of variables; beq—m length vector;
- $lb \leq x_h \leq ub$ each element of vector x_h should be higher than a lower bond (lb) and lower than an upper bound (ub).

For building the optimization model, we follow the steps: Identifying variables, setting the objective function, formulation of lower and upper bonds and definition of linear equations and inequalities systems. The *intlinprog* function requires the objective function modeling, setting the integer variables, bounds and constraints: f, $intcon$, A, Aeq, b, beq, ub, lb.

% intlinprog function call
[x,fval,exitflag,output] = intlinprog(f,intcon,A,b,Aeq,beq,lb,ub)

The advantage of this function is that it allows the imposing of some integer variables that characterize the electricity consumption.

The second optimization function is *fminimax* that is based on reaching the goal, considering it as daily average consumption. The problem of reaching the goal considering the minimization of the maximum variable can be formulated as:

$$\min_{h} \max (f(x_h) - goal) = \min_{h} \max \left(f(x_h) - \frac{T_c}{24} \right) \qquad (24)$$

3. Appliances Modeling and Optimization Results

First approach is to minimize the consumption peak. At first sight, it may be unattractive for consumers, but on long term it can sustainably reduce the electricity consumption payment due to the fact that by flattening the load curve overloading of the grids can be avoided. Second approach is to minimize the electricity payment that will significantly modify the load curve, but not in the flattening way due to the fact that most of the consumers will change their behavior by consuming when tariff rates are lower. Therefore, it may lead to new peaks that impose the need for a new tariff scheme in a chain reaction. After the new tariffs will be in force, the consumers again will tend to shift their appliances to operate at lower rates leading to other peaks that can be overcome only by new tariff schemes [24,25] that should discourage the consumption at peak hours.

For estimation of the payment, regardless the objective functions, we have considered a set of six three-level times of use (ToU) tariffs as in Figure 4. They encourage the consumption during night when the hourly rates are lower than the rates applied during the day. The implemented ToU tariffs ToU_B, ToU_C, ToU_D, and ToU_E are considered as in Irish Social Science Data Archive [4], while ToU_A and ToU_F are proposed by the authors considering the same rate structure as in [4] for testing purposes. ToU_A has similar rates without significant differences among rates, characterized by mildest slopes, while ToU_F has the highest differences between peak and off-peak rates, with sharp slopes. Tariffs, that represent a vector of 24 values that correspond to the 24 h, can be easily replaced with a more advanced tariff systems such as multiple-level ToU tariff, critical peak pricing tariff that is also based on different daily ToU tariffs or real-time tariff. The peak hours from the 1–24 h of the time intervals, 18 and 19 are the most expensive ones, while between 24–8, the electricity rate is less expensive.

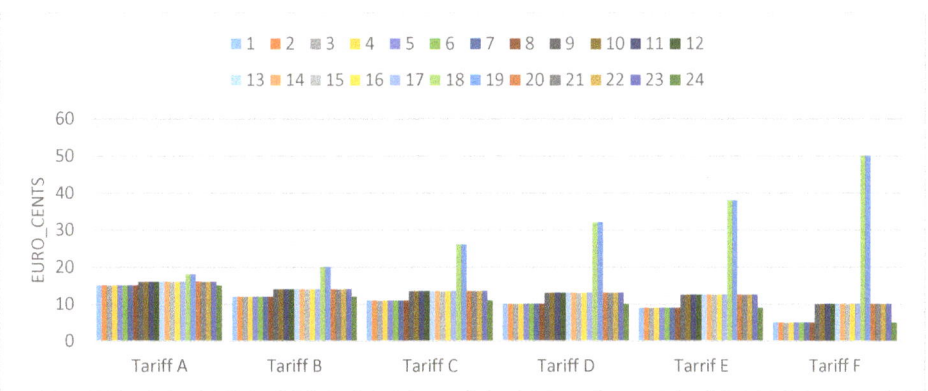

Figure 4. Hourly electricity time of use tariffs.

Before optimization, the peak is 2.55 kWh and the electricity payment is between 3.34 € (with ToU B) and 3.72 € (ToU_A).

The optimization results in case the objective function is minimization of the consumption peak are shown in Figure 5 (cumulative consumption) and Figure 6 (individual consumption).

The consumption of individual appliances is uniformly distributed during the 24 h so that the consumption peak is flattened based on the operation constraints of the programmable appliances and the consumer flexibility. Heating, car battery and water heater operate at night, due to the fact that they are programmable and do not involve consumer. In addition, the washing machine is programmed to operate in the afternoon even if the consumer is not at home. In the evening, non-programmable appliances such as: TV, vacuum cleaner and electric oven usually operate as they require the involvement of the consumer. After the optimization function is applied, we obtain the hourly schedule for each appliance for the entire day as in Figure 6.

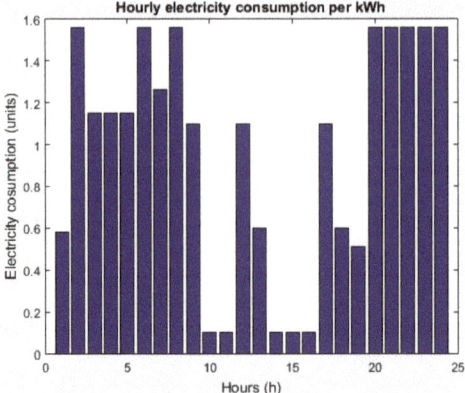

Figure 5. Daily consumption curve. Peak minimization with *intlinprog*.

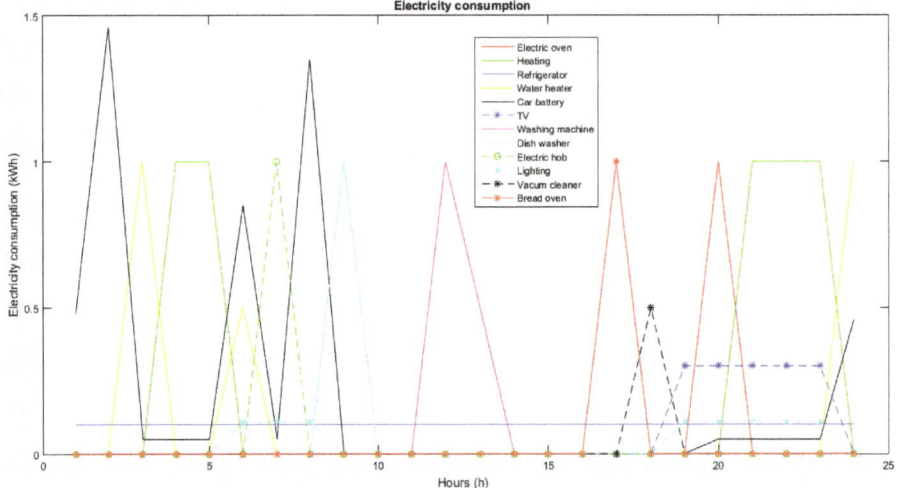

Figure 6. Individual consumption. Peak minimization with *inlinprog*.

In this scenario, the consumption peak is 1.56 kWh and the electricity payment varies between 2.196 € (Tou_F) and 3.632 € (ToU_A). Comparing to "without optimization case", the consumption peak reduces by around 39%, while the electricity payment reduces between 2.4% (ToU_A) and 31.1% (ToU_F).

In case the objective function is minimization of the electricity payment, the optimization results are presented in Figure 7 (cumulative consumption) and Figure 8 (individual consumption). Most of the appliances operate at lower tariff rate time intervals. Therefore, during the day only non-programmable appliances (such as refrigerator) operate. The consumption is moved from day and evening to night hours due to the ToU tariffs rates differences.

The consumption peak is 3 kWh and electricity payment is between 1.796 € (ToU_F) and 3.582 € (ToU_A). The peak consumption is almost double compared with the case when the objection function is minimization of the consumption peak. The electricity payment is less by 3.8% (ToU_A) and 50.2% (ToU_F) than in the initial case. The peak consumption is even higher compared with the no optimization case. Therefore, this approach is just a temporary solution that is not able to ensure sustainable development of the power system. It can just provide an overview on flexibility of the consumers and their mobility to react to different ToU tariffs. After the optimization function is applied, we obtain the hourly schedule for each appliance for the entire day as in Figure 8.

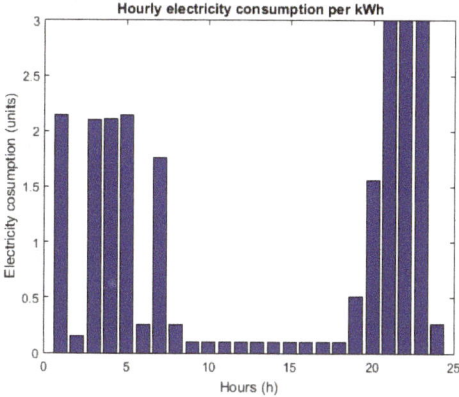

Figure 7. Daily consumption curve. Payment minimization with *intlinprog*.

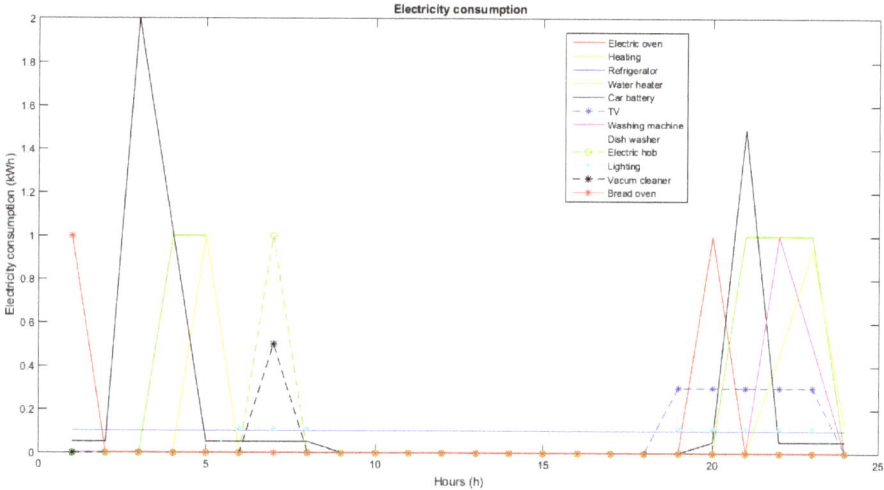

Figure 8. Individual consumption. Payment minimization with *inlinprog*.

Although the results obtained with *intlinprog* function are satisfactory, we also applied multi-objective *fminimax* Matlab function due to the fact that the hourly consumption of all appliances form an array of values that should be minimized.

The optimization results with *fminimax* function are presented in Figure 9 (cumulative consumption) and Figure 10 (individual consumption).

Comparing with *intlinprog* function, although the consumption peak is the same (1.56 kWh), the hourly schedule of the appliances does not comply with their characteristics and consumer's preferences due to the fact that no integer values could be obtained; with *fminimax* function, imposing integers is not possible.

Therefore, the minimization of the consumption peak as objective function and *intlinprog* as optimization function are the most appropriate combination to continue the simulations.

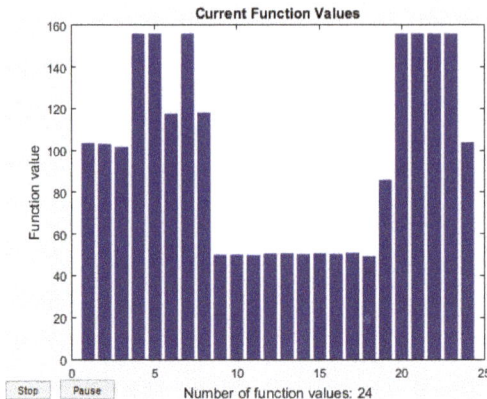

Figure 9. Daily consumption curve. Peak minimization with *fminimax*.

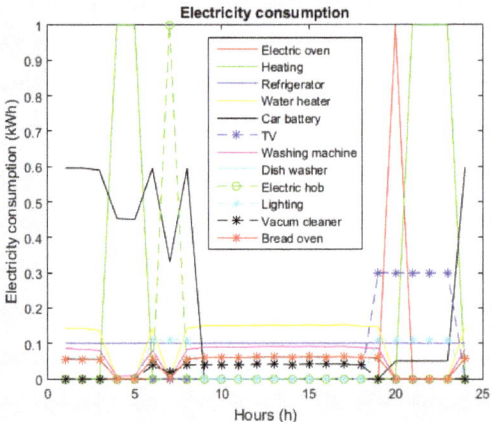

Figure 10. Individual consumption. Peak minimization with *fminimax*.

In addition to the twelve appliances, we have considered the RES distributed generation (PV) and a storage device. Thus, on top of the twelve appliances, we have added a PV with a total daily output of 2.3 kWh. Hourly minimum/maximum generated energy is around 0.05/0.2 kWh.

For simulations, the day-ahead PV forecast is required. Since it is predictable and mainly depends on solar irradiation, it can be done easily by ARIMA models or even feed-forward neural networks as in [26,27].

In case we consider the minimization of consumption peak, with PV the following results have been obtained: Consumption peak 1.56 kWh, electricity payment varies between 2.378 € (ToU_F) and 3.295 € (ToU_A). In Figure 11 the cumulative daily consumption curve and in Figure 12 the individual consumption curves for each appliance plus PV as negative consumption are shown.

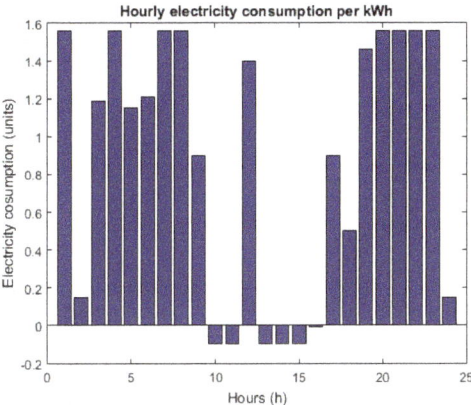

Figure 11. Daily consumption curve with PV. Peak minimization with *intlinprog*.

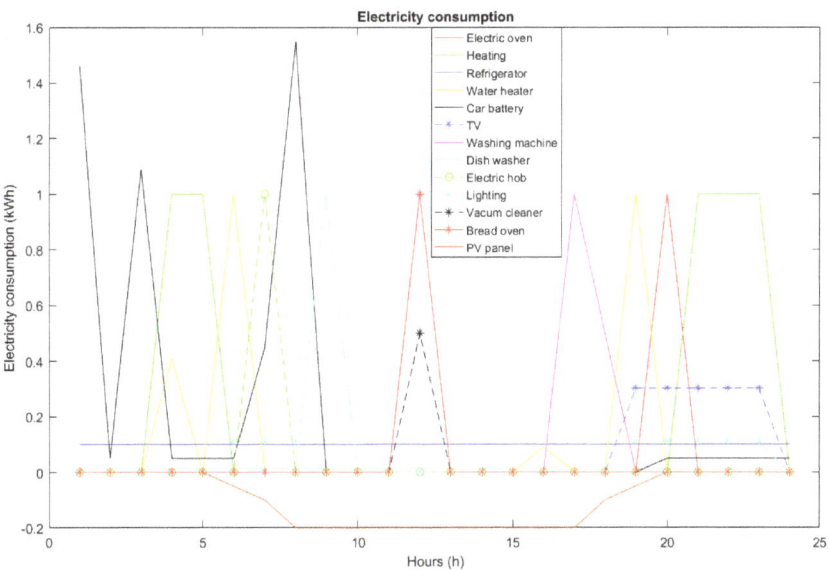

Figure 12. Individual consumption with PV. Peak minimization with *intlinprog*.

It can be noticed that the consumption peak remains unchanged, while the electricity payment reduces by 11.5% (ToU_A) and 34.1% (ToU_F) as a consequence of the local generation that reduces the consumption from the grid leading to less electricity payment.

Then, we have added the storage device with high efficiency (90%) that stores the energy generated by PV. It operates in charging/discharging alternative cycles considering the objective function of the optimization model. The maximum charging capacity is 0.1 kW, maximum consumed energy is 1 kWh, while for discharging and the maximum generated energy is 0.9 kWh. In Figure 13 (cumulative consumption) and Figure 14 (individual consumption), the optimization results with PV and SD are shown. The daily consumption curve indicates the reduction of the consumption peak, but at some time intervals consumption is fed by PV/ discharging of the SD (especially during 14 and 18 h).

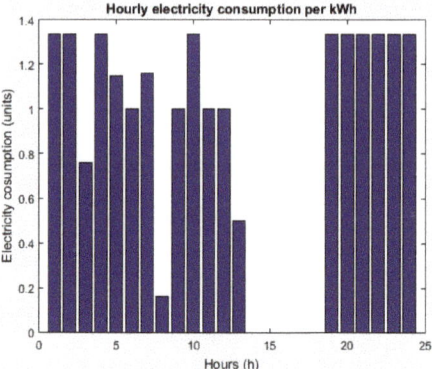

Figure 13. Daily consumption curve with PV & SD. Peak minimization with *intlinprog*.

Consumption of the appliances has been uniformly distributed considering their features, consumer' preferences and the objective: minimization of the consumption peak as in Figure 14.

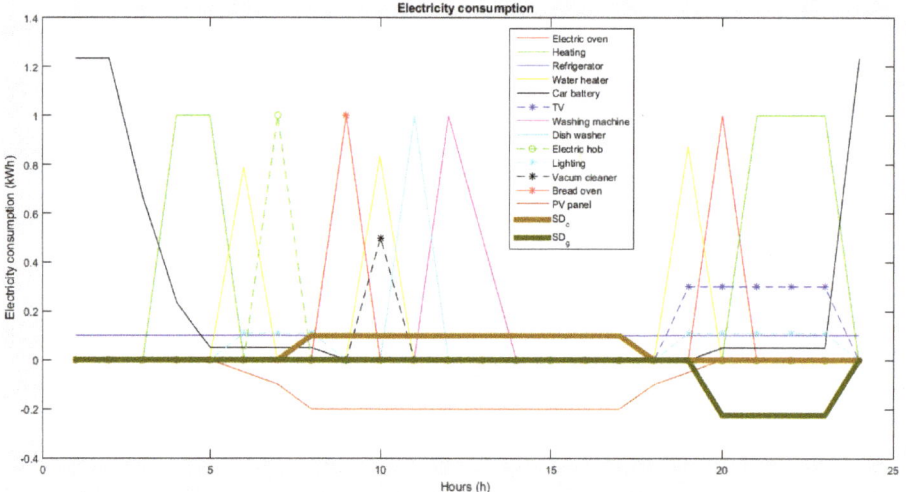

Figure 14. Individual consumption with PV & SD. Peak minimization with *intlinprog*.

In this case as a consequence of adding SD, the consumption peak decreases from 1.56 kWh to 1.34 kWh. Also, the electricity payment varies between 2.164 € (ToU_F) and 3.304 € (ToU_A).

Out of these analyses, the presence of the SD brings benefits from the consumption peak and electricity payment point of view, therefore both grid operators and consumers win when SD is installed, especially when ToU_F is implemented. In Table 2, the results of the comparative analysis with/without optimization based on the same ToU tariffs A and F, PV and SD scenarios in terms of consumption peak and payment are presented. The two ToU tariffs (A and F) give the extreme values in terms of consumption peak and payment. Tariffs ToU_E and ToU_F offer the best results in terms of electricity payment minimization (as in Table 3), especially in the scenario consumption peak minimization with PV and SD, applying *intlinprog* function.

Table 2. Comparative analysis with/without optimization, PV & SD scenarios, considering ToU_A and ToU_F.

Scenarios	Peak [kWh]	ToU_A €	ToU_F €	Peaks [%]	Payment ToU_A [%]	Payment ToU_F [%]
Without optimization	2.55	3.72	3.61	-	-	-
Peak min. *intlinprog*	1.56	3.63	2.20	−38.8	−2.4	−31.1
Payment min. *intlinprog*	3.00	3.58	1.80	17.7	−3.8	−50.2
Peak min. with PV. *intlinprog*	1.56	3.30	2.38	−38.8	−11.5	−34.1
Peak min. with PV & SD. *intlinprog*	1.34	3.30	2.16	−47.5	−11.2	−34.2

Table 3. Comparative analysis considering the six three-level ToU tariffs.

Scenarious/ToU	[kWh]/Euro Cents	ToU_A	ToU_B	ToU_C	ToU_D	ToU_E	ToU_F
Without Optimization	Peak	2.550	2.550	2.550	2.550	2.550	2.550
	Expenses	3.722	3.340	3.459	3.578	3.698	3.607
Min. Payment No PV	Peak	3.000	3.000	3.000	3.000	3.000	3.000
	Expenses	3.582	2.985	2.831	2.676	2.522	1.796
Min. Peak	Peak	1.56	1.56	1.56	1.56	1.56	1.56
	Expenses	3.632	3.095	2.993	2.891	2.790	2.196
Min. Peak With PV	Peak	1.56	1.56	1.56	1.56	1.56	1.56
	Expenses	3.295	2.853	2.825	2.797	2.769	2.378
Min. Peak With PV & SD	Peak	1.34	1.34	1.34	1.34	1.34	1.34
	Expenses	3.304	2.840	2.773	2.707	2.640	2.164

As in [24], the optimum capacity of SD could be established considering several consumption base case scenarios. This approach is based on the assumption that the residential consumers' goal is to minimize the electricity payment. Therefore, the electricity consumers' flexibility in terms of shifting appliances is rewarded by diminishing the payment using the SD.

In Figure 15, the daily consumption curves with *inlinprog* Matlab function with the two objective functions (minimization of consumption peak and minimization of electricity payment), then adding the PV and SD to the initial scenario are graphically depicted.

In the initial scenario with twelve appliances, we obtain the best results with minimization of consumption peak objective function, then, by adding the PV and SD the results significantly improved.

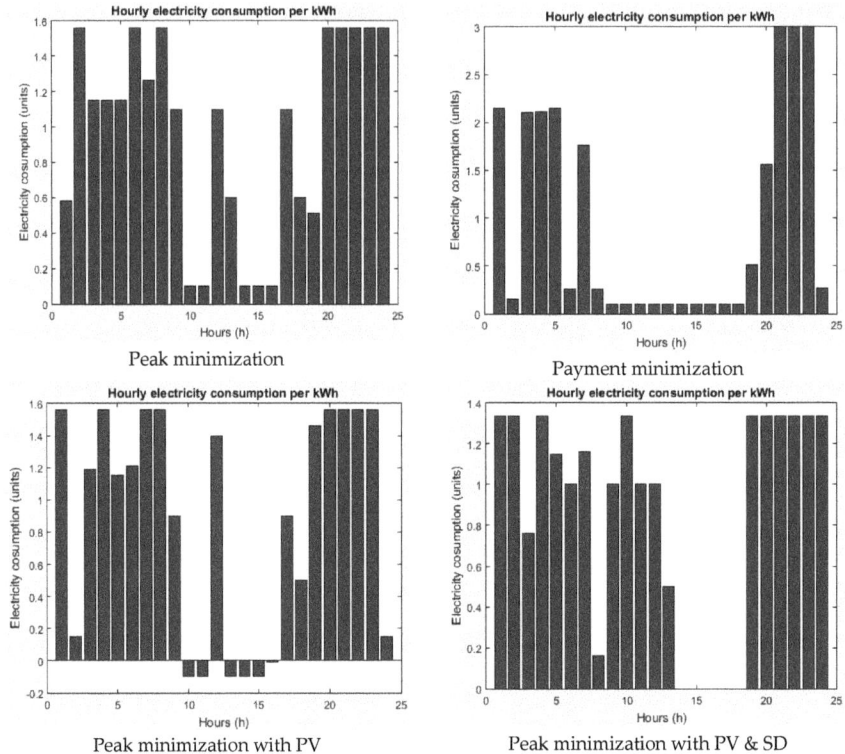

Figure 15. Daily consumption curves with *intlinprog*.

4. Conclusions

With smart grids technologies such as smart meters, sensors and actuators, the consumption optimization at the residential consumers is nowadays possible. The input data is gathered through a set of sensors based on Arduino components for monitoring and controlling the household appliances and collecting the data into a relational database running Oracle 12 Database. The optimization process considering the type of the appliances, reads the initial scheduler and optimizes their consumption based on two objective functions: Minimization of consumption peak and minimization of electricity payment. The results lead to the conclusion that only minimization of the consumption peak can bring sustainable development of the power systems. In addition, we performed simulations with two Matlab functions: *inlinprog* and *fminimax*; comparing the results, *intlinprog* function is the most appropriate optimization function for reaching our goal.

Then, on top of the twelve applications, we added PV panels and a storage device in order to investigate their effect on the consumption curve. In this scenario, the consumption peak reduced by 47.5% from 2.55 kWh (no optimization) to 1.34 kWh. Also, the daily electricity payment reduces by 11%. Although the highest electricity payment reduction is reached when the objective function is the minimization of the electricity payment (around 50%), the consumption peak in this case increased by 18%, that leads to non-uniformly loading of the power grid and additional stress that in the end new peaks may occur. Therefore, new advanced tariff scheme are needed in order to overcome the new peaks. Thus, it is demonstrated that on the long term only the minimization of the consumption peak lead to sustainable electricity payment minimization.

The presence of PV and SD is benefic for both grid operators and consumers due to the fact that consumption peak and electricity payment are diminished compared with initial scenario. Although the initial proposed scenarios reduce the consumption peak and electricity payment, presence of PV and SD will motivate the consumers to optimize the operation scheduling of their appliances since it leads to more savings.

In this paper, modeling of different types of appliances is performed, dividing the appliances in three categories based on their operation mode. Then, the Arduino based sensors controlling and monitoring of data generated by appliances are depicted, the processing of data being performed by several procedures and functions written in PLSQL language included in a proposed informatics solution architecture. Two objective functions: minimization of electricity payment and minimization of consumption peak are considered in order to identify the impact of PV and SD in different scenarios with six three-level ToU tariffs. Out of the two objective functions, the minimization of consumption peak proved to lead to a sustainable development due to the fact that by flattening the load curve overloading of the grids can be avoided. Second approach is to minimize the electricity payment that may lead to new peaks since the consumers will tend to shift their appliances to low rate tariffs. Out of simulations, the ToU tariff pattern that rewards and could motivate best the consumers that are willing to change the electricity consumption behavior is identified.

One limitation of our approach consists in assuming that the consumers are willing to involve in optimization process that is part of a complex demand response program (DR). However, this is not always the case. From different reasons such as (lack of willingness, refrain from revealing private information or participation denial), some consumers refuse to participate in the DR. In this case the DR should be improved, monitored and its performance periodically evaluated.

Another limitation consists in difficult implementation since sensors, storage devices and smart meters technologies are not yet implemented at large scale. However, this aspect will be overcome in the near future by the progress of new technologies with gradually decreasing investment cost and proper policy regulation.

Another limitation is related to the large volume of data flows that require real-time processing and big data technologies. The electricity suppliers or grid operators should develop or acquire new IC & T solutions that able to process large volume of semi- or unstructured data in order to rapidly react to certain circumstances. Nowadays, only few grid operators take advantage of big data technologies, but in short time perspective, the other smaller companies would benefit from cloud solutions that are more affordable and do not require expensive investment in hardware infrastructure.

The acquisition of the SD by grid operator can be a solution for their integration as planning and operation tool and improvement of the power system operation. This approach would incentive the electricity consumers to use SD. On one hand, the grid operator could benefit from the economy of scale by acquiring SD and should easily remotely control the SD with similar characteristics. On the other hand, the cost of SD is gradually decreasing. However, the investment decision should be taken based on a detailed feasibility study.

Author Contributions: All authors contributed equally to this work.

Funding: This research was funded by Executive Unit for Financing Higher Education, Research, Development and Innovation (UEFISCDI) grant number PN-III-P2-2.1-PTE-2016-0032, 4PTE/06/10/2016, PNIII—PTE 2016.

Acknowledgments: This paper presents some results of the research project: Informatics solutions for optimizing the operation of photovoltaic power plants (OPTIM-PV), project code: PN-III-P2-2.1-PTE-2016-0032, 4PTE/06/10/2016, PNIII—PTE 2016.

Conflicts of Interest: The authors declare no conflict of interest.

References

1. European Commission. Energy Union Package 2015. Available online: https://ec.europa.eu/energy/sites/ener/files/publication/FOR%20WEB%20energyunion_with%20_annex_en.pdf (accessed on 14 April 2018).

2. Final 10-year ETIP SNET R&I Roadmap (2017–2026). Available online: https://etip-snet.eu/pdf/Final_10_Year_ETIP-SNET_R&I_Roadmap.pdf (accessed on 14 April 2018).
3. ENTSO/E R&I Roadmap 2017–2026. Available online: http://riroadmap.entsoe.eu/wp-content/uploads/2016/06/entsoe_ri_roadmap_2017-2026.pdf (accessed on 14 April 2018).
4. ISSDA Irish Social Science Data Archive. Available online: http://www.ucd.ie/issda/data/commissionforenergyregulationcer (accessed on 5 June 2017).
5. Liu, T.; Chen, S.; Liu, Y.; Xu, Z.; Che, Y.; Duan, Y. SHE: Smart home energy management system for appliance identification and personalized scheduling. In Proceedings of the 2014 ACM International Joint Conference on Pervasive and Ubiquitous Computing, Seattle, WA, USA, 13–17 September 2014; ACM: New York, NY, USA, 2014; pp. 247–250.
6. Ye, C.; Miao, S.; Lei, Q.; Li, Y. Dynamic Energy Management of Hybrid Energy Storage Systems with a Hierarchical Structure. *Energies* **2016**, *9*, 395. [CrossRef]
7. Longe, O.M.; Ouahada, K.; Rimer, S.; Harutyunyan, A.N.; Ferreira, H.C. Distributed Demand Side Management with Battery Storage for Smart Home Energy Scheduling. *Sustainability* **2017**, *9*, 120. [CrossRef]
8. Pîrjan, A.; Oprea, S.-V.; Căruțașu, G.; Petroșanu, D.-M.; Bâra, A.; Coculescu, C. Devising Hourly Forecasting Solutions Regarding Electricity Consumption in the Case of Commercial Center Type Consumers. *Energies* **2017**, *10*, 1727. [CrossRef]
9. Wang, G.; Ciobotaru, M.; Agelidis, G.A. Power smoothing of large solar PV plant using hybrid energy storage. *IEEE Trans. Sustain. Energy* **2014**, *5*, 834–842. [CrossRef]
10. Pei, L.; Huang, Z.; Dong, D. A hierarchical management system for energy storage batteries. *Energy Storage Sci. Technol.* **2014**, *3*, 416–422. [CrossRef]
11. Wang, Y.; Huang, Y.; Wang, Y.; Li, F.; Zhang, Y.; Tian, C. Operation Optimization in a Smart Micro-Grid in the Presence of Distributed Generation and Demand Response. *Sustainability* **2018**, *10*, 847. [CrossRef]
12. Jayalakshmi, N.S.; Gaonkar, D.N.; Vikash Kumar, J.; Karthik, R.P. Battery-ultracapacitor storage devices to mitigate power fluctuations for grid connected PV system. In Proceedings of the 2015 Annual IEEE India Conference (INDICON), New Delhi, India, 17–20 December 2015; IEEE: Piscataway, NJ, USA, 2015.
13. Michalitsakos, P.; Mihet-Popa, L.; Xydis, G. A Hybrid RES Distributed Generation System for Autonomous Islands: A DER-CAM and Storage-Based Economic and Optimal Dispatch Analysis. *Sustainability* **2017**, *9*, 2010. [CrossRef]
14. Seo, H.-C. Development of Reclosing Method in a Distribution System with Distributed Generation and Battery Energy Storage System. *Energies* **2018**, *11*, 1407. [CrossRef]
15. Jurasz, J.; Piasecki, A. A simulation and simple optimization of a wind-solar-hydro micro power source with a battery bank as an energy storage device. *Energy Fuels* **2016**, *14*, 1–10. [CrossRef]
16. Hsu, Y.-L.; Chou, P.-H.; Chang, H.-C.; Lin, S.-L.; Yang, S.-C.; Su, H.-Y.; Chang, C.-C.; Cheng, Y.-S.; Kuo, Y.-C. Design and Implementation of a Smart Home System Using Multisensor Data Fusion Technology. *Sensors* **2017**, *17*, 1631. [CrossRef] [PubMed]
17. Oprea, S.-V.; Pîrjan, A.; Căruțașu, G.; Petroșanu, D.-M.; Bâra, A.; Stănică, J.-L.; Coculescu, C. Developing a Mixed Neural Network Approach to Forecast the Residential Electricity Consumption Based on Sensor Recorded Data. *Sensors* **2018**, *18*, 1443. [CrossRef] [PubMed]
18. Yang, L.; Lu, Y.; Xiong, L.; Tao, Y.; Zhong, Y. A Game Theoretic Approach for Balancing Energy Consumption in Clustered Wireless Sensor Networks. *Sensors* **2017**, *17*, 2654. [CrossRef] [PubMed]
19. Zheng, M.; Meinrenken, C.J.; Lackner, K.S. Smart households: Dispatch strategies and economic analysis of distributed energy storage for residential peak shaving. *Appl. Energy* **2015**, *147*, 246–257. [CrossRef]
20. Reihani, E.; Motalleb, M.; Ghorbani, R.; Saoud, L.S. Load peak shaving and power smoothing of a distribution grid with high renewable energy penetration. *Renew. Energy* **2016**, *86*, 1372–1379. [CrossRef]
21. Shirazi, E.; Jadid, S. Cost reduction and peak shaving through domestic load shifting and DERs. *Energy* **2017**, *124*, 146–159. [CrossRef]
22. Ampimah, B.C.; Sun, M.; Han, D.; Wang, X. Optimizing sheddable and shiftable residential electricity consumption by incentivized peak and off-peak credit function approach. *Appl. Energy* **2018**, *210*, 1299–1309. [CrossRef]
23. Uddin, M.; Romlie, M.F.; Abdullah, M.F.; Halim, S.A.; Bakar, A.H.A.; Kwang, T.C. A review on peak load shaving strategies. *Renew. Sustain. Energy Rev.* **2018**, *82*, 3323–3332. [CrossRef]

24. Oprea, S.V.; Bâra, A.; Cebeci, E.M.; Tor, O.B. Promoting peak shaving while minimizing electricity consumption payment for residential consumers by using storage devices. *Turk. J. Electr. Eng. Comput. Sci.* **2017**, *25*, 3725–3737. [CrossRef]
25. Oprea, S.V. Informatics solutions for electricity consumption optimization. In Proceedings of the IEEE 16th International Symposium on Computational Intelligence and Informatics, Budapest, Hungary, 19–21 November 2015; IEEE: Piscataway, NJ, USA, 2015; pp. 193–198.
26. Oprea, S.V.; Bara, A.; Carutasu, G.; Pirjan, A. Prosumers' Renewable Small-Size Generation Forecasting Analyses with ARIMA Models. In Proceedings of the ECAI 2016—International Conference–8th Edition Electronics, Computers and Artificial Intelligence, Ploiesti, Romania, 30 June–02 July 2016; IEEE: Piscataway, NJ, USA, 2016.
27. Lungu, I.; Bâra, A.; Căruțașu, G.; Pîrjan, A.; Oprea, S.V. Prediction intelligent system in the field of renewable energies through neural networks. *Econ. Comput. Econ. Cybern.* **2016**, *50*, 85–102.

© 2018 by the authors. Licensee MDPI, Basel, Switzerland. This article is an open access article distributed under the terms and conditions of the Creative Commons Attribution (CC BY) license (http://creativecommons.org/licenses/by/4.0/).

Article

Energy Management and Optimization of a PV/Diesel/Battery Hybrid Energy System Using a Combined Dispatch Strategy

Ali Saleh Aziz [1], Mohammad Faridun Naim Tajuddin [1,*], Mohd Rafi Adzman [1], Makbul A. M. Ramli [2] and Saad Mekhilef [3]

1. Centre of Excellence for Renewable Energy, School of Electrical Systems Engineering, Universiti Malaysia Perlis, Perlis 02600, Malaysia; iraq_1991@yahoo.com (A.S.A.); mohdrafi@unimap.edu.my (M.R.A.)
2. Department of Electrical and Computer Engineering, King Abdulaziz University, Jeddah 21589, Saudi Arabia; mramli@kau.edu.sa
3. Power Electronics and Renewable Energy Research Laboratory (PEARL), Department of Electrical Engineering, University of Malaya, Kuala Lumpur 50603, Malaysia; saad@um.edu.my
* Correspondence: faridun@unimap.edu.my; Tel.: +60-4988-5601

Received: 29 November 2018; Accepted: 22 January 2019; Published: 28 January 2019

Abstract: In recent years, the concept of hybrid energy systems (HESs) is drawing more attention for electrification of isolated or energy-deficient areas. When optimally designed, HESs prove to be more reliable and economical than single energy source systems. This study examines the feasibility of a combined dispatch (CD) control strategy for a photovoltaic (PV)/diesel/battery HES by combining the load following (LF) strategy and cycle charging (CC) strategy. HOMER software is used as a tool for optimization analysis by investigating the techno-economic and environmental performance of the proposed system under the LF strategy, CC strategy, and combined dispatch CD strategy. The simulation results reveal that the CD strategy has a net present cost (NPC) and cost of energy (COE) values of $110,191 and $0.21/kWh, which are 20.6% and 4.8% lower than those of systems utilizing the LF and CC strategies, respectively. From an environmental point of view, the CD strategy also offers the best performance, with CO_2 emissions of 27,678 kg/year. Moreover, the results show that variations in critical parameters, such as battery minimum state of charge, time step, solar radiation, diesel price, and load growth, exert considerable effects on the performance of the proposed system.

Keywords: combined dispatch (CD) strategy; optimization; HOMER; net present cost (NPC); sensitivity analysis

1. Introduction

Progressive energy demand growth and rapid depletion of fossil fuels have raised concerns of future energy supplies [1]. Moreover, the utilization of conventional energy sources from fossil fuels has resulted in tremendous increases in CO_2 emissions, which are the primary cause of global warming [2,3]. The atmospheric CO_2 concentration has increased by approximately 40% compared with that at the beginning of the industrial revolution [4]. These issues have led to the determination of alternative energy sources with the potential to reduce pollution and produce a sustainable energy supply [5]. The utilization of renewable energy sources (RESs) has grown exponentially over the last few years. By the end of 2017, approximately 26.5% of the total global electricity produced came from RES, as shown in Figure 1 [6]. RESs, such as solar, wind, hydropower, geothermal, and biomass, are easily replenished and environmentally friendly. However, the major drawback of RESs is the unpredictable and intermittent nature of power generation. This issue can be overcome by integrating different energy sources to produce a hybrid energy system (HES), which can solve reliability problems and provide an environmentally friendly solution [7–9]. Various power sources and storages, such as diesel generators, batteries, and supercapacitors, should be integrated into HESs to improve system stability and smooth out fluctuations [10,11].

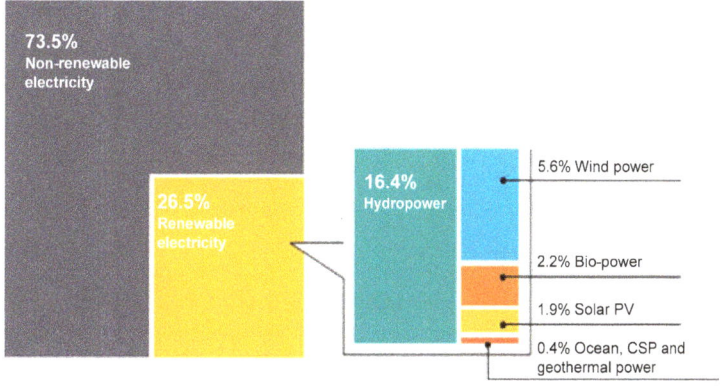

Figure 1. Energy mix of the global electricity generation by the end of 2017 [6].

The optimal design of HESs requires appropriate sizing of the components and a feasible energy management strategy. Selection of a suitable energy management strategy is critical because it determines the behavior of the system by controlling the flow of energy and deciding the priority of each component in the system [12]. The proper energy management strategy is therefore able to improve the power system stability, ensure the continuity of power supply, minimize the cost of energy (COE), and protect components against damage due to overloads. Moreover, in grid-connected systems, the energy management strategy is vital to control the flow of energy from and to the grid and for metering purposes [13,14]. Depending on the system configuration and the optimization objectives, different energy management strategies are performed based on various technical and economic criteria. These strategies can be less or more complicated, requiring the utilization of less or more complicated optimization algorithms [15].

The energy management strategy is conducted by selecting, installing, and programming a central controller to manage the flow of energy according to an optimized strategy. Various mathematical programs and optimization techniques are employed to design and plan energy management strategies. These strategies include off-grid and grid-connected HESs as presented in Figure 2 [14].

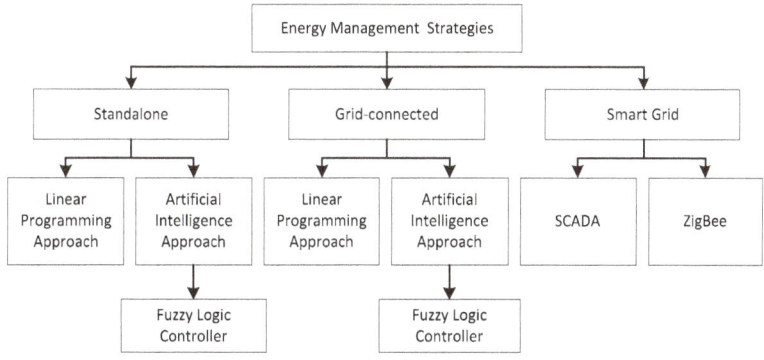

Figure 2. Commonly used approaches for energy management strategies [14].

Although the non-dispatchable RESs necessitate complex systems modeling, they are themselves simple to model without the need for control logic, in which the power is merely produced in direct response to the available renewable energy resources. On the other hand, the dispatchable components, such as the generator and battery, are more difficult to model because they must be controlled to achieve matching of demand and supply properly, and to compensate for the intermittency of the non-dispatchable RESs [16,17]. "Should a generator charge the battery or should excess solar production be responsible for charging the battery only and the generator used only to satisfy the load?" This question has led to some rules that manage the operation of generator and battery energy storage. The load following (LF) and cycle charging (CC) dispatch strategies are used to control generator operation and battery energy storage. Under the LF strategy, a generator produces only enough power to satisfy the load demand and does not charge the batteries. Batteries are charged only by excess electricity from RESs. Under the CC strategy, the generator operates at its maximum rated capacity whenever it is switched on to feed the load and charge the battery bank by the excess power [18,19]. In the CC strategy, a setpoint state-of-charge (SOC) can be applied. When a certain setpoint is selected, the system does not stop charging the battery until it reaches the specified setpoint [20–22]. The chosen dispatch strategies are applied only when dispatchable components such as generator and battery are operating simultaneously during a time step [23].

HOMER software that was developed by National Renewable Energy Laboratory (NREL), United States, is considered one of the most powerful tools for the optimization of HESs through performing a techno-economic analysis of the system with a project lifetime of a certain number of years [24]. Different technologies, such as PV, wind turbine, hydrokinetic, hydropower, biomass, conventional generator, battery, flywheel, supercapacitor, and fuel cell, can be addressed with HOMER. HOMER can model both the LF strategy and CC strategy to control the generator and energy storage operation. Both of these strategies make economic dispatch decisions; they choose the most economical way to serve the load at each time step. HOMER simulates hundreds or even thousands of all possible combinations of the energy system and then ranks the results according to the least-cost options that meet the load requirements [25–27].

Many studies throughout the world have used HOMER software to investigate the optimal design of proposed HESs using LF or CC strategies. Halabi et al. [28], Ansong et al. [29], Sigarchian et al. [30], Rezzouk and Mellit [31], and Madziga et al. [32] used the LF strategy to analyze the performance of suggested HESs to find the optimal configurations. Each study proposed certain components that differed from others and the simulation was conducted for a specific area. On the other hand, the authors in [9,33–37] carried out techno-economic analysis to investigate the optimum HES using the CC strategy in which different setpoints SOC were suggested.

Other studies have been conducted to investigate the optimum HES using both the LF strategy and the CC strategy to find the best one for the proposed HESs. Kansara and Parekh [38] performed dispatch analysis of a stand-alone wind/diesel/battery energy system and found that the system achieved better performance for high renewable penetration under the LF strategy; for low renewable penetration, the CC strategy was preferable. The optimal size of the PV/diesel/battery power system under the LF and CC control strategies was investigated in [39], and results showed that the optimum design for the CC strategy had a smaller COE than that of the LF strategy. The authors in [40] presented a techno-economic analysis of the PV/diesel/battery system and concluded that the LF provided better economic performance compared to the CC strategy. A comparison of the LF and CC control strategies for two different HESs was studied in [41], and it was found that for the PV/diesel HES, the CC strategy provided better performance in terms of net present cost (NPC), COE, and greenhouse gas (GHG) emissions, but not in terms of renewable penetration. For a wind/diesel system, the LF dispatch strategy demonstrated better techno-economic and environmental performance than the CC strategy. The authors in [42] investigated the best energy management strategy for HESs in three different places in India and found that the LF strategy achieved better techno-economic and environmental performance than the CC strategy for all locations. The optimum planning and design of the PV/diesel/battery for a school, a small dispensary, and 25 households using both LF and CC control strategies was examined by Fodhil et al. [43], who showed that the LF strategy was more cost effective for the dispensary, while the CC strategy was more cost effective for both the school and households. Moreover, the LF strategy had more PV penetration and less GHG emissions than the CC strategy in all cases. Shoeb and Shafiullah [44] investigated the performance of a PV/diesel/battery HES for standalone electrification of a rural area in southern Bangladesh. They showed that the CC strategy was more economical than the LF strategy. Ghenai and Bettayeb [45] analyzed the feasibility of different off-grid HESs using PV, diesel, battery, and fuel cell to supply electricity for a building in the University of Sharjah, UAE. They reported that for all cases, the use of the CC strategy for power generation was preferable to the LF strategy. Maatallah et al. [46] studied the techno-economic feasibility of a PV/wind/diesel/battery hybrid system for a case study in Bizerte, Tunisia and found that the LF strategy was more cost-effective than the CC strategy. The system under the LF strategy was able to save more than $27,000 during the project lifetime. A study on the viability of supplying electricity through an HES consisting of a PV, diesel generator, biomass plant, and batteries to rural areas in the northern region of Bangladesh was presented in [47]. It was found that the CC strategy was more economical than the LF strategy and it was capable of saving around $12,450 during the project lifetime. On the other hand, the LF strategy caused the cost of energy (COE) to increase from $0.188/kWh to $0.189/kWh.

From the previous works, it has been found that the optimum dispatch strategy is affected by various parameters, such as fuel price, fuel conversion efficiency of the generator, operation and maintenance (O&M) costs, renewable penetration, and battery and generator capacity. Moreover, LF and CC strategies are found to be simple and effective tools for the control of HESs, but they have some drawbacks. By using a generator to charge batteries during operation, the CC dispatch strategy provides more energy to the batteries. This approach shuts down the generator during the low-load period, which results in a reduction of the total operating time of the generator. However, the sequence of the battery charging/discharging process may damage or ruin the battery, which leads to increased replacement costs. On the other hand, when the generator satisfies the load demand without charging the batteries, the LF dispatch strategy ensures that the batteries remain charged by the RES output power only, and this maximizes the use of available renewable energy. Furthermore, in the LF strategy, the utilization of the battery is limited. Hence, a longer life of the battery can be achieved. However, the main drawback of the LF strategy is that for most of the time, the generator works at part load, and this increases the variable cost due to the lower off-design efficiency. In many cases, the system performance under one of these dispatch strategies is much better than that of the other. However, predicting the best strategy before running the model is difficult.

Based on the literature review and the research gap discussed above, the objective of this paper is to examine the techno-economic and environmental performance of an off-grid PV/diesel/battery HES supplying electricity to a small remote village in Iraq using a combined dispatch (CD) strategy. Combining the LF and CC dispatch strategies may help improve the system performance by allowing more efficient utilization of the generator and batteries. The system performance under this strategy is compared with the existing dispatch strategies (LF and CC), and HOMER software is utilized for analysis of the proposed systems. To the best knowledge of the authors, this is the first study investigating the techno-economic and environmental performance of an HES using the CD strategy.

The rest of the paper is organized as follows. Section 2 presents the methodology, which provides the assessment of site specifications, load profile, metrological data, system components, mathematical representation, and the control strategy. Section 3 presents the results obtained, including the optimization and sensitivity analysis. Finally, the conclusion is given in Section 4.

2. Methodology

Successful evaluation of any renewable energy project requires appropriate criteria to be applied in the selected area to ensure that the operational behavior of different scenarios can be analyzed accurately. The following analysis frameworks are adopted in the current work:

- Selected site specifications and load demand data
- Metrological data (solar radiation and temperature)
- Hybrid power system components
- Mathematical representation
- System operational control

2.1. Specification and Load Profile of the Selected Site

The remote rural village selected as a case study in this research is located at Muqdadiyah District, Diyala, Iraq. The village consists of 40 households with total number of people of around 150. Water-wells and hand pumps are the primary sources of water in the area. The village does not have access to grid power, and this situation may open new opportunities for utilization of stand-alone HESs to provide electricity for the area. Similar to most remote rural areas, the residential units of the village require a low electricity supply for electrical appliances and lighting. Figure 3 shows the total daily load profile of the village's households [48]. A 5% random variability is considered for time step to time step and day-to-day analyses to provide better reliability. During the first hours of the day (00:00–06:00), the power consumption is low since the people are asleep. From 06:00–07:00, the load profile shows an increase since most of the people get ready either to leave for schools or their work. Then the load profile decreases again, as most of the family members are outside, and continue until 14:00 when people come back to their homes. Starting from 14:00 the load profile shows a continues increment and the maximum is recorded from 19:00–21:00 since most people are present in their homes; then, the load decreases, which refers to the start of sleep time. A daily total energy demand of 145 kWh/day, a daily average of 6.04 kW, and power peak load of 18.29 kW are found.

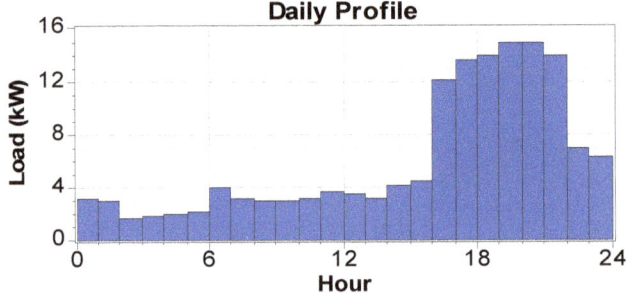

Figure 3. The total daily load profile of the village's households [49].

2.2. Solar Resource and Temperature

Solar radiation and ambient temperature are the two parameters with the most profound effects on the output PV power. In this framework, HOMER software uses the monthly average global horizontal radiation and ambient temperature as input parameters. Section 2.4 provides the mathematical expression of the effects of these parameters on the output PV power. The following points explain the solar radiation and ambient temperature data in the selected site, which were obtained from the NASA website [49].

- The solar radiation and clearance index data for the selected village (33° N latitude and 44° E longitude) are presented in Figure 4. The annual average solar radiation is 5.02 kWh/m²/day, maximum solar radiation (7.56 kWh/m²/day) is recorded in June, and the minimum solar radiation (2.62 kWh/m²/day) is recorded in December. The amount of solar radiation received by the area is relatively high, which indicates that the solar energy system is an attractive power source.

- The ambient temperature plays a vital role in the performance of PV modules. Therefore, accurate measurement of ambient temperature data is essential. The monthly average ambient temperature for the chosen area is illustrated in Figure 5. The summer season shows the highest ambient temperature, at 36.15 °C in July, and the winter season records the lowest ambient temperature, at 9.77 °C in January. In HOMER software, the ambient temperature is taken into consideration when calculating the output PV power.

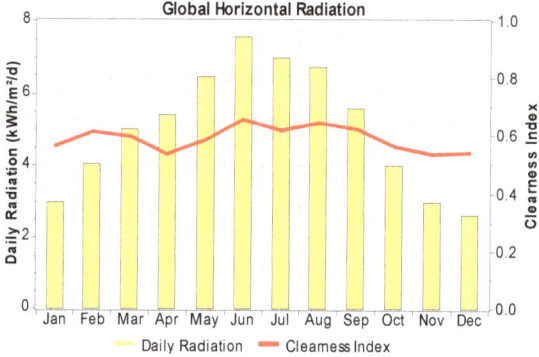

Figure 4. Monthly average solar global radiation and clearance index of the village.

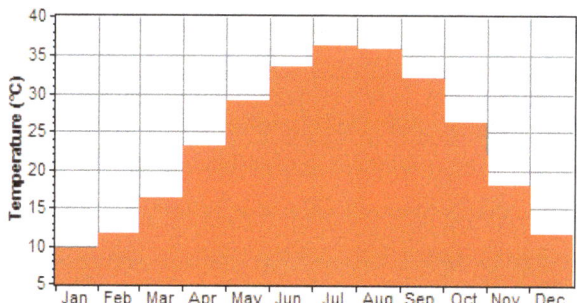

Figure 5. Monthly average temperature of the village.

2.3. System Components

In this research, the proposed HES consists of four components, i.e., the PV system, diesel generator, converter, and batteries. A schematic diagram of the proposed HES is illustrated in Figure 6. The techno-economic input parameters for all components in the HES are explained in detail in Table 1. Note that the technical parameters and costs of the components were obtained from several references [28,50–52]. Regarding the capacity, the optimizer lets HOMER find the optimal sizes for system components.

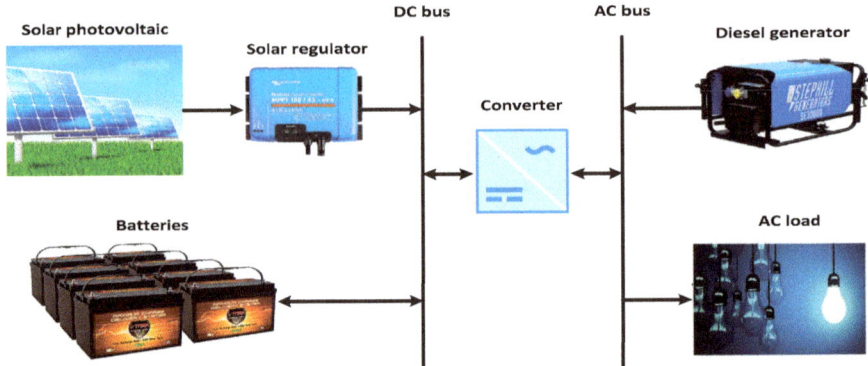

Figure 6. Schematic diagram of PV/diesel/battery hybrid energy system (HES).

Table 1. Input parameters and costs of different components.

Reference	Description	Specification
[50]	**1. PV system**	
	Tracking system	Fixed
	Nominal operating cell temperature	47 °C
	Temperature coefficient	−0.4%/°C
	Efficiency at standard test condition	18%
	Derating factor	80%
	Capital cost	$640/kW
	Operating and maintenance cost	$10/kW/year
	Cost of replacement	$640/kW
	Lifetime	25 years
[28]	**2. Diesel generator**	
	Cost of capital	$220/kW
	Cost of operating and maintenance	$0.03/kW/hour
	Cost of replacement	$200/kW
	Lifetime	15,000 h
[51]	**3. Batteries**	
	Model	Iron Edison Nickel Iron
	Nominal capacity	200 Ah (0.24 kWh)
	Nominal voltage	1.2 V
	Capital cost	$130
	Operating and maintenance cost	$1/year
	Replacement cost	$100
[52]	**4. Converter**	
	Efficiency	90% for inverter, 85% for rectifier
	Cost of capital	$550/kW
	Cost of operating and maintenance	$5/kW/year
	Cost of replacement	$450/kW
	Lifetime	15 years

2.4. Mathematical Representation

2.4.1. Output PV Power

The output PV power is known to be profoundly affected by the amount of the solar radiation available and the temperature. In this respect, the following expression is used in HOMER to compute the output PV power [53,54]:

$$P_{PV} = Y_{PV} f_{PV} \left(\frac{G_T}{G_{T,STC}} \right) [1 + \alpha_p (T_c - T_{c,STC})] \quad (1)$$

where Y_{PV} is the PV rated capacity under standard test conditions (kW), f_{PV} is the derating factor of PV, G_T is the solar radiation incident on the PV array in the current time step (kW/m²), $G_{T,STC}$ is the incident irradiance at standard test conditions (1 kW/m²), α_p is the temperature coefficient of power (%/°C), T_C is the temperature of the PV cell (°C) in the current time step, and $T_{C,STC}$ is the temperature of the PV cell under standard test conditions (25 °C).

The PV cell temperature is calculated using the following expression:

$$T_c = \frac{T_a + (T_{c,NOCT} - T_{a,NOCT})\left(\frac{G_T}{G_{T,NOCT}}\right)\left[1 - \frac{\eta_{mp,STC}(1-\alpha_p T_{c,STC})}{\tau\alpha}\right]}{1 + (T_{c,NOCT} - T_{a,NOCT})\left(\frac{G_T}{G_{T,NOCT}}\right)\left(\frac{\alpha_p \eta_{mp,STC}}{\tau\alpha}\right)} \tag{2}$$

where T_a is the ambient temperature (°C), $T_{C,NOCT}$ is the nominal operating cell temperature (NOCT) in (°C), $T_{a,NOCT}$ is the ambient temperature at which the NOCT is defined (20 °C), $G_{T,NOCT}$ is the solar radiation at which the NOCT is defined (0.8 kW/m²), $\eta_{mp,STC}$ is the maximum power point efficiency under standard test conditions (%), τ is the solar transmittance of any cover over the PV array (%), and α is the solar absorptance of the PV array (%).

2.4.2. Economic Model

Considering that the purpose of HOMER is to reduce the costs of system operation and determine the optimum system configuration, economics play a crucial role in this simulation. The optimum combination of the HES components is obtained on the basis of the NPC, which is the sum of all costs and revenues that take place throughout the lifetime of a project. To calculate the total NPC of a system, the following equation is used [55,56]:

$$NPC = \frac{C_{ann,tot}}{CRF(i, T_p)} \tag{3}$$

where $C_{ann,tot}$ is the total annualized cost ($/year), i is the annual real interest rate (%), T_P is the project lifetime (year), and CRF is the capital recovery factor, which is given by [55,56]:

$$CRF(i, n) = \frac{i(1+i)^n}{(1+i)^n - 1} \tag{4}$$

where n is the number of years.

Salvage costs (SC), which are the residual values of the system components by the end of the project lifetime, are taken into consideration in the NPC calculation. The following expression is utilized to find the SC [54]:

$$SC = C_{RC}\frac{T_{rem}}{T_{com}} \tag{5}$$

where C_{RC} is the cost of replacement ($), T_{rem} represents the remaining life of the component (year), and T_{com} refers to the component lifetime (year).

COE is the average cost per kWh of producing electricity, given by [53,56]:

$$COE = \frac{C_{ann,tot}}{E_{ann,tot}} \tag{6}$$

where $E_{ann,tot}$ is the total electrical load served (kWh/year).

2.5. Control Strategy

The two control methods of the hybrid PV/diesel/battery system are the LF and CC dispatch strategies. In this study, these two strategies are presented for PV/diesel/battery HES and compared with the proposed combined strategy. The implementation of these strategies in practice can be done using a suitable controller, such as microcontroller, PLC, FPGA, etc.

2.5.1. Load Following Strategy

Figure 7 shows the flowchart of the LF dispatch strategy for the PV/diesel/battery HES. The system operation of this model can be classified into three cases as follows:

➢ The first case is when the output PV power is equal to the load power. Here, the PV power meets the load demand, the batteries do not draw any energy, and the generator stays off. In this case, no excess power exists.

➢ The second case takes place when the output PV power is higher than the load power. The PV feeds the load resulting in excess power. In this case, the excess power will be damped if the battery is fully charged. In the case where the battery is not fully charged, the excess PV power is used to charge the battery. The generator also does not work in this case.

➢ The last case is when the PV power is lower than the load power. The two possible subcases are as follows:

- If SOC = SOC_{min}, the generator works to feed the net load (load minus renewable power). The generator provides only enough power to satisfy the net load without charging the battery. It is important to mention that if the minimum generator loading output power is higher than the net load, the generator works to feed the load and the excess power from the PV charges the battery.
- If SOC > SOC_{min}, a cost of discharging the battery is computed and compared with the cost of turning on the generator that operates only to serve the net load. If the battery discharging cost is higher than the cost of turning on a generator, then the battery would not be discharged while the generator runs and produce enough power to meet the load demand without charging the battery. Otherwise, the battery is discharged. The following equations explain the cost of each decision:
- The cost of discharging the batteries is calculated using the following equation [16,57]:

$$C_{disch} = C_{batt,wear}. \quad (7)$$

$C_{batt,wear}$ is the battery wear cost ($/kWh), which is given by [16,57]:

$$C_{batt,wear} = \frac{C_{batt,rep}}{N_{batt} Q_{life} \sqrt{\eta_{rt}}} \quad (8)$$

where $C_{batt,rep}$ is the battery replacement cost ($), N_{batt} is the number of batteries in the storage bank, Q_{life} is the single battery throughput (kWh), and η_{rt} is the battery round trip efficiency (%).

The cost of turning on the generator ($/kWh), in which the generator operates only to serve the net load, is calculated using the following expression [58]:

$$C_{gen} = \frac{F_{con} F_{price}}{L_{served} G_{output}} + \frac{C_{gen,rep}}{L_{served} G_{lifetime}} + \frac{C_{gen,O\&M}}{L_{served}} \quad (9)$$

where F_{con} is the fuel consumption (L/hour), F_{price} is the fuel price ($/L), L_{served} is the total load to be served, G_{output} is the generator output (kW), $C_{gen,rep}$ is the replacement cost of the generator ($/kWh), $G_{lifetime}$ is the generator lifetime, and $C_{gen,O\&M}$ is the O&M cost of the generator.

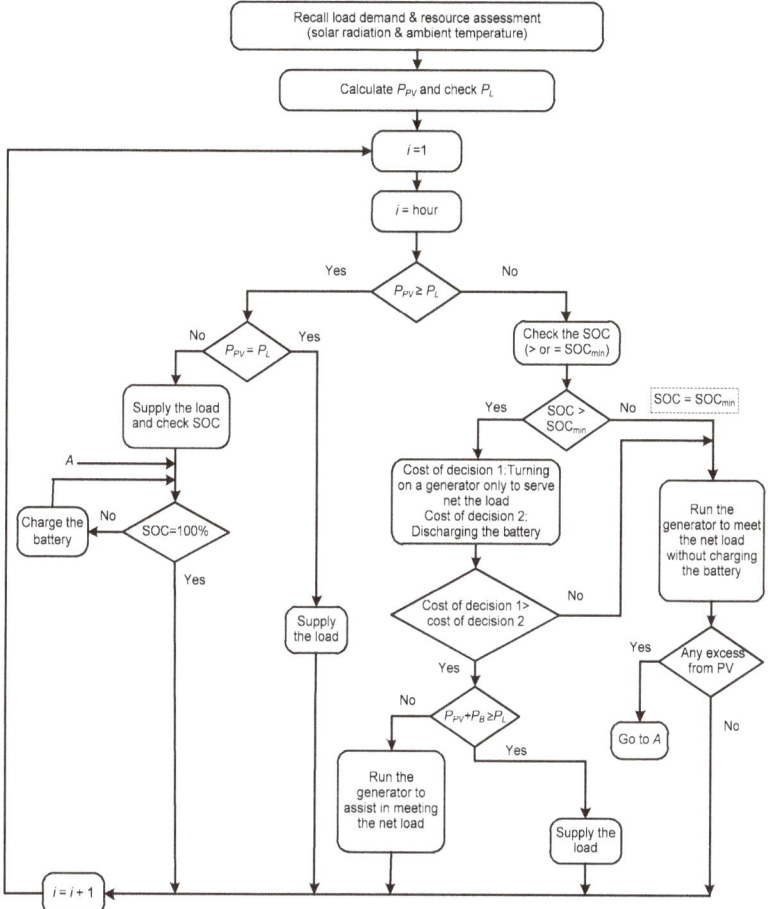

Figure 7. Load following (LF) dispatch strategy for the PV/diesel/battery HES.

2.5.2. Cycle Charging Strategy

The flowchart of the CC dispatch strategy for the PV/diesel/battery HES is shown in Figure 8. The operating strategy of this system is identical to that of the system utilizing LF dispatch. However, the strategy differs from the LF strategy in that when the generator is switched on, it runs at its maximum rated capacity to supply the net load and charge the battery with excess energy. The following equations explain the cost of each decision:

The cost of discharging the batteries is calculated using [16,57]:

$$C_{disch} = C_{batt,wear} + C_{batt,energy} \tag{10}$$

Battery energy cost, $C_{batt,energy}$ ($/kWh) is calculated in time step n using the following expression [57,58]:

$$C_{batt,energy,n} = \frac{\sum_{i=1}^{n-1} C_{cc,i}}{\sum_{i=1}^{n-1} E_{cc,i}} \tag{11}$$

where $C_{cc,i}$ is the cost of cycle charging in time step i ($), and $E_{cc,i}$ is the quantity of energy put into the batteries in time step i (kWh).

The cost of running the generator at maximum capacity to meet the net load and charge the battery is calculated using the following equation [58]:

$$C_{gen,ch} = C_{gen} + C_{cc} - C_{batt,energy} \qquad (12)$$

where C_{cc} here refers to the cost of cycle charge in the current time step, which is calculated using [58]:

$$C_{cc} = C_{gen,marg} + C_{batt,wear}. \qquad (13)$$

$C_{gen,marg}$ is the marginal cost of the generator ($/kWh), which is calculated using the following expression [58]:

$$C_{gen,marg} = \frac{F_{slope} F_{price}}{\eta_{rt}} \qquad (14)$$

where F_{slope} is the slope of fuel curve (L/kWh).

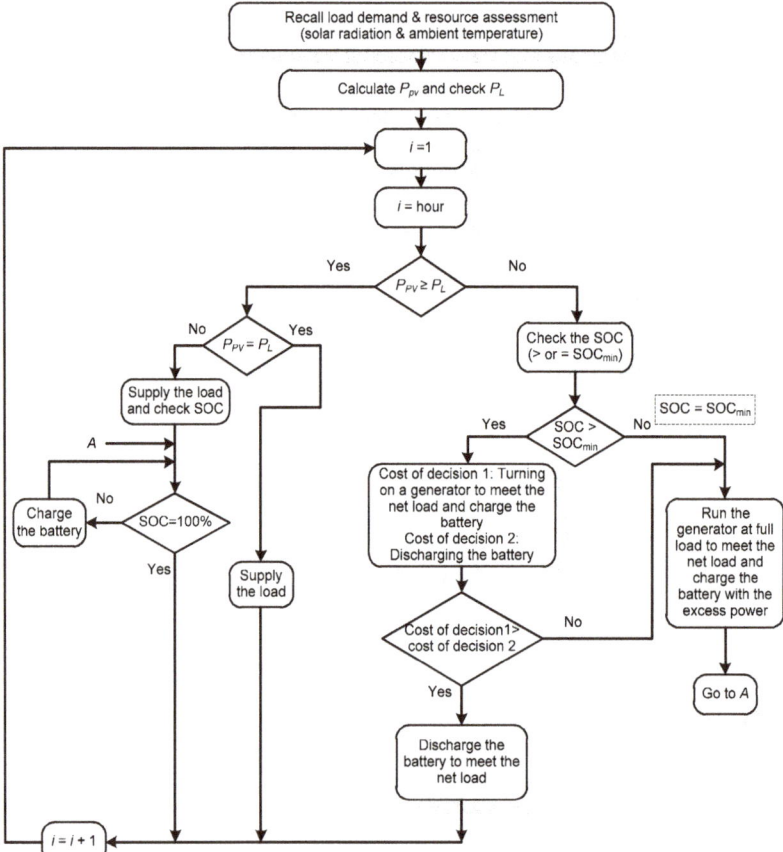

Figure 8. Cycle charging (CC) dispatch strategy for PV/diesel/battery HES.

2.5.3. Combined Dispatch Strategy

The decision of charging the battery by the diesel generator depends mainly on the future net load, which is the difference between the load demand and the output power from renewable sources. As predicting the future net load is challenging, the current net load is used in the CD strategy as a proxy for the future net load to investigate whether to charge the battery from the generator. This strategy tends to use the LF strategy when the net load is high and the CC strategy when the net load is low. During a low net load period, the CC strategy avoids the use of the generator. On the other hand, the LF strategy ensures continuous usage of the generator during high net load periods. The flexibility of the CD strategy based on high and low loads makes this model capable of producing better performance in the energy access scenarios compared with individual LF or CC strategies. Figure 9 shows the flowchart of the PV/diesel/battery HES with the CD strategy. As shown in the figure, three dispatch decisions are available in this strategy. At every time step, the CD strategy calculates the cost of each possible decision to obtain the least costly choice. This cost includes expenses in the current time step and the expected value of any change in the battery-stored energy, and a decision is made according to the tradeoff between the marginal cost of operating the generator and the battery wear cost. The details of the three cases are as follows:

➢ The first decision takes place when the generator provides only enough electricity to satisfy the net load without charging the battery.
➢ In the second decision, the generator meets the net load and charges the battery with the excess power.
➢ The third decision is feeding the net load by the battery alone.

The system operation of this model can be explained as follows:

➢ If the PV power is equal to or higher than the load demand, the power flow is the same for that of LF and CC strategies.
➢ If the PV power is less than the load power, the two possible subcases are as follows:

  - If SOC = SOC_{min}, the controller compares the cost of running the generator to produce only enough power to feed the net load without charging the battery with the cost of running the generator at its maximum capacity to serve the net load and charge the battery with the excess electricity. A decision is made based on the least cost dispatch decision.
  - If SOC > SOC_{min}, three decision costs are compared with each other, which are: the cost of turning on the generator that operates only to serve the net load without charging the battery, the cost of running the generator at its maximum capacity to serve the net load and charge the battery, and the cost of discharging the battery. The power flow solution is taken place according to the minimum cost decision.

3. Results and Discussion

The simulation results are presented and analyzed in this section. The HES was simulated under the three different dispatch strategies. First, the technical feasibility was examined to investigate the ability of the available energy to satisfy electric load demand throughout the year. Then, the economic viability and environmental effects of the proposed systems were investigated. Finally, a sensitivity analysis was performed to determine the impact of some critical parameters on the system performance. The simulation was performed using an annual discount rate of 7.8% and a 20-year project lifetime. Moreover, the system performance was set at a maximum capacity shortage of 1% and a battery minimum SOC of 25%.

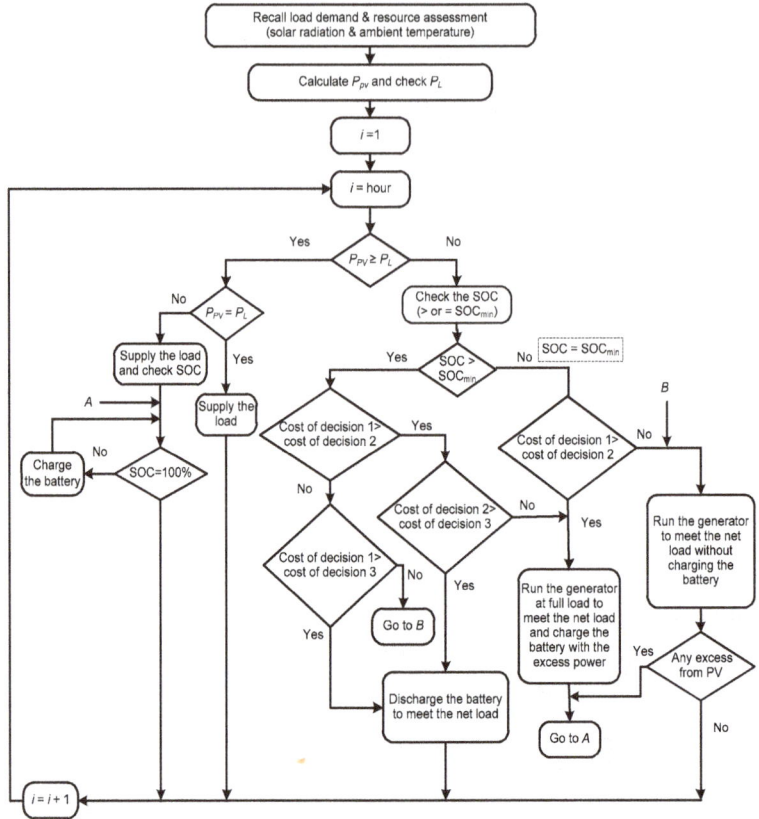

Figure 9. Combined dispatch (CD) strategy for PV/diesel/battery HES.

3.1. Optimization Results

A feasible system is one that can meet the load demands. In HOMER, the infeasible systems were excluded, while the feasible ones were filtered and presented by their NPC. A comparison of the optimized PV/diesel/battery systems under different dispatch strategies is provided in Table 2. The results show that best optimal combination of PV/diesel/battery HES under the CD strategy consisted of a 19.4 kW PV, a diesel generator with a 21 kW capacity, 220 batteries, and an 8.05 kW power converter. The system yielded an NPC of $110,191 and COE of $0.21/kWh, which were 20.6% and 4.8% lower than those of systems utilizing the LF and CC strategies, respectively. These results can be interpreted by the fact that the CD strategy obtained the most economical cost dispatch decision between LF and CC at every time step.

In the LF dispatch strategy, better renewable source utilization could be attained because the diesel generator contributed approximately 40.6% of the total energy production, the lowest compared with that in other strategies. This condition led to greater dependency on renewable components, which was in contrast to the CC dispatch strategy, wherein the generator produced a 73.8% share of the total energy production. The renewable fractions of LF, CC, and CD strategies were 44.7%, 18.4%, and 35.6%, respectively. The monthly average electric production for the system under the three dispatch strategies for one year is presented in Figure 10.

In terms of diesel generator performance, the diesel generator under the CD strategy yielded 2171 working hours; these values were lower than those of the LF and CC strategies, and, hence, the higher generator operation life was obtained under this strategy. The results show that the operating lifespans of LF, CC, and CD strategies were 4.08 years, 4.58 years, and 6.91 years, respectively. Moreover, it was found that the generator under the CD strategy consumed the lowest amount of diesel with 10,574 L/year, which was 1.7% and 23% lower than the system under the LF and CC strategies, respectively. Figure 11 shows the fuel consumption duration curve for the system under the three dispatch strategies for the entire year. It is evident that despite the fact that the generator working hours of the LF strategy were more than those of the CC strategy, the fuel consumption of the LF strategy was lower than that of the CC strategy. This is because the generator under the LF strategy produced only enough power to feed the load, in contrast to the CC strategy, in which the generator operated at its maximum rated capacity to supply the load and charge the battery; hence, more diesel would be consumed.

Battery energy storage is a critical component of standalone HESs because this parameter affects the reliability of meeting loads. Battery throughput (kWh/year) is a performance measure defined as the amount of energy that cycles through the storage bank in one year. Battery throughput affects the battery operational lifetime; the lower the annual battery throughput (lower charge/discharge cycles per year), the higher the battery lifetime [59]. Figure 12 shows the battery input/output energies under the three different strategies. Since the use of the battery in the LF strategy was limited, the LF strategy presented the lowest amount of input/output energy, followed by the CD strategy and the CC strategy, which showed the highest amount of input/output energy. These results show the LF strategy presented the lowest annual battery throughput, which was estimated to be 54.15 kWh, in contrast to the CC strategy, which had a yearly throughput of 79.9 kWh. The annual throughput of CD was calculated to be 70.76 kWh.

Battery autonomy refers to the number of hours a battery can support the critical load without charging during the main failure. Achieving some battery autonomy is necessary, especially during the rainy season when the output PV power cannot meet the load demand [60]. The simulation results indicate that the battery autonomy of the LF strategy was 8.94 h, which was the highest among those of other strategies. This result is due to the high nominal capacity of the system under this strategy. The battery autonomy values of the CC and CD strategies were 4.17 and 6.55 h, respectively.

Table 2. Optimization results of the HES for different dispatch strategies. NPC is net present cost; COE is cost of energy.

Item	Unit	Load Following	Cycle Charging	Combined Dispatch
PV	kW	28.5	10.2	19.4
Battery	-	300	140	220
Diesel generator	kW	21	21	21
Converter	kW	5.28	5.78	8.05
NPC	$	138,704	115,722	110,191
COE	$/kWh	0.264	0.22	0.21
Renewable fraction	%	44.7	18.4	35.6
PV production	kWh/year	42,884	15,314	29,116
Diesel generator production	kWh/year	29,255	43,187	34,080
Total production	kWh/year	72,139	58,501	63,196
Excess electricity	%	19	0.774	6.74
Generator number of starts	start/year	567	494	498
Generator operating hours	hour/year	3675	3277	2171
Generator operation life	year	4.08	4.58	6.91
Fuel consumption	L/year	10,757	13,887	10,574
Batteries throughput	kWh/year/battery	54.15	79.9	70.76
Batteries nominal capacity	kWh	72	33.6	52.8
Batteries autonomy	hour	8.94	4.17	6.55

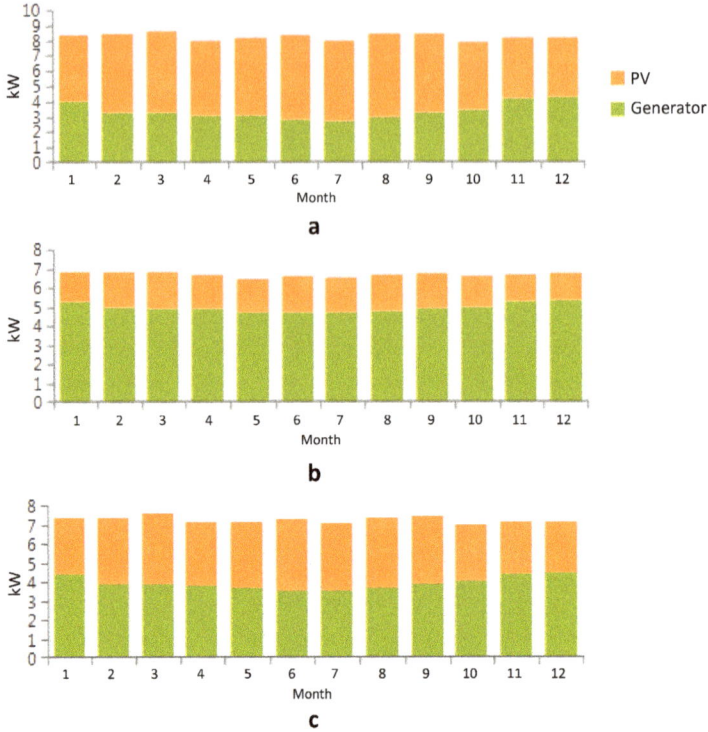

Figure 10. Monthly average electric production for the HES under (**a**) load following (LF) strategy, (**b**) CC strategy, and (**c**) CD strategy.

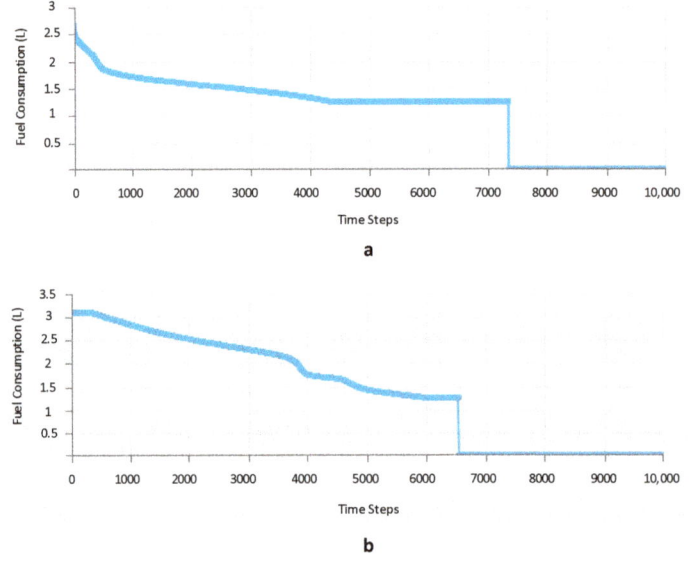

Figure 11. *Cont.*

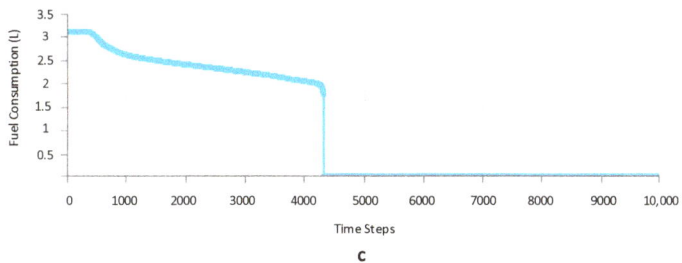

Figure 11. Fuel consumption duration curve for the HES under (**a**) LF strategy, (**b**) CC strategy, and (**c**) CD strategy.

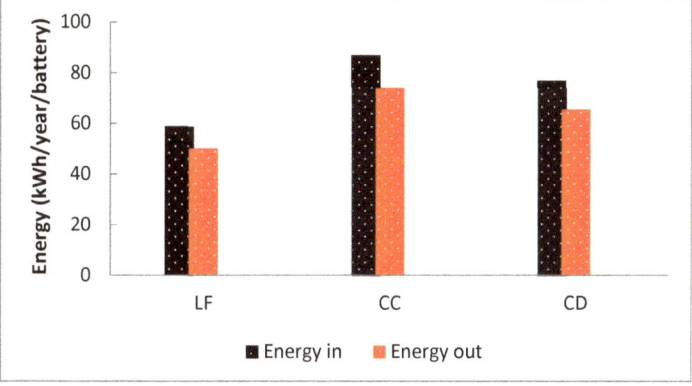

Figure 12. Input/output energy of battery for different dispatch strategies.

3.2. Economic Analysis of the System under the Three Dispatch Strategies

A cost summary of the system under all dispatch strategies is shown in Figure 13. The following bullets describe the cost analysis for each configuration:

- The lowest capital cost was achieved by using the CC strategy ($32,537), followed by the CD strategy ($50,042). The system under CC had a capital cost of $64,779, which was the highest among the systems compared, because of the large sizes of renewable components.
- Once installed, the PV, converter, and batteries under all strategies were inexpensive to operate and maintain compared with the diesel generator, which contributed most of the total O&M cost. The CD strategy was found to have the lowest O&M cost, which was $18,172, i.e., 37.7% and 21.9% lower than those of LF and CC, respectively.
- The replacement cost of the LF, CC, and CD strategies were $9005, $8457, and $5135, respectively. The CD strategy had the lowest cost because it depended less on the diesel generator, which led to making this strategy have the highest generator lifetime as compared to other strategies.
- The fuel cost of the CD strategy was $42,016, which was lower than that of the other strategies, because of the relatively low fuel consumption in this strategy. The fuel costs of the LF and CC strategies were $42,742 and $55,182, respectively.
- The salvage costs over the project lifetime of the LF, CC, and CD strategies were estimated to be $−6974, $−3727, and $−5174, respectively; these costs originated from the remaining life of the system components.

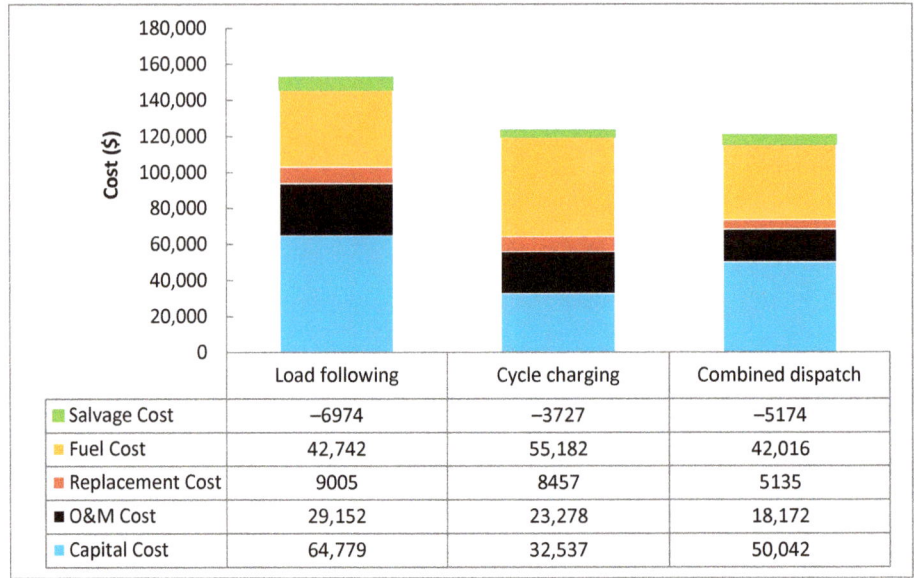

Figure 13. Cost summary of PV/diesel/battery under different dispatch strategies. O&M is operation and maintenance.

3.3. Environmental Assessment

Consumption of diesel has a negative impact on the environment and is harmful to human health, since different kinds of gaseous pollutants are emitted during the process. These emissions include nitrogen oxide (NO_x), sulfur dioxide (SO_2), particulate matter (PM), unburned hydrocarbon (UHC), carbon monoxide (CO), and carbon dioxide (CO_2) [61]. The HESs have positive effects on the environment. Combining PV with the diesel and battery is capable of reducing the emissions significantly [31]. The releasing of annual emissions for the HES under different control strategies has been determined and demonstrated to compare the cases from the environmental standpoint. Table 3 presents the greenhouse emissions of PV/diesel/battery with LF, CC, and CD strategies. It is obvious that the system with CD was the most environmentally friendly option by having the lowest amount of CO_2 emissions (27,678 kg/year), in contrast to the system with the CC strategy, which had CO_2 emissions of 36,352 kg/year. The CO_2 emissions of the system with the LF strategy were calculated to be 28,157 kg/year. These results were achieved according to the resulting fuel consumption presented in Table 2, which indicates that the diesel generator under the CD strategy yielded a fuel consumption of 10,574 L/year, lower than that of the LF and CC strategies. As a result, minimum greenhouse emissions were obtained under this strategy.

Table 3. Pollutant emissions for different control strategies. PM is particulate matter; UHC is unburned hydrocarbon.

Pollutant Emissions (kg/year)	LF	CC	CD
NO_x	167	215	164
SO_2	68.9	89	67.8
PM	1.08	1.39	1.06
UHC	7.74	10	7.61
CO	177	229	174
CO_2	28,157	36,352	27,678

3.4. Energy Balance of the Optimized Systems

In order to have a better understanding of the system operation, details of the system's power flow and the battery status during 24 h in April for LF, CC, and CD strategies are provided in this section. The energy flow and the battery status for each strategy are explained in the following points:

- LF strategy: Figure 14 shows the energy balance and battery SOC of the proposed HES under LF strategy. At the start of the day (00:00–05:30), the battery discharged its power to supply the load alone. At 05:30, the PV array began producing power, but it was not enough to meet the load; therefore, the batteries continued to discharge their power for half an hour. At 06:00, the batteries reached their minimum SOC, so the generator worked to meet the load while the batteries were charged by the excess power from PV until 07:00. For the period between 07:00 and 13:30, PV power was sufficient to fulfill all demand and charge the battery. From 13:30 to 16:30 the generator restarted and worked together with the PV. From 16:30 to 18:30, the batteries discharged to help the PV and the generator in satisfying the load. At 18:30, the PV power became zero, and the load was met by the generator and the batteries for the rest of the day.
- CC strategy: The energy balance and battery SOC of the proposed HES under CC strategy is depicted in Figure 15. For the first few hours (00:00–05:30), the load was served by the battery, similar to the LF strategy. From 05:30 to 07:00, both the PV and battery shared to satisfy the load. From 07:00 to 10:00, the generator produced power so that both the PV and generator could serve the load. Since it was a cycle-charging scheme, the excess power of the generator was used to charge the battery. The load was satisfied by the PV, which also charged the batteries from 10:00 to 12:30. At 12:30, the PV output showed a decrease in its output power; therefore, the batteries discharged their power to compensate for the required power and continued to do so until 15:30. At 15:30, the generator worked with the PV to feed the load and charge the batteries. The PV output fell to zero at 18:30 and the generator continued to feed the load for the remaining time of the day.
- CD strategy: Figure 16 presents the energy balance and battery SOC of the proposed HES under CD strategy. As in the LF and CC strategies, the load was served by the battery alone from 00:00 until 05:30. From 05:30 until 15:30, the PV supplied the load and charged the battery during the high solar radiation hours, while the batteries discharged their power to share with the PV in satisfying the load during the low solar radiation hours. Around 15:30, the generator operated at its maximum rated capacity to share in meeting the load with the PV and charged the batteries by the excess power. This case represented the CC strategy, which continued until 18:30. After 18:30, the PV output became zero, and the generator produced only enough power to satisfy the load without charging the battery. This case referred to the LF strategy and continued until 19:30. The CC strategy took place again from 19:30 to 21:30. In the period from 21:30 to 22:00, the generator and the batteries shared to feed the load. At 22:00, the generator turned off, and the load was served by the batteries for the remaining time.

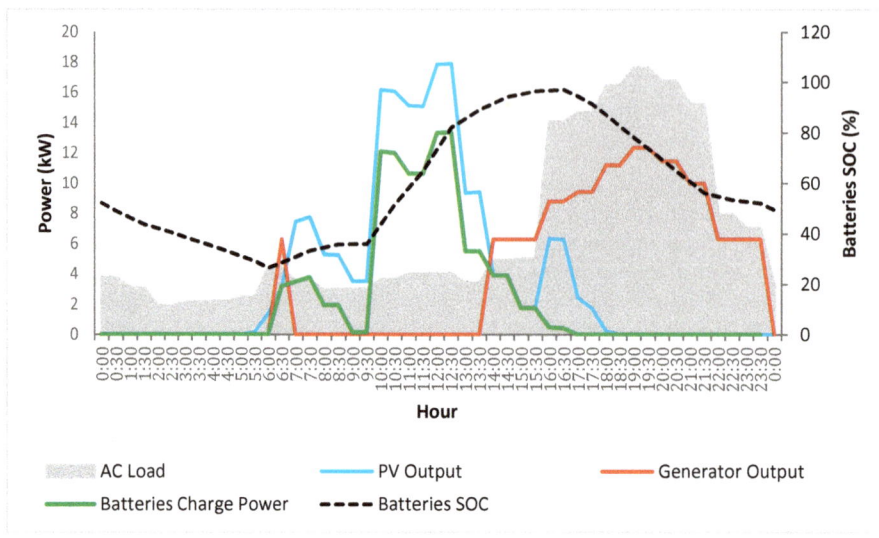

Figure 14. Energy balance and battery SOC of PV/diesel/battery under LF strategy.

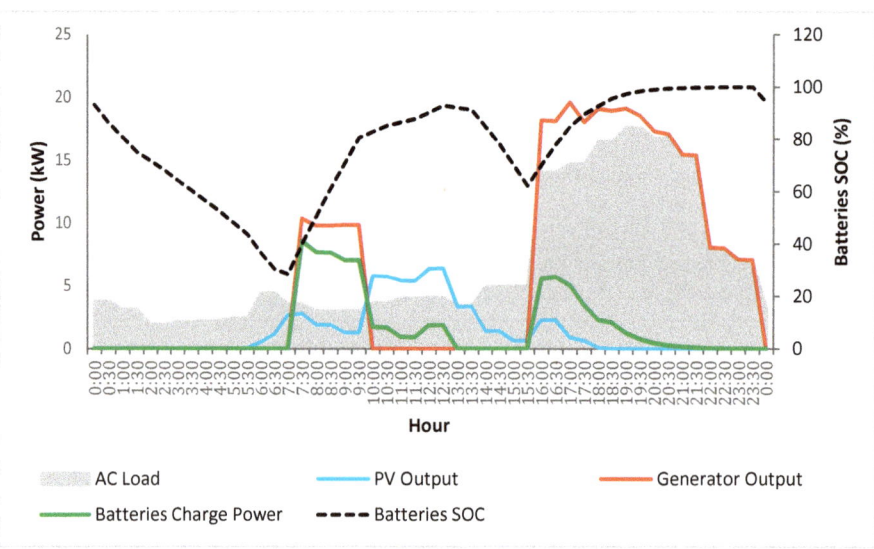

Figure 15. Energy balance and battery SOC of PV/diesel/battery under CC strategy.

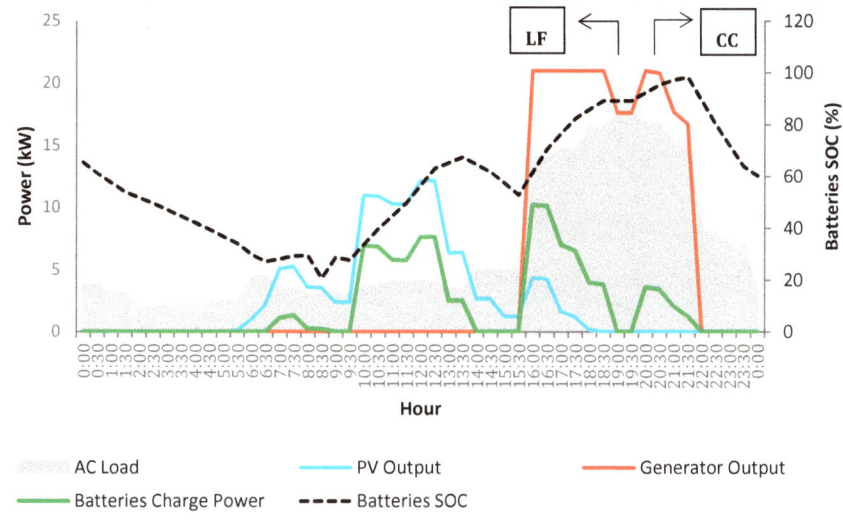

Figure 16. Energy balance and battery SOC of PV/diesel/battery under CD strategy.

3.5. Sensitivity Analysis

In this part of the study, sensitivity analysis was conducted to investigate the effect of important parameters on the system performance. The parameters considered for sensitivity analysis were battery minimum SOC, time step, solar radiation, diesel price, and load growth.

3.5.1. Battery Minimum SOC

Battery SOC_{min} is the lowest acceptable battery charge level. The battery level is never drawn below the SOC_{min}, which is given as a percentage of the total capacity. In this part, the impact of the SOC_{min} on the system performance was investigated. The variations in SOC_{min} were set to 15%, 20%, 25%, 30%, 35%, and 40. Figure 17 shows a graph of the total NPC and CO_2 emissions cost as a function of SOC_{min} variation. As can be seen in the Figure, when the SOC_{min} increased from 15% to 40%, CO_2 emissions increased from 25,468 kg/year to 34,159 kg/year. This result indicated that an increase in SOC_{min} would increase the system's dependence on the diesel generator to supply the load demand and charge the battery, resulting in greater CO_2 emissions. Moreover, NPC increased from $107,637 to $114,274. However, researchers have recommended that the SOC_{min} not be set to an extremely low value to avoid damaging the storage bank by excessive discharge [62–65].

3.5.2. Time Step

The system operation was simulated by performing energy balance calculations in every time step for 8760 h in a year. A comparison between the output power of the available energy sources and load demand was conducted at every time step, and the flow of energy from and to every component in the system was investigated. For a system that included a generator and a battery, a decision was made at each time step on the operation of the generator and whether to charge or discharge the battery. Sensitivity analysis was performed in this section to evaluate the effect of time step variations from 5 min to 1 h on the system performance. As shown in Figure 18, the best economic performance was achieved by setting the time step to 5 min, which gave an NPC of $109,982. This can be explained by the fact that reducing the time step led to an increase in the number of times of cost comparison, which led to a minimum possible cost. At the same time, the most environmentally friendly system could also be obtained when the time step was set to 5 min; in such a case, the CO_2 emissions were 26,257 kg/year.

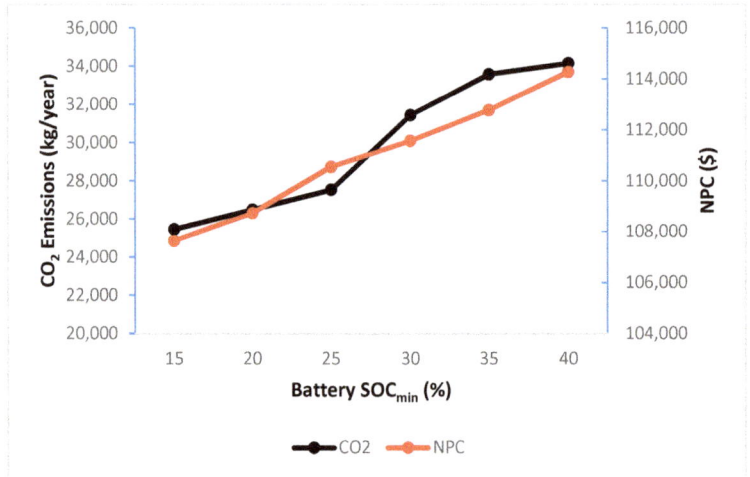

Figure 17. Impact of battery minimum SOC on CO_2 emissions and NPC.

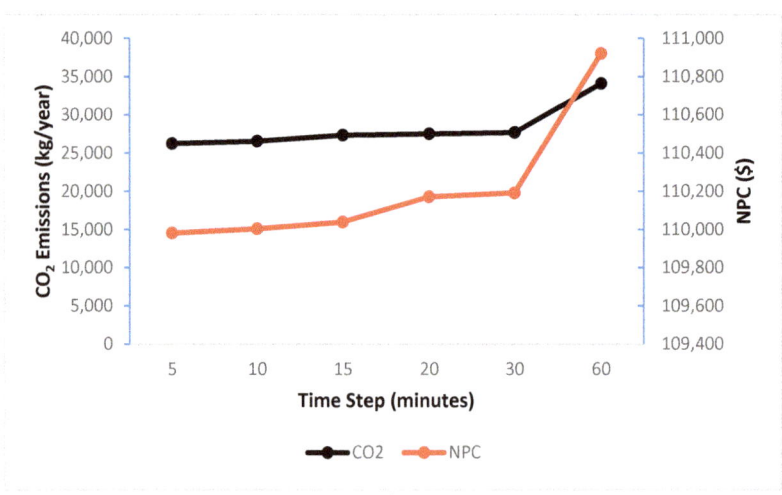

Figure 18. Impact of time step on CO_2 emissions and NPC.

3.5.3. Solar Radiation

Global solar radiation intensity and the efficiency of the solar panels play a significant role in the harvesting of solar energy in the HESs. The power produced by the PV panels increases when the solar radiation increases and vice-versa; hence, the variation of solar radiation can significantly affect the performance of the system. In this subsection, the annual average global solar radiation was varied between 4 kWh/m^2/day and 6 kWh/m^2/day. Figure 19 shows the graph of NPC and CO_2, with the variation of solar radiation. The NPC and CO_2 of the system decreased from $113,792 to $107,627 and from $35,038/kWh to $26,874/kWh, respectively, when solar radiation increased from 4 kWh/m^2/day to 6 kWh/m^2/day. The decrements of NPC and CO_2 occurred because the increase of solar radiation increased the output power of PV; hence, the generator operation hours and fuel consumption were reduced.

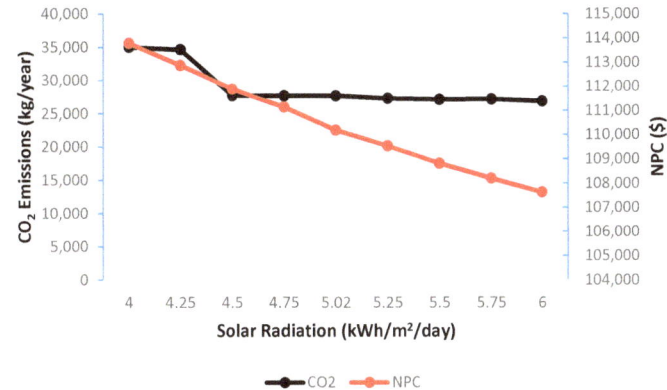

Figure 19. Impact of solar radiation on CO_2 emissions and NPC.

3.5.4. Diesel Fuel Price

It is well known that the fuel price fluctuates continuously. The fuel price fluctuations encourage the use of renewable energy technologies, which offer stabilization of electricity costs [59]. The current price of diesel in Iraq is about $0.4/L with some variability from time to time. To investigate the effect of diesel fuel variation on the system performance, a sensitivity analysis was done by varying the price between $0.25/L and $0.55/L. The impact of diesel price variation on the NPC and CO_2 is depicted in Figure 20. It is obvious that with the rise in diesel fuel price from 0.25/L and $0.55/L, CO_2 emissions decreased from 34,683 kg/year to 26,578 kg/year while NPC increased from $91,051 to $126,989. These results can be explained by the fact that a rise in diesel price led to making the utilization of the diesel generator less competitive. Therefore, the generator working hours were reduced, which affected the optimal power flow of the system.

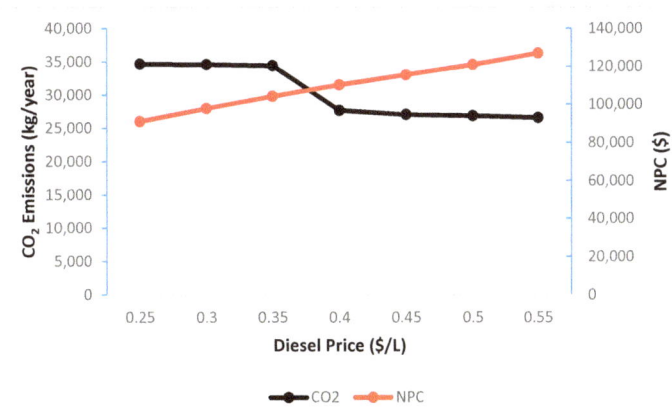

Figure 20. Impact of diesel price on CO_2 emissions and NPC.

3.5.5. Load Consumption Growth

The output power of an energy system depends on the load requirements that should be satisfied all the time. If the load consumption increases, the energy production should imperatively be increased. The load demand is usually varied from time to time. In this subsection, different load consumptions (145, 150, 155, 160, 165, 170, 175, and 180 kWh/day) were considered to investigate their impact on the system performance. The effect of load consumption growth on the NPC and CO_2 is presented in Figure 21. The results indicate that NPC and CO_2 of the system increased by 22.6% and 52.3%, respectively, when the load consumption increased from 145 kWh/day to 180 kWh/day. The increments of NPC and CO_2 were mainly due to the increase of the required capacity of the different components of the system, including the diesel generator to increase the energy productions of the system.

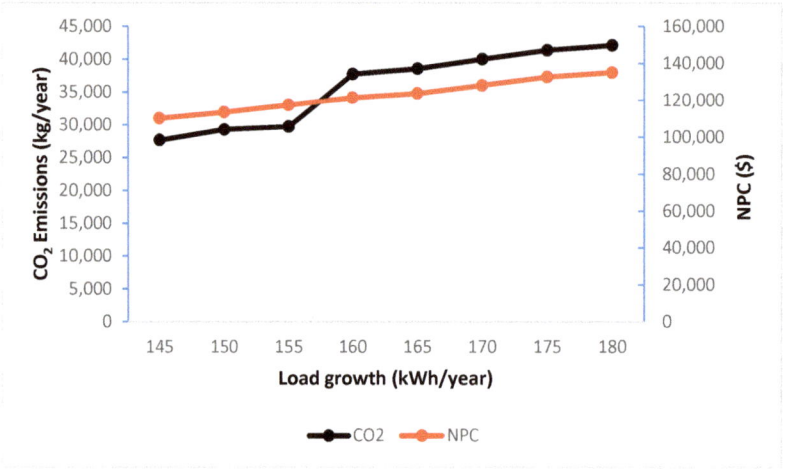

Figure 21. Impact of load growth on CO_2 emissions and NPC.

4. Conclusions

Optimization of energy sources is a critical key in the assessment of the feasibility of HESs. This study presents an optimal plan and design by providing a systematic techno-economic and environmental evaluation of a PV/diesel/battery off-grid configuration for a rural area in Iraq. Three different control strategies were proposed in this study, i.e., LF, CC, and CD. HOMER software was used to evaluate the overall analyses, including the optimization and sensitivity. The cost analysis results show that the combination of 19.4 kW PV, a diesel generator with a 21 kW capacity, 220 batteries, and an 8.05 kW power converter with the CD strategy was the optimal solution for this case study by having an NPC of $110,191 and a COE of 0.21 $/kWh, which were 20.6% and 4.8% less expensive than the systems under the LF and CC strategies, respectively. Moreover, minimum fuel consumption made the CD dispatch strategy the most suitable option for the environment, considering its CO_2 emissions of 27,678, which were 1.7% and 23.9% lower than those of the LF and CC strategies, respectively. Furthermore, sensitivity analysis showed that variations in some important parameters, such as battery minimum SOC, time step, solar radiation, diesel price, and load growth, had significant effects on system performance. This study could play a vital role in decision making towards better energy management strategies.

Author Contributions: A.S.A. contributed theoretical approaches, simulation, and preparing the article; M.F.N.T. and M.R.A. contributed on supervision; M.A.M.R. and S.M. contributed on article editing.

Funding: This research was funded by the Ministry of Education (MOE) Malaysia, grant number FRGS/1/2015/TK10/UNIMAP/03/2.

Acknowledgments: The authors would like to thank the Universiti Malaysia Perlis and the Ministry of Higher Education (MOHE) Malaysia for providing the facilities and financial support (Fundamental Research Grant Scheme (FRGS) under a grant number of FRGS/1/2015/TK10/UNIMAP/03/2).

Conflicts of Interest: The authors declare no conflicts of interest.

References

1. Mehrpooya, M.; Mohammadi, M.; Ahmadi, E. Techno-economic-environmental study of hybrid power supply system: A case study in Iran. *Sustain. Energy Technol. Assess.* **2018**, *25*, 1–10. [CrossRef]
2. Owusu, P.A.; Asumadu-Sarkodie, S. A review of renewable energy sources, sustainability issues and climate change mitigation. *Cogent. Eng.* **2016**, *3*, 1167990. [CrossRef]
3. Ramli, M.A.; Hiendro, A.; Twaha, S. Economic analysis of PV/diesel hybrid system with flywheel energy storage. *Renew. Energy* **2015**, *78*, 398–405. [CrossRef]
4. Moreira, D.; Pires, J.C. Atmospheric CO_2 capture by algae: Negative carbon dioxide emission path. *Bioresour. Technol.* **2016**, *215*, 371–379. [CrossRef] [PubMed]
5. Kim, H.; Bae, J.; Baek, S.; Nam, D.; Cho, H.; Chang, H. Comparative analysis between the government micro-grid plan and computer simulation results based on real data: The practical case for a South Korean Island. *Sustainability* **2017**, *9*, 197. [CrossRef]
6. REN21. *Renewables 2018 Global Status Report*; REN21 Secretariat: Paris, France, 2018; ISBN 978-3-9818911-3-3.
7. Shahzad, M.K.; Zahid, A.; Rashid, T.U.; Rehan, M.A.; Ali, M.; Ahmad, M. Techno-economic feasibility analysis of a solar-biomass off grid system for the electrification of remote rural areas in Pakistan using HOMER software. *Renew. Energy* **2017**, *106*, 264–273. [CrossRef]
8. Shi, B.; Wu, W.; Yan, L. Size optimization of stand-alone PV/wind/diesel hybrid power generation systems. *J. Taiwan Inst. Chem. Eng.* **2017**, *73*, 93–101. [CrossRef]
9. Nag, A.K.; Sarkar, S. Modeling of hybrid energy system for futuristic energy demand of an Indian rural area and their optimal and sensitivity analysis. *Renew. Energy* **2018**, *118*, 477–488. [CrossRef]
10. Arul, P.; Ramachandaramurthy, V.K.; Rajkumar, R. Control strategies for a hybrid renewable energy system: A review. *Renew. Sustain. Energy Rev.* **2015**, *42*, 597–608. [CrossRef]
11. Strnad, I.; Prenc, R. Optimal sizing of renewable sources and energy storage in low-carbon microgrid nodes. *Electr. Eng.* **2017**, *100*, 1661–1674. [CrossRef]
12. Cheng, L.; Wang, W.; Wei, S.; Lin, H.; Jia, Z. An improved energy management strategy for hybrid energy storage system in light rail vehicles. *Energies* **2018**, *11*, 423. [CrossRef]
13. Li, H.; Eseye, A.T.; Zhang, J.; Zheng, D. Optimal energy management for industrial microgrids with high-penetration renewables. *Prot. Control Mod. Power Syst.* **2017**, *2*. [CrossRef]
14. Olatomiwa, L.; Mekhilef, S.; Ismail, M.; Moghavvemi, M. Energy management strategies in hybrid renewable energy systems: A review. *Renew. Sustain. Energy Rev.* **2016**, *62*, 821–835. [CrossRef]
15. Vivas, F.; Heras, A.D.L.; Segura, F.; Andújar, J. A review of energy management strategies for renewable hybrid energy systems with hydrogen backup. *Renew. Sustain. Energy Rev.* **2018**, *82*, 126–155. [CrossRef]
16. Farret, F.A.; Simões, M. *Godoy Integration of Renewable Sources of Energy*; John Wiley & Sons, Inc.: Hoboken, NJ, USA, 2018.
17. Zambrana, M.N.F.V. System Design and Analysis of a Renewable Energy Source Powered Microgrid. M.Sc. Thesis, University of Maryland, College Park, MD, USA, 2018.
18. Sawle, Y.; Gupta, S.; Bohre, A.K. Optimal sizing of standalone PV/Wind/Biomass hybrid energy system using GA and PSO optimization technique. *Energy Procedia* **2017**, *117*, 690–698. [CrossRef]
19. Hosseinalizadeh, R.; Shakouri, G.H.; Amalnick, M.S.; Taghipour, P. Economic sizing of a hybrid (PV–WT–FC) renewable energy system (HRES) for stand-alone usages by an optimization-simulation model: Case study of Iran. *Renew. Sustain. Energy Rev.* **2016**, *57*, 1657. [CrossRef]
20. Olatomiwa, L.; Mekhilef, S.; Huda, A.S.N.; Sanusi, K. Techno-economic analysis of hybrid PV-diesel-battery and PV-wind-diesel-battery power systems for mobile BTS: The way forward for rural development. *Energy Sci. Eng.* **2015**, *3*, 271–285. [CrossRef]

21. Ajlan, A.; Tan, C.W.; Abdilahi, A.M. Assessment of environmental and economic perspectives for renewable-based hybrid power system in Yemen. *Renew. Sustain. Energy Rev.* **2017**, *75*, 559–570. [CrossRef]
22. Hsu, D.; Kang, L. Dispatch Analysis of Off-Grid Diesel Generator-Battery Power Systems. *Int. J. Emerg. Electr. Power Syst.* **2014**, *15*. [CrossRef]
23. Zobaa, A.F.; Bansal, R.C. *Handbook of Renewable Energy Technology*; World Scientific Pub. Co.: Singapore, 2011.
24. Jung, T.; Kim, D.; Moon, J.; Lim, S. A scenario analysis of solar photovoltaic grid parity in the Maldives: The case of Malahini resort. *Sustainability* **2018**, *10*, 4045. [CrossRef]
25. Brenna, M.; Foiadelli, F.; Longo, M.; Abegaz, T. Integration and optimization of renewables and storages for rural electrification. *Sustainability* **2016**, *8*, 982. [CrossRef]
26. Jiang, F.; Xie, H.; Ellen, O. Hybrid energy system with optimized storage for improvement of sustainability in a small town. *Sustainability* **2018**, *10*, 2034. [CrossRef]
27. Mazzola, S.; Astolfi, M.; Macchi, E. A detailed model for the optimal management of a multigood microgrid. *Appl. Energy* **2015**, *154*, 862–873. [CrossRef]
28. Halabi, L.M.; Mekhilef, S.; Olatomiwa, L.; Hazelton, J. Performance analysis of hybrid PV/diesel/battery system using HOMER: A case study Sabah, Malaysia. *Energy Convers. Manag.* **2017**, *144*, 322–339. [CrossRef]
29. Ansong, M.; Mensah, L.D.; Adaramola, M.S. Techno-economic analysis of a hybrid system to power a mine in an off-grid area in Ghana. *Sustain. Energy Technol. Assess.* **2017**, *23*, 48–56. [CrossRef]
30. Sigarchian, S.G.; Paleta, R.; Malmquist, A.; Pina, A. Feasibility study of using a biogas engine as backup in a decentralized hybrid (PV/wind/battery) power generation system—Case study Kenya. *Energy* **2015**, *90*, 1830–1841. [CrossRef]
31. Rezzouk, H.; Mellit, A. Feasibility study and sensitivity analysis of a stand-alone photovoltaic–diesel–battery hybrid energy system in the north of Algeria. *Renew. Sustain. Energy. Rev.* **2015**, *43*, 1134–1150. [CrossRef]
32. Madziga, M.; Rahil, A.; Mansoor, R. Comparison between three off-grid hybrid systems (solar photovoltaic, diesel generator and battery storage system) for electrification for Gwakwani village, South Africa. *Environments* **2018**, *5*, 57. [CrossRef]
33. Upadhyay, S.; Sharma, M. Development of hybrid energy system with cycle charging strategy using particle swarm optimization for a remote area in India. *Renew. Energy* **2015**, *77*, 586–598. [CrossRef]
34. Rajbongshi, R.; Borgohain, D.; Mahapatra, S. Optimization of PV-biomass-diesel and grid base hybrid energy systems for rural electrification by using HOMER. *Energy* **2017**, *126*, 461–474. [CrossRef]
35. Amutha, W.M.; Rajini, V. Cost benefit and technical analysis of rural electrification alternatives in southern India using HOMER. *Renew. Sustain. Energy. Rev.* **2016**, *62*, 236–246. [CrossRef]
36. Sen, R.; Bhattacharyya, S.C. Off-grid electricity generation with renewable energy technologies in India: An application of HOMER. *Renew. Energy* **2014**, *62*, 388–398. [CrossRef]
37. Singh, A.; Baredar, P.; Gupta, B. Techno-economic feasibility analysis of hydrogen fuel cell and solar photovoltaic hybrid renewable energy system for academic research building. *Energy Convers. Manag.* **2017**, *145*, 398–414. [CrossRef]
38. Kansara, B.U.; Parekh, B.R. Dispatch, control strategies and emissions for isolated wind-diesel hybrid power system. *Int. J. Innov. Technol. Explor. Eng.* **2013**, *2*, 152–156.
39. Eu-Tjin, C.; Huat, C.K.; Seng, L.Y. Control strategies in energy storage system for standalone power systems. In Proceedings of the 4th IET Clean Energy and Technology Conference, Kuala Lumpur, Malaysia, 14–15 November 2016.
40. Markovic, M.; Nedic, Z.; Nafalski, A. Microgrid solutions for insular power systems in the outback of Australia. *Masz. Elektr. Zesz. Probl.* **2016**, *109*, 103–107.
41. Kansara, B.U.; Parekh, B.R. Penetration of renewable energy resources based dispatch strategies for isolated hybrid systems. *Int. J. Electr. Electron. Eng. Res.* **2013**, *3*, 121–130.
42. Saraswat, S.K.; Rao, K.V.S. 10 kW solar photovoltaic—Diesel hybrid energy system for different solar zones of India. In Proceedings of the 2016 International Conference on Emerging Technological Trends (ICETT), Kollam, India, 21–22 October 2016.
43. Fodhil, F.; Hamidat, A.; Nadjemi, O. Energy control strategy analysis of hybrid power generation system for rural Saharan community in Algeria. In *Artificial Intelligence in Renewable Energetic Systems Lecture Notes in Networks and Systems*; Springer: Cham, Switzerland, 2017; pp. 108–120.
44. Shoeb, M.; Shafiullah, G. Renewable energy integrated islanded microgrid for sustainable irrigation—A Bangladesh perspective. *Energies* **2018**, *11*, 1283. [CrossRef]

45. Ghenai, C.; Bettayeb, M. Optimized design and control of an off grid solar PV/hydrogen fuel cell power system for green buildings. *IOP Conf. Ser. Earth Environ. Sci.* **2017**, *93*, 012073. [CrossRef]
46. Maatallah, T.; Ghodhbane, N.; Nasrallah, S.B. Assessment viability for hybrid energy system (PV/wind/diesel) with storage in the northernmost city in Africa, Bizerte, Tunisia. *Renew. Sustain. Energy Rev.* **2016**, *59*, 1639–1652. [CrossRef]
47. Islam, M.S.; Akhter, R.; Rahman, M.A. A thorough investigation on hybrid application of biomass gasifier and PV resources to meet energy needs for a northern rural off-grid region of Bangladesh: A potential solution to replicate in rural off-grid areas or not? *Energy* **2018**, *145*, 338–355. [CrossRef]
48. Hassan, Q.; Jaszczur, M.; Abdulateef, J. Optimization of PV/wind/diesel hybrid power system in HOMER for rural electrification. *J. Phys. Conf. Ser.* **2016**, *745*, 032006. [CrossRef]
49. Surface Meteorology and Solar Energy. Available online: https://eosweb.larc.nasa.gov/cgi-bin/sse/homer.cgi (accessed on 15 August 2018).
50. Garni, H.Z.A.; Awasthi, A.; Ramli, M.A. Optimal design and analysis of grid-connected photovoltaic under different tracking systems using HOMER. *Energy Convers. Manag.* **2018**, *155*, 42–57. [CrossRef]
51. Bimble Solar. Available online: https://www.bimblesolar.com/NiFe200ah (accessed on 5 September 2018).
52. Aziz, A.S.; Tajuddin, M.F.N.; Adzman, M.R. Feasibility analysis of PV/wind/battery hybrid power generation: A case study. *Int. J. Renew. Energy Res.* **2018**, *8*, 661–671.
53. Nacer, T.; Hamidat, A.; Nadjemi, O.; Bey, M. Feasibility study of grid connected photovoltaic system in family farms for electricity generation in rural areas. *Renew. Energy* **2016**, *96*, 305–318. [CrossRef]
54. Mamaghani, A.H.; Escandon, S.A.A.; Najafi, B.; Shirazi, A.; Rinaldi, F. Techno-economic feasibility of photovoltaic, wind, diesel and hybrid electrification systems for off-grid rural electrification in Colombia. *Renew. Energy* **2016**, *97*, 293–305. [CrossRef]
55. Alharthi, Y.; Siddiki, M.; Chaudhry, G. Resource Assessment and Techno-Economic Analysis of a Grid-Connected Solar PV-Wind Hybrid System for Different Locations in Saudi Arabia. *Sustainability* **2018**, *10*, 3690. [CrossRef]
56. Alam, M.; Bhattacharyya, S. Decentralized renewable hybrid mini-grids for sustainable electrification of the off-grid coastal areas of Bangladesh. *Energies* **2016**, *9*, 268. [CrossRef]
57. Rendall, C.O. Economic Feasibility Analysis of Microgrids in Norway: An Application of HOMER Pro. Master's Thesis, Norwegian University of Life Sciences, Ås, Norway, 2018.
58. HOMER Energy Index. Available online: https://www.homerenergy.com/products/pro/docs/3.11/index.html (accessed on 25 September 2018).
59. Maheri, A. Multi-objective design optimisation of standalone hybrid wind-PV-diesel systems under uncertainties. *Renew. Energy* **2014**, *66*, 650–661. [CrossRef]
60. Tharani, K.L.; Dahiya, R. Choice of battery energy storage for a hybrid renewable energy system. *Turk. J. Electr. Eng. Comput. Sci.* **2018**, *26*, 666–676. [CrossRef]
61. Fazelpour, F.; Soltani, N.; Rosen, M.A. Economic analysis of standalone hybrid energy systems for application in Tehran, Iran. *Int. J. Hydrogen Energy* **2016**, *41*, 7732–7743. [CrossRef]
62. Wang, Y.; Jiao, X.; Sun, Z.; Li, P. Energy Management strategy in consideration of battery health for PHEV via stochastic control and particle swarm optimization algorithm. *Energies* **2017**, *10*, 1894. [CrossRef]
63. Bordin, C.; Anuta, H.O.; Crossland, A.; Gutierrez, I.L.; Dent, C.J.; Vigo, D. A linear programming approach for battery degradation analysis and optimization in offgrid power systems with solar energy integration. *Renew. Energy* **2017**, *101*, 417–430. [CrossRef]
64. Rivera-Barrera, J.; Muñoz-Galeano, N.; Sarmiento-Maldonado, H. SoC estimation for lithium-ion batteries: Review and future challenges. *Electronics* **2017**, *6*, 102. [CrossRef]
65. Galad, M.; Spanik, P.; Cacciato, M.; Nobile, G. Comparison of common and combined state of charge estimation methods for VRLA batteries. In Proceedings of the ELEKTRO, Strbske Pleso, Slovakia, 16–18 May 2016.

© 2019 by the authors. Licensee MDPI, Basel, Switzerland. This article is an open access article distributed under the terms and conditions of the Creative Commons Attribution (CC BY) license (http://creativecommons.org/licenses/by/4.0/).

Article

A Two-Step Approach to Solar Power Generation Prediction Based on Weather Data Using Machine Learning

Seul-Gi Kim, Jae-Yoon Jung and Min Kyu Sim *

Department of Industrial & Management Systems Engineering, Kyung Hee University, 1732 Deogyeong-daero, Giheung-gu, Yongin-si, Gyenggi-do 17104, Korea; nysg6190@khu.ac.kr (S.-G.K.); jyjung@khu.ac.kr (J.-Y.J.)
* Correspondence: mksim@khu.ac.kr; Tel.: +82-31-201-2537

Received: 4 February 2019; Accepted: 9 March 2019; Published: 12 March 2019

Abstract: Photovoltaic systems have become an important source of renewable energy generation. Because solar power generation is intrinsically highly dependent on weather fluctuations, predicting power generation using weather information has several economic benefits, including reliable operation planning and proactive power trading. This study builds a model that predicts the amounts of solar power generation using weather information provided by weather agencies. This study proposes a two-step modeling process that connects unannounced weather variables with announced weather forecasts. The empirical results show that this approach improves a base approach by wide margins, regardless of types of applied machine learning algorithms. The results also show that the random forest regression algorithm performs the best for this problem, achieving an R-squared value of 70.5% in the test data. The intermediate modeling process creates four variables, which are ranked with high importance in the post-analysis. The constructed model performs realistic one-day ahead predictions.

Keywords: renewable energy; solar power generation prediction; smart grid; photovoltaic power; machine learning

1. Introduction

A smart grid is an electrical grid system that manages energy-related operations, including production, distribution, and consumption. Efficient smart grid operations are aided by reliable power supply planning. Supply planning on renewable energy operations, such as sunlight, wind, tides, and geothermal energy, involves a unique (unique class) class of prediction problem because these natural energy sources are intermittent and uncontrollable, due to fluctuating weather conditions [1]. (This paper is the expanded version of the cited conference paper.)

The photovoltaic geographic information system (PVGIS) [2] provides climate data and the performance assessment tools of photovoltaic (PV) technology mainly for Europe and Africa. Based on historical averages, PVGIS offers a practical guideline for expected solar radiance in geological locations. Also, many studies are conducted to predict the level of future solar irradiance or PV power generation in solar plants using weather information.

Sources of weather information include both measured weather records and weather forecasts. This study finds that most previous studies have focused on exploiting only single source and that few studies have attempted to utilize both information sources. Thus, this study proposes a novel two-step prediction process for PV power generation using both weather records and weather forecasts. This study demonstrates the philosophy of data-driven modeling with as much relevant data as possible to improve model performance.

Popular prediction methods for solar irradiance or PV power generation can be largely divided into three categories [3]. The first category is physical methods that predict the future solar position and the resulting irradiance without relying on other climate data. Though the prediction of the solar position can be significant, this approach is likely to overlook other relevant climatic conditions. For example, the sky condition of clouds or rain blocks solar irradiance. The second category is statistical methods, which can be further divided into classical methods and modern statistical-learning based methods (also known as machine learning). With rapid developments of statistical learning methods over the last decade, many studies have adopted this data-driven approach to developing PV prediction models [4]. Lastly, hybrid methods [5,6] apply not only statistical methods but also other methods, such as mathematical optimization or signal processing.

Since many studies using statistical learning methods have appeared, a paper reviewing these studies is also published [4]. This review paper classifies the line of studies according to adopted machine learning algorithms. However, no review study has attempted to discuss data sources of the predictive studies in our knowledge. Needless to say, which data source is used in a data-driven approach is crucial to the model performance, so this study briefly reviews the sources of predictors used in existing papers.

First, there is a group of studies that use recorded weather observations as key predictors. In the case of using current weather as predictors, an implied hypothesis is that future irradiance and PV generation are related to the current weather. Studies in this stream adopt methods, such as neural networks [7], heterogeneous regressions [8], and deep belief network [9]. When the time span of recorded weather observations is expanded, time-series analysis approaches are adopted, such as autoregressive moving average (ARMA) [10], autoregressive integrated moving average (ARIMA) [11–13], and a few variants of recurrent neural networks (RNNs) [14,15]. These studies have shown significant predictability. However, using only actual weather records is likely to be a suboptimal strategy.

Instead, utilizing weather forecasts that reliable weather agencies announce in punctual manners has certain benefits. Thus, a greater number of studies adopt weather forecasts as primary predictors. These studies [16–23] model future PV power generation using announced weather forecasts targeted for the future time. Nonetheless, weather forecasts have some issues in terms of data quality. First, they are not exactly accurate, and the weather agencies typically announce values under concerns of risk averseness [24]. It may limit the performance of resulting predictive models that rely only on weather forecasts. Second, weather forecasts by weather agencies tend to include fewer variables compared to weather records. For example, the Korea Meteorological Administration (KMA) (The KMA is the central administrative body of the Republic of Korea that oversees weather-related affairs.) announces forecasts only for the surface temperature of the ground, while the KMA observes and records 10 cm-, 20 cm-, and 30 cm- underground temperatures as well. Lastly, due to the concerns about inaccuracy, several variables in weather forecasts are announced in less fine units, often in the form of categorical variables instead of numerical variables. Regarding the quality of data alone, weather observation is, therefore, a richer and more accurate data source.

Due to the pros and cons of weather observations and weather forecasts, we believe that these two data sources should be utilized in a complementary manner. In fact, a few studies [24,25] use both observations and forecasts for prediction. Bacher et al. [25] propose an adaptive linear time series model whose autoregressive component for recent solar irradiation is supplemented by an exogenous input of weather forecasts. Interestingly, they report that recent weather records are more important when the forecasting horizon is less than two hours. On the other hand, weather forecasts begin contributing more when the forecasting horizon becomes longer than two hours. Detyniecki et al. [24] adopt a fuzzy decision tree learning that takes both weather forecast and weather observation into their input.

This study first built a base model that uses weather forecasts to predict solar power generation. The focus was then moved to the existence of a set of the variables, which we call auxiliary variables, that are not included in weather forecasts but are observed by weather agencies. In particular, the solar radiation among the auxiliary variables is known as a significant predictor for solar power generation [8,26]. Therefore, an auxiliary model identifies the relationship between weather forecast variables and the auxiliary variables, then the main model for solar power generation uses both weather forecast variables and the auxiliary variables generated by the auxiliary model. In the language of statistical learning, the base model aims to identify a regression function that relates the weather forecast variables and the solar power generation. The main model additionally incorporates the identified relationship between the weather forecast variables and the auxiliary variables into the process of training another function. The auxiliary variables can be understood as latent variables—not directly observable but can be inferred from attainable variables of weather forecasts.

Figure 1 presents a graphical abstract of the models proposed in this study. Suppose the prediction target is for time $t + 1$ and the prediction is made at time t. Weather forecast $\hat{\mathbf{F}}_{t+1|t}$ contains weather forecast variables announced by weather agency at time t, targeted for the weather at time $t + 1$. The hat notation implies that this vector contains forecasted values. The base model f predicts power generation at time t, y_{t+1}, from the weather forecast $\hat{\mathbf{F}}_{t+1|t}$. Weather observation $\mathbf{O}_{t+1}$ contains variables actually observed at time $t + 1$ but not forecasted by the weather agency prior to the time $t + 1$. Therefore, the auxiliary model identifies the best regression function g^* from a parametrized family of g. Lastly, the main model uses $\hat{\mathbf{O}}_{t+1} = g^*(\hat{\mathbf{F}}_{t+1|t})$ from the auxiliary model along with the original forecast $\hat{\mathbf{F}}_{t+1|t}$ in order to predict power generation y_{t+1}.

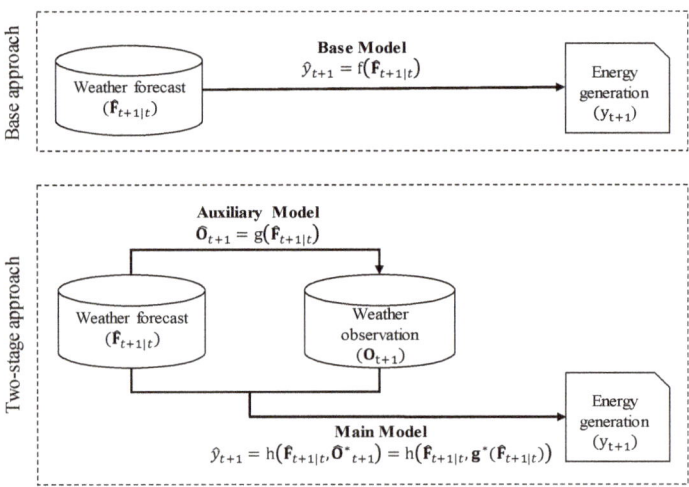

Figure 1. The base model and the proposed two-step approach for solar power generation prediction based on weather data.

In building the three prediction models, this study tests multiple machine learning algorithms that have been frequently used for predictive analytics [4]. The tested algorithms include linear regression, support vector regression (SVR) [27], classification and regression tree (CART) [28], k-nearest neighbors (k-NN) [29,30], adaptive boosting (AdaBoost), random forest regression (RFR) [31], and artificial neural network (ANN) [7,32,33].

The study contributes to the research lines in the following ways:

- This study proposes an approach to expanding predictors for the prediction of solar power generation. It exemplifies a practical application to include relevant but delayed climatic data that are not available in real-time.

- Many practical applications, including renewable energy operations, call for predictions using weather information as predictors. The proposed approach can be applied to predictions for renewable energy operations, such as wind, tide, and geothermal power production.
- Generally, identifying latent variables and incorporating them in the prediction process often enhance the model performance. The proposed two-step approach does so with various machine learning algorithms.
- In applications of machine learning methods, identifying latent variable structures in prior often enhances the performance of resulting models. This study indirectly investigates how much each machine learning algorithm gains benefit from the intermediate latent variable identification process.

2. Materials and Methods

This section describes the data set and methods used to develop the models. Section 2.1 describes the sources of the data and the preprocessing steps and Section 2.2 briefly explains the machine learning algorithms used in this study. Section 2.3 finally formulates the prediction problems that the three proposed models aim to solve.

2.1. Data Collection and Preprocessing

A solar power generation data from the Yeongam Photovoltaic Power Plant in South Korea were collected from a publicly available database (http://www.data.go.kr) provided by the government. The weather-related data were provided by the KMA. Solar elevation information was obtained from a database by the Stellarium®.

The variables in the dataset can be divided into four categories as listed in Table 1. First, hourly power generation data were collected. The hourly data excluded daylight-free hours (00:00–08:00 and 20:00–24:00) and were collected over three years from 2013-01-01 to 2015-12-31. Second, weather forecast data were collected. This study collected all available variables for the same period where power generation data were available during the corresponding period announced for the same period. The constructed models predicted future power generation amounts using weather forecast data announced for the future period. The KMA announced short-term weather forecasts at the city or district level for each three-hour period from 02:00 each day. We used the forecast data announced at 11:00 targeted for 09:00, 12:00, 15:00, and 18:00 of the following day (corresponding to 22, 25, 28, and 31 hours after the announcement, respectively). Solar elevation data were collected from an open source program called Stellarium (www.stellarium.org). Specifically, we estimated the solar elevation (0°–90°) for the same period using the latitude 34.751702 and longitude 126.458533, which is the geographical location of the power plant. The position of solar affects how much solar radiation energy is collected at the ground, along with other weather conditions, such as rain, snow, cloud, and the density of air. Third, actual weather records were obtained. This study included all-weather observation variables that were not included in weather forecasts. In the step of auxiliary modeling, this study built a prediction model for the weather observation variables (Radiation, VaporPressure, SurfaceTemperature, and AtmosphericPressure) using weather forecast data, called *auxiliary variables* in this research.

The preprocessing task prepared the data into the structure suitable for quantitative modeling. Categorical variables in the weather forecasts, such as RainfallType, SkyType, and WindDirection, were converted to multiple binary variables through one-hot coding. A week index variable (Weeknum) was created to reflect seasonal changes. This variable assigns index sequentially from the first week to the last week of each year. To include information about the time of the day, the variable TimeZone was used to indicate three-hour intervals.

Table 1. Dependent and Independent Variables [1].

	Source	Variable Name	Description
Dependent variable	Power plant (y)	Generation	Solar power generation (kWh)
Independent variable	Weather forecast ($\hat{F}$)	RainfallType	0: none, 1: rain, 2: rain/snow, 3: snow
		SkyType	1: sunny, 2: a little cloudy, 3: cloudy, 4: overcast
		WindDirection	1: west, 2: east, 3: south, 4: north
		WindSpeed	Wind speed (m/s)
		Humidity	Humidity (%)
		Temperature	Temperature (°C)
	Weather observation (O)	Elevation	Solar Elevation (0°–90°) by Stellarium®
		Radiation	Radiation (MJ/m^2)
		VaporPressure	Vapor pressure (hPa)
		SurfaceTemperature	Surface temperature (°C)
		AtmosphericPressure	Atmospheric Pressure (hPa)
	Derived variables	Weeknum	Weekly index (1–53)
		TimeZone	1: 09:00–12:00, 2: 12:00–15:00, 3: 15:00–18:00, 4: 18:00–21:00

[1] The total number of available observations is 4380 (1095 days × 4 observations/day).

2.2. Machine Learning Methods

This subsection briefly describes the machine learning methods tested in this study. The methods include popular supervised learning methods in the research line [4]. Linear regression is a simple but effective modeling technique where the linear relationship between independent variables and a dependent variable is to be identified. SVR is a variant of linear regression where prediction error that is smaller than some threshold is ignored in order to minimize the effect of outliers. Kernel functions, such as polynomial and radial basis functions, help the SVR perform the non-linear separation [26,34]. ANN is becoming an increasingly popular method for non-linear regression due to its effectiveness in data prediction. To find the relationship between input and output nodes, multi-layered hidden nodes are connected and their weights are updated through the error backpropagation algorithm [6,31,32,35,36]. CART, also known as recursive partitioning, splits an entire data set into two groups by searching for the best split condition that can reduce the sum of squared errors (SSE) mostly. This binary partitioning occurs recursively until each leaf node reaches to have enough impurity [27,37]. k-NN is a non-parametric method used for classification and regression [28,29]. For each instance, the predicted value is based on the weighted average value of the k neighborhood instances where each of the weight is commonly given as an inverse value of the distance between the target instance and each instance of the k nearest neighbors. Since k-NN treats input variables indiscriminately, this study scales and normalizes all the input instances in a preprocessing step.

As a representative ensemble learning method, AdaBoost fits additional copies of the decision tree but with the weights adjusted to the error of the current prediction. By subsequently focusing more on difficult instances, the learning mechanism boosts weaker learners to produce powerful "committees" [3]. Another powerful ensemble implementation is RFR, which consists of a collection of decision trees that are built from each bootstrapping sampling of the entire data set [30]. Averaging values from each tree, RFR generates a prediction value.

In the post-analysis, this study measures the Gini importance or the mean decrease in impurity (MDI) as an important measure to investigate the effect of each predictor. The Gini importance is defined as the total decrease in node impurity, averaged over all trees of the ensemble. By sorting the predictors using the Gini importance, the contribution of each predictor can be evaluated.

2.3. Problem Statements

In this study, we aimed to build a model that predicts solar power generation one day ahead of the actual operation. The base model identified the best function f^* in which the predictors were limited to the weather forecast variables.

$$f^* = \text{argmin}_f L(y_{t+1}, \hat{y}_{t+1}) \text{ s.t. } \hat{y}_{t+1} = f(\hat{\mathbf{F}}_{t+1|t}) \tag{1}$$

where $\hat{\mathbf{F}}_{t+1|t}$ is a vector of weather forecast variables available at day t and targeted for day $t + 1$ (The hat notation emphasizes that this quantity is forecasted.), y_{t+1} is a quantity of power generation at day $t + 1$, and L is a cost function where this study adopted the measure of mean squared error (MSE) as a popular choice.

Though a few variables in weather observation were missing in weather forecasts, this study aimed to fully exploit weather information for building prediction process. That is, the weather observation variables were predicted using weather forecast variables. This auxiliary model aimed to find the best performing function g^*, such as

$$g^* = \text{argmin}_g L(\mathbf{O}_{t+1}, \hat{\mathbf{O}}_{t+1}) \text{ s.t. } \hat{\mathbf{O}}_{t+1} = g(\hat{\mathbf{F}}_{t+1|t}) \tag{2}$$

where $\mathbf{O}_{t+1}$ is a vector of weather observation variables that are known to be related to solar power but not included in weather forecast [7,25].

Finally, the main model aimed to exploit the two previous models by including both $\hat{\mathbf{O}}_{t+1} = g^*(\hat{\mathbf{F}}_{t+1|t})$ and $\hat{\mathbf{F}}_{t+1|t}$ as predictors. The main model identified the best function h^* such that

$$h^* = \text{argmin}_h L(y_{t+1}, \hat{y}_{t+1}) \text{ s.t. } \hat{y}_{t+1} = h(\hat{\mathbf{F}}_{t+1|t}, g^*(\hat{\mathbf{F}}_{t+1|t})) \tag{3}$$

where g^* is obtained from the auxiliary model.

The base model provides a baseline for comparisons to the main model, which includes generated predictors. Since predictive relationships are complex and difficult to grasp, this study tests several machine learning algorithms, such as linear regression, SVR, CART, k-NN, AdaBoost, and RFR, which are suitable for the structure of the data and the problem. Before applying the machine learning algorithms, proper scaling is performed. Specifically, distance-based methods, including k-NN and SVR, need standardization so-called z-score normalization, in order to carry comparable importance in model generation process [34]. To calculate z-score, each variable x is subtracted by its mean μ and divided by its standard deviation σ, that is, $z = (x - \mu)/\sigma$. ANN needs a min-max scaling to a bounded range, such as between 0 and 1, in these experiments. The normalized value can be calculated by $(x - min(x))/(max(x) - min(x))$. This step is necessary so that all variables are in a comparable range before fed into a network [34]. Tree-based methods, such as AdaBoost, CART, and RFR, do not need scaling since they bisect each variable in a non-parametric manner [34]. Linear regression does not need to scale the data, either. By optimizing parameters under the train set, prediction models based on each machine learning algorithm are built with the machine learning package in Python, scikit-learn [38].

3. Results

This section presents the results of the methods described in the previous section. The results identify (1) which machine learning method produces the best-performing model, (2) whether the predicted values for auxiliary variables created during the auxiliary modeling step have significant forecasting performance for solar power generation, and (3) how much each independent variable among weather forecast and weather observation contributes to prediction performance. Section 3.1

explains the setting of experiments, Section 3.2 presents the performance of the auxiliary model formulated as Equation (2), and Section 3.3 compares performances of the base model in Equation (1) and the main model in Equation (3).

3.1. Measures for Model Comparison

To build models, the data for three years were split to a training set (30 months; from 2013-01-01 to 2015-06-30) and a test set (6 months; from 2015-07-01 to 2015-12-31). Using the train set, five-fold cross-validation was performed to find the best model for each prediction algorithm. The random search technique was adopted to search the proper parameter set of the best model.

An error measure of the mean squared error (MSE) was employed in choosing the best model among candidates. Specifically, the MSE measures an average value of squares of errors, formulated as:

$$MSE = \frac{1}{N} \sum_{i=1}^{N} (y_i - \hat{y}_i)^2 = RMSE^2 \tag{4}$$

where y_i is the i-th actual value, $\hat{y}_i$ is the predicted value for y_i, N is the number of samples, and RMSE implies the square root of MSE. Along with the MSE, this paper presents two other error measures, the R-squared value and the adjusted R-squared value. The R-squared value R^2, also known as the coefficient of determination, is the proportion of the variance of the dependent variable that is explained by the independent variables.

$$R^2 = 1 - \frac{\sum_{i=1}^{N}(y_i - \hat{y}_i)^2}{\sum_{i=1}^{N}(y_i - \bar{y})^2} \tag{5}$$

where $\bar{y}$ is the mean of the actual values of y. The adjusted R-squared value, denoted R^2_{adj}, penalizes the number of independent variables used to generate the predicted value, after measuring the proportion of the variance explained by independent variables.

$$R^2_{adj} = 1 - (1 - R^2)\frac{N-1}{N-p-1} \tag{6}$$

where p is the total number of the independent variables in the model.

3.2. Performance of Auxiliary Model

The proposed approach of this study features a two-step process, of which the first step predicts the observed variables (O) using the forecast variables ($\hat{F}$). The intermediate result created by this auxiliary model with RFR is presented in Table 2. Among four auxiliary variables, the prediction made on the first three variables, Radiation, VaporPressure, and SurfaceTemperature, are highly accurate with R^2 higher than 97%. The other variable AtmosphericPressure also has generally acceptable accuracy. Having these auxiliary variables is equivalent to having another set of weather forecast when predicting the future solar power generation.

Table 2. Performance of the auxiliary model on the test set.

	RMSE	R^2
Radiation	0.128	97.0
VaporPressure	0.743	99.3
SurfaceTemperature	1.252	98.7
AtmosphericPressure	4.288	72.4

[1] See Table A2 for selected hyperparameters to generate the models.

3.3. Performance of Base and Main Models

For the base model and the main model, popular machine learning algorithms in the line of studies are applied. Table 3, sorted by MSE in the main model, presents performances in the test set. It can be seen that R^2 for the test ranges from 57.9% to 70.1% in the base model, and from 67.2% to 70.5% in the main model. In the base model, RFR outperforms the others by large margins. Other methods exhibit similar performance except that k-NN performs poorly. In the main model, RFR still performs the best, but the margin is narrowed as other methods gain more from the two-step prediction process employed in the main model.

Table 3. Performance of the base model and the main model in the test set.

Algorithm	Base Model		Main Model		Improvement	
	RMSE	R^2	RMSE	R^2	RMSE	R^2
AdaBoost	669.5	0.604	609.2	0.672	60.3 (9.0%)	0.068
Linear Reg.	635.5	0.643	608.6	0.673	26.9 (4.2%)	0.030
CART	619.2	0.661	607.9	0.673	11.3 (−1.8%)	0.012
SVR	689.9	0.579	605.7	0.676	84.2 (12.2%)	0.097
ANN	606.0	0.675	597.4	0.684	−8.6 (1.4%)	0.009
k-NN	630.2	0.649	596.4	0.686	−35.8 (5.7%)	0.037
RFR	581.5	0.701	577.5	0.705	4.0 (−0.7%)	0.004

[2] Algorithms are ordered by R^2 of the main model, [3] See Tables A1 and A3 for selected hyperparameters to generate the models.

Figure 2 emphasizes the improvements in accuracy from utilizing the two-step process. By incorporating auxiliary variables (**O**), each algorithm experiences an improvement as much as 9.7% (R^2 of SVR). The best performing algorithm, RFR, gains 0.4% improvement in R^2.

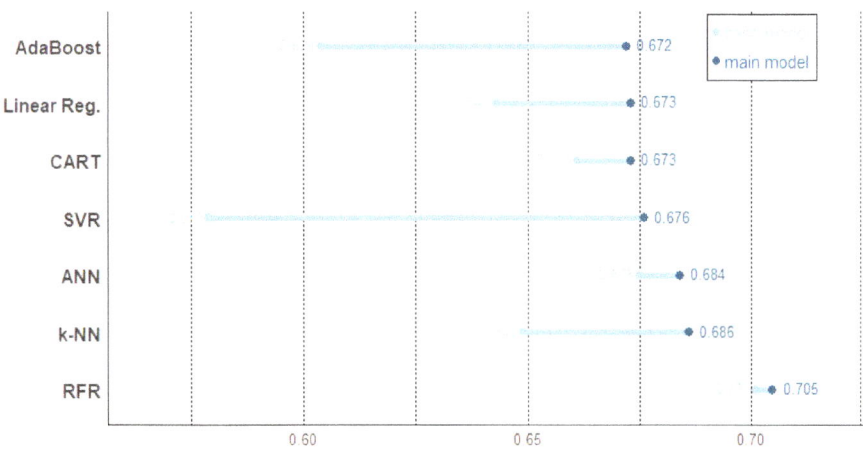

Figure 2. Improved prediction performance in terms of R^2 from the base model to the main model.

Figure 3 presents a time-series plot of predicted and actual values in a month (August 2015) of the test set. The predicted values are produced by the best RFR model with a parameter fitted by learning the train set. Overall, the predicted values track the fluctuations of actual power generation well, except for a series of under-predictions for the peak hour in early days in August and a big over-prediction on the peak hours on 22 August which is unavoidable by an unpredicted weather event. The day was unexpectedly foggy and heavily clouded (the maximally clouded day in August).

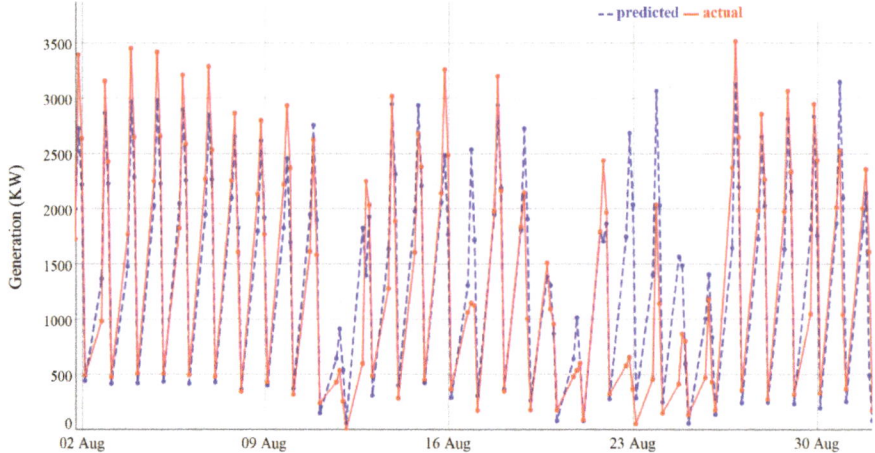

Figure 3. Actual values and predicted values by RFR from 1 August to 31 August 2015.

In the experiments on prediction models so far, all available variables in Table 1 are used. The advantages of the two-step approach are validated under this untouched setting. As a post analysis, the necessity of each predictor is assessed using a classical variable selection method, called *backward elimination*. Backward elimination starts with all variables, and a single variable is removed in each step until doing so would reduce the overall performance of the model. A performance measure R^2_{adj} is used for this process, which penalizes the number of predictors so that a more concise model is promoted. Figure 4 presents which predictor is removed at each step.

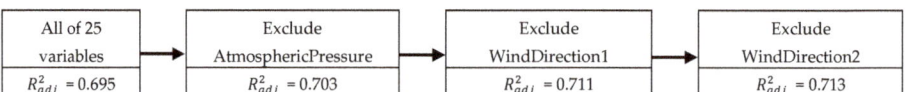

Figure 4. Backward elimination process with the RFR model.

The process begins with 25 predictors, including binary dummy variables, generated from categorical variables. This original model has a R^2_{adj} value of 0.695. Excluding AtmosphericPressure would enhance the R^2_{adj} value to 0.703, yielding a prediction model with 24 predictors. This weather variable for sea-level pressure turns out to be secondary to direct weather variables. Then, excluding WindDirection1 (West) and WindDirection2 (East) would enhance the R^2_{adj} value to 0.711 and 0.713, respectively. Winds blowing from North or South carry more information compared to the winds blowing from East or West. No further removal is beneficial in terms of R^2_{adj}. This process ensures some redundant, highly correlated or ineffective predictors to be removed. After the backward elimination process, the final model contains the smallest number of essential variables, but still achieves high prediction accuracy.

3.4. Importance of Variables

The above experiments demonstrate that the proposed two-step approach to solar power generation prediction improves the performance compared to the base model, regardless of the tested algorithms. Another way to validate its benefits is to measure whether the auxiliary variables are indeed pivotal components in the main model. Determining the necessary predictors, the importance of each variable in the final model is examined in the next subsection. Because the RFR model performs best, we adopt the Gini importance or mean decrease in impurity (MDI) as an important measure. The MDI is defined as the total decrease in node impurity, averaged over all trees of the ensemble. By sorting the predictors using the important measure, the contribution of each predictor is ranked.

Figure 5 presents the Gini importance of each variable in the main model with RFR. This figure supports the hypothesis for the benefit in the two-step process. One of the auxiliary variable, Radiation, is the most important variable with the importance of 43.7%. Other auxiliary variables, such as SurfaceTemperature and VaporPressure, are ranked in the upper half among all variables. The top four important variables consist of how much solar radiation is emitted (Radiation), from which solar position (Elevation), at what time of the day (TimeZone5 and TimeZone3). The condition of the air (Humidity) and temperatures (SurfaceTemperature and Air-Temperature) also affect solar power generation. Sky condition of overcast (SkyType4) also plays a role.

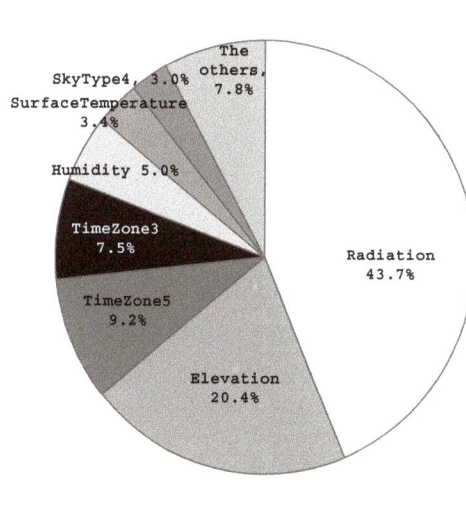

Variable	Gini
Radiation*	43.7%
Elevation	20.4%
TimeZone5	9.2%
TimeZone3	7.5%
Humidity	5.0%
Surface Temperature*	3.4%
SkyType4	3.0%
VaporPressure*	1.7%
TimeZone4	1.2%
(Air)-Temperature	1.0%
Weeknum	1.0%
WindSpeed	0.8%
TimeZone2	0.6%
RainfallType0	0.5%
SkyType1	0.4%
SkyType3	0.4%
RainfallType1	0.1%
WindDirection4	0.1%
WindDirection3	0.1%
SkyType2	0.0%
RainfallType3	0.0%
RainfallType2	0.0%

* indicates an auxiliary variable.

Figure 5. Gini importance of variables in the main model based on RFR.

4. Discussion

For the prediction of solar power generation during operations, weather forecast variables are readily available. On the other hand, the auxiliary variables are not available to use. The auxiliary modeling step utilizes the historical data to identify the relationship between available forecasts and suitable predictions for the auxiliary variables. The generated predictions for the auxiliary variables using available forecasts are highly accurate (see Table 2). The main model utilizes the predicted values for the auxiliary variables along with available forecasts. On comparing the base models and the main models with popular machine learning algorithms, it is shown that the main models successfully improved the performance of the base models (see Table 3, Figure 2, and Figure 3). Among the tested machine learning algorithms, the models generated by RFR outperform the other models. The relative importance for each predictor is identified (see Figure 5) after the removal of a few variables (see Figure 4).

The results can be interpreted as follows:

- For predicting the solar power generation, the forecasts for the amount of solar radiation is the most important among the others, in terms of the Gini importance. The forecast for solar radiation is not directly available from the weather agency but can be indirectly generated by the proposed auxiliary model. Next, the position of the solar relative to the ground (Elevation) carries important information, and the operation time of the day affects the power generation. Since solar elevation can be accurately forecasted by astrophysics and the time of the day (TimeZone) is deterministic, the future information for these two variables are attainable accurately. The condition of the atmosphere (Humidity and VaporPressure) and the temperatures (SurfaceTemperature and Air-Temperature) also affect the power generation.

- Forecasts for auxiliary variables are not readily available during actual power generation operations, but their values are later realized and highly correlated to the solar power generation. This relationship is captured by the auxiliary model, and the main model exploits this information to outperform the base model regardless of the machine learning methods applied. This approach, regarded as identification of latent variables, enhances the performances of solar power prediction.
- On comparing the different machine learning methods, models with higher capacity, such as RFR, k-NN, and ANN, perform relatively well. RFR, the best performing method, is characterized by its ensemble approach with multiple randomized trees and known for its robustness in the test data set. It is generally known that RFR is especially suitable when multiple categorical variables are involved, as in our case. The main results support the robustness and good performance of RFR.

5. Conclusions

This study proposes a two-step approach to solar power generation prediction to fully exploit the information contained in the weather data. Specifically, the predicted values for auxiliary variables contribute greatly to enhancing prediction performance.

Studies in this line present a wide range of errors from 3% to 38% [4]. The large over-prediction on 22 August due to an unexpected weather event (see Figure 3) indicates how the error distribution can be highly skewed. Skewed errors result in lower overall accuracy, especially for the power plant located in areas of unpredictable weather. In particular, the power plant of this study is located in a landfill area in the southwestern part of the Korean peninsula, which is surrounded by three seas and 70% of whose total area is mountainous, making weather predictions very difficult. To aid actual operations, it would be meaningful in future studies, especially for areas with low weather predictability, to present confidence intervals of the predicted value, as well as the predicted values themselves.

This paper exemplifies a practical application of feature extraction such that latent variables, relevant but delayed weather data in this study, are identified prior to the main modeling. This study validates that the process of latent structure identification improves the solar power generation problem and aids PV plant operations.

Furthermore, other renewable energy operations, such as wind, tide, and geothermal power production, can also be benefitted from the proposed approach. More generally, it can also be applied to other fields that require predicting future weather conditions.

Author Contributions: S.-G.K. and J.-Y.J. designed the research purposes and methodology; S.-G.K. developed the software that is validated and further developed by M.K.S.; S.-G.K. and M.K.S. analyzed the data; S.-G.K. and M.K.S. wrote the original draft; M.K.S. and J.-Y.J. reviewed and finalized the draft. This research project was conducted under the supervision of J.-Y.J.

Funding: This work was supported by the National Research Foundation of Korea (NRF) grant funded by the Korea government (MSIT) (NRF-2017H1D8A2031138).

Conflicts of Interest: The authors declare no conflict of interest.

Appendix A Candidates and Optimal Values of Proposed Models

This appendix section presents tables for illustrating hyperparameter tuning process for each model proposed in this study.

Table A1. Hyperparameter candidates and optimal values for the base model.

Method	Set of Considered Hyperparameters (The Selected Value is in the Gray Box)	Description
AdaBoost	n_estimators: 30, 31, ..., **94**, ..., 99, 100	The maximum number of estimators at which boosting is terminated.
	learning_rate: 0.01, 0.05, **0.1**, 0.3, 0.5, 0.7, 1.0	Learning rate of shrinking the contribution of each regressor.
	loss: linear, square, **exponential**	The loss function to use when updating the weights after each boosting iteration.
Linear Reg.	fit_intercept = True	Whether to calculate the intercept for this model.
CART	max_depth: 1, 2, ..., **6**, ..., 29, 30	The maximum depth of the tree.
	max_features: 1, 2, ..., **14**, ..., 20, 21	The number of features to consider when looking for the best split.
	min_sample_split: 2, 3, **4**, ..., 10, 11	The minimum number of samples required to split an internal node.
	min_sample_leaf: 1, 2, ..., **8**, 9, 10, 11	The minimum number of samples required to be at a leaf node.
SVR	C: 0.1, 0.25, 0.3, 0.5, 0.75, **1**	Penalty parameter C of the error term.
ANN	hidden_layer_sizes: 100, 105, ..., **265**, ..., 295, 300	The number of neurons in a single hidden layer.
	activation: logistic, tanh, **ReLU**	The activation function for the hidden layer.
	learning_rate: **constant**, invscaling, adaptive	Learning rate schedule for weight updates.
	max_iter: 1000, **2000**, 3000	The maximum number of iterations.
k-NN	n_neighbors: 5, 6, ..., **18**, 19, 20	The number of neighbors to use by default.
	algorithm: auto, ball_tree, kd_tree, **brute**	The algorithm used to compute the nearest neighbors.
	weights: **uniform**, distance	Weight function used in the prediction.
RFR	n_estimators: 5, 6, ..., **15**, ..., 19, 20	The number of trees in the forest.
	max_depth: 1, 2, ..., **13**, ..., 29, 30	The maximum depth of the tree.
	max_features: 1, 2, ..., **9**, ..., 20, 21	The number of features to consider when looking for the best split.
	min_samples_split: 2, 3, ..., **10**, 11	The minimum number of samples required to split an internal node.
	min_sample_leaf: 1, 2, ..., **9**, 10, 11	The minimum number of samples required to be at a leaf node.

Table A2. Hyperparameter candidates and optimal values for the auxiliary models with the random forest regression method.

Radiation							Vapor Pressure						
n_estimators							n_estimators						
5	6	...		15	...	19	20	5	6	...		...	18	19	20
max_depth							max_depth						
1	2	...		15	...	29	30	1	2	...			...	29	30
max_features							max_features						
1	2	...		13	...	16	17	1	2	...		12	...	16	17
min_samples_split							min_samples_split						
2	3	...			...	10	11	2	3	4	...		...	10	11
min_sample_leaf							min_sample_leaf						
1	2	...			...	10	11	1	2	...			...	10	11

Surface Temperature							Atmospheric Pressure						
n_estimators							n_estimators						
1	2	...			...	19	20	1	2	...		17	...	19	20
max_depth							max_depth						
1	2	...		23	...	29	30	1	2	...	10		...	29	30
max_features							max_features						
1	2	...		12	...	16	17	1	2	...	9		...	16	17
min_samples_split							min_samples_split						
2	3	4	...		...	10	11	2	3	...			...	10	11
min_sample_leaf							min_sample_leaf						
1	2	...			...	10	11	1	2	...		7	...	10	11

Table A3. Hyperparameter candidates and optimal values for the main model.

Method	Set of Considered Hyperparameters (The Selected Value is in the Gray Box)	Description
AdaBoost	n_estimators: 30, 31, …, **90**, …, 99, 100	The maximum number of estimators at which boosting is terminated.
	learning_rate: 0.01, **0.05**, 0.1, 0.3, 0.5, 0.7, 1.0	Learning rate of shrinking the contribution of each regressor.
	loss: linear, square, **exponential**	The loss function to use when updating the weights after each boosting iteration.
Linear Reg.	fit_intercept = True	Whether to calculate the intercept for this model.
CART	max_depth: 1, 2, …, **6**, …, 29, 30	The maximum depth of the tree.
	max_features: 1, 2, …, **19**, 20, 21	The number of features to consider when looking for the best split.
	min_sample_split: 2, 3, 4, **5**, …, 10, 11	The minimum number of samples required to split an internal node.
	min_sample_leaf: 1, 2, …, **7**, …, 10, 11	The minimum number of samples required to be at a leaf node.
SVR	C: **0.1**, 0.25, 0.3, 0.5, 0.75, 1	Penalty parameter C of the error term.
ANN	hidden_layer_sizes: 100, 105, …, **290**, 295, 300	The number of neurons in a single hidden layer.
	activation: logistic, tanh, **ReLU**	The activation function for the hidden layer.
	learning_rate: **constant**, invscaling, adaptive	Learning rate schedule for weight updates.
	max_iter: 1000, **2000**, 3000	The maximum number of iterations.
k-NN	n_neighbors: 5, 6, …, **10**, …, 19, 20	The number of neighbors to use by default.
	algorithm: auto, ball_tree, kd_tree, **brute**	The algorithm used to compute the nearest neighbors.
	weights: **uniform**, distance	Weight function used in the prediction.
RFR	n_estimators: 5, 6, …, **19**, 20	The number of trees in the forest.
	max_depth: 1, 2, …, **27**, 28, 29, 30	The maximum depth of the tree.
	max_features: 1, 2, …, **10**, …, 20, 21	The number of features to consider when looking for the best split.
	min_samples_split: 2, 3, …, **7**, …, 10, 11	The minimum number of samples required to split an internal node.
	min_sample_leaf: 1, 2, …, **9**, 10, 11	The minimum number of samples required to be at a leaf node.

References

1. Kim, S.; Jung, J.-Y.; Sim, M. Machine Learning Methods for Solar Power Generation Prediction based on Weather Forecast. In Proceedings of the 6th International Conference on Big Data Applications and Services (BigDAS2018), Zhengzhou, China, 19–22 August 2018.
2. Suri, M.; Huld, T.; Dunlop, E.D. Geographic aspects of photovoltaics in Europe: Contribution of the PVGIS website. *IEEE J. Sel. Top. Appl. Earth Obs. Remote Sens.* **2008**, *1*, 34–41. [CrossRef]
3. Antonanzas, J.; Osorio, N.; Escobar, R.; Urraca, R.; Martinez-de-Pison, F.J.; Antonanzas-Torres, F. Review of photovoltaic power forecasting. *Sol. Energy* **2016**, *15*, 78–111. [CrossRef]
4. Voyant, C.; Notton, G.; Kalogirou, S.; Nivet, M.L.; Paoli, C.; Motte, F.; Fouilloy, A. Machine learning methods for solar radiation forecasting: A review. *Renew. Energy* **2017**, *1*, 569–582. [CrossRef]
5. Abedinia, O.; Raisz, D.; Amjady, N. Effective prediction model for Hungarian small-scale solar power output. *IET Renew. Power Gener.* **2017**, *11*, 1648–1658. [CrossRef]
6. Abuella, M.; Chowdhury, B. Improving Combined Solar Power Forecasts Using Estimated Ramp Rates: Data-driven Post-processing Approach. *IET Renew. Power Gener.* **2018**, *12*, 1127–1135. [CrossRef]
7. Chaouachi, A.; Kamel, R.M.; Nagasaka, K. Neural network ensemble-based solar power generation short-term forecasting. *J. Adv. Comput. Intell. Intell. Inform.* **2010**, *14*, 69–75. [CrossRef]
8. Hossain, M.R.; Oo, A.M.; Ali, A.S. Hybrid prediction method of solar power using different computational intelligence algorithms. In Proceedings of the Power Engineering Conference (AUPEC), Christchurch, New Zealand, 26 September 2012.
9. Li, L.L.; Cheng, P.; Lin, H.C.; Dong, H. Short-term output power forecasting of photovoltaic systems based on the deep belief net. *Adv. Mech. Eng.* **2017**, *9*, 1687814017715983. [CrossRef]
10. David, M.; Ramahatana, F.; Trombe, P.J.; Lauret, P. Probabilistic forecasting of the solar irradiance with recursive ARMA and GARCH models. *Sol. Energy* **2016**, *133*, 55–72. [CrossRef]
11. Pedro, H.T.; Coimbra, C.F. Assessment of forecasting techniques for solar power production with no exogenous inputs. *Sol. Energy* **2012**, *86*, 2017–2028. [CrossRef]
12. Phinikarides, A.; Makrides, G.; Kindyni, N.; Kyprianou, A.; Georghiou, G.E. ARIMA modeling of the performance of different photovoltaic technologies. In Proceedings of the 39th Photovoltaic Specialists Conference (PVSC), Tampa, FL, USA, 16–21 June 2013.
13. Hassan, J. ARIMA and regression models for prediction of daily and monthly clearness index. *Renew. Energy* **2014**, *68*, 421–427. [CrossRef]
14. Alzahrani, A.; Shamsi, P.; Dagli, C.; Ferdowsi, M. Solar irradiance forecasting using deep neural networks. *Procedia Comput. Sci.* **2017**, *114*, 304–313. [CrossRef]
15. Abdel-Nasser, M.; Mahmoud, K. Accurate photovoltaic power forecasting models using deep LSTM-RNN. *Neural Comput. Appl.* **2017**, 1–4. [CrossRef]
16. Sharma, N.; Gummeson, J.; Irwin, D.; Shenoy, P. Cloudy computing: Leveraging weather forecasts in energy harvesting sensor systems. In Proceedings of the 7th Annual IEEE Communications Society Conference, Sensor Mesh and Ad Hoc Communications and Networks (SECON), Boston, MA, USA, 21 June 2010.
17. Sharma, N.; Sharma, P.; Irwin, D.; Shenoy, P. Predicting solar generation from weather forecasts using machine learning. In Proceedings of the 2nd IEEE International Conference, Smart Grid Communications (SmartGridComm), Brussels, Belgium, 17–20 October 2011.
18. Amrouche, B.; Le Pivert, X. Artificial neural network based daily local forecasting for global solar radiation. *Appl. Energy* **2014**, *130*, 333–341. [CrossRef]
19. Zamo, M.; Mestre, O.; Arbogast, P.; Pannekoucke, O. A benchmark of statistical regression methods for short-term forecasting of photovoltaic electricity production, part I: Deterministic forecast of hourly production. *Sol. Energy* **2014**, *105*, 792–803. [CrossRef]
20. Gensler, A.; Henze, J.; Sick, B.; Raabe, N. Deep Learning for solar power forecasting-An approach using AutoEncoder and LSTM Neural Networks. In Proceedings of the 2016 IEEE International Conference, Systems, Man, and Cybernetics (SMC), Budapest, Hungary, 9 October 2016.
21. Andrade, J.R.; Bessa, R.J. Improving renewable energy forecasting with a grid of numerical weather predictions. *IEEE Trans. Sustain. Energy* **2017**, *8*, 1571–1580. [CrossRef]
22. Leva, S.; Dolara, A.; Grimaccia, F.; Mussetta, M.; Ogliari, E. Analysis and validation of 24 hours ahead neural network forecasting of photovoltaic output power. *Math. Comput. Simul.* **2017**, *131*, 88–100. [CrossRef]

23. Persson, C.; Bacher, P.; Shiga, T.; Madsen, H. Multi-site solar power forecasting using gradient boosted regression trees. *Sol. Energy* **2017**, *150*, 423–436. [CrossRef]
24. Detyniecki, M.; Marsala, C.; Krishnan, A.; Siegel, M. Weather-based solar energy prediction. In Proceedings of the 2012 IEEE International Conference, Fuzzy Systems (FUZZ-IEEE), Brisbane, Australia, 10 June 2012.
25. Bacher, P.; Madsen, H.; Nielsen, H.A. Online short-term solar power forecasting. *Sol. Energy* **2009**, *83*, 1772–1783. [CrossRef]
26. Sharma, S.; Jain, K.K.; Sharma, A. Solar cells: In research and applications—A review. *Mater. Sci. Appl.* **2015**, *6*, 1145. [CrossRef]
27. Mori, H.; Takahashi, A. A data mining method for selecting input variables for forecasting model of global solar radiation. In Proceedings of the 2012 IEEE PES, Transmission and Distribution Conference and Exposition (T&D), Orlando, FL, USA, 7–10 May 2012.
28. Voyant, C.; Paoli, C.; Muselli, M.; Nivet, M.L. Multi-horizon solar radiation forecasting for Mediterranean locations using time series models. *Renew. Sustain. Energy Rev.* **2013**, *28*, 44–52. [CrossRef]
29. Pedro, H.T.; Coimbra, C.F. Nearest-neighbor methodology for prediction of intra-hour global horizontal and direct normal irradiances. *Renew. Energy* **2015**, *80*, 770–782. [CrossRef]
30. Lee, K.; Kim, W.J. Forecasting of 24 hours Ahead Photovoltaic Power Output Using Support Vector Regression. *J. Korean Inst. Inf. Technol.* **2016**, *14*, 175–183. [CrossRef]
31. Almeida, M.P.; Perpinan, O.; Narvarte, L. PV power forecast using a nonparametric PV model. *Sol. Energy* **2015**, *115*, 354–368. [CrossRef]
32. Song, J.J.; Jeong, Y.S.; Lee, S.H. Analysis of prediction model for solar power generation. *J. Digit. Converg.* **2014**, *12*, 243–248. [CrossRef]
33. Yona, A.; Senjyu, T.; Funabshi, T.; Sekine, H. Application of neural network to 24-hours-ahead generating power forecasting for PV system. *IEEJ Trans. Power Energy* **2008**, *128*, 33–39. [CrossRef]
34. Kuhn, M.; Johnson, K. *Appl. Predict. Model*, 1st ed.; Springer: New York, NY, USA, 2013.
35. Voyant, C.; Soubdhan, T.; Lauret, P.; David, M.; Muselli, M. Statistical parameters as a means to a priori assess the accuracy of solar forecasting models. *Energy* **2015**, *90*, 671–679. [CrossRef]
36. Kalogirou, S.A. Artificial neural networks in renewable energy systems applications: A review. *Renew. Sustain. Energy Rev.* **2001**, *5*, 373–401. [CrossRef]
37. Breiman, L. *Classification and Regression Trees*, 1st ed.; Routledge: New York, NY, USA, 1984.
38. Pedregosa, F.; Varoquaux, G.; Gramfort, A.; Michel, V.; Thirion, B.; Grisel, O.; Blondel, M.; Prettenhofer, P.; Weiss, R.; Dubourg, V.; Vanderplas, J. Scikit-learn: Machine learning in Python. *J. Mach. Learn. Res.* **2011**, *12*, 2825–2830.

© 2019 by the authors. Licensee MDPI, Basel, Switzerland. This article is an open access article distributed under the terms and conditions of the Creative Commons Attribution (CC BY) license (http://creativecommons.org/licenses/by/4.0/).

Article

An Adaptable Engineering Support Framework for Multi-Functional Energy Storage System Applications

Claudia Zanabria [1,*], Filip Pröstl Andrén [1] and Thomas I. Strasser [1,2]

1. Center for Energy—Electric Energy Systems, AIT Austrian Institute of Technology, 1210 Vienna, Austria; Filip.Proestl-Andren@ait.ac.at (F.P.A.); Thomas.Strasser@ait.ac.at (T.I.S.)
2. Institute of Mechanics and Mechatronics, Vienna University of Technology, 1040 Vienna, Austria
* Correspondence: claudia.zanabria@ait.ac.at

Received: 17 September 2018; Accepted: 31 October 2018; Published: 12 November 2018

Abstract: A significant integration of energy storage systems is taking place to offer flexibility to electrical networks and to mitigate side effects of a high penetration of distributed energy resources. To accommodate this, new processes are needed for the design, implementation, and proof-of-concept of emerging storage systems services, such as voltage and frequency regulation, and reduction of energy costs, among others. Nowadays, modern approaches are getting popular to support engineers during the design and development process of such multi-functional energy storage systems. Nevertheless, these approaches still lack flexibility needed to accommodate changing practices and requirements from control engineers and along the development process. With that in mind, this paper shows how a modern development approach for rapid prototyping of multi-functional battery energy storage system applications can be extended to provide this needed flexibility. For this, an expert user is introduced, which has the sole purpose of adapting the existing engineering approach to fulfill any new requirements from the control engineers. To achieve this, the expert user combines concepts from model-driven engineering and ontologies to reach an adaptable engineering support framework. As a result, new engineering requirements, such as new information sources and target platforms, can be automatically included into the engineering approach by the expert user, providing the control engineer with further support during the development process. The usefulness of the proposed solution is shown with a selected use case related to the implementation of an application for a battery energy storage system. It demonstrates how the expert user can fully adapt an existing engineering approach to the control engineer's needs and thus increase the effectiveness of the whole engineering process.

Keywords: energy management system; energy storage system; semantic web technologies; rules; ontology; engineering support; smart grid architecture model; model driven architecture; IEC 61850; IEC 61499

---

1. Introduction

The reduction of CO_2 emissions is motivating the integration of renewables into power grids. As a consequence, a higher penetration of distributed generators such as Photovoltaic (PV), wind turbines, and small hydro power stations is taking place [1]. A side effect of this is the perturbation of the power system stability and quality. Those issues were addressed by different studies [2], which encourage the use of Battery Energy Storage Systems (BESS). Moreover, BESS can also support services related to the minimization of supply costs and market integration, among others [3]. Consequently, BESS will play a multi-functional role in the near future. The BESS is often accompanied by an Energy Management Systems (EMS) where the BESS's services and functions are hosted. Hence, the EMS should exchange information with stakeholders and Intelligent Electronic

Devices (IEDs) spread out over the electric grid. Thereby, the EMS development process should consider interoperability across systems as well as evolving requirements of smart grid systems. In this context, the realization of EMS involves challenging tasks, such as alignment with smart grid information models, conflicts resolution within a multi-functional system, as well as handling a diversity of tools for EMS validation. Different approaches handle these issues to different degrees as demonstrated by Zanabria et al. [4]. Nevertheless, a full flexibility within a rapid prototyping context is still not covered by the mentioned approaches. This motivates the open issues addressed in this work. An outlook of them is given below.

At the design phase, control engineers gather documents that encapsulate information regarding IED capabilities, network topologies, control applications structure, etc. Those documents are considered to provide important input for the EMS design. Thereby, a methodology that supports an automated treatment of this information to support the development process is necessary. Santodomingo et al. [5] process models based on the Smart Grid Architecture Model (SGAM) [6] and Use Cases defined with the IEC 62559 [7] standard to support network operators in selecting adequate technical solutions for their business needs. The referred study shows attempts to benefit from available information sources (SGAM models) in an automated way.

Another approach, the so called Energy Management System Ontology (EMSOnto) [8] also proposes a resolution of conflicts within an EMS [9]. Nevertheless, inconsistencies detected within BESS applications imply not only conflicts across Use Cases (UC) but also others such as incompatibility between a BESS and a service, misconfigured units, wrong write/red access permissions, etc. This motivates to look for mechanisms that enable a customized identification of inconsistencies based on evolving engineering requirements.

What is more, due to control engineering practice and legacy solutions a variety of tools, programming languages and documents are likely to be employed during the validation phase. Current approaches [4] do not answer those requirements since models and code generated were tied to a specific platform. However, higher flexibility is expected in terms of software artifacts generation. For instance, EMSOnto automatically generates code to be employed in a power system emulator (e.g., MATLAB/Simulink). Here, the compatibility with only one specific platform is provided. On the other hand, the rapid engineering methodology called Power System Automation Language (PSAL) generates models compatible with IEC 61499 [10]. However, also here, the generation of models in other specific platforms rather than IEC 61499 is not guaranteed.

EMSOnto is one of the more suitable approaches that addresses the above mentioned issues in a holistic manner. EMSOnto guides and supports control engineers during the design, proof-of-concept and implementation stages of BESS services and applications. Consequently, this work considers EMSOnto as reference framework to demonstrate how a flexible and automated engineering process can be attained. By applying the outcomes of the current work to EMSOnto, an improved version of it will be reached. Moreover, the proposed methodology may also be used to reinforce other modern approaches such as PSAL and SGAM.

Since EMSOnto is taken as reference approach, the role of a new actor, the so called EMSOnto expert user, is introduced to answer the exposed open issues. Thus, concepts from Model-Driven Engineering (MDE) and ontologies are combined to offer an adequate solution. The evaluation of the new capabilities of EMSOnto is performed by the realization of a selected use case example. As a result, an acceleration in the realization of EMS applications is gained as well as a further identification of inconsistencies. Moreover, the achieved engineering process is flexible enough to be aligned with different software platforms.

This paper is organized as follows: Section 2 outlines the motivation of this work, the open issues to be addressed, and the role of control engineers in the scope of EMSOnto as well. In Section 3 mechanisms to enhance the engineering process are introduced and discussed. A use case example is used in Section 4 to demonstrate how the control engineer's requirements are precisely addressed.

Thereafter, in Section 5, a UC example is realized with EMSOnto to evaluate and analyze its new features. Finally, Section 6 addresses the conclusions and discusses future needs.

2. ESS Application Development Process Using Modern Engineering Approaches

This section defines the scope of the current work and outlines the engineering process of multi-functional storage systems. Furthermore, an overview of different engineering approaches that support the development process of BESS applications is also presented. Subsequently, issues not yet addressed by those approaches are highlight. Since EMSOnto is considered to be the most adequate approach, the engineering process under the basis of EMSOnto is presented. Furthermore, the open issues and research goals under the frame of EMSOnto are defined.

2.1. Realization of Multi-Functional ESS Applications

BESS are being used to enhance the power quality and to maintain the power stability, among others. This involves frequency and voltage control, self-consumption, peak-shaving, etc. [3,11,12]. Those functionalities are usually embedded within an external system (e.g., EMS) that controls active and reactive power of the battery to facilitate the referred services. Thereby, a participation of different stakeholders is required, such as network operators, Energy Market Operator (EMO), or Distributed Energy Resources (DER), as outlined in Figure 1. As an example, the UC Primary Control Reserve (PCR) [11] controls the active power of a BESS to support the regulation of the grid frequency. Thereby a communication of the EMS with a smart meter, network operator and a BESS is needed. On the other hand, the UC minimization of costs with peak-shaving [12] would require a communication with a PV generator and households to measure the active power generated and consumed.

The realization of EMS applications often follows certain stages, such as specification, design, implementation, and proof-of-concept evaluation [13]. In the specification stage a definition of requirements to be satisfied by an EMS is carried out. As a sequel, a conceptual design of EMS is elaborated, this implies the definition of control strategies and communication and component architectures. The validation of the proposed control design is mainly carried out in offline power system simulators (e.g., DIgSILENT/PowerFactory) and controller platforms (e.g., IEC 61499) [14]. Often the validation process entails an iterative refinement of the control application design. Subsequently, a prototype is realized and implemented within a real hardware controller.

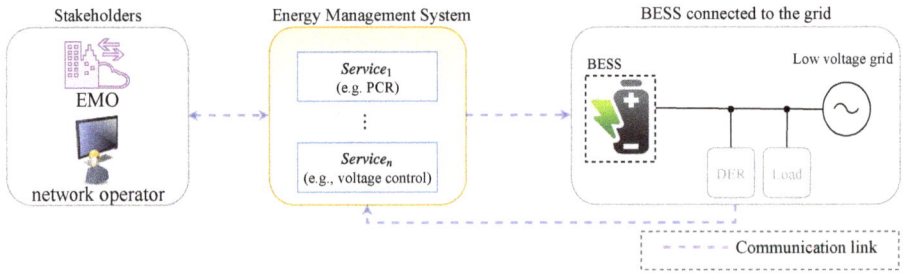

Figure 1. Framework of the BESS control applications.

2.2. Application Engineering Using Modern Approaches

The engineering process of EMS can nowadays be supported by using modern development approaches [4]. Each of them tackles different specific issues. An analysis on how those approaches support the EMS development process is addressed as follows.

An approach called IntelliGrid (IEC 62559) [15] is mainly used to specify use cases and was suggested by EPRI in 2003 to face the complexity involved within power systems. IntelliGrid is the most used standard to describe smart grid use cases. The main outcome of IntelliGrid is a set of use case templates that guide engineers during the definition of business cases, use cases and requirements. Those templates are supported with a narrative and visual representation. The process involving the realization of IntelliGrid templates is basically a manual work. This means that information sources supporting the EMS specification (e.g., IED's models, component and communication architectures) are not automatically processed to fill out the templates. Besides this, the consistency of the information collected by the referred templates is not verified. This may lead to unattended inconsistent data at the specification stage.

On the other hand, the Unified Modeling Language (UML) is the most prominent general-purpose language to specify and design use cases in an object-oriented approach [16]. UML provides different structure (class, component, etc.) and behavior (sequence, activity, etc.) diagrams to allow a complete modeling of use cases. Nowadays, UML is getting more attention also in the domain of power system to model smart grid solutions [17]. A different approach (SGAM) conceived to model smart grid projects is recommended by [4]. SGAM provides a set of concepts, coordinated viewpoints and a method to map use cases within a smart grid model. An implementation of SGAM is achieved by the SGAM-Toolbox (SGAM-TB), a UML-based domain specific language available as an extension of the Enterprise Architecture (EA) tool [18]. A drawback of both UML and SGAM is that the information gathered at the specification phase is not automatically processed to achieve important knowledge for the design stage. This does not guarantee a consistency between specification and design stages.

A holistic approach called PSAL has as main purpose the modeling of SGAM UCs and at the same time provide automated support for rapid prototyping. To do this, PSAL provides a formal domain specific language to describe smart grid systems. PSAL stands out from the previously mentioned approaches due to the rapid generation of software artifacts (code and communication configurations) offered. However, PSAL does not identify inconsistencies within the planned design. This, in turn, is tackled by EMSOnto, an approach focused on the identification of conflicts and on the rapid prototyping of multi-functional BESS applications. Nevertheless, EMSOnto lacks flexibility due to the restricted set of inconsistencies detection offered. On the other hand, as already mentioned, EMSOnto and PSAL address the rapid generation of software artifacts. However, since the needed type of software artifacts depend on the UCs to be implemented and on needed legacy solutions, it should be possible to support multiple target platforms. This feature is neither covered by PSAL nor EMSOnto. Currently, the generated models and code using PSAL are aligned to the automation standard IEC 61499 and the communication model IEC 61850. EMSOnto currently only provides generation of code for the system simulator MATLAB/Simulink.

The mentioned open issues entail the setting of the following research goals: (i) benefit from information sources to automate the realization of design tasks, (ii) facilitate the deduction of knowledge to be customizable to control engineer's requirements, and (iii) generate software artifacts to support the whole engineering process. To meet those goals, EMSOnto is taken as the reference framework due to its completeness and holistic understanding. Hence, the solution proposed in this work has been aligned to EMSOnto but should still be applicable not only to EMSOnto but also to other existent and upcoming approaches. The next paragraphs outline EMSOnto and formulate the already mentioned open issues and research goals under the EMSOnto basis.

2.3. EMSOnto Development Process

An overview of the work performed by control engineers on the frame of EMSOnto is shown in Figure 2. At the specification phase, engineers study the requirements to be fulfilled by the EMS and starts a rough definition of the behavior and the structure of the EMS control applications. At this stage, different approaches such as SGAM, Unified Modeling Language (UML) and Systems Modeling Language (SysML) for system specification are employed [6,16].

The next step is the design of the EMS, which should be congruent with the specification already defined. To this end, EMSOnto offers spreadsheet templates (EMS-templates) that collect and structure the EMS's information. These spreadsheets have headlines with the attributes of functions and variables created in the scope of the EMS. Moreover, models of IEDs and UCs derived from smart grids standards (e.g., Common Information Model, IEC 61850) are available within the UC and IED repository. Those pre-built models ease the population of EMS-templates and thus compatibility with existent information models is achieved.

Once the EMS-templates are completely filled a reasoner engine is executed resulting in the derivation of inferred data. The resulted data is queried to identify inconsistencies in the design, a report form gathers those inconsistencies and is handled and analyzed by control engineers. This analysis may lead to adjustments within the planned design. Therefore, at this stage EMS-templates are amended. This process is repeated until control engineers are satisfied with the resulted design. In consequence, a consistent and less error-prone EMS is achieved before any implementation. To support the proof-of-concept a set of software artifacts are made available to control engineers.

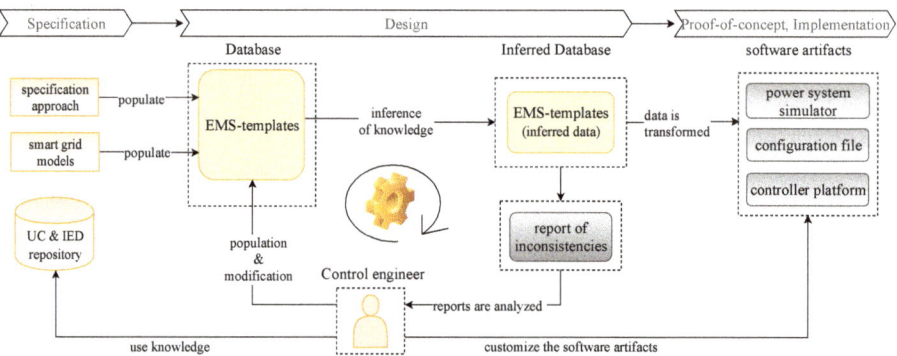

Figure 2. EMSOnto in practice by control engineers.

2.4. Open Issues

EMSOnto was put in practice by control engineers in the scope of a large range of UCs [8]. Those realizations raised a series of requirements understood as open issues within EMSOnto. They are detailed as follows:

- *Information sources are not exploited:* At the design stage control engineers dispose documents that support the design of EMS. This may correspond to files describing IED capabilities, smart grid use cases, communication networks, information models, etc. (e.g., IEC 61850, IntelliGrid, SGAM) [15,18,19]. Since those files contain requirements and important knowledge for the design process, control engineers manually need to import selected data into the EMS-templates. This repetitive manual work is time consuming and exposed to human errors. Hence, an automatic exchange between EMS-templates and other information sources is sought.
- *Restricted inference:* EMSOnto supports the identification of conflicts between use cases. However, this is not the only kind of inconsistency that would harm the suitable operation of EMS. For instance, the setting up of IED registers with a wrong unit value would also impact the correct operation. Besides this, the inference of important data to support the design of EMS's control strategies is also missed. Since knowledge to be inferred depends on engineer's needs a flexible customization of EMSOnto to enlarge inferred knowledge is desired.
- *Limited generation of software artifacts:* EMS-templates can be automatically transformed into models and code compliant with a specific power system simulator (i.e., MATLAB/Simulink). Nevertheless, software platforms to be targeted depend on best practices established for testing and validation. In the power system domain, those platforms involve controller platforms, co-simulation platforms, communication network simulators, etc. (e.g., IEC 61499, Mosaik,

OMNET++) [20–22]. Therefore, generation of software artifacts, compatible with a large set of platforms, should be guaranteed.

Summarizing, taking all the open issues into account an adaptable engineering support framework for ESS applications is required which is introduced and discussed in the following sections.

3. Mechanisms to Automate and Increase Flexibility of EMSOnto

This section introduces the role of a new actor called EMSOnto expert, whose main focus is to offer mechanisms to overcome the aforementioned gaps. A detailed list of actions performed by the expert is exposed as well as the foundations of the proposed solution.

3.1. EMSOnto Expert Participation

The main key of EMSOnto is the conception of an ontology (EMS-ontology) which gives a common understanding of EMS control applications and the process to be controlled by them (e.g., BESS, smart meter). A formal definition of the ontology is reached by Description Logics (DL) [23], a formal language to model real systems, where the classification of common individuals is done by concepts and the relations between concepts are established by roles. DL languages comprises an assertional component (*ABox*) and a terminological component (*TBox*). A *TBox* defines concepts, roles, and constraints between them. In turn, an *ABox* asserts the membership of individuals in concepts and roles. On that basis, the EMS-*ABox* is populated by the EMS-templates and the EMS-*TBox* provides a knowledge representation of EMS applications. The terms EMS-ontology and EMS-*Tbox* are used interchangeably as well as EMS-templates and EMS-*ABox*. In this manuscript, concepts are indicated with the font typewriter. A new actor, the EMSOnto expert, is introduced to expand the features of EMSOnto. The actions taken by the expert are depicted in Figure 3.

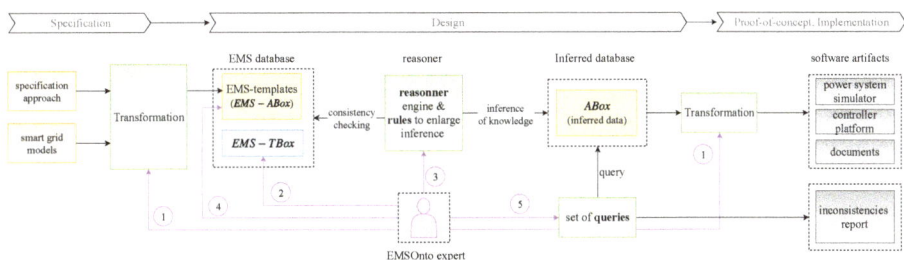

Figure 3. Tasks to be performed by EMSOnto expert.

(1) Transformation across information: Control engineers make use of documents which contain specifications regarding the EMS design (e.g., SGAM models), this information is gathered within the EMS-templates. Since EMS-templates are aligned to the EMS-*TBox*, the referred gathering process involves a matching between the information source, collected from specification approaches or smart grid models, and concepts and roles defined within the EMS-*TBox*. For instance, EMS's structure represented by UML use case diagrams comprise the concepts UseCase, LogicalActor, etc. [16]. Hence, a transformation of a UseCase instance requires a correspondence with the concept UC defined within EMS-*TBox*. This is possible since the UC and the UseCase represent the same information. A similar mechanism is applied for an automatic generation of software artifacts from EMS-templates. In such a case, a transformation from EMS-*ABox*'s individuals into models, code and text compliant with specific platform environments is performed.

(2) Adjustment of the EMS-TBox: Since the EMS-*ABox* should be consistent with the *TBox*, the gathering of new assertions may impact the *TBox*. This applies to knowledge not modeled in the *TBox* but required in the transformation from or to other models. On the other hand, the inference of implicit knowledge may also require an upgrade of the *TBox*. For instance, the detection of "*BESS that*

do not support the active power required by certain service" would require to create the concept Pmax which symbolizes the upper limit of an active power's value.

(3) *Creation of new rules*: The reasons that motivated an extension of the EMS-*TBox* could require an improvement on the expressivity afforded by the EMS-*TBox*. This issue is tackled by rules which are used to improve the intelligence own by ontologies. Rules are composed of a premise and a conclusion, satisfaction of the premise results in the derivation of implicit data or conclusions. As a result, any requirement involving complex inferences that are not achievable by the setting up of axioms within the EMS-*TBox* would be resolved by rules.

(4) *Upgrade of EMS-templates*: As explained before, EMS-*TBox* is vulnerable to be extended due to inference of knowledge or software artifacts generation. Since an *ABox* should be consistency with a *TBox* any modification to the EMS-*TBox* would also impact the EMS-*ABox*. This means that, instantiations of new concepts and roles would require an upgrade of EMS-templates.

(5) *Set-up new queries*: The detection of inconsistencies is based on a series of questions to be asked to the inferred EMS-*ABox*. Implementation of this is based on the formulation of query templates that model well-defined questions. Thus, the inferred data resulted from action 3 is queried and processed to collect important knowledge for the inconsistencies report, as shown in Figure 3. Besides that, queries could also target inference of essential knowledge to support the EMS design.

3.2. Transformation Mechanisms and Techniques

The tasks related to transformation of information across domains will be performed by MDE techniques [24], where models are key players. Thereby, a model compliant with a UML class representation is envisaged to represent the EMS-ontology. This model will be the essential part of the model transformations carried out along this work.

3.2.1. Model-Driven Engineering in Power System Domain

The basics principles necessary to achieve the contemplated transformations between source and target data are covered by MDE approach [24]. This section outlines the concepts within MDE, benefits of the approach as well as power system applications where MDE was implemented.

The main focus of MDE is the automation of a system development process. On that basis, MDE analyzes a development process in an object-oriented manner, resulting in the conception of models. Interoperability between models is reached by mapping rules. Which are defined under Model-to-Model (M2M) or Model-to-Text (M2T) transformations, also known as code generation.

Implementation of MDE by the Object Management Group (OMG) goes under the name of Model Driven Architecture (MDA). Within MDA, the phases of a development process are supported by a Platform Independent Model (PIM) and a Platform Specific Model (PSM). PIM defines system functionality and is characterized by an independence with a specific platform solution. PIM is deployed into a concrete platform (PSM) and usually PSM is translated into software artifacts (e.g., C++, Java code). M2M supports transformation from PIM to PSM and M2T from PSM into code.

An automated engineering method to support the complete development process of smart grid control applications is provided by Pröstl Andrén et al. [10]. Power System Automation Language (PSAL) is a Domain Specific Language (DSL) and supports the design of UCs compliant with SGAM. MDE techniques are used to generate executable code and communication configuration scripts from UCs specifications. This is a holistic approach that covers the full engineering process.

On the other hand, Andrén et al. [25] proposes an automatic generation of IEC 61499 compliant control applications from IEC 61850 descriptions. This approach implements M2M and M2T transformations. However, the obtained IEC 61499 system is not aligned to SGAM. Another study [26] designs a BESS control application within a power system emulator (Matlab/Simulink) for validation purposes and uses MDE to replicate the validated application into a controller platform (IEC 61499). This study proposes a rapid prototyping that covers proof-of-concept and implementation. However, other stages of the engineering process such as specification and design were not considered.

Furthermore, [18] Dänekas et al. proposes the SGAM-Toolbox (SGAM-TB) by following the MDA approach. The SGAM-TB is implemented as an extension to Enterprise Architect (EA) and therefore code generation is also possible. However, this code is not formatted to specific needs of control engineers.

Summarizing, apart from EMSOnto, PSAL is the only approach that offers a holistic approach. However, some gaps at the function and information layer of PSAL are covered by EMSOnto. Those gaps regard the identification of inconsistencies and the lack of a mechanism to gather unlimited data. Those topics were tackled in the current EMSOnto. The new version of EMSOnto is formulated with the aim of giving flexibility at all the stages of the engineering process. This flexibility is not addressed by any solution already presented and is going to conduct the research of this work.

3.2.2. UML Representation of EMS-Ontology

Since models are the main concept behind MDE, the conception of models across the full development process is a main task to be supported by the EMSOnto expert. In this light, a model of EMS-ontology is achieved and exposed in this section.

OMG recommends the formulation of models by OMG specifications such as UML and SysML, among others [27]. UML, a widely used standard for describing structure and behavior of complex systems is selected to represent the models. On the other hand, the expressiveness needed by the EMS-$TBox$ is given by $SROIQ(D)$, a logic defined within DL concept that provides complex role inclusion axioms, inverse roles constructors, qualify cardinality restrictions among others [28]. Thus, mapping rules to relate $SROIQ(D)$ elements to UML metamodel elements are established to conceive a UML model of the EMS-$TBox$, see Table 1. A DL concept is transformed into a UML class, a concept subsumption into a UML generalization and so on, a practical implementation of those conversions is given in [29]. It is worth mentioning that the full $TBox$ cannot be expressed by a UML class diagram since not all the $TBox$ axioms can be transformed into UML elements (e.g., transitivity axioms, composition role constructor). The resulted UML class diagram is presented in Figure 4 and goes under the name of EMS-Data Model (DM). Only a simplify model is depicted. InformationFlow is a new concept derived from the EMSOnto extension, it will be introduced later in the paper.

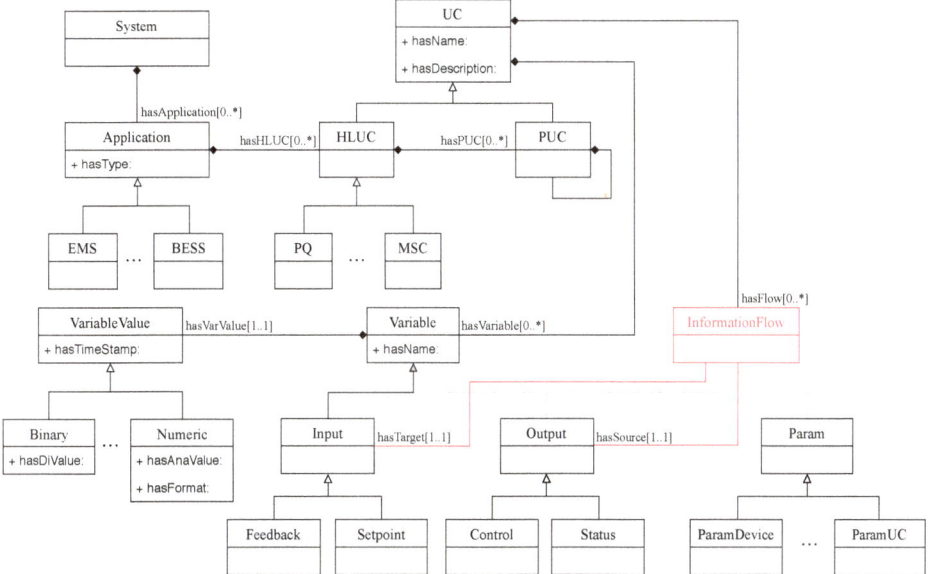

Figure 4. UML class diagram representing the EMS-$TBox$ (reduced model).

Table 1. Matching between elements of $SROIQ(D)$ and UML metamodel.

DL ($SROIQ(D)$)	UML
concept	class
concept subsumption ($\sqsubseteq$)	generalization
data type	datatype
role	association, composition, aggregation
concrete role	attribute

4. Analysis of a Use Case Example by an EMSOnto Expert

Different mechanisms to enlarge the capabilities of EMSOnto were proposed above. This section outlines the role of the EMSOnto expert regarding the implementation of a specific UC.

4.1. Use Case to Be Analyzed by the EMSOnto Expert

The UC Customer Energy Management System (CEMS) was implemented using EMSOnto as a basis [8]. SGAM is the preferred approach among smart grid specification approaches as demonstrated in [4]. The SGAM information layer represent system's components as well as their information flows. Moreover, it combines domain-axis and zone-axis to structure smart grid control applications. For the CEMS in this UC, a detailed model is provided as an SGAM description, as seen in Figure 5. This shows the main CEMS's actors (i.e., BESS, CEMS, Meter, etc.) and the scope of the CEMS, which is located at the customer side and at the station level.

The CEMS is embedded with Frequency-Watt (FW) and Self-Consumption (SelfC) services. Both services control the active power injected/subtracted by the BESS. To this aim, the signals State of Charge (SoC) (SoC_{bat}) and active power (P_{bat}) are retrieved from the BESS. In turn, the frequency (F_{grid}) and active power (P_{grid}) at the Point of Common Coupling (PCC) are taken from a smart meter. Generation of a PhotoVoltaic (PV) DER and the consumption of a household impact the behavior of the CEMS. Thus, they are also considered during design stage. After execution of the SelfC and FW services, a calculated active power setpoint (P_{ref}) is sent to the BESS.

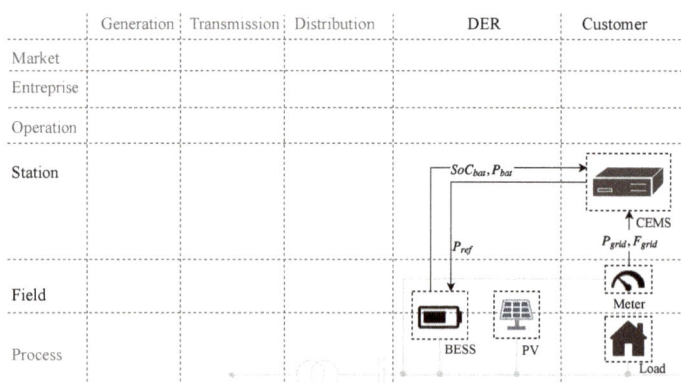

Figure 5. SGAM information layer representation of the CEMS (UC example) [8].

4.2. Requirements from Control Engineers

The CEMS is implemented with EMSOnto. Control engineers in charge of the implementation employs specific tools for power systems emulator (MATLAB/Simulink) and controller platforms (IEC 61499 platform) [20,30]. Furthermore, not only specific tools but specific documents (IEC 61850 SCL files, SGAM models) were used for the description of IED's functionalities and control applications architecture. This is due to common practices of control engineers. Nevertheless, other tools and frameworks are also available to support the engineering process of power systems (e.g., Modelica and IEC 61131-3) [31,32] as well as other information models for smart grid networks (e.g., CIM) [33]. The fact of using specific tools to evaluate the proposed methodology would enable a concrete understanding of the actions taken by the EMSOnto expert.

A series of requirements that motivates an expansion and customization of EMSOnto were figured out by control engineers. These requirements are bound to specific tools, documents and information models as exposed below.

(R1) Benefiting from SGAM models: CEMS is roughly described at the specification stage by means of the SGAM-TB. The information added at that stage is reused at design time to fill the EMS-templates. Therefore, an automatic import of SGAM models into EMS-templates is expected with the aim of reducing the manual work conducted by control engineers.

(R2) Discovering CEMS constraints: The controlled process (i.e., the BESS) is surrounded by constraints that characterizes technical limitations of electrical devices and/or safety restrictions. Those constraints need to be considered during design of control strategies. Therefore, an understanding of them is absolutely required during design time. Hence, an automatic retrieval of constraints from the process is expected.

(R3) SoC estimator function generation: In the scope of multi-functional BESS, the battery provides support not only to an individual service, but more than one. In this context, a solution to track the delivered active power of the BESS to a specific service is performed by control engineers. This consists on the calculation of SoC per service by a function called *SoC_estimator*. As a result, this function is implemented whenever a UC controls a BESS. Hence, an automatic generation of the *SoC_estimator* function is sough in the scope of R3.

(R4) Generation of configuration files: EMS and IEDs need to be parametrized and customized to specific requirements and initial state values. A selection of parameters is gathered in a sort of configuration file that is filled out before any execution or simulation of the mentioned systems. These configuration files support the validation of the EMS through the proof-of-concept and implementation phases. R4 seeks for an automatic generation of such files, which should expose the parameters of functions to be deployed within an EMS as well as the IED's parameters.

(R5) Identification of batteries mismatching services: A service that request BESS support would ask to fill certain requirements such as capacity (kWh), maximal power provision (kW), activation time (s) [34,35]. Thus, not all the BESS would pass the pre-qualification procedures defined by network operators. BESS able to participate in Primary Control Reserve (PCR) should provide certain power. A report that identifies BESS that do not meet this requirement is expected.

(R6) Generation of IEC 61499 models: The design of CEMS is validated within a power system emulator as part of the proof-of-concept. Following this step, the implementation of it is performed within a controller platform fully compliant with IEC 61499. On the other hand, some components of the CEMS (e.g., communication interfaces, control logic) were directly tested in the controller platform. This motivates an automatic generation of the IEC 61499 control application either from EMS-templates or from the MATLAB/Simulink project.

(R7) Benefiting from IEC 61850 ICD files: Power system information models hold data important for the IDE and UC repository. In the scope of CEMS, Information Capability Device (ICD) files are handled to control engineers. ICD files are XML-based, which serialize information regarding functions supported by an IED and are often supplied by the IED manufacturer. Thus, R7 seeks for an automated import of ICD files into the repository.

(R8) Identification of misconfigured units: An EMS controlling and monitoring an IED (e.g., BESS) should be aligned to the units of the IED. In other words, a control signal targeting a setpoint should respect the units of the value to be targeted. Otherwise, a wrong value would be set. This applies also to a status been monitored by feedback. For instance, a BESS is providing its SoC status (SoC_{bat}), which has the value of 80%. An EMS retrieving this value may expect raw data between the range 0–1 (0.8). This mismatch between units would cause a wrong behavior of the EMS. Hence, an identification of such inconsistencies is fundamental.

4.3. Analysis Phase: Analysis of Requirements

An analysis of the mentioned requirements determines the actions to be performed by the EMSOnto expert. This analysis is structured under a succession of steps. The first step corresponds to investigate where the transformation across data models takes place. This corresponds to R1, R4, R6 and R7. Those requirements involve different data models that need to be interrelated with EMS-DM. Therefore, data models of SGAM, IEC 61850, IEC 61499 and Matlab/Simulink are studied and an alignment of EMD-DM to the referred data models is sought. Following this, classes and associations not considered within EMS-DM are created and added to it, affecting the EMS-$TBox$.

The next step is to identify requirements clearly connected to the inference of knowledge. R2, R3, R5 and R8 correspond to this category. A complete understanding of the knowledge to be inferred in the scope of those requirements is needed. Thereby mathematical models representing constraints within a control application (R2) are investigated as well as models that drive the operation of $SoC\_estimator$ function (R3). Furthermore, an understanding of technical limitations within a BESS are also needed (R5) just as information models that conduct the structure of sources and targets defined within information flows (R8). In the aftermath, the $TBox$ is again extended to cope with R2, R3, R5 and R8. Eventually, the creation of queries and rules could also take place.

4.4. Realization Phase: Implementations Performed by EMSOnto Expert

The actions taken by EMSOnto expert are highlighted and enumerated in Figure 6. The first step is dedicated to the creation of new concepts, roles, and constraints within the EMS-$TBox$. The expressivity of EMS-$TBox$ depends on the ontology language used to implement it (e.g., OIL, DAML+OIL, OWL 2 [36]). In case that a higher expressivity is needed the implementation of rules is carried out as part of the next step.

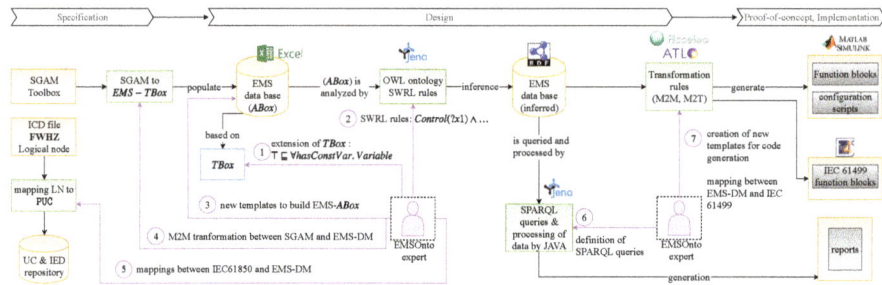

Figure 6. Design and Implementation of CEMS UC— automated by EMSOnto expert.

As a sequel, population of the EMS-$TBox$ is done against assertions of the CEMS use case, hence, CEMS knowledge is collected within EMS-templates. EMS-templates are complemented with the information from SGAM models and ICD files. To achieve this, a mapping between the SGAM model and the EMS-DM is performed, as well as a transformation between IEC 61850 models (ICD files) and EMS-DM. When a complete $ABox$ is reached the inference of knowledge takes place. Queries are built upon the inferred knowledge to allow the identification of inconsistencies. Finally, the EMSOnto expert establishes M2M and M2T transformation to generate software artifacts as depicted in Figure 6. The implementation of the aforementioned actions are carried out by Semantic Web Technologies, ontology editors and programming frameworks. A detail of the aforementioned actions is addressed below.

4.4.1. Action 1: Extending the EMS-Ontology

Ontology Web Language (OWL) 2, a popular ontology language recommended by the World Wide Web Consortium (W3C) is chosen to implement the expressivity required by EMS-$TBox$ ($SROIQ(D)$

logic) [37]. OWL 2 is based on $SROIQ$ and furthermore it supports data types. Thus, a complete implementation of the TBox is reached. On that basis, concepts, roles, and axioms not considered within the EMS-TBox are investigated and addressed in this section.

R1 seeks to benefit from UML behavior and structure diagrams specified with the SGAM-TB. Because of that, a concordance between SGAM-TB and EMS-DM should be reached. The EMS-DM focuses on function and information aspects of smart grid control applications. Thereby, the business, communication, and component SGAM layers are out of the scope. The function analysis of SGAM-TB involves different concepts, such as High-Level Use Cases (HLUC), Primary Use Cases (PUC), Logical Actors (LA) [38]. Those concepts are already included in the EMS-DM. However, the concept Information Object (IO) that represents communication between LAs and PUCs was not yet addressed. Hence, it is added to the EMS-DM under the name of InformationFlow, which stands for the UC's information that relates one Output to one Input as exposed in Table 2 and Figure 4.

Table 2. Terminological axioms to align EMS-DM with SGAM model.

Concepts, Roles, OWL Axioms	Description
$UC \sqsubseteq \exists hasFlow.InformationFlow \sqcap \ldots$, $InformationFlow \sqsubseteq\ \leq 1hasSource.Output \sqcap\ \leq 1hasTarget.Input \sqcap \exists hasSource.Output \sqcap \exists hasTarget.Input$	InformationFlow represents information exchanged between UCs. Hence, the role hasFlow relates the concepts UC and InformationFlow. InformationFlow owns one Output as source and one Input as target. A formal representation of this requires the use of qualified number restrictions constructor ($\leq$ n R.C, R is a role and C is a concept) [23].

An investigation of mathematical models for IEDs is carried out in the scope of R2. For simplicity, a single-output and single-input process is represented in state space as shown in Equation (1). Where u is the input variable, y is the process output and x the state vector with dimension $n \times 1$ that represent the entire state of the process at any time. In the scope of our study, the process is represented by electrical devices connected along the low voltage distribution grid, such as BESS, PV, load, etc. These processes are frequently surrounded by constraints on control variables (u), state variable (x), or output (y), as described by Equation (2). A control application (i.e., CEMS) that intends to manipulate outputs of the process (y) needs to respect the referred constraints.

From an IED's perspective, an instantiation of previous data with EMS-TBox concepts is achievable: Setpoint (u), Manipulated (y), State $\{x_1, \ldots, x_n\}$, Constraint $\{u_{max}, \ldots\}$. In this work, only constraints on setpoint variables (u_{min}, u_{max}) are inferred, the others (e.g., y_{max}) follow a similar mechanism. On that basis, the CEMS infers constraints from the IED's setpoints. To achieve this, the concept Constraint is further developed as depicted in Table 3, where Variables defined within a Constraint are modeled as well as the relation between Constraints.

$$x' = Ax + Bu, \quad y = Cx + Du, \quad x = \begin{bmatrix} x_1 & \ldots & x_n \end{bmatrix}^T \qquad (1)$$

$$x_{min} \leq x \leq x_{max}, \quad u_{min} \leq u \leq u_{max}, \quad y_{min} \leq y \leq y_{max} \qquad (2)$$

Table 3. Concepts, roles and axioms to infer CEMS's constraints.

Concepts, Roles, OWL Axioms	Description
$Constraint \sqsubseteq \exists hasConstVar.Variable \sqcap \exists IsConsLinkCons.Constraint$, $hasVarConst \sqcap hasConstVar \sqsubseteq T$, $trans(IsConsLinkCons)$	A Constraint owns Variables, this relation is represented by hasConstVar. The inverse role of hasConstVar is given by hasVarConst. The role IsConsLinkCons relates Constraints and the transitivity property (trans) is assigned to it.
$Variable \sqsubseteq \exists hasVarConst.Constraint \sqcup \ldots$, $hasVarConst \circ IsConsLinkCons \sqsubseteq hasVarConst$	A Variable belongs to a Constraint is represented by hasVarConst. A Variable can inherit Constraints from other Variables, this inference is achieved by complex role inclusion axioms ($R_1 \circ R_2 \sqsubseteq R_3$, being R_1, R_2 and R_3 roles).

The generation of the *SoC_estimator* function is contemplated in R3. Thereby, an analysis of data that conforms such function is carried out. Considering a HLUC controlling the active power supplied by a BESS, the calculation of HLUC's SoC (SoC_{hluc}) follows the logic in Equation [3]. SoC_{hluc} depends on the SoC initially assigned to the HLUC (SoC_{ini}), the current charged into the BESS (I_{bat}), as well as the total capacity of the BESS (Q_{bat}) and a portion capacity assigned to the HLUC (Q_{hluc}). The value I_{bat} is calculated from the active power (P_{bat}) and the nominal voltage (V_{nom}) of the battery. A modeling of the mentioned variables results in the following instantiations: `State{Vnom, SoC}`, `ParamDevice{SoCini, CAh}` and `ParamUC(CAh_UC)`. Nevertheless, a further representation of those variables is required, resulting in new concepts. For instance, the concept `Vnom` represents nominal voltages and is subsumed by `State`. A detail of all those new concepts is given in Table 4.

On the other hand, the creation of new roles facilitates the generation of *SoC_estimator*. Hence, `ControlBESS` relates a UC controlling a BESS and the concrete role `hasI_O` facilitates a remote control of internal variables defined within the scope of a UC as detailed in Table 4.

$$SoC_{hluc} = SoC_{ini} + \int \frac{I_{bat}}{Q_{bat}Q_{hluc}}, \quad I_{bat} = \frac{P_{bat}}{V_{nom}} \quad (3)$$

Table 4. Extension of EMS-*TBox* to generate *SoC_estimator* function.

Concepts, Roles, OWL Axioms	Description
$Vnom \sqcup SoC \sqcup P \sqcup I \sqcup \cdots \sqsubseteq State$	Vnom and SoC represent a nominal voltage (e.g., V_{nom}) and a state of charge (e.g., SoC_{hluc}) respectively. P models an active power (e.g., P_{bat}) and I a current value (e.g., I_{bat}).
$SoCini \sqcup CAh \sqcup \cdots \sqsubseteq ParamDevice$	SoCini represents an initial SoC (e.g., SoC_{ini}), CAh models the full capacity of an energy storage device (e.g., Q_{bat}).
$CAh\_UC \sqcup \cdots \sqsubseteq ParamUC$	Capacity assigned to a UC is represented by CAh_UC (e.g., Q_{hluc}).
$Application \sqsubseteq \exists IsConnectedTo.Application \ldots,$ $HLUC \sqsubseteq \exists ControlBESS.BESS \ldots$	The role IsConnectedTo relates two applications and the role ControlBESS relates a HLUC that is connected to a BESS.
$hasI\_O \; keyfor \; Internal$	hasI_O gathers information about whether or not a variable of type Internal is assigned to an Input or Output. Thereby, variables affected by this role are those subsumed by Internal (Param, State, ...).

4.4.2. Action 2: Enlargement of Inference Process

The inference of data reached by OWL ontologies can be extended by Semantic Web Rule Language (SWRL). This concept is standardized by W3C as language for expressing rules and provides powerful deductible reasoning mechanisms compared to OWL. SWRL is based on OWL DL, Datalog and Horn-like rules [39]. As a result, the syntax of SWRL is represented by atoms, rule body (antecedent) and rule head (consequent): $Atom \wedge Atom \wedge \cdots \rightarrow Atom$. If the conditions on the antecedent hold, then also the conditions in the consequent hold. On that basis, SWRL rules *r1* and *r2* are designed to support the identification of CEMS's constraints, as shown in Table 5. Where *r1* deduces Variables assigned to Constraints and *r2* infers relations between Constraints.

Table 5. Rules to infer CEMS's constraints.

SWRL Rules	Description
r1: $Control(?x1) \wedge IsAssignedTo(?x1,?s) \wedge Setpoint(?s)$ $\wedge hasVarConst(?s,?c1) \wedge Constraint(?c1) \rightarrow \mathbf{hasVarConst}(?x1,?c1)r$	A Control inherits constraints assigned to the Setpoint that it targets. A setpoint variable ?s owns the constraint ?c1, if ?s is controlled by ?x1, then ?x1 inherits the constraint ?c1.
r2: $Constraint(?c1) \wedge hasConstVar(?c1,?x1) \wedge IsVarLinkVar(?x1,?x2)$ $\wedge Variable(?x1) \wedge Variable(?x2) \wedge hasVarConst(?x2,?c2)$ $\rightarrow \mathbf{IsConsLinkCons}(?c1,?c2)$	The role IsConsLinkCons is established when a relation between Costraints is detected.

On the other hand, certain rules that requires the creation of new instances, in the consequent that these do not appear in the antecedent, are not possible by SWRL. In those cases, SPARQL Update queries are used instead [40]. SPARQL Update is a standard language that executes updates to triples in a graph store. It uses the data operators INSERT and DELETE for inserting and removing triples.

Following that, mechanisms to generate the *SoC_estimator* function use *r3* to identify when a HLUC is controlling a BESS by means of ControlBESS. As a sequel, *r4* attaches a new PUC (*SoC_estimator*) to the identified HLUCs. Finally, *r5* is responsible for the connections across the BESS, the HLUC and the created PUC (*SoC_estimator*). Those rules are detailed in Table 6.

Table 6. Rules to support the inference of the function: *SoC_estimator*.

SWRL Rules and SPARQL Update Query	Description
r3: EMS(?y) ∧ HLUC(?z) ∧ hasHLUC(?y,?z) ∧ BESS(?x) ∧ IsConnectedTo(?x,?y) ∧ hasControl(?z,?x1) ∧ Control(?x1) ∧ IsAssignedTo(?x1,?x2) ∧ P(?x2) ∧ hasVariable(?x,?x2) → ControlBESS(?z,?x)	If an EMS contains a HLUC that controls active power (P) of a BESS. Then, such HLUC and BESS are bound by ControlBESS.

r4: PREFIX CEMS :< http : //.../CEMS# >
 INSERT{
   ?x1 **CEMS:hasPUC** ?x3.
   ?x3 rdf : type CEMS : PUC.
   ?x3 **CEMS:hasType** "SoC_estimator"^^xsd : string.
   ?x3 CEMS : hasName ?x3N }
 WHERE{
   ?x1 rdf : type CEMS : HLUC.
   ?x1 CEMS : ControlBESS ?x2.
   ?x1 CEMS : hasName ?x1N.
   BIND(URI(CONCAT("SoC_", STR(?x1))) as ?x3).
   BIND(CONCAT(STR(?x1N),"_SoC") as ?x3N)}
	A PUC of type 'SoC_estimator' is added to a HLUC that controls a BESS. The name assigned to the new PUC is a concatenation of HLUC's name and the string '_SoC'. For instance, a PUC called FW_SoC is assigned to a HLUC(FW) named FW.

r5: PREFIX CEMS :< http : //.../CEMS# >
 INSERT{
   ?x5 **CEMS:IsAssignedTo** ?x4. }
 WHERE{
   ?x1 CEMS : ControlBESS ?x2.
   ?x1 CEMS : hasPUC ?x3.
   ?x3 CEMS : hasType "SoC_estimator"^^xsd : string.
   ?x3 **CEMS:hasFeedback** ?x4.
   ?x4 CEMS : hasType "Vnom"^^xsd : string.
   ?x2 **CEMS:hasStatus** ?x5.
   ?x5 CEMS : hasType "Vnom"^^xsd : string }
	A BESS's Status is assigned to a Feedback of the function PUC(*SoC_estimator*). For simplicity, only the inference of relations between BESS's V_{nom} and PUC(*SoC_estimator*) are shown. Thus, since V_{nom} is needed to calculate SoC_{hluc}, a role IsAssignedTo representing the relation of Status(V_{nom}) and PUC(*SoC_estimator*)'s Feedback is established.

4.4.3. Action 3: Setting up of New EMS-Templates

This section addresses all the modifications carried out along EMS-templates to keep a consistency with the extended EMS-*TBox*. The extended headlines are highlight with blue color.

The detection of CEMS's constraints would need to gather instances of these general statements: a UC (s) contains a Setpoint (x), which in turn is constrained by Constraint (C_1). Moreover, a Constraint (C_1) is related to a Constraint (C_2). Thereupon, an extension of the spreadsheet template collecting the UC's attributes is performed and resulted in Table 7. Moreover, the field Const_Description is included to gather constraint's descriptions.

Table 7. Extension of EMS-templates to collect constrains of a variable.

UC	Variable	Description	Type	Const.	Const_Description	IsConsLink.
s	x	variables's description	Setpoint	C_1	$x_{min} \leq x \leq x_{max}$	C_2

The vocabulary defined to model the *SoC_estimator* involves variables of BESS and UCs. The current version of EMS-templates contemplates generic models of BESS and UCs. Nevertheless, those models need to be extended to achieve the implementation of the *SoC_estimator* as shown in Table 8. For instance, the concept CAh_UC that represents the capacity assigned to a UC is now included in the UC(*UC_generic*), just as the CAh concept in the UC(*BESS*). Besides this, a complete model of PUC(*SoC_estimator*) is also required, see Table 9. All the extended models are made available in the IED and UC repository. Hence, control engineers are free to edit all the fields except Type.

Table 8. BESS and UC models concerned by UC (*SoC_estimator*).

UC	Variable	Description	I_O	Type	Value	Format	Unit
BESS	CAh	total capacity of the battery	Status	CAh		double	Ah
	Vnom	nominal voltage of the battery	Status	Vnom		double	V
UC_generic	SoCini	initial SoC of the use case	Status	SoCini		double	%
	CAh	capacity assigned to a UC	Status	CAh_UC		double	Ah

Table 9. Parameters and states relevant to UC(*SoC_estimator*).

UC	Variable	Description	I_O	Type	Value	Format	Unit
SoC_estimator	SoC_UC	SoC of a UC	Status	SoC		double	%
	I	current charged into the battery	Status	I		double	A
	U_bat	voltage of the battery	Feedback	Vnom		double	V
	CAh_bat	total capacity of a battery	Feedback	CAh		double	Ah
	P_UC	active power set by a UC	Feedback	P		double	kW
	SoC_ini	initial SoC of the UC	Feedback	SoCini		double	%
	CAh_UC	capacity assigned to a UC	Feedback	CAh_UC		double	Ah

4.4.4. Action 4: Mapping Between SGAM-TB and EMS-DM

The exchange of data between SGAM models and EMS-templates is addressed using MDA techniques. Hence, a model of SGAM-TB is investigated and proposed in this section.

A UML class representation of SGAM-TB is suggested by Neureiter et al. [38]. Such representation motivates the SGAM-TB model proposed in this work. It is worth highlighting that only function and information layers are modeled. The function layer of SGAM-TB involves description of main functionalities within the concept HLUC, further details of HLUCs are given in PUC. Actors involved within those PUCs are modeled by the LogicalActor. In turn, the information layer includes the InformationObject a concept representing any information flow. Which is structured by a data model (DataModel). The aforementioned concepts are combined to conceive a UML class representation of SGAM-TB, see Figure 7a. Considering this, classes, attributes, and other components of SGAM-TB model are transformed into EMS-DM components, as shown in Figure 7b. For a straightforward interpretation only the mapping between classes is depicted.

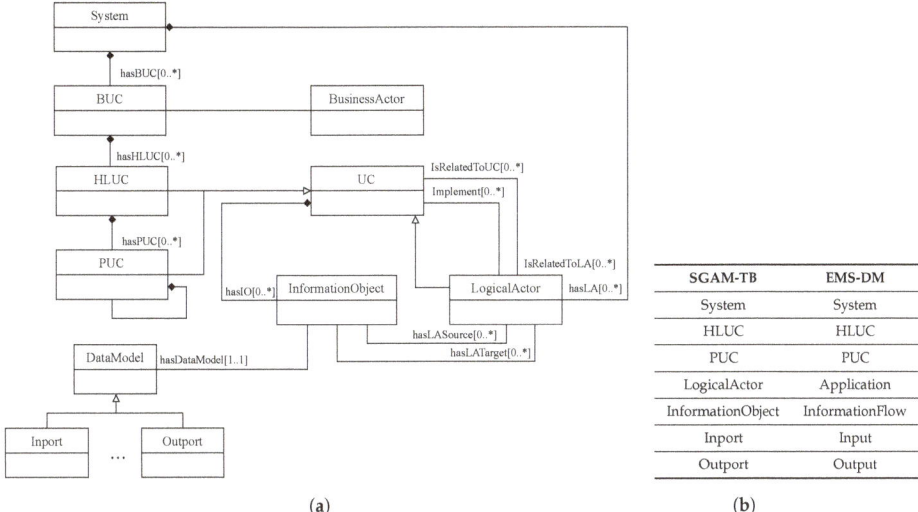

Figure 7. Proposed foundations to achieve a mapping between SGAM-TB and EMS-DM. (**a**) UML class diagram of SGAM-TB; (**b**) matching between SGAM-TB and EMS-DM.

4.4.5. Action 5: Mapping Between IEC 61850 an EMS-DM

The IEC 61850 approach is object-oriented and has been lately used in power system domain to improve interoperability between IEDs. The abstraction of services provided by an IED begins with a Logical Node (LN), which models services for protection, measurement, control, etc. Each LN is defined as a class within the IEC 61850-7-4 specification [41]. Data Objects (DO) refine the definition of a LN, they are specified by Common Data Classes (CDC) in IEC 61850-7-3 [42].

On the other hand, IEC 61850 defines a language to configure IEDs under the name of System Configuration Language (SCL) in IEC 61850-7-6. An SCL specifies the capabilities of an IED by LN classes, which are of type LNodeType. Each of them contains DOs of type DOType which in turns contain DAs of type DAType. A LNodeType should correspond to a LN as well as a DOType to a CDC.

The instantiation of the mentioned classes is exemplified by the LNodeType:$MMXU$, see Figure 8. This notation (class:individual) is employed to assign an individual to a class. The mentioned LNodeType represents a calculation of current, voltage, frequency, etc. Thereby, it contains among others the DO:Hz for frequency measurement. This DO contains DA:mag to abstract the mean value of the frequency.

On the basis of SCL representation, the mapping between EMS-DM and IEC 61850 is carried out as follows: a LNodeType is transformed into a UC and a DO into a Variable. The type of Variable (i.e., Feedback, Control, etc.) to be generated is obtained from the attribute *cdc* within DOType. Hence, a MV (Measured Value) results in a Status assigned with a Numeric value. The DAs elements of a DOType (i.e., t, mag, units, etc.) are mapped into the attributes of Numeric (i.e., hasTimeStamp, hasAnaValue, hasUnits, etc.). Those mappings are referenced by color match in Figure 8.

In the previous example a DO:Hz representing a MV resulted in a Status. However, a DO may belong to the other types such ASG (Analogue Setting) or SPS (Single Point Status) among others. Depending on that, a generation of either a Setpoint or Status, etc. is performed. It is worth highlighting that not all the DAs of a DOType can be mapped into the EMS-DM. For instance, the DA:q that represents the quality of the information is not mapped. To accomplish that, an extension of attributes assigned to Variable is required. As a summary, any IED functionality defined within a LN is translated into a UC and upload into the repository.

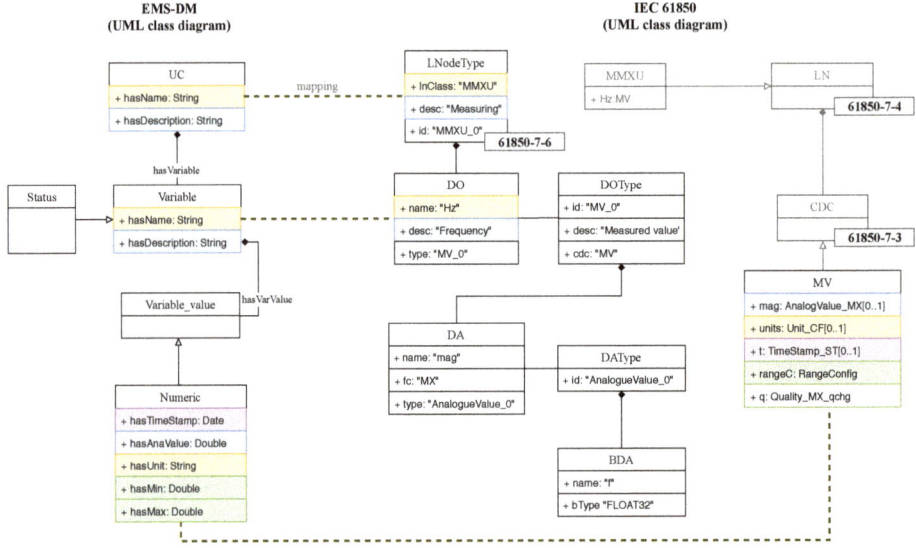

Figure 8. Mapping between IEC 61850 (LN) and EMS-DM (UC).

4.4.6. Action 6: Generation of Inconsistency Reports

The identification of certain inconsistencies may require the extension of the EMS-*TBox* as well as definition of new rules and queries besides adjustments to the EMS-templates. However, the fulfillment of R5 and R8 requires just the elaboration of new queries, those are addressed as follows. In the scope of R5, constraints on a BESS are analyzed. The active power value of a battery is limited as shown in Equation [4]. The identification of services that attempt to control P_{bat} with values that exceed the technical limits requires the modeling of *Pmax*. Besides this, the power demanded by a service is also needed (*BessSize*). Those concepts were already included in the EMS-*TBox*.

A formalization of competency questions is achieved by DL or SPARQL queries [43]. This depends on the type of question, usually elaborated questions need the expressivity of SPARQL syntax. In turn, simple questions can be formulated by DL. On that basis, R5 implies DL queries to answer questions about BESS's restrictions and UC's parameters, see Table 10.

$$P_{min} < P_{bat} < P_{max} \qquad (4)$$

Table 10. Querying the EMS-*ABox* to identify batteries mismatching services.

Query	Question	DL Query
Q_I	What is the variable of type *Pmax* defined within a BESS(*battery*)?	$Pmax \sqcap \exists hasParam^-.BESS\{battery\}$
Q_{II}	What is the active power to be required by a service HLUC (*Service$_n$*)?	$BessSize \sqcap hasParam^-.HLUC\{Service_n\}$

On the other hand, since R8 investigates misconfigured units, an analysis of the information flow is required. An information flow is conformed by a source and a target. A source can be of type Control and Output. In turn, a target is of type Setpoint and Feedback. Units of the source should match the units configured at the target. Enough information is already provided within the EMS-templates to conclude mistaken units. The SPARQL query in Table 11 investigates and compares units within a Setpoint and Control.

Table 11. SPARQL query to investigate misconfigured units.

Query	Question	SPARQL Query
Q_{III}	What are the setpoints of a HLUC? What is the unit configured within a setpoint? What is the unit of a control variable targeting certain setpoint? What are the units that mismatch?	$SELECT\ ?control\ ?setpoint\ ?unit\_ct\ ?unit\_sp$ $WHERE\{$ $?HLUC\ \ CEMS:hasSetpoint\ ?setpoint;$ $\qquad rdf:type\ \ CEMS:HLUC.$ $?setpoint\ \ CEMS:hasUnit\ ?unit\_sp.$ $?control\ \ CEMS:IsAssignedTo\ ?setpoint;$ $\qquad rdf:type\ \ CEMS:Control;$ $\qquad CEMS:hasUnit\ ?unit\_ct$ $FILTER(!regex(STR(?unit\_ct),\ STR(?unit\_sp))\}$

4.4.7. Action 7: Software Artifacts Generation

This section tackles the issues exposed in R4 and R6. Therefore, an automatic generation of configuration files and software artifacts compliant with IEC 61499 is pursued.

The achievement of a configuration file from EMS-templates is carried out by M2T transformations. Hence, the model source is given by EMS-DM and the text target is defined by control engineer's practice. As an example, a MATLAB script that supports the proof-of-concept of CEMS is sought. Such script should expose parameters within a HLUC. Hence, a M2T is implemented by Acceleo templates, an open source framework for code generation, see Figure 9. That Acceleo template investigate all HLUCs (i.e., functions) included within an EMS. Afterwards, the parameters (Param) assigned to the identified HLUCs are extracted and customized.

```
[for (app: Application| asystem.hasApplication -> filter(EMS) ) separator('\n')]
  %[app.eClass().name/] [app.hasName /];
  [for (hluc:HLUC | app.hasHluc ) separator('\n')]
    %HLUC [hluc.hasName/]
    %Parameters of the Hluc:
    %======================================
    [for (var:Param | hluc.hasVariable->filter(ParamHluc) )]
      [app.hasName.concat('.').concat(hluc.hasName).concat('.').concat('Param').concat('.').
      concat(var.hasName).concat('=') /][ var.hasVarValue-> filter(Analog).hasAnaValue/];
    [/for] ...
```

Figure 9. Acceleo template to generate configuration file of HLUCs.

The proof-of-concept of CEMS is also supported by generation of MATLAB/simulink models. A generation of MATLAB/Simulink applications from EMS-templates is based on M2M transformation where the PIM is given by EMS-DM and the PSM by a MATLAB/Simulink model. As first step, a conception of MATLAB/Simulink model should be reached, this model is proposed within the MATLAB Simulink Integration Framework for Eclipse (MASSIF) project [44]. The root of that model is SimulinkModel, which contains Block and SubSystem. A SubSystem is characterized by properties and ports of type Outport and InPort. Those ports may be related to a SingleConnection. This configuration is quite similar to EMS-DM structure. Hence, the concepts: Application^{EMS} and UC^{EMS} are transformed into SubSystem^{Sim}, Input^{EMS} into InPort^{Sim}, InformationFlow^{EMS} into SingleConnection^{Sim} and so on, as shown in Table 12. To increase the understanding of classes transformation the syntax class^{Model} is used.

An automatic generation of IEC 61499 applications from EMS-templates requires a model representing IEC 61499 applications. A UML class representation of it is reached from an analysis of the standard IEC 61499 [45], see Figure 10. As it can be noticed the depicted model share many similarities with the EMS-DM, thus a matching between them is possible as shown in Table 12. A UC^{EMS} is mapped to a FB^{61499}, which is defined by a FBType^{61499}. The FBType can take the form of one single block (BasicFB) or an arrangement of blocks (FBNetwork), see Figure 10. On the other hand, Applications^{EMS} is converted into Application^{61499}, Inputs^{EMS} into InputVarss^{61499} and so on.

Model-to-model transformation across MATLAB/Simulink, IEC 61499 and EMS-DM can be implemented by following the mappings referenced in Table 12. This gives a flexibility to transform EMD-DM instances (i.e., EMS-templates) into specific platforms important to the proof-of-concept and implementation phases. Transformation rules are carried out using ATL Transformation Language (ATL), a M2M transformation language that implements unidirectional transformations within the Eclipse platform [46]. To illustrate this, the rules to transform UC^{EMS} into $FBType^{61499}$ are shown in Figure 11. Where a navigation through the components of FBType (i.e., InputVars, OutputVars, InternalVars) enables a match to EMS-DM elements (i.e., Input, Output, Param).

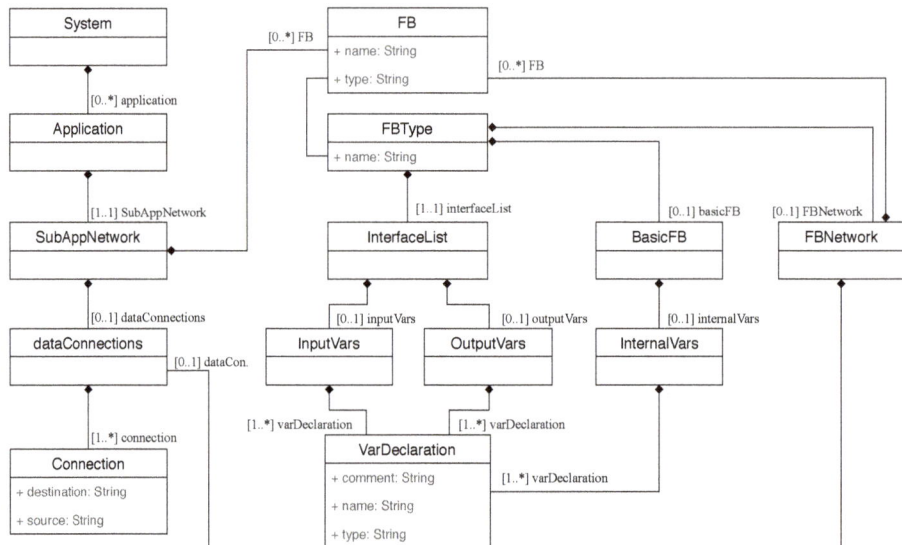

Figure 10. UML class diagram representation of IEC 61499 (PSM).

```
rule UC2FBType {
from
s:EMS!UC
to
t:iec61499!FBTypeType(
    name <- s.hasName,
    interfaceList <- SinterfaceList),
    SinterfaceList:iec61499!InterfaceListType (
        inputVars <- SInputVarsType,
        outputVars <- SOutputVarsType,
    SInputVarsType: iec61499!InputVarsType(
        varDeclaration <- s.hasVariable -> select(e e.oclIsTypeOf(EMS!Input)) ->
        collect(e thisModule.Input2VarDeclaration(e))), ...

lazy rule Input2VarDeclaration {
from
s:EMS!Input(s.isContainsElement())
to
t: iec61499!VarDeclarationType(
    comment <- s.hasDescription,
    name <- s.hasName ,
    type <- thisModule.TypeConversion(s.hasFormat)) }
```

Figure 11. ATL rule transformation from UC^{EMS} into $FBType^{61499}$ (reduced model).

Table 12. Mapping between EMS, MATLAB/Simulink and IEC 61499 data models.

EMS-DM	MATLAB/Simulink	IEC 61499
System	SimulinkModel	System
UC, HLUC, PUC	SubSystem	FB, FBType
Application	SubSystem	Application
Input	Inport	InputVars
Output	Outport	OutputVars
InformationFlow	SingleConnection	Connection
Param	Property	InternalVars

5. CEMS Implemented with the Extended EMSOnto

This section shows how the above presented improvements of the EMSOnto are used by control engineers during the realization of a CEMS.

5.1. EMS-Templates

The previously introduced CEMS is specified using SGAM-TB, which is available as an extension of the design tool Sparx Systems EA, thus UML and SysML support the CEMS design. Hence, the structure of the CEMS is specified under UML use case diagrams, the resulting model is depicted in Figure 12. From such a representation it is deduced that the CEMS implements $SelfC$ and FW, which are associated with a $Meter$ and a $BESS$. Moreover, $SelfC$ implements the functions $PI\_Control$ and $Limit\_SoC$ just as FW implements $Linar\text{-}Control$ and $Limit\_SoC$. All those details define the CEMS structure and should be considered at the design phase. To support this, an automatic transfer of information from SGAM models to EMS-templates is executed. This involves UML use case, sequence, and activity diagrams. The resulted EMS-template from a use case diagram is shown in Table 13. This represents a first version of the EMS-$ABox$ and the starting point of the design process.

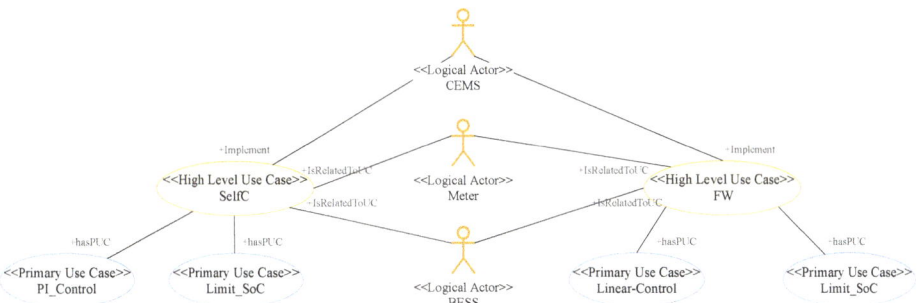

Figure 12. CEMS's structure represented by a UML use case diagram.

Table 13. Spreadsheet generated automatically from CEMS's UC diagram.

System	Appl.	Application Description	Type	HLUC	HLUC Description	PUC
Sys	CEMS	customer energy management system	-	SelfC	power from the grid is avoided	PI_Control
			-	FW	active power is injected to support frequency regulation	Limit_SoC
	BESS	model of a BESS	BESS	-	-	-
	Meter	smart meter connected at PCC point	Meter	-	-	-

5.2. UC and IED Repository

At the design phase, ICD files provided by the IED's manufacturer are available to the control engineers. Selected ICD files are imported into the UC repository to benefit from IED and UC models.

A reduced model of an ICD file containing the LN:*FWHZ* is illustrated in Figure 13. A detail description of that LN is available in IEC 61850-90-7 [47]. The XML file shows variables within the LN:*FWHZ*, which are represented as DO elements. Attributes of those variables are contained within the DOType identified under *ASG_0*. Those attributes are of type *AnalogogueValue_0* and own values of type *FLOAT32*. With that in mind, it is understood that LN:*FWHZ* contains the variables *Wgra*, *HzStr* and *HzStop*. Each of them owns a series of attributes: *setMag*, *units*, *minVal* and *maxVal* which are of type *FLOAT32*. The exposed LN:*FWHZ* is transformed into PUC(*FWHZ*), afterward this PUC is load into the UC & IED repository, thus accessible to control engineers.

Since the operation followed by the LN:*FWHZ* is similar to PUC(*Linear-Control*), it is worth setting the PUC's type to *FWHZ*. As a result, all the variables related to LN:*FWHZ* are imported within *Linear-Control*, see Table 14. However not all the attributes' values are collected (e.g., min, max, unit), this is because those values are set during real-time operation of IEDs.

```xml
<LNodeType lnClass="FWHZ" id="FWHZ_0">
    <DO name="Wgra" type="ASG_0" desc="active power gradient in percent of frozen ...
    <DO name="HzStr" type="ASG_0" desc="delta frequency between start frequency ...
    <DO name="HzStop" type="ASG_0" desc="delta frequency between stop frequency ...
</LNodeType>
<DOType cdc="ASG" id="ASG_0" desc="Analogue setting">
    <DA dchg="true" fc="SP" name="setMag" bType="Struct" type="AnalogueValue_0"/>
    <DA dchg="true" fc="CF" name="units" bType="Struct" type="Unit_0"/>
    <DA dchg="true" fc="CF" name="minVal" bType="Struct" type="AnalogueValue_0"/>
    <DA dchg="true" fc="CF" name="maxVal" bType="Struct" type="AnalogueValue_0"/>
</DOType>
<DAType id="AnalogueValue_0">
    <BDA name="f" bType="FLOAT32"/>
</DAType>
```

Figure 13. ICD file containing a reduced model of LN:*FWHZ*.

Table 14. PUC (*Linear-Control*) generated automatically from repository's models (LN:*FWHZ*).

PUC	Variable	description	Type	Format	Min	Max	Unit
Linear-Control	Wgra	active power gradient in percent of frozen active power value per Hz	Setpoint	FLOAT32			
	HzStr	delta frequency between start frequency and nominal frequency	Setpoint	FLOAT32			
	HzStop	delta frequency between stop frequency and nominal frequency	Setpoint	FLOAT32			

5.3. Constraints of the CEMS

The inference of CEMS's constraints entails a collection of technical limitations regarding the process to be controlled (i.e., DERs, IEDs). This means that IED's constraints are documented within the extended EMS-templates. Once this process is complete, IED models are upload to the repository and CEMS's constraints are automatically resolved.

As an example the BESS's constraints are inferred. The BESS model is built by control engineers, see Table 15. As it can be noticed, the BESS owns the variables active power (*Pbat*) and apparent power (*Sbat*). The active power is set by Setpoint(*sp_Pref*). This statement is exemplified by the role IsAssignedBy. Moreover, constraints on the BESS are detailed. Hence, C1 represents a constraint on *Pbat* and C2 a constraint on *Sbat*. Since *Sbat* and *Pbat* are constrained by $Pbat^2 + Qbat^2 \leq Sbat^2$ the relation IsConsLinkCons(C1, C2) is established. The resulted BESS is upload to the repository. Next, a reasoner engine is executed, and new knowledge is provided to the control engineers. One of those results corresponds to PUC(*Limit_SoC*) being limited by the constraints C1 and C2. This implicit data is highlight with blue color in Table 16. The role IsAssignedTo is the inverse of IsAssignedBy.

Table 15. Extended BESS model apprised by control engineers

UC	Variable	Description	Type	IsAssignedBy	Const.	Const. Description	IsConsLink.
BESS	Pbat	active power	State	sp_Pref	C1	$P_{min} \leq P \leq P_{max}$	C2
	Sbat	apparent power	State		C2	$P^2+Q^2 \leq S^2$	-

Table 16. Constraints allocated to the PUC (*Limit_SoC*).

PUC	Variable	Description	Type	IsAssignedTo	Const.	Const. Description
Limit_SoC	ct_Pref	signal to control the charging/discharging of the BESS	Control	sp_Pref	C1	$P_{min} \leq P \leq P_{max}$
					C2	$P^2+Q^2 \leq S^2$

5.4. Inconsistencies Report

Inconsistencies reports are generated from querying the CEMS's inferred data. The original EMSOnto already supports reports of conflicts between UCs [8]. On top of that, the extended EMSOnto allows the identification of further inconsistencies such as BESS mismatching a service (I_I) and misconfigured units (I_{II}). During the design process the report shown in Table 17 was provided to control engineers. The report shows I_{II} as an identified mismatch, which led to a correction of the unit configured within Control(*ct_Pref*), a control signal targeting the setpoint (*sp_Pref*) of the BESS.

Table 17. Inconsistencies reports handled to control engineers.

Inconsistency	Detected	Conclusion Derived from Queries.	Control Engineer Analysis
Mismatches between a BESS and a service/I_I	X	The technical limitations of the BESS are not violated.	-
Units are misconfigured/I_{II}	✓	The unit of the control variable (*ct_Pref*) is set to W and the unit configured in a setpoint (*sp_Pref*) is kW.	Correction of the unit at the control level is required.

5.5. SoC Estimator Function

Since *FW* and *SelfC* are controlling the BESS they need to estimate how much power was delivered by the BESS. This is reached by instantiating the function *SoC_estimator* and attaching those instances to HLUC{*FW*, *SelfC*}. EMSOnto provides an automatic creation of PUC(*FW_SoC*) and PUC(*SelfC_SoC*) as shown in Table 18. The PUCs are instances of *SoC_estimator* and therefore share the same attributes (e.g., *CAh_UC*, *I*). This new information is highlighted by blue color.

Table 18. SoC of the functions HLUC {*FW*, *SelfC*} is estimated by adding new PUCs.

HLUC	PUC	Description	Type	Variable	Description
FW	FW_SoC	state of charge of a HLUC	SoC_estimator	CAh_UC	capacity assigned to a UC
SelfC	SelfC_SoC	state of charge of a HLUC	SoC_estimator	I	current charged into the battery

5.6. Software Artifacts Generation

An automatic generation of the MATLAB script shown in Figure 14 is performed. Only the parameters' values are generated. Other attributes such as format and unit are dismissed. Moreover, a refinement of the code was not necessary and it was used as it was generated. This script supports the proof-of-concept of the CEMS simulated within a MATLAB/Simulink platform [8].

On the other hand, the validated CEMS is transformed into IEC 61499 models, the ensemble of function blocks focused on HLUC(*FW*) is shown in Figure 15. A refinement of the obtained model consisting on the replacement of function blocks is carried out. Furthermore, two clients simulating the communication with real BESS and meters are generated but not configured (*BESS* and *Meter*).

```
%Application CEMS;
%HLUC FW
%Parameters of the Hluc:
%=====================================
CEMS.FW.Param.Priority=1.0;
CEMS.FW.Param.CAh=0.3;
CEMS.FW.Param.SoC=10.0;
CEMS.FW.Param.BessSize=100.0;
CEMS.FW.Param.Ena=1.0;
CEMS.FW.Param.SoCini=1.0;
```

Figure 14. Script generated to configure HLUC(*FW*) presented in a MATLAB/Simulink model.

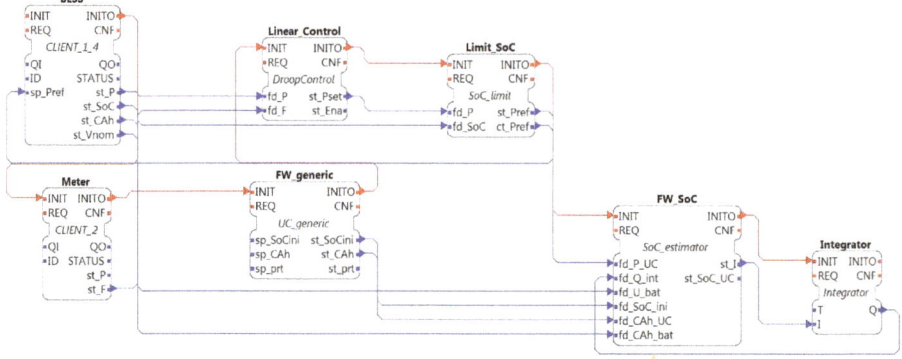

Figure 15. IEC 61499 model issued from EMS-templates (only *FW* is represented).

5.7. Evaluation of Requirements and Open Issues

This section exposes how the open issues stated in Section 2.4 are addressed by analyzing the implementation of the selected use case example (CEMS).

Documents available at design phase are exploited: The previous experiments regarding the CEMS demonstrate an automatic processing of documents intended to support the CEMS design. In other words, knowledge structured under information models (IEC 61850) and specification approaches (SGAM) are collected to populate EMS-templates.

SGAM models provide a high amount of information regarding the specification of an EMS. Since the design should implement a solution according to EMS specifications, SGAM models were imported within EMS-templates. As a result, at the design phase control engineers starts with a design totally aligned with the requirements defined by stakeholders. Another benefit of this, is that potential manual errors, susceptible to be introduced by manual input of EMS-templates, are avoided. Furthermore, it has been demonstrated that by model-driven engineering techniques it is possible to employ SGAM models in a computerized way to support the whole engineering process. This corresponds to the use of SGAM knowledge for the generation of software artifacts. Nevertheless, the presented experiments did not consider other specification approaches such as Intelligrid and UML. Getting information derived from such approaches would imply to carry out a formal data representation as well as transformation rules and maybe extensions on the EMS-*TBox* as it was done with SGAM. The fact of proving an alignment between SGAM and EMSOnto sets the basis for other possible alignments.

Another executed experiment corresponds to the import of IEC 61850 SCL files into the IED and UC repository. IEC 61850 suggests not only power system components but also IED functionality. In that regard, the PUC(*Linear-Control*) is successfully generated from the LN *FWHZ*. Only the LN's attributes required by engineers were covered in the experiments. However, since the EMS-*TBox* does not contain all the classes and attributes defined within IEC 61850 a full import of the LN's attributes would involve a further extension of the EMS-*TBox*.

The successful import of SCL files attempts to demonstrate that power utility information models can be used to feed the UC and IED repository. This means that the same mechanisms used to import IEC 61850 models are expected for other models such as energy market communication models formalized in IEC 62325 and also smart meter models defined within IEC 62056 among others [48].

Enlargement of inferred knowledge: Inferences provided by EMSOnto were extended. The inferences performed in this work support control engineers by deducing the constraints to be respected. Besides this, the identification of inconsistencies that may harm the correct operation of control applications was carried out. On top of that, new functions (*SoC_Estimator*) were created to accelerate and support the design of control applications.

Since different sort of inferences were achieved this may lead to think that EMSOnto is flexible to implement any kind of inference. However, this strictly depends on the expressivity provided by the ontology ($SROIQ(D)$) and also the rules. Although $SROIQ(D)$ is well defined, the expressivity for rules is determined by rule languages. In this work, SWRL and SPARQL were enough to address the control engineer's requirements. Nevertheless, other requirements may need to use the Rule Interchange Format (RIF), which provides the means to insert primary logic programming languages [37], or even to extend ontologies with probabilistic inference and temporal logic [49].

Flexible generation of software artifacts: CEMS experiments demonstrate that generated code and models support control engineers not only in the proof-of-concept but also during the implementation phase. To make this possible, mappings between a power system emulator (MATLAB/Simulink), a controller platform (IEC 61499 platform) and Emsonto are conceived as well as text templates for code generation (MATLAB script). This evaluation lays the groundwork for implementations to other specific platforms (IEC 61131-3, PowerFactory).

The final realization of control applications requires the participation of engineers for the customization of generated models as well as the definition of the control algorithm's behavior. In the use case example, communication interfaces were generated (*BESS* and *Meter* clients), although not configured. Consequently, the generation of a full control application ready to be tested is not fully automated. Indeed, other approaches cover those gaps. For instance, PSAL supports the design and implementation of communication and component layers [10]. Therefore, the cooperation between different engineering and development approaches is encouraged.

6. Conclusions

In the near future, a high integration of BESS within the distribution grid may very well become a reality since they can provide a broad range of services. Those services are likely to be deployed within an EMS which in turn will control the active and reactive power of a BESS. In that context, this work seeks an automated and adaptable approach that would support the engineering process of such EMS. Therefore, available engineering approaches are studied to realize what are the open issues to be addressed. Among the studied approaches, EMSOnto—a holistic approach for multi-functional BESS—is selected as the reference methodology.

The identified open issues are handled by the EMSOnto expert, whose main role is the customization of the EMSOnto according to control engineer's requirements. As a solution, the expert proposes a set of actions based on model-driven engineering techniques and ontologies. Those are summarized as the extension of the EMS-*TBox*, setting up of new rules and queries, extension of EMS-templates, and finally the conception of data models and transformation rules. All those actions are exemplified by a selected use case example showing the realization of a CEMS, which encapsulates a series of control engineering requirements. The realization of the CEMS allows to evaluate the efficacy and limitations of the proposed solutions. The experiments were tied to specific tools and standards (MATLAB/Simulink, IEC 61850, IEC 61499, SGAM) to show the EMSOnto expert's actions. These experiments set foundations for further customizations of EMSOnto with the aim of encompassing other standards and tools. An analysis of the tasks performed by the EMSOnto expert, after collecting and understanding the control engineer's requirements, is outlined, and discussed:

1. Define an ontology/data model of the EMS under study
2. Integrate rules and queries to the ontology
3. Propose a methodology to gather knowledge from the EMS
4. Design data models for specific software platforms, IEDs, DERs, etc.
5. Elaborate transformation rules for code/text and model generation

From the abovementioned steps, only steps 4 and 5 should be carried out to exploit information sources available at the specification stage. This is demonstrated by the CEMS example, where EMS-templates are fed with SGAM models. Hence, a consistency between specification and design phases was reached. Similarly, the abovementioned steps can also be applied to other specification and design approaches such as IntelliGrid, PSAL, etc. On the other hand, by following steps 1, 2 and 3 the inference of implicit knowledge is achieved. It was demonstrated by detecting inconsistencies within multi-functional BESS (e.g., misconfigured units) and by deducing functions at the design time (e.g., *SoC_estimator*). The proposed methodology can enable inference within other modern approaches (e.g., PSAL) as well. Besides this, a customized generation of software artifacts is achieved by performing steps 1, 4 and 5. As a result, EMSOnto is now able to support the implementation stage with generated code aligned to a controller platform (IEC 61499). In the same way, other approaches can also benefit from the referred steps to achieve compatibility with different specific platforms.

From the CEMS use case example, it is worth noticing that limitations of inferences depend on the expressivity of the ontology (e.g., OWL ontology) and the intelligence provided by rule languages. Consequently, it is encouraged to choose the right ontology language for the handling of the knowledge to be inferred. Moreover, the validation of EMS involves a dynamic process. Therefore, the identification of certain inconsistencies at that stage would require the time notion, which in a future work can be achieved by extending the expressiveness provided by ontologies with temporal logic [50]. On the other hand, with the current approaches, it is common to document the planned design (e.g., PSAL, SGAM, EMSOnto). Modifications to the stipulated design occurs often during proof-of-concept and implementations stages. However, since the modified information is not tracked then the generated software artifacts would not be aligned with the final validated design. To handle that issue, reverse engineering along the engineering process is encouraged [51]. Moreover, the validation of multi-functional BESS applications often involves tools such as communication networks and co-simulation frameworks. This means that the communication and component domains should be considered by the selected approach. However, this is not always the case since the mentioned domains are not modeled within EMSOnto. Thereby, as future work it is encouraged to combine different approaches, for instance, combining EMSOnto with PSAL where a formal semantic for communication interfaces and components is addressed.

Author Contributions: C.Z. wrote the paper and carried out the conceptualization, investigation and validation of the work. F.P.A. and T.I.S. participated in the conceptualization of the proposed approach and reviewed the final manuscript. T.I.S. supervised the overall work.

Funding: This work is partly supported by the Austrian Ministry for Transport, Innovation and Technology (bmvit) and the Austrian Research Promotion Agency (FFG) under the ICT of the Future Programme in the MESSE project (FFG No. 861265).

Conflicts of Interest: The authors declare no conflict of interest.

References

1. Alizadeh, M.; Parsa Moghaddam, M.; Amjady, N.; Siano, P.; Sheikh-El-Eslami, M. Flexibility in Future Power Systems with High Renewable Penetration: A Review. *Renew. Sustain. Energy Rev.* **2016**, *57*, 1186–1193. [CrossRef]
2. Koller, M.; Borsche, T.; Ulbig, A.; Andersson, G. Review of Grid Applications with the Zurich 1MW Battery Energy Storage System. *Electr. Power Syst. Res.* **2015**, *120*, 128–135. [CrossRef]
3. *EERA Joint Programme on Smart Grids-Sub-Programme 4-Electrical Energy Technologies*; Technical Report D4.3 Integration of Storage Resources to Smart Grids: Possible Services, D4.4 Control Algorithms for Storage Applications in Smart Grid; EERA: Brussels, Belgium, 2014.
4. Zanabria, C.; Pröstl Andrén, F.; Strasser, T.I. Comparing Specification and Design Approaches for Power Systems Applications. In Proceedings of the 2018 IEEE PES Transmission and Distribution Conference and Exhibition—Latin America, Lima, Peru, 18–21 September 2018; p. 5.
5. Santodomingo, R.; Uslar, M.; Goring, A.; Gottschalk, M.; Nordstrom, L.; Saleem, A.; Chenine, M. SGAM-Based Methodology to Analyse Smart Grid Solutions in DISCERN European Research Project. In Proceedings of the 2014 IEEE International Energy Conference (ENERGYCON), Dubrovnik, Croatia, 13–16 May 2014; pp. 751–758. [CrossRef]
6. Working Group Sustainable Processes (SG-CG/SP). *CEN-CENELEC-ETSI Smart Grid Coordination GroupSustainableProcesses*; Technical Report; CEN-CENELEC-ETSI: Brussels, Belgium, 2012.
7. International Electrotechnical Commission (IEC). *IEC 62559-2 Use Case Methodology-Part2: Definition of the Templates for Use Cases, Actor List and Requirement List*; IEC: Geneva, Switzerland, 2015.
8. Zanabria, C.; Pröstl Andrén, F.; Kathan, J.; Strasser, T.I. Rapid Prototyping of Multi-Functional Battery Energy Storage System Applications. *Appl. Sci.* **2018**, *8*, 1326. [CrossRef]
9. Zanabria, C.; Pröstl Andrén, F.; Kathan, J.; Strasser, T. Approach for Handling Controller Conflicts within Multi-Functional Energy Storage Systems. In Proceedings of the CIRED-Open Access Proceedings Journal, Glasgow, UK, 12–15 June 2017; pp. 1575–1578.
10. Andrén, F.P.; Strasser, T.I.; Kastner, W. Engineering Smart Grids: Applying Model-Driven Development from Use Case Design to Deployment. *Energies* **2017**, *10*, 374. [CrossRef]
11. Tayyebi, A.; Bletterie, B.; Kupzog, F. Primary Control Reserve and Self-Sufficiency Provision with Central Battery Energy Storage Sytens. In Proceedings of the NEIS Conference, Hamburg, Germany, 21–22 September 2017; pp. 21–22.
12. Riffonneau, Y.; Bacha, S.; Barruel, F.; Ploix, S. Optimal Power Flow Management for Grid Connected PV Systems with Batteries. *IEEE Trans. Sustain. Energy* **2011**, *2*, 309–320. [CrossRef]
13. Boehm, B.; Turner, R. *Balancing Agility and Discipline: A Guide for the Perplexed, Portable Documents*; Addison-Wesley Professional: Boston, MA, USA, 2003.
14. Andrén, F.; Lehfuss, F.; Strasser, T. A Development and Validation Environment for Real-Time Controller-Hardware-in-the-Loop Experiments in Smart Grids. *Int. J. Distrib. Energy Resour. Smart Grids* **2013**, *9*, 27–50.
15. Gottschalk, M.; Uslar, M.; Delfs, C. *The Use Case and Smart Grid Architecture Model Approach: The IEC 62559-2 Use Case Template and the SGAM Applied in Various Domains*; Springer: New York, NY, USA, 2017.
16. Weilkiens, T. *Systems Engineering with SysML/UML: Modeling, Analysis, Design*; Elsevier: Amsterdam, The Netherlands, 2011.
17. Tornelli, C.; Radaelli, L.; Rikos, E.; Uslar, M. *WP 4 Fully Interoperable Systems Deliverable R4.1: Description of the Methodology for the Detailed Functional Specification of the ELECTRA Solutions*; Technical Report; ELECTRA IRP; 2015. Available online: http://www.electrairp.eu/index.php?option=com_attachments&task=download&id=441 (accessed on 12 March 2017).
18. Dänekas, C.; Neureiter, C.; Rohjans, S.; Uslar, M.; Engel, D. Towards a Model-Driven-Architecture Process for Smart Grid Projects. In *Digital Enterprise Design & Management*; Springer: New York, NY, USA, 2014; pp. 47–58.
19. Higgins, N.; Vyatkin, V.; Nair, N.K.C.; Schwarz, K. Distributed Power System Automation With IEC 61850, IEC 61499, and Intelligent Control. *IEEE Trans. Syst. Man Cybern. Part C* **2011**, *41*, 81–92. [CrossRef]
20. Zhabelova, G.; Vyatkin, V.; Dubinin, V. Towards Industrially Usable Agent Technology for Smart Grid Automation. *IEEE Trans. Ind. Electron.* **2014**, *62*, 2629–2641. [CrossRef]

21. Schütte, S.; Scherfke, S.; Sonnenschein, M. MOSAIK—Smart Grid simulation API—Toward a Semantic Based Standard for Interchanging Smart Grid Simulations. In Proceedings of the 1st International Conference on Smart Grids and Green IT Systems, Porto, Portugal, 21 April 2012; SciTePress—Science and Technology Publications: Porto, Portugal, 2012; pp. 14–24. [CrossRef]
22. Bhor, D.; Angappan, K.; Sivalingam, K.M. Network and Power-Grid Co-Simulation Framework for Smart Grid Wide-Area Monitoring Networks. *J. Netw. Comput. Appl.* **2016**, *59*, 274–284. [CrossRef]
23. Kroetzsch, M.; Simancik, F.; Horrocks, I. A Description Logic Primer. *arXiv* **2013**, arXiv:1201.4089.
24. Brambilla, M.; Cabot, J.; Wimmer, M. Model-Driven Software Engineering in Practice. *Synth. Lect. Softw. Eng.* **2012**, *1*, 1–182. [CrossRef]
25. Andrén, F.; Strasser, T.; Kastner, W. Model-Driven Engineering Applied to Smart Grid Automation Using IEC 61850 and IEC 61499. In Proceedings of the Power Systems Computation Conference, Wroclaw, Poland, 18–22 August 2014; pp. 1–7.
26. Zanabria, C.; Pröstl Andrén, F.; Kathan, J.; Strasser, T. Towards an Integrated Development of Control Applications for Multi-Functional Energy Storages. In Proceeding of the IEEE 21st International Conference on Emerging Technologies and Factory Automation (ETFA), Berlin, Germany, 6–9 September 2016; pp. 1–4.
27. Morales-Trujillo, M.E.; Oktaba, H.; Piattini, M. The making of an OMG standard. *Comput. Stand. Interfaces* **2015**, *42*, 84–94. [CrossRef]
28. Horrocks, I.; Kutz, O.; Sattler, U. The Even More Irresistible SROIQ. *Kr* **2006**, *6*, 57–67.
29. Brockmans, S.; Volz, R.; Eberhart, A.; Löffler, P. Visual modeling of OWL DL ontologies using UML. In Proceeding of the International Semantic Web Conference, Hiroshima, Japan, 7–11 November 2004; Springer: Berlin/Heidelberg, Germany, 2004; pp. 198–213.
30. Andren, F.; Brundlinger, R.; Strasser, T. IEC 61850/61499 Control of Distributed Energy Resources: Concept, Guidelines, and Implementation. *IEEE Trans. Energy Convers.* **2014**, *29*, 1008–1017. [CrossRef]
31. Franke, R.; Wiesmann, H. Flexible Modeling of Electrical Power Systems—The Modelica PowerSystems Library. In Proceedings of the 10th International Modelica Conference, Lund, Sweden, 10–12 March 2014; pp. 515–522. [CrossRef]
32. Buscher, M.; Kube, M.; Piech, K.; Lehnhoff, S.; Rohjans, S.; Fischer, L. Towards Smart Grid-Ready Substations: A Standard-Compliant Protection System. In Proceeding of the 2016 Power Systems Computation Conference (PSCC), Genoa, Italy, 20–24 June 2016; pp. 1–6. [CrossRef]
33. Fiaschetti, L.; Antunez, M.; Trapani, E.; Valenzuela, L.; Rubiales, A.; Risso, M.; Boroni, G. Monitoring and Controlling Energy Distribution: Implementation of a Distribution Management System Based on Common Information Model. *Int. J. Electr. Power Energy Syst.* **2018**, *94*, 67–76. [CrossRef]
34. Divya, K.; Østergaard, J. Battery Energy Storage Technology for Power Systems—An Overview. *Electr. Power Syst. Res.* **2009**, *79*, 511–520. [CrossRef]
35. Braam, F.; Diazgranados, L.M.; Hollinger, R.; Engel, B.; Bopp, G.; Erge, T. Distributed Solar Battery Systems Providing Primary Control Reserve. *IET Renew. Power Gen.* **2016**, *10*, 63–70. [CrossRef]
36. Slimani, T. Ontology Development: A Comparing Study on Tools, Languages and Formalisms. *Indian J. Sci. Technol.* **2015**, *8*. [CrossRef]
37. Hitzler, P.; Krotzsch, M.; Rudolph, S. *Foundations of Semantic Web Technologies*; CRC Press: Boca Raton, FL, USA, 2009.
38. Neureiter, C.; Uslar, M.; Engel, D.; Lastro, G. A Standards-Based Approach for Domain Specific Modelling of Smart Grid System Architectures. In Proceeding of the 2016 11th System of Systems Engineering Conference (SoSE), Kongsberg, Norway, 12–16 June 2016; pp. 1–6. [CrossRef]
39. Horrocks, I.; Patel-Schneider, P.F.; Boley, H.; Tabet, S.; Grosof, B. SWRL: A Semantic Web Rule Language Combining OWL and RuleML. *W3C Member Submission*, 21 May 2004; p. 79.
40. Paul, G.; Alexandre, P.; Axel, P. *SPARQL 1.1 Update*; W3C: Cambridge, MA, USA, 2013; Volume 21.
41. IEC. *Communication Networks and Systems for Power Utility Automation. Part 7-4: Basic Communication Structure—Compatible Logical Node Classes and Data Object Classes*; IEC: Geneva, Switzerland, 2010.
42. IEC. *Communication Networks and Systems for Power Utility Automation. Part 7-3: Basic Communication Structure-Common Data Classes*; IEC: Geneva, Switzerland, 2011.
43. Gearon, P.; Passant, A.; Polleres, A. *SPARQL 1.1 Query Language*; W3C Recommendation; W3C: Cambridge, MA, USA, 2013; Volume 21.

44. Massif: MATLAB Simulink Integration Framework for Eclipse. 2016. Available online: https://github.com/viatra/massif (accessed on 12 March 2017).
45. International Electrotechnical Commission. *IEC 61499-1/Ed.2: Function Blocks—Part 1: Architecture*; Standard; IEC: Geneva, Switzerland, 2012.
46. van Amstel, M.; Bosems, S.; Kurtev, I.; Ferreira Pires, L. Performance in Model Transformations: Experiments with ATL and QVT. In *Theory and Practice of Model Transformations*; Lecture Notes in Computer Science; Springer: Berlin/Heidelberg, Germany; Zurich, Switzerland, 2011; pp. 198–212.
47. International Electrotechnical Commission (IEC). *IEC/TR 61850-90-7—Communication Networks and Systems for Power Utility Automation—Part 90-7: Object Models for Power Converters in Distributed Energy Resources (DER) Systems*; IEC: Geneva, Switzerland, 2013.
48. Uslar, M.; Specht, M.; Dänekas, C.; Trefke, J.; Rohjans, S.; González, J.M.; Rosinger, C.; Bleiker, R. *Standardization in Smart Grids*; Power Systems; Springer: Berlin/Heidelberg, Germany, 2013.
49. Gayathri, K.; Easwarakumar, K.; Elias, S. Probabilistic Ontology Based Activity Recognition in Smart Homes Using Markov Logic Network. *Knowl. Based Syst.* **2017**, *121*, 173–184. [CrossRef]
50. Baader, F.; Borgwardt, S.; Lippmann, M. Temporal query entailment in the description logic SHQ. *Web Semant.* **2015**, *33*, 71–93. [CrossRef]
51. Brunelière, H.; Cabot, J.; Dupé, G.; Madiot, F. MoDisco: A Model Driven Reverse Engineering Framework. *Inf. Softw. Technol.* **2014**, *56*, 1012–1032. [CrossRef]

© 2018 by the authors. Licensee MDPI, Basel, Switzerland. This article is an open access article distributed under the terms and conditions of the Creative Commons Attribution (CC BY) license (http://creativecommons.org/licenses/by/4.0/).

Article

Feasibility Analysis of Behind-the-Meter Energy Storage System According to Public Policy on an Electricity Charge Discount Program

Byuk-Keun Jo [1,2], Seungmin Jung [3] and Gilsoo Jang [1,*]

1. School of Electrical Engineering, Korea University, Anam Campus, 145 Anam-ro, Seongbuk-gu, Seoul 02841, Korea; chisjolin@korea.ac.kr
2. Regional Cooperation Division, Korea Energy Agency, 388 Poeun Dae-ro, Suji-gu, Yongin-si, Gyunggi-do 16842, Korea
3. Department of Electrical Engineering, Hanbat National University, 125 Dongseo-daero, Yuseong-gu, Daejeon 34158, Korea; seungminj@hanbat.ac.kr
* Correspondence: gjang@korea.ac.kr; Tel.: +82-2-3290-3246

Received: 29 November 2018; Accepted: 21 December 2018; Published: 1 January 2019

Abstract: Energy storage systems are crucial in dealing with challenges from the high-level penetration of renewable energy, which has inherently intermittent characteristics. For this reason, various incentive schemes improving the economic profitability of energy storage systems are underway in many countries with an aim to expand the participation rate. The electricity charge discount program, which was introduced in 2015 in Korea, is one of the policies meant to support the economic feasibility of demand-side energy storage systems. This paper quantitatively evaluated the impact of the electricity charge discount program on the economic feasibility of behind-the-meter energy storage systems. In this work, we first summarized how electricity customers can benefit from behind-the-meter energy storage systems. In addition, we represented details of the structure that make up the electricity charge discount program, i.e., how the electricity charge is discounted through the discount scheme. An optimization problem that establishes a charge and discharge schedule of an energy storage system to minimize each consumer's electricity expenditure was defined and formulated as well. The case study results indicated that the electricity charge discount program has improved the profitability of behind-the-meter energy storage systems, and this improved profitability led to investment in behind-the-meter energy storage systems in Korea. As a result of the electricity charge discount program, Korea's domestic demand side energy storage system market size, which was only 27 billion dollars in 2015 in Korea, has grown to 825 billion dollars in 2018.

Keywords: energy storage system; electricity charge discount program; peak reduction; economic feasibility analysis; policy effectiveness evaluation

1. Introduction

In the 2016 Paris Climate Change Agreement, South Korea committed to a 37% reduction target by 2030 [1]. In order to achieve this, Korea announced the 3020 Renewable Energy Implementation Plan on December 2017 [2,3]. The renewable portfolio standard (RPS) was strengthened, while the proportion of renewable energy generation was increased, and the aging coal plant shut down plan was begun in Korea [2,4]. However, integrating renewable energy resources with the power system can negatively affect the power system, due to the inherent intermittency of renewable energy resources and rapid variation of renewable energy generation [4]. Particularly in an isolated power system with relatively low power system inertia, like Korea, the negative impact of intermittent renewable energy resources may have greater leverage [5]. As one of the various solutions to relieve the impact of renewable energy resources integration, global leaders—such as the U.S. and Germany—are promoting the expansion of energy storage system (ESS) applications [6]. Korea has also been supporting ESS businesses and technologies. Plans to disseminate 1.7 GW ESS by 2020 through the research and development (R&D) investment of $5.7 billion and the establishment of institutional incentives were announced in 2010 [7]. A renewable energy certificate (REC) is given to ESSs that are linked to renewable energy, such as wind and photovoltaic power [8]. In the transmission and distribution networks, ESS has provided the primary and secondary frequency regulation service and substitution for upgrade of decrepit distribution network facilities. The ESS supply business is proceeding annually with government subsidies, and electricity charge discount programs are provided for using ESS to reduce consumers' electricity bills.

ESS is a solution to various challenges associated with power systems. ESS applications improve the flexibility and reliability of the power system [9,10]. First of all, ESS does not require external excitation to start. This makes ESS suitable for providing a black start service. Case studies of black start services of ESSs in medium voltage distribution networks are discussed in Manganelli et al. [11]. Further, upgrades of transmission and distribution network facilities can be deferred through ESS installation [12–14].

ESS is more useful for providing frequency regulation services than conventional generators due to the fast ramping capability of ESS [15,16]. The determination scheme of the optimal cost-effective size of ESS to provide a primary frequency reserve has been studied [17,18]. In the New York Independent System Operator's market, an ESS is more profitable on load frequency control service than energy arbitrage on the customer side [19].

According to Moore and Shabani [6], the ESS capacity should be 10 to 20 percent of the total intermittent renewable energy generation in order to avoid power system disturbance caused by the integration of renewable energy generation. Mitigating the fluctuation of intermittent renewable energy generation through ESSs has been studied in Zhao et al., Beaudin et al., Barton and Infield and Suberu et al. [20–23]. The determination of the size and location of ESS installation for integrating wind power generation has also been studied in Nguyen and Lee and Dvorkin et al. [24,25]. Nguyen and Lee [24] examined a method of dispatching wind power generation while minimizing the size of ESS connected with wind power and Dvorkin et al. [25] conducted a study for the determination of the size and location of ESS installation so as to maximize profit through energy arbitrage through surplus power from renewable generation variation.

Numerous studies have been carried out on ESS applications as demand-side energy resources, and these have mainly focused on reducing each customer's electricity expenditure. The charge and discharge scheduling of behind-the-meter ESS for a smart home's electricity bill minimization is studied in Longe et al. [26]. The proposed energy scheduling and distributed storage (ESDS) algorithm of Longe et al. [26] aims to reduce electricity bills while increasing customer satisfaction. Studies on charge and discharge scheduling algorithms further considered the operation and wear-out costs of batteries [27,28]. The proposed scheduling schemes are established in consideration of the fact that repetitive charge and discharge of ESS degrade each battery's lifetime. In Babacan et al. [29], the scheduling algorithm of the behind-the-meter ESS aiming to minimize each customer's electricity expenditure, as well as mitigate the fluctuation of net demand load with solar photovoltaic generation, is proposed and demonstrated. In addition, an optimization scheme for sizing the behind-the-meter ESS in peak shaving application for a base charge reduction of electricity bills is proposed and demonstrated in Martins et al. [30].

ESS applications on the demand side bring economic benefits to electricity customer first. Further, large amounts of demand-side ESSs improve public value, such as a power system's flexibility and reliability [31,32]. The increase of ESS capacity on the demand side in the power system provides controllability of the demand loads that increases the demand-side management capability, resolves the fluctuation of distributed generation and uncertainty of loads, and reduces tangible and intangible costs for power system operation. For these reasons, governments in various countries around the world have implemented policies encouraging the dissemination of ESS [6,33]. These governments have initiated subsidies, released loan support funds, and introduced incentive programs in tax credit and electricity tariff systems.

Subsidies to supplement the low profitability of ESS with high initial investment costs are the most direct way to improve the economic feasibility of ESS. In Germany, one-third of the installation costs for ESS with solar panels are subsidized. In Japan, since the Great East Japan Earthquake, federal government and municipal governments have enhanced subsidy programs for the purpose of widely spreading household emergency power sources. In California, USA, the Self-Generation Incentive Program (SGIP) offers subsidies for advanced ESSs [33]. In Korea, more than 85 MW, 210 MWh ESSs have been installed through government subsidies of more than $53.6 billion.

These governments have also enforced mandatory installation schemes. In California, the major power suppliers in the state are obliged to install 1325 MW of ESSs by 2020, and they are currently issuing orders to install ESSs, which are equivalent to 1% of the peak time power requirement for small businesses [33]. This installation obligation will be gradually strengthened from 2.25% of the electricity supply in 2014 to 5% in 2020. In Korea, ESS has been mandatory in the public sector since 2017. All of the public buildings in Korea with a contract power of 1000 kW or more should install an ESS [34]. The capacity of installed ESS should be more than five percent of the building's contract power. ESS has to be preferentially considered as a back-up power source of public buildings when back-up power is installed at public buildings. In addition, the states of Oregon and Massachusetts have also enforced mandatory programs for ESS [33].

Financial incentives are introduced and initiated in Korea for improving the economic effectiveness of behind-the-meter ESSs. Up to a six percent investment tax credit is given to electricity customers who install an ESS with a high efficiency certification at their own sites. Through the ESS electricity charge discount program (ECDP), electricity customers can earn benefits through not only load shifting and peak reduction, but also using the electricity charge discount. Institutional incentive and technological advancement in the market environment have improved the economic feasibility of ESS applications and driven investment in ESS projects.

The aim of this paper is to analyze the impact of the Korea's public policies on enhancing the economic feasibility of behind-the-meter ESS to disseminate ESS to the demand side. A behind-the-meter ESS is an ESS placed at the bottom of the receiving end. It is electrically connected to the electricity customer, not to the utility. The incentive policy for improving the economic feasibility of behind-the-meter ESS leads to the expansion of demand-side customers as a whole.

This paper conducts a sensitivity analysis of the behind-the-meter ESS's economic feasibility according to the ESS ECDP. Since the release of ESS ECDP in 2015, the duration of ESS ECDP has been extended, and the discount rate has been increased. After 2020, the incentives of ESS ECDP will be gradually decreased and finally expire in 2026. This paper attempts to verify that the ESS ECDP affects the profitability of behind-the-meter ESS and that increased profitability through ECDP leads to increased scale of investment in ESS on the demand side.

ESS ECDP is described in Section 2 and the cost-benefit structure of behind-the-meter ESS is modeled and formulated in Section 3. Section 4 quantitatively analyzes the sensitivity of the economic feasibility according to ESS ECDP in stages. Finally, the impact of public policy on demand-side ESS is reviewed in Section 5. This paper provides governments with the impact of public policy on the ESS as a reference for decision making regarding ESS diffusion.

2. Energy Storage System (ESS) Electricity Charge Discount Program (ECDP)

2.1. Benefits of Behind-the-Meter ESS

The benefits of ESS on the demand side are generated through the sale of stored power, demand response, and electricity bill reduction [30]. Since December 2014 in Korea, consumers can sell less than 1000 kW of electric power from their behind-the-meter ESSs to the Korea Electric Power

Corporation (KEPCO). In this case, less than 50% of the total annual stored energy is allowed to be sold to the KEPCO, and the settlement of sold power is calculated using the system marginal price (SMP). However, on average, the SMP has been lower than the electricity usage rate at peak time since 2015 in Korea. In addition, following the initiation of ESS ECDP, the self-consumption of stored power in ESS at the peak load time can earn discounts on the base charge as well as electricity bill reduction through peak reduction and load shifting. Therefore, consumers are more motivated to utilize self-consumption than sell stored power.

In the case of the demand response, ESS is not used as a sole resource, as it is used with various resources such as light, electric motors, and other facilities. In addition, since the reduced peak load by ESS is already reflected in the baseline decision, which is the basis of the demand response settlement calculation, ESS is not useful as a demand response resource. For these reasons, electricity bill management is the dominant factor for giving customers benefits through ESS rather than the cases of the sale of stored power or demand response.

In Korea, a time of use (TOU) rate is applied to electricity customers, excluding households. The TOU rates consist of the base charge, which is charged according to the peak load, and the usage charge, which is charged according to the electricity usage. In the TOU rates system, one's electricity bill can be reduced through peak reduction and load shifting with the ESS charge–discharge scheme. The base charge is saved by discharging the ESS to lower the peak load. The usage charge is reduced by load shifting, which means that ESS charge with low-cost electricity at light load time and discharge at peak load time then replaces expensive electricity purchases. Consequently, the ESS operator would establish an optimal ESS charge–discharge schedule that maximizes the benefits of the ESS by determining whether to use the limited ESS resources for peak reduction for base charge savings or for load shift in order to reduce usage charge.

2.2. ESS Electricity Charge Discount Program

The ESS ECDP has been implemented in Korea since 2015 to encourage expansion of the installation of ESS on the demand side. It is designed to establish a structure to give back to ESS investors according to the contribution of behind-the-meter ESS to the power system such as reduced peak demand, increased power system flexibility, and improved load management capability. The ESS ECDP consists of base charge discount and usage charge discount for ESS charge. If an ESS reduces the peak load through discharging at peak load time, then the base charge of the electricity bill is discounted by reflecting the peak load reduction. The electricity rate for electricity usage to charge ESS at light load time is discounted.

The effectiveness of the ESS ECDP depends on the discount period and the discount rate. All of the discounts are available during the discount period, and therefore, upon expiration of the discount period, a behind-the-meter ESS can only save base charge through peak load reduction and reduce usage charge through energy arbitrage, i.e., load shifting. Since it was initiated, the range and effectiveness for the ESS ECDP have been strengthened in stages. The ESS ECDP began with a 10% discount on the electricity usage for charging ESS in 2015, and it was extended to base charge discount so as to return the peak reduction contribution of ESS to the ESS owner with base charge reduction benefit in 2016. On 1 January 2017, the Korean government implemented a policy to temporarily increase the discount rate of ECDP for the purpose of enhancing the effectiveness of the incentive system, and in May, announced a one-year extension of the applicable period of the increased discount rate. With the initiation of the increased discount rate of ECDP, a weight to the discount rate depending on the battery storage capacity has been introduced. This weight is determined by the ratio of ESS battery storage capacity to contract power of the electricity customer. Table 1 summarizes the range of ECDP, discount rate, and period in four stages, and these are depicted by year in Figure 1. Table 2 describes the weights according to the ratio of contract power to battery storage capacity.

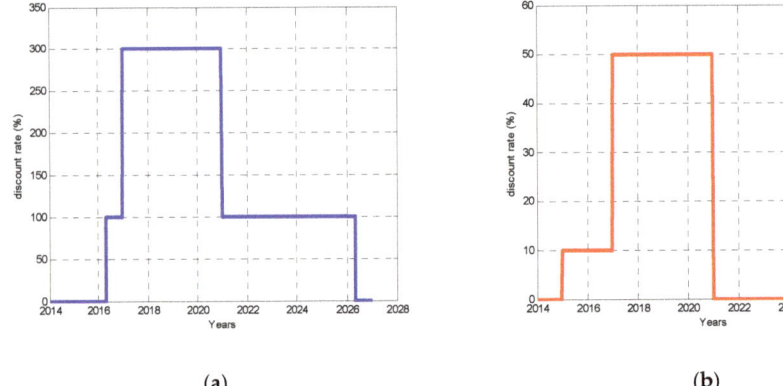

Figure 1. (a) Discount rate of base charge discount. (b) Discount rate of usage charge discount.

Table 1. Summary of ESS ECDP in Korea.

Stages	ESS ECDP	Discount Rate		Discount Period
		Base	Charging	
Stage 1	Charging rate discount was initiated	-	10%	1.1.2015–12.31.2017
Stage 2	Base charge discount was initiated	100%	-	4.1.2016–3.31.2026
Stage 3	Temporarily enhanced discount rate was additionally applied with a weight factor	300%	50%	1.1.2017–12.31.2019
Stage 4	Applicable period of additional discount was extended	300%	50%	1.1.2017–12.31.2020

Table 2. Weight of ESS ECDP.

Battery Capacity/Contract Power (%)	Weight
More than 10%	120%
Between 5 and 10%	100%
Less than 5%	80%

Before the initiation of ESS ECDP, the benefits of behind-the-meter ESS relied on load shifting and peak reduction. These were not enough to attract investors to invest in demand-side ESSs. However, the ESS ECDP significantly increases the profitability of behind-the-meter ESSs.

Figure 2 illustrates schematically how ESS benefits consumers, i.e., the profit-making structure via peak reduction and load shifting through behind-the-meter ESS. The green line shown in Figure 2a is a typical load pattern of the commercial building with a contract power of 3000 kW. A 250 kW, 1000 kWh ESS is applied to the building in order to reduce electricity bills, and the schedule of charge and discharge for this ESS is drawn in green and red bars, respectively. The orange line is a net load pattern that reflects the ESS charge and discharge on the original load pattern.

First of all, as shown in Figure 2a, the ESS reduces the original peak of 1898 kW to 1715 kW through discharging at the peak time period. This peak reduction also reduces the base charge of the electricity bill.

The electric power charged in the ESS in a light load time period, which is low in its electricity usage rate, is used by discharging at a peak load time period, in which an electricity usage rate is expensive. ESS provides benefits as much as the difference in electricity usage charges for these two time periods. A discount for electricity usage to charge ESS improves the profitability of arbitrage through load shifting.

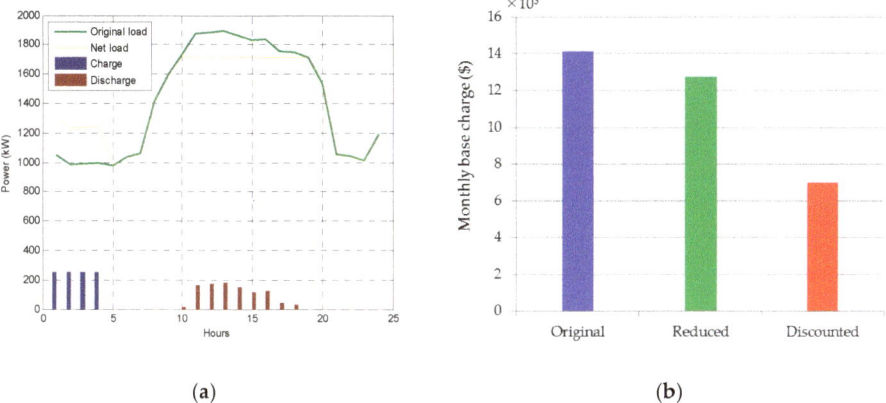

Figure 2. (**a**) Peak reduction and load shifting through behind-the-meter ESS. (**b**) Base charge reduction.

Figure 2b provides information on the benefit generation structure through peak reduction. The base charge of the electricity bill for the original peak, the blue bar, is reduced to the green bar via ESS discharging at peak time. In addition, ECDP provides additional discounts on the base charge by reflecting the contribution of the peak load reduction. The red bar in Figure 2b is the base charge, which is the final payment of the customer. Compared to the original base charge, the final base charge was reduced by about 50.3%. This is a reduction of about 45.0% additional savings compared to 9.6%, which is the base charge reduction through the peak load reduction of the ESS without ECDP. Details in the improvement of profitability of behind-the-meter ESSs due to ECDP and the resulting expansion of ESSs on the demand side are discussed in Section 4.

3. Problem Definition and Formulation

This chapter defines and formulates the cost-benefit structure of a behind-the-meter ESS as well as the optimization problem for electricity bill minimization. In this chapter, in terms of minimizing the electricity bill, the cost is defined as the electricity charge for charging the ESS. In addition, the benefits are the base charge and usage charge reduction, and discounts for electricity charge through the ESS ECDP. The optimization problem, which establishes a schedule of behind-the-meter ESS for minimizing electricity bills, is organized so as to analyze the effectiveness of ESS ECDP on the profitability of behind-the-meter ESSs. Furthermore, the correlation between ESS profitability and diffusion effects is derived to assess the extent to which economic improvements in behind-the-meter ESS affected the spread of demand-side ESSs.

3.1. Base Charge Reduction

The base charge for electricity is charged monthly, depending on the peak load, in kW, regardless of electricity usage, in kWh. It is calculated as the product of the annual peak load and the unit base rate. If the annual peak load is under 30% of the contract power, the base charge is calculated based on 30% of the contract power. As a short peak power occurring in summer or winter becomes the basis for the base rate calculation of the following year, thorough peak load management is required.

With the ESS, the base charge can be lowered by simultaneously discharging at the time when the peak load occurs. When the load exceeds a predetermined level, the ESS starts to discharge and the load no longer increases. However, due to the storage capacity limit of ESSs, it is important to set an appropriate level of the peak load control value by considering the forecasted load.

Meanwhile, peak reduction through a ESS not only leads to savings in the base charge, but also has the effect of improving the customer's load factor, which is defined in Equation (1). A higher load factor, i.e., thorough peak management, improves the customer's economic feasibility [35].

$$Load\ factor,\ LF = \sum_{t=(T_l+T_m+T_p)} P_t^l \times \frac{1}{P_{Tmax}^l}(T_l + T_m + T_p) \tag{1}$$

The base charge savings through peak reduction are represented as follows:

$$\pi^b = \left(P_{Tmax}^l + P_{Tmax}^c - P_{Tmax}^d\right)\mu^b \tag{2}$$

The effectiveness of the base charge reduction through ESS differs depending on each electricity customer's load factor. Due to the limited storage capacity of the ESS, the base charge reduction through peak load reduction can be better for electricity customers with a low load factor than those with a high load factor.

3.2. Usage Charge Reduction

The usage charge reduction stems from load shifting, a type of arbitrage transaction. This is to reduce the electricity bill based on the difference between the two rates by charging the ESS at the light load time with a low electricity rate and discharging the ESS at the peak load time with a high electricity rate. The usage charge reduction through load shifting is formulated in Equation (3):

$$\pi^u = \mu^l \sum_{d=D}\left(\sum_{t=T_l}\left(P_{d,t}^c - P_{d,t}^d\right) + \mu^m \sum_{t=T_m}\left(P_{d,t}^c - P_{d,t}^d\right) + \mu^p \sum_{t=T_p}\left(P_{d,t}^c - P_{d,t}^d\right)\right) \tag{3}$$

3.3. Elecricity Charge Discount

ESS ECDP is applied in two ways, as described above, with a discount on the base charge and usage rate for charging the ESS. The base charge discount is calculated by multiplying the electricity base rate and peak load reduction as in Equation (4). The peak load reduction is predetermined by the electricity company, KEPCO, in Korea as in Equation (5). Discounts on usage charges are granted only for electricity usage to charge ESS at light load times. The discount for usage charge is represented as Equation (6).

$$\pi_d^b = \Phi\mu^b\delta^b\omega \tag{4}$$

$$\Phi = \frac{\sum_{d=D}\sum_{t=T_p}\left(P_{d,t}^d - P_{d,t}^c\right)}{3} \tag{5}$$

$$\pi_d^u = \mu^l\delta^c\omega \sum_{d=D}\sum_{t=T_l}\left(P_{d,t}^c\right) \tag{6}$$

3.4. Electricity Bill Minimization

The purpose of behind-the-meter ESSs is to minimize electricity bills, which can be achieved by maximizing benefits through ESSs. Since ESS has a capacity limit, the ESS operator should resolve the optimization problem of allocating limited stored energy considering all the benefit-making options: peak reduction, arbitrage, and ECDP. The problem of electricity bill minimization through behind-the-meter ESS is defined as Equation (7). Then, the decision variables of the cost minimization problem are the kW powers of ESS charging and discharging.

$$\min.C = \left(P^l_{Tmax} + P^c_{Tmax} - P^d_{Tmax}\right)\mu^b$$
$$+ \sum_{d=D}\left(\mu^l \sum_{t=T_l}\left(P^c_{d,t}(1-\delta^c\omega) - P^d_{d,t} + P^l_{d,t}\right) + \mu^m \sum_{t=T_m}\left(P^c_{d,t} - P^d_{d,t} + P^l_{d,t}\right)\right. \quad (7)$$
$$\left. + \mu^p \sum_{t=T_p}\left(P^c_{d,t} - P^d_{d,t} + P^l_{d,t}\right)\right) - \tfrac{1}{3}\mu^b\delta^b\omega \sum_{d=D_w}\sum_{t=T_p}\left(P^d_{d,t} - P^c_{d,t}\right)$$

subject to

$$0 \leq P^c \leq P^c_{max}$$
$$0 \leq P^d \leq P^d_{max}$$
$$P^d_{max} = P^c_{max}$$
$$(1-\varepsilon)E \leq \sum_{t=1}^{T}\left(P^c_t - P^d_t\right) \leq \varepsilon E, \quad \forall t$$
$$\sum_{t=T_{n-1}}^{T_n} P^d_t \leq \sum_{t=1}^{T_{n-1}}\left(P^c_t - P^d_t\right), \quad \forall t$$

3.5. Economic Feasibility Evaluation

The effectiveness of the ESS ECDP on the profitability of behind-the-meter ESS is analyzed through economic feasibility evaluation in this paper. In order to evaluate the economic feasibility of behind-the-meter ESS, the cost and profit of behind-the-meter ESS are defined and represented as follows:

cost:

$$C = \sum_{d=D}\sum_{t=T}\left(\mu_t P^c_{d,t}\right) - \pi^u_d = \sum_{d=D}\left(\mu^l(1-\delta^c\omega)\sum_{t=T_l}P^c_{d,t} + \mu^m\sum_{t=T_m}P^c_{d,t} + \mu^p\sum_{t=T_p}P^c_{d,t}\right) \quad (8)$$

and profit:

$$\pi = \left(P^l_{Tmax} + P^c_{Tmax} - P^d_{Tmax}\right)\mu^b + \sum_{d=D}\left(\mu^l\sum_{t=T_l}P^d_{d,t} + \mu^m\sum_{t=T_m}P^d_{d,t} + \mu^p\sum_{t=T_p}P^d_{d,t}\right) \quad (9)$$
$$+ \tfrac{1}{3}\mu^b\delta^b\omega\sum_{d=D_w}\sum_{t=T_p}\left(P^d_{d,t} - P^c_{d,t}\right)$$

3.6. Correlation Coefficient

The correlation between the economic feasibility of behind-the-meter ESS and the installation capacity of ESS on the demand side is analyzed in order to verify the effectiveness of the ESS ECDP, which is a public policy aiming to expand ESS diffusion. The closer the relationship of economic feasibility of ESS and installed capacity of demand-side ESS, the closer the correlation coefficient is to 1. A positive correlation coefficient means positive correlation and a negative correlation coefficient means inverse correlation. The correlation coefficient between the economic feasibility of behind-the-meter ESS and the dissemination of demand side ESS is carried out in this work.

4. Case Studies

In this section, the impact of the ESS ECDP on the improvement of the economic feasibility of the behind-the-meter ESS is quantitatively analyzed. Furthermore, the way in which this economic improvement led to actual ESS dissemination on the demand side is evaluated. All the simulations and calculations in this paper are conducted with MATLAB 7.10.0 (R2010) of MathWorks in Natick, Massachusetts, U.S.A. Also, all the figures provided in this paper are also drawn through same program.

4.1. Load Analysis and ESS Scheduling

A complex center with 2500 kW of contract power was selected as a sample for the evaluation of economic feasibility. Figure 3 shows the 24-h load patterns and its distribution for each hour of the actual load of the building for a year. In winter, the load pattern is different from the other seasons because of the different peak time period applied to electricity charge and heating demand in the evening and at night. Therefore, the loads are analyzed separately for spring, summer, autumn, and winter. The peak load usually occurs as the flow of outside visitors increases around lunch and dinner time. Due to the characteristics of this building, which has many night shifts, the load at all times is kept above a certain level, except for in the early morning hours. This building is subject to the electricity tariff, A-level high voltage with option II, as summarized in Tables 3 and 4.

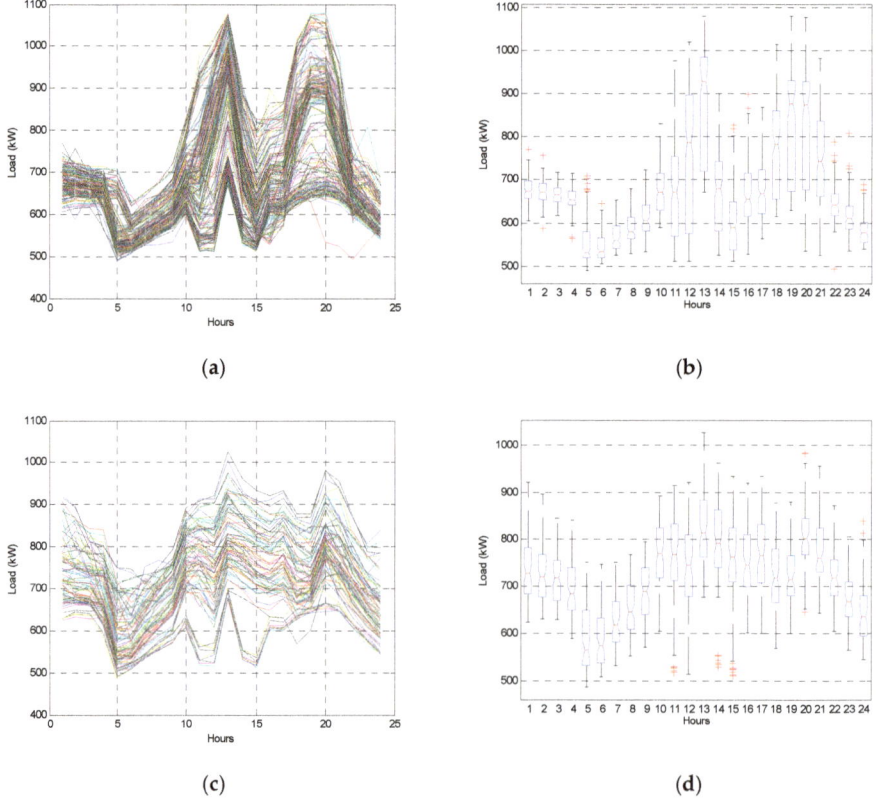

Figure 3. (**a**) Daily loads in spring, summer, and fall. (**b**) Distributions of loads in spring, summer, and fall. (**c**) Daily loads in winter. (**d**) Distributions of loads in winter.

Table 3. Electricity Tariff for a Building with A-level High Voltage, Option II.

Base Charge ($/kW)	Usage Rate (cent/kWh)			
	Time	Summer	Spring, Fall	Winter
7.43	Light Load Time	5.01	5.01	5.63
	Medium Load Time	9.73	7.02	9.75
	Peak Load Time	17.06	9.76	14.88

Table 4. Time Table for TOU Price in Korea.

Time	Summer, Spring, Fall	Winter
Light Load Time	23:00–09:00	23:00–09:00
Medium Load Time	09:00–10:00	09:00–10:00
	12:00–13:00	12:00–17:00
	17:00–23:00	20:00–22:00
Peak Load Time	10:00–12:00	10:00–12:00
	13:00–17:00	17:00–20:00
		22:00–23:00

4.2. ESS Scheduling

An overview of the behind-the-meter ESS in this work is given in Table 5. Excluding pumped-storage hydro power plants and uninterruptible power supplies, more than 99% of ESSs in Korea are lithium-ion battery ESS. In addition, in order to meet the High-efficiency Appliance Certification criteria for ESS in Korea, the ESS should have a round-trip efficiency of over 89%. Given the service characteristics of ESS applications, behind-the-meter ESS for peak reduction and load shifting is suitable to have a storage time of about two to five hours [36]. A 250 kW, 1000 kWh ESS capable of sustaining the rated power for four hours is considered for this study.

Table 5. Specification of ESS and Conditions of Analysis.

Battery type	Lithium-ion
Rated power	250 kW
Storage capacity	1000 kWh
Round-trip efficiency	89%
Life cycles	3000 cycles (≈12 years)
Depth of discharge (DOD)	95%
Maintenance cost	0.2% of capital cost, annually

The optimization problem for the ESS charge–discharge scheduling was solved, and the results are shown in Figure 4. Since the base charge saving and base charge discount through peak reduction are more profitable than arbitrage through load shifting, the ESS is mainly discharged at the peak time period. In addition, since the maximum load occurring at the light load time is not considered in the calculation of the base charge, the ESS charged from midnight to its rated power without considering any additional calculations for charging. If the ESS charge is made at a light load time, there is no variation in the economics of how and when it is charged. Thus, it is more effective for the operator to set the charging schedule in consideration of the operating aspect and the life cycle of the ESS.

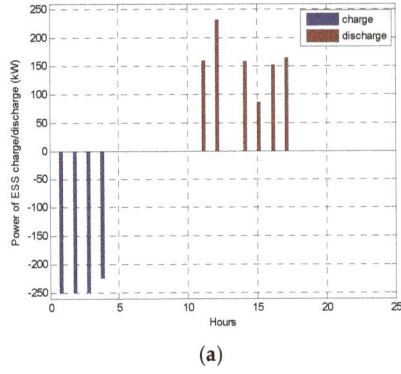
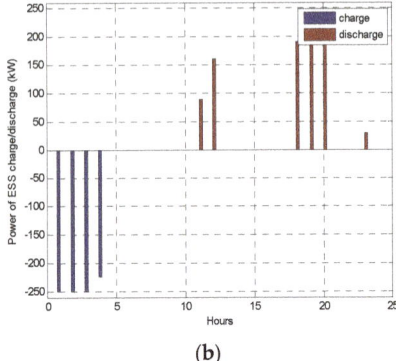

Figure 4. (a) ESS charge-discharge schedule in spring, summer, and fall; (b) and in winter.

4.3. Economic Feasibility of ESS without Subsidies or Fixed Cost

Since the initiation of the ESS ECDP in Korea, discounts are strengthened in stages. In this work, the economic feasibility of ESS ECDP of each stage is analyzed.

The cases for analysis are summarized in Table 6. Case 1 is the case without discount before the ESS ECDP initiation. Cases 2 to 5 are cases of ESS ECDP in stages, as shown Table 1 after ESS ECDP initiation. Cases 4 and 5 reflect a temporarily raised discount rate for base charge discount. In Cases 4 and 5, a discount rate of 300% is initially applied, and then a discount rate of 100% is applied after the end of the 300% discount period.

Table 6. Controlled Cases for ESS's Economic Feasibility Analysis.

Discounts	Case I	Case II	Case III	Case IV	Case V
Base charge discount	-	-	100%	300%, 100%	300%, 100%
Period of base charge discount (years)	-	-	10	3, 7	4, 6
Usage charge discount for the ESS charge	-	10%	10%	50%	50%
Period of usage charge discount (years)	-	3	3	3	4

In this study, the impact of the improved profitability of behind-the-meter ESS through the ESS ECDP on the dissemination of demand side ESSs in Korea is analyzed. Subsidies that have been provided to behind-the-meter ESSs should be considered so as to analyze the correlation of economic feasibility with the dissemination of ESS in real cases. In order to consider the cash discount rate in the case study, the discount rate is assumed to be 3%. In this work, the capital cost of ESS is fixed at $625,000.

In order to assess only the effect of the ESS ECDP, controlled cases without other considerations, such as cost degradation and governmental subsidy, were analyzed. Economic feasibility evaluation is conducted in terms of payback period, benefit-cost ratio, internal rate of return (IRR), and net present value (NPV). The results are carried out in Table 7. Cases 1, 2, and 3 did not achieve payback during the lifetime of the ESS. Thus, in these cases, the ESS project is not profitable. In Cases 4 and 5, ECDP's incentives ensure that the ESS project is economically viable.

Table 7. Results of economic feasibility analysis in controlled cases.

Results	Case I	Case II	Case III	Case IV	Case V
Payback years	-	-	-	6.85	5.73
B/C ratio	0.483	0.489	0.861	1.03	1.08
IRR (%)	−14.2	−14.1	−3.6	1.21	3.28
NPV (thousand US dollars)	−389	−385	−105	24.6	63.5

4.4. Economic Feasibility of ESS in Real Cases

Regarding the real cases, subsidies and price decrease are considered. The value of subsidies for behind-the-meter ESS is given as 80% of the total cost of ESS installation since 2012, and it is decreased by year. The real cases in this work including subsidies are summarized in Table 8. Further, details in the price decrease of ESS in the domestic market in Korea are shown as Figure 5. In this work, the actual effect of the market environment and public policy including the ESS ECDP on economic feasibility are evaluated. The results are shown in Tables 9 and 10. Table 9 represents the results of analysis of ESS economic feasibility in private markets excluding government subsidies for ESS private investment and ESS economics comparison. Table 10, on the other hand, shows the economic feasibility evaluation results of ESS considering government subsidies.

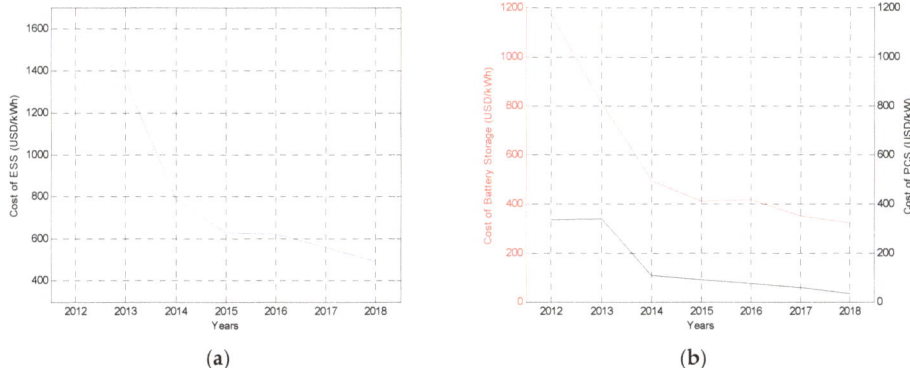

Figure 5. (a) Total cost of behind-the-meter ESS installation. (b) Cost of power conditioning system (PCS) and battery storage of ESS.

According to Table 9, which is the result of the ESS economic feasibility evaluation in the private market excluding government subsidies, ESS is worth the investment only in Cases 4 and 5 that were implemented the ESS ECDP with a 300% base charge discount and 50% usage charge discount. Until then, a behind-the-meter ESS is not profitable.

In Table 10, even though ECDP was introduced in Case 2, the economic feasibility of Case 2 is lower than that of Case 1. This is because the government subsidy rate for behind-the-meter ESS decreased from 70% to 50% in Case 2. It can be concluded that the economic feasibility of the ESS project is more dependent on subsidies than institutional incentives before the incentive programs are fully implemented. In Case 3, compared to Case 2, the subsidy rate was maintained, and the effect of improving the economic efficiency was significantly increased by introducing the base charge discount.

In Case 4 and 5 of real case studies, the profitability of a behind-the-meter ESS drastically increased regardless of whether government subsidies were given or not. Therefore, it can be inferred that private capital investment in behind-the-meter ESS would have been influenced. In other words, it is analyzed in Section 4.5 whether improved economic feasibility leads to the diffusion of demand-side ESSs.

Table 8. Real Cases for ESS's Economic Feasibility Analysis.

Parameters	Case I	Case II	Case III	Case IV	Case V
Base charge discount	-	-	100%	300%, 100%	300%, 100%
Period of base charge discount (years)	-	-	10	3, 7	4, 6
Usage charge discount for ESS charge	-	10%	10%	50%	50%
Period of usage charge discount (years)	-	3	3	3	4
Governmental subsidy for ESS (only for subsidy program)	70%	50%	50%	30%	30%
Cost of ESS with 0.25 MW, 1 MWh (thousand US dollars)	786.6	629.5	622.3	558.9	558.9

Table 9. Results of Real Cases in Private Market: Without Subsidies.

Results	Case I	Case II	Case III	Case IV	Case V
Payback years	-	-	-	5.72	4.63
B/C ratio	0.38	0.49	0.86	1.15	1.21
IRR (%)	−18.4	−14.2	−3.50	5.76	8.29
NPV (thousand US dollars)	−584	−390	−102	104	143

Table 10. Results of Real Cases in Government-Driven Market: With Subsidies.

Results	Case I	Case II	Case III	Case IV	Case V
Payback years	7.52	-	4.99	3.01	3.00
B/C ratio	1.28	0.971	1.73	1.65	1.73
IRR (%)	5.75	−0.65	16.4	25.7	30.3
NPV (thousand US dollars)	79.2	−11.0	273	306	345

4.5. Corealations between Economic Feasibility and Dissemination of a Behid-the-Meter ESS

Figure 6 shows the capacity of domestic ESS installation on demand side according to the investor by year. The light blue bar in Figure 6, which is marked as "Subsidy," means the capacity of ESS installed by governmental subsidy program. Except for "Subsidy," no others are given government subsidies. The green bar labeled "Public Sector" refers to the ESS installed by the public organizations in line with the government's demand-side ESS diffusion policy. "Manufacturer's Demonstration," which is orange bar, represents a leading investment by ESS makers to take advantage of future market share. The dark red one labeled "Private Investment" is the amount of ESS installed for consumer's economic profit, such as electricity bill management, and a government subsidy is not given here. Finally, the correlation coefficient of IRR and ESS investment is shown in Table 11.

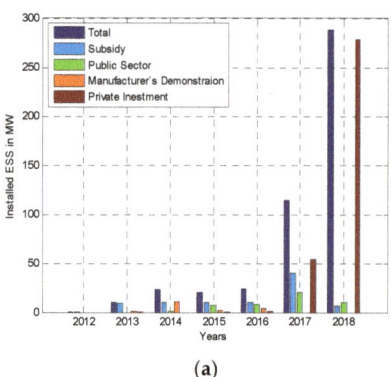
(a)

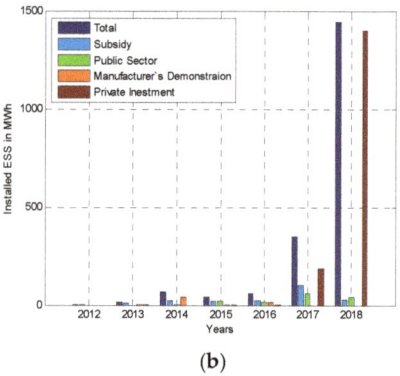
(b)

Figure 6. Korea's annual domestic ESS market size by source of investment in (a) MW, and (b) MWh.

Table 11. Correlations between IRR and ESS Investment.

Correlation Coefficient (with IRR)	Total		Private Only	
	kW	kWh	kW	kWh
	0.8093	0.7513	0.7431	0.7113

5. Discussion

Based on the actual case of installing 250 kW and 1000 kWh ESS at the contract power of a 2500 kW complex center, we examined whether the ESS ECDP, the incentive policy for the expansion of the demand side ESS, acted as intended by the Korean government. As shown in Table 7, an ESS's economic feasibility is improved as the discount rate is raised and the discount period is extended. According to the results of case studies shown in Table 9, without government subsidies, incentives of a 300% base rate discount and 50% charge rate discount should be given for positive profitability. On the other hand, Case 1 of Table 10, which receives 70% of government subsidies, is profitable because it can save a considerable amount of investment cost without the incentive. However, this is not constrained by the limited government budget, and therefore cannot lead to continued ESS market expansion.

Considering that it usually takes less than a year of 1 MWh-sized behind-the-meter ESS installation processes, the improved profitability of behind-the-meter ESS due to the ESS ECDP enhancement is immediately connected to the demand side ESS dissemination of the following year. As shown in Figure 6, the amount of ESS installation on the demand side has increased noticeably from 2017, where in Case 4, the incentive of the ESS ECDP was strengthened and the behind-the-meter ESS had positive profitability. The extension of the ESS ECDP in May 2017 further improved the economic efficiency of behind-the-meter ESSs, as in Case 5, resulting in the explosion of demand-side ESSs in 2018.

We can see in Figure 6 that the private sector investment contributed more to the growth of the Korean domestic demand side ESS market than any other sectors. We can also infer from this study that it is effective to build a market structure that can secure return on investment in order to form the demand side ESS market as a private led market.

As a result, the ESS ECDP for improving the economic feasibility of behind-the-meter ESSs in order to diffuse the ESS on the demand side in Korea has directly attracted investment in demand-side ESSs. The ESS ECDP initiated in 2015 to promote private sector investment in the behind-the-meter ESS has resulted in the growth of Korea's demand-side ESS market, which was only $27 billion in 2015, to $825 billion in 2018. In addition, after the advanced ESS ECDP initiation in 2017, which provides an IRR of at least 5% as in Case 4 of Table 9, the private sector ESS investment, which had not existed before 2015, had risen to $105 billion in 2017 and $783 billion in 2018. These were about 54% and 95% of the total market, respectively. These results suggest that it is possible to promote the private investment on the domestic demand side ESS with public policy on behind-the-meter ESS.

6. Conclusions

This paper shows that public policy to support the profitability of behind-the-meter ESS in Korea drives increased installation in demand-side ESSs. In order to clarify this, we first described the ESS ECDP, the representative public policy to enhance the profitability of behind-the-meter ESSs in Korea. In addition, we defined the cost–benefit structure of demand side ESS and formulated the cost minimization problem for electricity bill reduction. Finally, we clarified how ESS ESCP affects the profitability of a behind-the-meter ESS and how the improvement of profitability of a behind-the-meter ESS influences increased investment in behind-the-meter ESSs. Since the Korean government has continued subsidies and incentives to support the profitability of behind-the-meter ESSs, investment in demand-side ESSs has increased in Korea. In particular, the investment has led by the private sector and amounted to $783 billion, about 95% of the total domestic demand side ESS market for the seven months from January to July 2018.

The results of this paper can be used as reference for policymaking in countries considering ESS diffusion through institutional effort to expand renewable energy supply. When the countries intending to promote the spread of ESS diffusion through public policy determined the level of incentive degree in view of the targeted amount of ESS installation, the results represented in this paper can be referenced.

Previous studies aiming to maximize the economic efficiency of ESS resources on the demand side have mainly focused on the optimization of the charging–discharging schedule of ESSs in combination with other distributed resources such as photovoltaic generation and electric vehicles. However, this paper quantitatively evaluates the impact of policy decision-making on the economic performance of a behind-the-meter ESS, thereby broadening the approach to the study of the demand-side ESSs. Furthermore, studies on the effects of public policy promoting ESS applications on the supply side, such as renewable portfolio standards (RPS), can be considered in further study.

Author Contributions: Conceptualization, B.-K.J. and G.J.; Methodology, B.-K.J.; Software, B.-K.J.; Validation, B.-K.J., S.J. and G.J.; Formal Analysis, B.-K.J.; Investigation, B.-K.J.; Resources, B.-K.J.; Data Curation, B.-K.J.; Writing—Original Draft Preparation, B.-K.J.; Writing—Review and Editing, S.J. and G.J.; Visualization, B.-K.J.; Supervision, G.J.; Project Administration, S.J. and G.J.; Funding Acquisition, S.J.

Funding: This work was supported by Korea Electric Power Corporation Grant (R18XA06-40), and National Research Foundation Grant (No. 2018R1C1B5030524) funded by the Korean government.

Conflicts of Interest: The authors declare no conflict of interest.

Nomenclature

$P_{d,t}^l$	Load power on day d at time t in kW
$P_{d,t}^d$	Discharging power of ESS on day d at time t in kW
$P_{d,t}^c$	Charging power of ESS on day d at time t in kW
E	Storage capacity of the ESS in KWh
π^b	Base charge savings through peak load reduction in US dollars
π^u	Usage charge savings through load shifting in US dollars
π_d^b	Base charge discount amount in US dollars
μ^b	Unit rate for base charge of electricity bill in US dollars per kW
μ^p	Unit rate for electricity usage at peak load time period in US dollars per kWh
μ^m	Unit rate for electricity usage at medium load time period in US dollars per kWh
μ^l	Unit rate for electricity usage at light load time period in US dollars per kWh
D	Days of the month
D_w	Weekdays of the month
Y	Life years of the ESS
T_p	Peak load time period
T_m	Medium load time period
T_l	Light load time period
T^{max}	The time of peak load occurrence
ω	Weight of discount rate
δ^b	Base charge discount rate
δ^u	Usage charge discount rate for ESS charging
ε	Depth of discharge of battery storage of the ESS

References

1. Intended Nationally Determined Contribution (INDC). Available online: http://www4.unfccc.int/Submissions/INDC/Published%20Documents/Republic%20of%20Korea/1/INDC%20Submission%20by%20the%20Republic%20of%20Korea%20on%20June%2030.pdf (accessed on 30 June 2015).
2. Cornot-Grandolphe, S. *South Korea's New Electricity Plan: Cosmetic Changes or a Breakthrough for the Climate?* Institut Francais des Relations Internationales (ifri): Paris, France, 2018.
3. Renewables Now. Available online: https://renewablesnow.com/news/south-korea-eyes-20-renewables-share-by-2030-594850/ (accessed on 18 December 2017).
4. Oilprice.com. Available online: https://oilprice.com/Latest-Energy-News/World-News/South-Korea-Starts-Phasing-Out-Old-Coal-Plants.html (accessed on 28 February 2018).
5. Ulbig, A.; Borsche, T.S.; Andersson, G. Impact of Low Rotational Inertia on Power System Stability and Operation. *IFAC Proc.* **2014**, *47*, 7290–7297. [CrossRef]

6. Moore, J.; Shabani, B. A Critical Study of Stationary Energy Storage Policies in Australia in an International Context: The Role of Hydrogen and Battery Technologies. *Energies* **2016**, *9*, 674. [CrossRef]
7. Lee, S.I. *Plans for Energy Storage System Market Creation*; Korea Energy Economics Insitute: Ulsan, Korea, 2015.
8. Renewable Portfolio Standards (RPS). Available online: http://www.energy.or.kr/renew_eng/new/standards.aspx (accessed on 1 January 2019).
9. Denholm, P.; Jorgenson, J.; Hummon, M.; Jenkin, T.; Palchak, D. *The Value of Energy Storage for Grid Applications*; National Renewable Energy Laboratory: Denver West Parkway Golden, CO, USA, 2013.
10. Berrada, A.; Loudiyi, K.; Zorkani, I. Valuation of energy storage in energy and regulation markets. *Energy* **2016**, *115*, 1109–1118. [CrossRef]
11. Manganelli, M.; Nicodemo, M.; D'Orazio, L.; Pimpinella, L.; Carmen Falvo, M. Restoration of an Active MV Distribution Grid with a Battery ESS: A Real Case Study. *Sustainability* **2018**, *10*, 2058. [CrossRef]
12. Schoenung, S.M.; Eyer, J. *Benefit/Cost Framework for Evaluating Modular Energy Storage*; Sandia National Laboratories: Albuquerque, NM, USA, 2008.
13. Eyer, J. *Electric Utility Transmission and Distribution Upgrade Deferral Benefits from Modular Electricity Storage*; Sandia National Laboratories: Albuquerque, NM, USA, 2009.
14. Oprea, S.; Bara, A.; Uta, A.I.; Pirjan, A.; Carutasu, G. Analyses of Distributed Generation and Storage Effect on the Electricity Consumption Curve in the Smart Grid Context. *Sustainability* **2018**, *10*, 2264. [CrossRef]
15. Kottick, D.; Blau, M.; Edelstein, D. Battery energy storage for frequency regulation in an island power system. *IEEE Trans. Energy Conv.* **1993**, *8*, 455–459. [CrossRef]
16. Zhang, F.; Hu, Z.; Xie, X.; Zhang, J.; Song, Y. Assessment of the Effectiveness of Energy Storage Resources in the Frequency Regulation of a Single-Area Power System. *IEEE Trans. Power Syst.* **2017**, *32*, 3373–3380. [CrossRef]
17. Mercier, P.; Cherkaoui, R.; Oudalov, A. Optimizing a Battery Energy Storage System for Frequency Control Application in an Isolated Power System. *IEEE Trans. Power Syst.* **2009**, *24*, 1469–1477. [CrossRef]
18. Oudalov, A.; Chartouni, D.; Ohler, C. Optimizing a Battery Energy Storage System for Primary Frequency Control. *IEEE Trans. Power Syst.* **2007**, *22*, 1259–1266. [CrossRef]
19. Walawalkar, R.; Apt, J.; Mancini, R. Economics of electric energy storage for energy arbitrage and regulation in New York. *Energy Policy* **2007**, *35*, 2558–2568. [CrossRef]
20. Zhao, H.; Wu, Q.; Hu, S.; Xu, H.; Rasmussen, C.N. Review of energy storage system for wind power integration support. *Appl. Energy* **2015**, *137*, 545–553. [CrossRef]
21. Beaudin, M.; Zareipour, H.; Schellenberglabe, A.; Rosehart, W. Energy storage for mitigating the variability of renewable electricity sources: An updated review. *Energy Sustain. Dev.* **2010**, *14*, 302–314. [CrossRef]
22. Barton, J.P.; Infield, D.G. Energy storage and its use with intermittent renewable energy. *IEEE Trans. Energy Conv.* **2004**, *19*, 441–448. [CrossRef]
23. Suberu, M.Y.; Mustafa, M.W.; Bashir, N. Energy storage systems for renewable energy power sector integration and mitigation of intermittency. *Renew. Sustain. Energy Rev.* **2014**, *35*, 499–514. [CrossRef]
24. Nguyen, C.; Lee, H. Optimization of Wind Power Dispatch to Minimize Energy Storage System Capacity. *J. Electr. Eng. Technol. JEET* **2014**, *9*, 1080–1088. [CrossRef]
25. Dvorkin, Y.; Fernandez-Blanco, R.; Kirschen, D.S.; Pandzic, H.; Watson, J.; Silva-Monroy, C.A. Ensuring Profitability of Energy Storage. *IEEE Trans. Power Syst.* **2017**, *32*, 611–623. [CrossRef]
26. Longe, O.M.; Ouahada, K.; Rimer, S.; Harutyunyan, A.N.; Ferreira, H.C. Distributed Demand Side Management with Battery Storage for Smart Home Energy Scheduling. *Sustainability* **2017**, *9*, 120. [CrossRef]
27. Yan, L.; Baek, M.; Park, J.; Park, Y.; Roh, J.H. An Optimal Energy Storage Operation Scheduling Algorithm for a Smart Home Considering Life Cost of Energy Storage System. *J. Electr. Eng. Technol. JEET* **2017**, *12*, 1369–1375. [CrossRef]
28. Choi, Y.; Kim, H. Optimal Scheduling of Energy Storage System for Self-Sustainable Base Station Operation Considering Battery Wear-Out Cost. *Energies* **2016**, *9*, 462. [CrossRef]
29. Babacan, O.; Ratnam, E.L.; Disfani, V.R.; Kleissl, J. Distributed energy storage system scheduling considering tariff structure, energy arbitrage and solar PV penetration. *Appl. Energy* **2017**, *205*, 1384–1393. [CrossRef]
30. Martins, R.; Hesse, H.C.; Jungbauer, J.; Vorbuchner, T.; Musilek, P. Optimal Component Sizing for Peak Shaving in Battery Energy Storage System for Industrial Applications. *Energies* **2018**, *11*, 2048. [CrossRef]
31. Staffell, I.; Rustomji, M. Maximising the value of electricity storage. *J. Energy Storage* **2016**, *8*, 212–225. [CrossRef]

32. Sioshansi, R.; Denholm, P.; Jenkin, T.; Weiss, J. Estimating the value of electricity storage in PJM: Arbitrage and some welfare effects. *Energy Econ.* **2009**, *31*, 269–277. [CrossRef]
33. Winfield, M.; Shokrzadeh, S.; Jones, A. Energy policy regime change and advanced energy storage: A comparative analysis. *Energy Policy* **2018**, *115*, 572–583. [CrossRef]
34. Current Status and Prospects of Korea's Energy Storage System Industry 2017 KOTRA. Available online: http://www.investkorea.org/kotraexpress/2017/07/Industry.html (accessed on 31 December 2018).
35. Uddin, M.; Romlie, M.F.; Abdullah, M.F.; Halim, S.A.; Bakar, A.H.A.; Kwang, T.C. A review on peak load shaving strategies. *Renew. Sustain. Energy Rev.* **2018**, *82*, 3323–3332. [CrossRef]
36. Pearre, N.S.; Swan, L.G. Technoeconomic feasibility of grid storage: Mapping electrical services and energy storage technologies. *Appl. Energy* **2015**, *137*, 501–510. [CrossRef]

© 2019 by the authors. Licensee MDPI, Basel, Switzerland. This article is an open access article distributed under the terms and conditions of the Creative Commons Attribution (CC BY) license (http://creativecommons.org/licenses/by/4.0/).

MDPI
St. Alban-Anlage 66
4052 Basel
Switzerland
Tel. +41 61 683 77 34
Fax +41 61 302 89 18
www.mdpi.com

Sustainability Editorial Office
E-mail: sustainability@mdpi.com
www.mdpi.com/journal/sustainability

www.ingramcontent.com/pod-product-compliance
Lightning Source LLC
LaVergne TN
LVHW071937080526
838202LV00064B/6623